AF248849

Soil Biology

Volume 15

Series Editor
Ajit Varma, Amity Institute of Microbial Sciences,
Amity University Uttar Pradesh, Noida, UP, India

Chandra Shekhar Nautiyal • Patrice Dion
Editors

Molecular Mechanisms of Plant and Microbe Coexistence

Foreword by V.L. Chopra

 Springer

Editors
Professor Dr. Chandra Shekhar Nautiyal
National Botanical Research Institute
Rana Pratap Marg
Lucknow 226001
India
csn@nbri.res.in

Professor Dr. Patrice Dion
Département de phytologie
Pavillon Charles-Eugène Marchand
1030, avenue de la Médecine
Université Laval
Québec (Québec) G1V 0A6
Canada
patrice.dion@fsaa.ulaval.ca

ISBN 978-3-540-75574-6   e-ISBN 978-3-540-75575-3
DOI:10.1007/978-3-540-75575-3

Soil Biology ISSN: 1613–3382

Library of Congress Control Number: 2008923850

Cover design: WMXDesign GmbH, Heidelberg, Germany

Printed on acid-free paper

5 4 3 2 1 0

springer.com

SCIENCE OF LIFE

Pollens of flowers, sperms of animals
Coded with messages internal
Busy in the acts of creation
Millions perish in action
The few successes are the great Creations!
Also succeed microbes many
Which disrupt the inner harmony
Of organisms big and mighty!
Bodies, anti bodies, biochemical wars
Evolutionary wisdom of immune systems
Nature's beauty opens one petal
For the researcher with knowledge tool
One by one petals open
Trillions more waiting to be seen

Y.S. Rajan

Foreword

Coexistence between microorganisms and plants has moulded civilization from the time humans began to rely extensively on cultivated crops for food. It is becoming increasingly clear that microbes are the basic of the biosphere. The interactions of plant roots with soil microorganisms are spatially and temporally complex. These interactions are fundamental to soil carbon dynamics and availability of inorganic and organic N to plants. The primary source of mineral nutrients for plants is the decomposition of organic matter by soil microbes.

Molecular Mechanisms of Plant and Microbe Coexistence reflects on the world of microbes in the soil that cause a soil to become biologically alive to enable plants to withstand better the rigors of life. For these reasons, our challenge is to understand the molecular processes that are crucial for establishing successful plant-microbe coexistence. This will not only lead to new scientific discoveries, but will provide the bases for new strategies for combating infectious diseases, producing novel and environment friendly, microbial inoculants for protecting plants from disease, for promoting healthy growth and ensuring precise regulation of nutrient supply to plants.

The book has assembled contributions from leading authorities of varied backgrounds on topics relating to interactions between plant and microbe coexistence research and dealing with contemporary scientific findings and cutting-edge technologies. Studying plant-microbe associations is important for all of these practical reasons and also for the goal of improved agricultural productivity. Exploitation of beneficial plant-microbe coexistence in the rhizosphere can promote plant health and have significant implications for low input sustainable agriculture. A major difficulty faced by plant biologists and microbiologists is that many groups of microbes that inhabit rhizosphere are not cultivable in the laboratory. Recent developments in molecular biology are shedding light on microbial diversity of rhizosphere. Understanding the complexity of this environment and how the microbial communities adapt and respond to alterations in the physical, chemical and biological properties of the rhizosphere remains a significant challenge for plant and microbial biologists. Contributors have focused on recent and exciting findings in the biology of plant and microbe coexistence and speculated on the technical and conceptual developments that will drive innovative research to open new vistas.

The practical utility of understanding plant-microbe coexistence is obvious. Genomic technologies can elucidate the role of novel genes in plant-microbe coexistence in the rhizosphere. The understanding of the molecular signaling processes and the functions they regulate within the rhizosphere will play a pivotal role in promoting beneficial plant and microbe coexistence, in overcoming existing limitations, and in designing strategies for the generation of novel inoculant consortia with applications in sustainable environmental biotechnology. The availability of new and powerful technologies for studying plant microbial coexistence in the soil guarantees a better understanding of these processes, which will not only facilitate their successful applications in biotechnology but will also provide new insights into sustaining plant and animal productivity, maintaining or enhance water and air quality, and supporting human health and habitation.

I sincerely hope that this volume will be of great interest to a wide spectrum of biologists.

Member of Planning Commission   V.L. Chopra
Government of India
Yojna Bhavan, Sansad Marg
New Delhi 110001
India

Preface

The coexistence of plant and microbes has had major effects on the development of civilization since humans began to rely extensively on cultivated crops for food, soon after the emergence of agriculture, dating back to about 10,000 years before the present (BP). The refinement of our knowledge on this system is brought about by the interaction between existing social values and the integrity of our measuring tools- with interplay of scientific and philosophical rationale. Thus, the wisdom gained and practices adopted have been passed down through generations. The ecological competence of traditional farmers is now reflected in the resurgence of organic agriculture. The available ancient literature includes the four Vedas, Susruta Samhita, Charaka Samhita, Krishi-Parashara, and Surapala's Vriskshayurveda. This literature is most likely to have been composed between 8000 and 1000 BP. Indeed, ancient civilizations often worshipped the soil as the foundry of life itself. We are becoming increasingly aware that microbes are the basis of the biosphere. The interactions of plant roots with soil microorganisms are spatially and temporally complex; however, it is becoming increasingly apparent that these interactions are fundamental to soil carbon dynamics and the availability of inorganic and organic N to plants. The primary source of mineral nutrients for plants is the decomposition of organic matter by humble soil microbes. The present book on molecular mechanisms of plant and microbe coexistence thus reflects upon the world of microbes in the soil that cause a soil to become biologically alive and assist other organisms in withstanding the rigors of life. We know that microbes inhabited the Earth long before multicellular life appeared and diversified. Following their appearance, multicellular plants and animals coexisted, interacted and coevolved with microbes. The evolution and activity of all plants and animals has thus been influenced by microorganisms, often as a result of intimate interactions.

We are at a time when the confluence of technological advances and the explosion of knowledge on plant-microbe coexistence will enable significant advances over the next decade. Thus, mechanisms controlling the multiple interactions between plant roots, other organisms and the soil environment are currently arousing great scientific interest. As plant and microbe coexistence research is a true interdisciplinary field of biological and soil sciences, the volume deals with new scientific findings and cutting-edge technologies, including molecular biological and functional genomic approaches. The present book has been organized in four sections, covering molecular mechanisms of plant and microbe coexistence from the point of view of populations, genomes, molecules and methods, respectively. Opening the first section, Chap. 1 provides an overview of plant-associated soil microorganisms and places the other book chapters into perspective. Chapter 2 deals with the role of microbial diversity in enhancing soil and plant health. Chapter 3 seeks to examine the structure of soil microbial communities in the light of evolution. Particular emphasis is placed on the defining role of plant-microbe mutualistic symbioses. Chapter 4 deals with new techniques, based on the genomics of rhizosphere colonization that offer opportunities for greatly expanding our knowledge of signaling, recognition and interaction in the root zone. Chapter 5 describes how the arbuscular mycorrhizal fungi, by maintaining belowground endosymbiosis with the roots of vascular land plants, influence the interactions between their host plant and aboveground insects.

The second section of the book describes the molecular processes that are crucial in establishing successful plant-microbe coexistence. Chapter 6 relates to macromolecular structure and evolutionary genomics, examining how these relate to the evolution of function in transcript RNA and protein molecules. This approach, involving the definition of rooted phylogeny of proteomes and fold architectures, is leading to fundamental understandings on genome coexistence. Chapter 7 highlights the main features of the currently known complete genomes of nitrogen-fixing symbiotic rhizobia, comparisons between these genomes revealing their evolution. Chapter 8 discusses the pathogen stress-induced epigenetic changes occurring in plants as a result of the production of a plant-derived signal, named the systemic recombination signal (SRS), which triggers the destabilization of the somatic and meiotic cell genomes and leads to heritable changes in response to stress. Functional genomics and proteomics are two rapidly expanding research areas. Chapter 9 deals eloquently with recent advances in functional genomics and proteomics of plant-associated microbes which will form the basis for new microbial inoculant-based strategies to combat infectious plant diseases promote plant growth and regulate nutrient supply to plants. Chapter 10 discusses how recent advances in molecular genetics will contribute to our knowledge of biocontrol process, and suggest new avenues for solving intriguing problems that the biocontrol industry faces, like inconsistency in performance of *Trichoderma* spp. under field conditions.

The third section relates to studies indicating a significant role for signaling in successful plant-microbe coexistence. Dedicated studies are needed to unravel the function and mechanism of signaling during the different stages of plant-microbe coexistence. Many Gram-negative, plant-associated bacteria use *N*-acyl homoserine lactone (AHL)-mediated quorum sensing to regulate traits involved in symbiotic, pathogenic or surface-associated relationships with their corresponding host plant. Chapter 11 presents an overview of the diverse phenomena regulated by quorum sensing in representative groups of these bacteria and illustrates the regulatory complexity often associated with these signaling networks. Chapter 12 then synthesizes eloquently the information available on the various types of signaling interactions that occur in the rhizosphere between microorganisms and plants. The investigation of host proteins interacting with viral proteins is a very promising approach to dissect the molecular basis of viral infections and to understand how viruses integrate in the complex structural and regulatory networks controlling plant growth; these plant-viral relationships are examined in Chap. 13. Given the major effects of rhizodeposition on composition and activities of microbial communities inhabiting rhizosphere soil, Chap. 14 discusses the state-of-the-art of studies on microbial activity and microbial diversity in the rhizosphere soil.

The fourth section explores questions related to the extent of diversity within naturally occurring microbial communities and addresses the challenge of studying as yet uncultivable prokaryotes. Techniques as described in methods-based Chaps. 15–18 offer opportunities for greatly expanding our knowledge of interaction of various eukaryotes in the root zone. Chapter 15 sets in motion as an example siderotyping as a particularly promising method for the characterization and identification at the species level of fluorescent and non-fluorescent *Pseudomonas*. Offering simplicity and rapidity of execution, siderotyping could advantageously replace a phenotypic numerical analysis. Chapter 16 deals with recent advances towards understanding oomycetes, which have been facilitated by the development of genomics databases and proteomics-based strategies. These new tools have usefully complemented traditional methods of gene cloning and classical genetics. Advanced molecular methods that can be used to analyze rhizosphere and soil-derived nucleic acids have been described in Chap. 17, along with examples where the use of these approaches has contributed significantly to our understanding of microbial life in soil and of microbial interactions with plants. Chapter 18 concentrates on in-depth morphotyping and molecular methods to characterize the ectomycorrhizal fungi. It is anticipated that these methods will allow us to understand the interplay of genes and functions in an ecosystem.

We could not have completed this book without the unflinching cooperation of our invaluable contributors who are authorities from varied background and, despite being heavily occupied, were always willing to accede to our demands to focus on exciting sub disciplines via simple schematic diagrams or at times even sharing their unpublished work. While editing the chapters we have taken care that personal style of the contributors is not influenced by our own. Without the legendary elephantine cool patience and composed posture of Professor Ajit Varma, Series Editor, this book could have remained indefinitely as an idea in our minds. We wish to thank Dr. Dieter Czeschlik and Dr. Jutta Lindenborn, Springer Heidelberg, for excellent feedback and professional support throughout the book preparation process. Jutta deserves special recognition of her very kind and supportive nature. Our thanks are also extended to Puneet S. Chauhan for help in compiling the chapters and those who have participated in the production of this book, whose indispensable help has significantly improved the quality of individual chapters and of the book as a whole, but the mistakes that are left remain our responsibility. Shekhar in particular is grateful to Professor Y.S. Rajan for writing a poem exclusively for the book despite the fact he is not a biologist but has worked on Indian satellite, launching and space application programmes. Shekhar would like to convey his indebtedness to his wife Manju and daughters Shikha and Isha, and Patrice to his family for their all-time support, encouragement and understanding in the face of lost evenings, weekends, holidays and for the break they gave to us to write uninterrupted. Finally, it is hoped that the excitement and significant opportunities presented in this volume about our newfound understanding of the relationships and the challenges that this brings for studying plant microbial coexistence will stimulate readers to push the field forward to new frontiers.

Lucknow, India   Chandra Shekhar Nautiyal
Québec City, Canada   Patrice Dion
October 2007

Contents

Contributors

Acosta, José L.
Centro de Investigación Sobre Fijación de Nitrógeno, Universidad Nacional
Autónoma de México, Cuernavaca, Morelos, México 62210,
jlacosta@ccg.unam.mx

Ascher, J.
Dipartimento della Scienza del Suolo e Nutrizione della Pianta,
Universita' degli Studi di Firenze, 50144 Firenze, Italy, judith.ascher@unifi.it

Azul, Anabela Marisa
Centre for Functional Ecology, Department of Botany, University of Coimbra,
Coimbra, Portugal, amjrazul@ci.uc.pt

Bhatia, C.R.
17, Rohini, Plot No. 29–30, Sector 9-A, Vashi, Navi Mumbai 400703, India,
neil@bom7.vsnl.net.in

Bledsoe, Caroline S.
Department of Land, Air and Water Resources, University of California,
Davis, California 95616-8627, USA, csbledsoe@ucdavis.edu

Boyko, Alex
Department of Biological Sciences, University of Lethbridge, Lethbridge,
Alberta T1K 3M4, Canada, oleksander.boyko@uleth.ca

Braeken, Kristien
Centre of Microbial and Plant Genetics, Katholieke Universiteit Leuven,
Kasteelpark Arenberg 20, B-3001 Leuven, Belgium,
Kristien.Braeken@biw.kuleuven.be

Bustos, Patricia
Centro de Investigación Sobre Fijación de Nitrógeno, Universidad Nacional
Autónoma de México, Cuernavaca, Morelos, México 62210
paty@ccg.unam.mx

Caetano-Anolles, Gustavo
Department of Crop Sciences, 332 NSRC, 1101 West Peabody Drive,
University of Illinois at Urbana-Champaign, Urbana, IL 61801, USA,
gca@uiuc.edu

Castillo, Santiago
Centro de Investigación Sobre Fijación de Nitrógeno, Universidad Nacional
Autónoma de México, Cuernavaca, Morelos, México 62210, iago@ccg.unam.mx

Ceccherini, M.T.
Dipartimento della Scienza del Suolo e Nutrizione della Pianta,
Universita' degli Studi di Firenze, 50144 Firenze, Italy,
mariateresa.ceccherini@unifi.it

Chauhan, Puneet Singh
National Botanical Research Institute, Rana Pratap Marg,
Lucknow 226001, India, puneetnbri@rediffmail.com

Daniels, Ruth
Hasselt University, Biomedical Research Institute. Agoralaan,
Building A, B-3590 Diepenbeek, Belgium, ruth.daniels@uhasselt.be

Dávila, Guillermo
Centro de Investigación Sobre Fijación de Nitrógeno,
Universidad Nacional Autónoma de México, Cuernavaca, Morelos,
México 62210, davila@ccg.unam.mx

Digilio, Maria Cristina
Dipartimento di Entomologia e Zoologia agraria "Filippo Silvestri" - Università
degli Studi di Napoli "Federico II" Via Università 100 - 80055 Portici (NA) Italy,
digilio@unina.it

Dion, Patrice
Département de phytologie, Pavillon Charles-Eugène-Marchand, 1030,
Avenue de la Médecine, Université Laval, Québec, Canada G1V 0A6,
patrice.dion@fsaa.ulaval.ca

Fernández, José L.
Centro de Investigación Sobre Fijación de Nitrógeno,
Universidad Nacional Autónoma de México, Cuernavaca, Morelos,
México 62210, jlfernan@ccg.unam.mx

Fischer-LeSaux, Marion
UMR 077 de Pathologie Végétale, Centre INRA d'Angers, Beaucouzé, France,
mlesaux@angers.inra.fr

Fong, Audrey M. V. Ah
Department of Plant Pathology and Center for Plant Cell Biology,
University of California, Riverside 92521, USA, audreya@ucr.edu

González, Ismael Hernández
Centro de Investigación Sobre Fijación de Nitrógeno,
Universidad Nacional Autónoma de México, Cuernavaca, Morelos,
México 62210, ishernan@ccg.unam.mx

Gonzalez, Victor
Centro de Ciencias Genómicas, Universidad Nacional Autónoma de México Av.
Universidad N/C, Col. Chamilpa CP 62210. Apdo. Postal 565-A, Cuernavaca,
Morelos, México, vgonzal@ccg.unam.mx

Gruffaz, Christelle
Laboratoire de Microbiologie et Génétique, Université Louis Pasteur,
28, rue Goethe, 67083 Strasbourg Cedex, France,
E-mail: christelle.gruffaz@gem.u-strasbg.fr

Guerri, G.
Dipartimento della Scienza del Suolo e Nutrizione della Pianta,
Universita' degli Studi di Firenze, 50144 Firenze, Italy, giulia.guerri@unifi.it

Guerrieri, Emilio
Istituto per la Protezione delle Piante CNR Via Università 133,
80055 Portici (NA), Italy guerrieri@ipp.cnr.it

Hampp, Rüdiger
Botanical Institute, Physiological Ecology of Plants, University of Tübingen,
Auf der Morgenstelle 1, 72076 Tübingen, Germany,
ruediger.hampp@uni-tuebingen.de

Judelson, Howard S.
Department of Plant Pathology, University of California, Riverside, California
92521, USA, howard.judelson@ucr.edu

Kovalchuk, Igor
Department of Biological Sciences, University of Lethbridge, Lethbridge,
Alberta T1K 3M4, Canada, igor.kovalchuk@uleth.ca

Lozano, Luis
Centro de Investigación Sobre Fijación de Nitrógeno,
Universidad Nacional Autónoma de México, Cuernavaca, Morelos,
México 62210, llozano@ccg.unam.mx

Mabood, Fazli
Department of Plant Science, McGill University, Macdonald Campus,
21,111 Lakeshore Road, Ste Anne-de-Bellevue, Que., Canada H9X 3V9
fazli.mabood@elf.mcgill.ca

MacFarlane, Stuart A.
Plant Pathology Department, Scottish Crop Research Institute,
Invergowrie, Dundee DD2 5DA, UK, Stuart.MacFarlane@scri.ac.uk

Meyer, J.M.
Département Génétique Moléculaire, Génomique et Microbiologie,
UMR 7156 Université Louis-Pasteur/CNRS, Strasbourg, France,
meyer@gem.u-strasbg.fr

Michiels, Jan
Centre of Microbial and Plant Genetics, Katholieke Universiteit Leuven,
Kasteelpark Arenberg 20, B-3001 Leuven, Belgium,
jan.michiels@agr.kuleuven.ac.be

Morris, Melissa H.
Department of Land, Air and Water Resources, University of California,
Davis California 95616-8627, USA, mhmorris@ucdavis.edu

Mukherjee, P.K.
Nuclear Agriculture and Biotechnology Division, Bhabha Atomic
Research Centre, Mumbai, India, prasunmukherjee1@gmail.com

Mukhopadhyay, A.N.
Sangini, 151 Akansha, Udhay-II, Raibareilly Road, Lucknow, India,
mukhopadhyayamar@indiatimes.com

Nannipieri, Paolo
Dipartimento della Scienza del Suolo e Nutrizione della Pianta,
Universita' degli Studi di Firenze, 50144 Firenze, Italy,
paolo.nannipieri@unifi.it

Nautiyal, Chandra Shekhar
National Botanical Research Institute, Rana Pratap Marg,
Lucknow 226001, India, csn@nbri.res.in, nautiyalnbri@lycos.com

Ndayizeye, Maxime
Centre of Microbial and Plant Genetics, Katholieke Universiteit Leuven,
Kasteelpark Arenberg 20, B-3001 Leuven, Belgium

Pietramellara, G.
Dipartimento della Scienza del Suolo e Nutrizione della Pianta,
Universita' degli Studi di Firenze, 50144 Firenze, Italy,
giacomo.pietramellara@unifi.it

Renella, G.
Dipartimento della Scienza del Suolo e Nutrizione della Pianta,
Universita' degli Studi di Firenze, 50144 Firenze, Italy,
giancarlo.renella@unifi.it

Santamaría, Rosa I.
Centro de Investigación Sobre Fijación de Nitrógeno,
Universidad Nacional Autónoma de México, Cuernavaca, Morelos,
México 62210, rosa@ccg.unam.mx

Schrey, Silvia
Botanical Institute, Physiological Ecology of Plants, University of Tübingen,
Auf der Morgenstelle 1, D-72076 Tübingen, Germany

Smith, D.L.
Plant Science Department, McGill University/Macdonald Campus,
21,111 Lakeshore Road, Ste. Anne de Bellevue, Quebec, Canada H9X 3V9
donald.smith@mcgill.ca

Srivastava, Suchi
National Botanical Research Institute, Rana Pratap Marg,
Lucknow 226001, India, ssnbri@rediffmail.com

Suz, Laura M.
Departamento de Micología, Real Jardín Botánico-CSIC, Madrid, Spain,
laura.martinez@ctfc.es

Tarkka, Mika[1,2]
[1]Botanical Institute, Physiological Ecology of Plants, University of Tübingen,
Auf der Morgenstelle 1, D-72076 Tübingen and [2]UFZ, Helmholtz-Centre
for Environmental Research, Department of Soil Ecology,
Theodor-Lieser-Strasse 4, D-06120 Halle, Germany

Thies, Janice E.
Department of Crop and Soil Sciences, Cornell University, Ithaca, NY
jet25@cornell.edu

Uhrig, Joachim
University of Cologne, Department of Botany III, Gyrhofstr. 15,
D-50931 Cologne, Germany
joachim.Uhrig@uni-koeln.de

Valori, F.
Dipartimento della Scienza del Suolo e Nutrizione della Pianta,
Universita' degli Studi di Firenze, 50144 Firenze, Italy, federico.valori@unifi.it

Vanderleyden, Jos
Centre of Microbial and Plant Genetics, Katholieke Universiteit Leuven,
Kasteelpark Arenberg 20, B-3001 Leuven, Belgium,
jos.vanderleyden@biw.kuleuven.be

Woo Jin Jung
Division of Applied Bioscience and Biotechnology, College of Agriculture
and Life Science, Chonnam National University, Gwangju 500–757,
South Korea, woojung@chonnam.ac.kr

Part I
Coexistence Between Populations

Chapter 1
Plant Associated Soil Micro-organisms

Mika Tarkka, Silvia Schrey, and Rüdiger Hampp(⊠)

1.1 Micro-organisms of the Rhizosphere

1.1.1 The Rhizosphere

Roots constitute important plant organs for water and nutrient uptake. However, they also release a wide range of carbon compounds of low molecular weight. These can amount to between 10% and 20% of total net fixed carbon (Rovira 1991) and form the basis for an environment rich in diversified microbiological populations, the rhizosphere (Hiltner 1904). The rhizosphere has been defined as a narrow zone of soil which is influenced by living roots. Bacteria are an important part of micro-organisms inhabiting this ecological niche. Abundance and turnover of rhizobacteria are regulated by microfaunal grazers such as protozoa. Consequently, beneficial effects of protozoa on plant growth have been related to nutrients released from consumed bacterial biomass. This has been termed 'microbial loop' (Bonkowski 2004) and works as follows: organic compounds released from roots stimulate bacterial growth. Bacteria can solubilize nutrients from the mineral soil layer, but will sequester them. Consumption of bacteria by soil protozoa and nematodes will then liberate nutrients, which in due course will become available for plants. Fungi form another important part of the rhizosphere. Most terrestrial plants develop symbiotic structures (mycorrhiza) with soil-borne fungi. In these interactions the fungal partner provides the plant with improved access to water and soil nutrients due to more or less complex hyphal structures, which emanate from the root surface and extend far into the soil. The plant, in return, supplies carbohydrates for fungal growth and maintenance (Hampp and Schaeffer 1998; Smith and Read 1997). Due to leakage and the turnover of mycorrhizal structures, these solutes are

R. Hampp
Botanical Institute, Physiological Ecology of Plants, University of Tübingen,
Auf der Morgenstelle 1, 72076 Tübingen, Germany
e-mail: ruediger.hampp@uni-tuebingen.de

C.S. Nautiyal, P. Dion (eds.) *Molecular Mechanisms of Plant and Microbe Coexistence*. Soil Biology 15, DOI: 10.1007/978-3-540-75575-3
© Springer-Verlag Berlin Heidelberg 2008

also released into the mycorrhizosphere where they can be accessed by the other micro-organisms. It has been shown that microbial communities within the rhizosphere are distinct from those of non-rhizosphere soil (Curl and Truelove 1986; Whipps and Lynch 1986).

1.1.2 Fungi: Symbionts, Saprotrophs, Pathogens

Plants and soil communities are mainly linked by the provision of photoassimilates by the plant. In addition, plants supply organic matter by litter such as leaves, or by root exudates.

For nutrient recycling and supply to plants, symbiotic and saprotrophic fungi are essential components of the rhizosphere.

1.1.2.1 Arbuscular Mycorrhiza (AM)

With regard to symbiotic root fungus interactions, two major types exist; the arbuscular mycorrhiza (AM) and the ectomycorrhiza (ECM). They differ in morphological features and in the type of fungi. AM are typical for most herbaceous plants, including crop plants and also for the majority of tropical tree species (Janos 1987). In addition, AM constitutes the most ancient form of mycorrhiza documented by fossil findings (Redecker et al. 2000), which can explain its global occurrence.

The fungi involved are obligate biotrophs. They belong to the order of the Glomales (Glomeromycota). Typically, they form extra- and intraradical mycelia, as well as inter- and intracellular hyphae, coiled hyphae, arbuscules, vesicles, auxiliary cells close to or within the cortex of the host root. Most of these structures increase the effective surface area for solute exchange (see also Smith and Read 1997).

Colonization of fine roots by AM fungi starts with the formation of appressoria on the surface of epidermal cells and is then followed by the development of penetration hyphae. After successful epidermal penetration, hyphae invade the apoplast of the root cortex. Typical arbuscules are formed as intracellular terminal structures of trunk hyphae.

Gallaud (1905) described two major structural classes of arbuscular mycorrhizae, which he named *Arum-* and *Paris*-type, after the plants in which they were first described (for review see Smith and Smith 1990). In the *Arum*-type the fungus spreads relatively rapidly in the cortex via intercellular air spaces (Brundrett and Kendrick 1990). Short side-branches of the fungus penetrate the cortical cells and ramify dichotomously to produce characteristic arbuscules. Hyphal coils may be formed, but they are usually not a major component of the intraradical mycelium. In the *Paris*-type, colonization of the roots is characterized by extensive development of intracellular coiled hyphae, which spread directly from cell to cell within the cortex. From these coils arbuscules can be developed, and there is very little, if any, intercellular growth. As a consequence, the growth rate of the infection units within the root is much slower than for the *Arum*-type (Smith and Read 1997).

Investigations of Barrett (1958) indicated that the host, not the endophyte, determines the structural class of AM mycorrhiza. A similar conclusion was drawn by Gerdemann (1965), who showed that the same fungus can form a *Paris*-type mycorrhiza in *Liriodendron sp.* and an *Arum*-type in *Zea mays* plants.

1.1.2.2 Ectomycorrhiza (ECM)

ECM establishes with fine roots of autotrophic trees and shrubs, especially of the families *Betulaceae, Pinaceae, Fagaceae, Salicaceae* and *Dipterocarpaceae* (Read 1991; Smith and Read 1997). The fungal partners belong to the basidiomycetes and ascomycetes. Typically, hyphae form a mantle of varying thickness around the fine roots. From there, hyphae or more specialised hyphal aggregates (rhizomorphs) radiate into the substrate in order to exploit nutrients and water. Mantle hyphae also extend into the apoplast of the root cortex. Here, they form highly branched networks, which establish a large surface area for solute exchange. This structure is called the Hartig net and constitutes the interface for the exchange of photoassimilates, soil water and nutrients between the host plant and its fungal partner.

Communities of ECM trees are dominating in the boreal and temperate plant biomes and are also important in certain tropical rain forest environments (Read 1993). In these diverse plant formations, ECM fungi are best adapted to mobilise the sparse heterogeneous resources in phosphorus and especially in nitrogen from the litter layer. This function is ensured by a high diversity of fungi, which has been estimated between 5000 and 6000 species (Molina et al. 1992). This high biodiversity of ECM fungi corresponds to a broad range of capabilities for the uptake of specific forms of organic and inorganic nitrogen and phosphorus, allowing the development of tree vegetations with low plant species diversity despite the above-mentioned heterogeneity and limitation of nutrient resources (Read 1993).

Due to competition for photoassimilates, mycorrhiza-forming fungi can become protective for their source plant by preventing pathogenic fungi or nematodes from root colonization (Graham 2001). However, some can also become highly parasitic (Jonsson et al. 2001).

1.1.2.3 Saprotrophs

While interactions between wood-decay fungi themselves have been reviewed by Boddy (2000), there is only little information about their interaction with mycorrhiza-forming fungi (Leake et al. 2002). In a forest ecosystem both types of fungi rely on a large supply of organic carbon, either from photosynthesis (symbiotic fungi) or from wood and other litter (saprophytes). Both types of fungi can explore large volumes of soil due to their ability to form hyphal aggregates (rhizomorphs), which allow for long distance solute transport (Finlay and Read 1986a,b; Boddy 1993, 1999). Except for the difference in carbon source, their strategies for nutrient acquisition are similar, which can cause competition (Leake et al. 2002). This competition is not limited to N and P but also includes organic compounds, as (ecto) mycorrhizal fungi can produce a variety of extracellular enzymes which can degrade a wide range of soil organics (for literature see Leake et al. 2002). The ability of (ecto)mycorrhizal fungi to efficiently acquire N and P from organic and inorganic pools in the soil brings them in direct competition with wood decomposer fungi, which require the same nutrients. It can thus be assumed that saprotrophs play mainly a role as primary colonizers, i.e. growth on lignocellulose-rich plant litter, or on recalcitrant organic residues, which cannot be accessed by symbiotic fungi (Persson et al. 1980; see also Sect. 2.1.2).

1.1.2.4 Fungal Networking

Hyphal strands have been shown to connect neighboring trees and can thus establish a large network for assimilate transfer according to source sink gradients (Perez-Moreno and Read 2004). Molecular proof for such networks comes from DNA analysis of roots and associated fungi (Saari et al. 2005). Hyphal connections also exist between fine roots of seedlings and of adult trees (Matsuda and Hijii 2004). This could help to compensate for shadowing and thus increase seedling performance under limiting light (Booth 2004). As mentioned above, ECM is typical for trees and shrubs, but also some herbaceous plants form this type of mycorrhiza. The latter have possibly an important function in bridging forest gaps by spreading ECM fungi, as well as perpetuating fungal inocula when, e.g. after fire, tree seedlings start to re-establish (Dickie et al. 2004; Richard et al. 2005).

1.1.3 Plant Beneficial Bacteria

The release of carbon by plant roots results in greater microbial populations and activity in the rhizosphere than in the bulk soil. The rhizosphere/bulk soil ratio for Gram negative bacteria reaches from 2 to 20 and for actinomycetes from 5 to 10 (Morgan et al. 2005). The diversity and structure of bacterial communities is plant specific and varies over time (Smalla et al. 2001; Barriuso et al. 2005), and these microbes can have a negative, neutral or beneficial effect to plant fitness. Detrimental effects are caused by bacterial pathogens and parasites and bacteria that produce phytotoxic substances. The occurrence of pathogenic bacteria is however low in healthy plant populations. This is due to plant defence systems, which are selective and could cause the enrichment of plant beneficial microbes within the rhizosphere. The plant beneficial bacteria include saprophytes that degrade the organic litter, plant growth promoting rhizobacteria (PGPR), and antagonists of plant root pathogens (Barea et al. 2005). This section deals with the plant beneficial rhizosphere bacteria.

Plant growth promoting rhizobacteria (PGPR) are usually in contact with the root surface, and increase plant growth by three major mechanisms: i) by improved mineral nutrition ii) by phytohormone production and/or iii) by disease suppression (Weller 1988; Lucy et al. 2004; Haas and Defago 2005). The PGPR must be able to colonise the root and to be present in sufficient numbers to exert their functions. *Pseudomonas* and *Bacillus* are the most commonly investigated PGPR, and often the dominating bacterial groups in the rhizosphere (Marilley and Aragno 1999; Morgan et al. 2005). Diverse PGPR strains have been used successfully for crop inoculations, including members of *Azospirillum, Azotobacter, Bacillus, Enterobacter, Pseudomonas, Serratia* and *Xanthomonas* (see Lucy et al. 2004 for a comprehensive list).

Two groups of PGPR exist: those that are involved in nutrient cycling and plant growth stimulation (biofertilizers), and those that are involved in the biological control of plant pathogens (biopesticides). Bacteria may support the plant growth by the mobilisation of inorganic nutrients, by nitrogen fixation and by the production of phytohormones including auxins, cytokinin and volatile substances such as butanediol (Barea et al. 2005). In soils with low phosphate (P), P-solubilising bacteria release phosphate ions from low-soluble inorganic P crystals and from organic phosphate sources. These bacteria exude organic acids that solubilise the inorganic P crystals and exude enzymes that split the organophosphates (Vessey 2003). Although many P solubilising bacteria have been characterized, their relative importance in the PGPR effect is uncertain. However, if the phosphate ions are released in an area rich with mycorrhizal fungal hyphae, the hyphae may transport the P to the plants and the PGPR effect is mostly detectable (Artursson et al. 2005; Barea et al. 2005; see Sect. 1.2.1).

Nitrogen fixing bacteria significantly improve nitrogen availability in the soil. In this process the bacteria, called diazotrophs, convert atmospheric nitrogen (N_2) into ammonia compounds that can be used by other organisms, including plants. *Klebsiella* strains have been isolated from the rhizosphere of a variety of plants and these bacteria are often called associative nitrogen fixers, since they are diazotrophs that colonise the root surface (Haahtela et al. 1986). Recent data suggests that some of the *Klebsiella* species fix nitrogen as plant endophytes (Iniguez et al. 2004). The nitrogen fixation capacity occurred only with one variety out of four tested by Iniguez et al. (2004), indicating a strong specificity for this interaction.

Azospirillum spp. are diazotrophs, freely living in the soil or in association with roots (Bashan et al. 2004). The inoculation of roots with *Azospirillum* spp. often promotes plant growth, not only due to the nitrogen fixation but also due to the ability of the bacteria to produce phytohormones (Steenhoudt and Vanderleyden 2000). In general, PGPR often enhance plant growth through the production of plant growth regulators (Lucy et al. 2004). The auxin type phytohormones produced by the *Azospirillum* spp. induce root branching and thus improve plant nutrient uptake from the soil (Dobbelaere et al. 1999). The growth of the plants may also be induced by bacterial cytokinin production (Lucy et al. 2004). Garcia de Salamone et al. (2001) detected five *Pseudomonas fluorescens* PGPR strains which promoted plant growth through the production of cytokinins identified as dihydrozeatin riboside and isopentenyl adenosine. Recently, volatile compounds that promote plant growth were isolated form bacterial cultures by Ruy et al. (2003). The authors observed that two *Bacillus* spp. and an *Enterobacter cloacae* strain induce *Arabidopsis thaliana* growth in Petri dish assays with a separate compartment for the bacteria. The growth promoting volatile was identified as 2,3-butanediol. Both *Bacillus subtilis* GB03 and *Bacillus amyloliquefaciens* IN937a produced this substance, whereas it was undetectable in other PGPR strains that did not trigger enhanced growth via volatile emissions.

The second major PGPR mechanism is to reduce the incidence of plant disease. Infectious diseases are often caused by soilborne organisms including both bacteria and fungi. The soils where soilborne diseases are infrequent are called suppressive soils, and it has been shown that the disease suppression is often caused by specific bacterial and fungal populations (Weller et al. 2002). Recent research highlights three major mechanisms for the disease suppression: antagonism, direct pathogen-agonist-interactions, and induced systemic resistance (Compant et al. 2005).

Antagonists are naturally occurring organisms that express traits that enable them to interfere with pathogen growth, survival and infection. Bacteria antagonistic to plant pathogens represent an important part of the rhizosphere communities. The proportion of antagonistic strains is up to 35% of the culturable bacteria (Opelt and Berg 2004). The *Burkholderia cepacia* complex (former *Pseudomonas cepacia*) is a group of nine closely related bacterial species that have useful properties in the natural environment (Chiarini et al. 2006), and they have emerged as powerful biocontrol agents (Bevivino et al. 1998). The Gram-negative rods from *Stenotrophomonas maltophilia* (earlier *Xanthomonas maltophilia*) are also typical rhizosphere inhabitants, and a focus of scientific interest due to their potential for biological control (Dunne et al. 1998; Nakayama et al. 1999). Actinomycetes also form a group of important biocontrol agents (Whipps 2001), which is discussed later in this review (Sect. 1.3.2). The most thoroughly investigated group of PGPR antagonists are still the fluorescent pseudomonads (Haas and Defago 2005). These bacteria produce diverse antagonistic secondary metabolites that suppress the growth of other organisms. As an example, the extracellular pigment pyoverdin is an efficient siderophore (iron carrier), and the production of pyoverdin by pseudomonads in iron-poor soils is an effective way to suppress the growth of non-producers by depriving the pathogens from iron (Kloepper et al. 1978). A great variety of pyoverdines have been identified from the *Pseudomonas* spp. Pseudomonads also produce metal chelating agents with proposed properties other than iron scavenging. Pyochelin, e.g. effectively binds copper and zinc, and possesses strong antimicrobial activity (Cornelis and Matthijs 2002). However, the antimicrobial effect of pyochelin, and of some other siderophores can be explained by their effective metal chelating activity (Haas and Defago 2005).

Direct antibiosis is used by several PGPR as a mechanism for biocontrol. Antibiosis by PGPR pseudomonads is often caused by the production of several antimicrobial substances. These chemicals not only suppress fungi but also are often toxic against bacteria (Compant et al. 2005). From antimicrobial compounds produced by pseudomonads, the mode of action has been partly determined for six classes of substances thus far. These include the electron transport inhibitors phenazines, phloroglucinols, which cause membrane damage in *Pythium* spp. and are phytotoxic at higher conditions, pyrrolnitrin, acting as a fungicide, cyclic lipopeptides that have surfactant properties against fungi and plants or chelate cations, and finally HCN, which is a potent inhibitor of metalloenzymes. A comprehensive list of the antibiotics, producer strains, target organisms and effects on the host plants have been covered in a review by Raaijmakers et al. (2002). Siderophore and antibiotics production has been observed in other PGPR isolates as well, including *Bacillus* and *Stenotrophomonas* spp. (Compant et al. 2005), and *Streptomyces* spp. (see Sect. 1.3.2).

The second group of antagonistic compounds are lytic enzymes, such as cell wall hydrolases that attack pathogens. The ability to degrade fungal cell walls by chitinases is shared by many biocontrol PGPR including *Pseudomonas, Serratia* and *Streptomyces* spp. (Whipps 2001). In addition to chitinases, some bacterial strains produce beta-glucanases and proteases (Dunne et al. 2000). Synergism between the action of cell wall degrading agents and antibiotics was observed by Woo et al. (2002). The authors showed that the pre-treatment of plant pathogenic fungi with cell wall degrading enzymes made them more susceptible to the antifungal substance syringomycin.

Another important mechanism of biocontrol is the degradation of virulence factors (Compant et al. 2005). Albicidins are a family of phytotoxins and antibiotics which play an important role in the pathogenesis of sugarcane leaf scald disease. The *albA* gene from *Klebsiella oxytoca* encodes a protein which inactivates albicidin by heat-reversible binding (Zhang and Birch 1997). In contrast to the mechanism in *K. oxytoca,* an esterase produced by *Pantoea dispersa* is able to degrade the albicidins rendering them inactive (Zhang et al. 1998).

Bacterial cells sense their population density through a cell-cell communication system and trigger expression of particular genes when the density reaches a threshold. This type of gene regulation, which controls diverse biological functions including virulence, is known as quorum sensing. Certain PGPR are able to quench the quorum sensing capacity of the neighbouring bacteria by degrading autoinducer signals, thereby blocking expression of several virulence genes (Dong et al. 2000).

Inoculation of plants with some PGPR elicits a phenomenon known as induced systemic resistance (ISR; Van Loon et al. 1998). The ISR allows the plants to endure pathogen attacks that without bacterial pre-inoculation could be lethal. The effect is systemic, e.g. root inoculation with the biocontrol PGPR yields the whole plant non-susceptible (Haas and Defago 2005). Thus far, *Pseudomonas, Burkholderia* and *Bacillus* spp. have been shown to elicit ISR (Barka et al. 2000; Brooks et al. 1994; Ryu et al. 2004), and the search for effective substances is in progress. Root treatment of *Phaseolus vulgaris* with *Pseudomonas putida* BTP1 leads to significant reduction of the disease caused by the pathogen *Botrytis cinerea* on leaves. Ongena et al. (2005) isolated the molecular determinant of *P. putida* mainly responsible for the induced systemic resistance and identified it as a polyalkylated benzylamine structure. Exposure to butanediol, the volatile that induces the growth of *Arabidopsis* seedlings (Ruy et al. 2003), decreases disease severity by the bacterial pathogen *Erwinia carotovora* in the same plant (Ryu et al. 2004). Transgenic lines of *Bacillus subtilis* that emitted reduced levels of 2,3-butanediol, decreased *Arabidopsis* protection against pathogen infection compared with seedlings exposed to volatiles from wild-type bacterial lines.

Environmental pollution with metals and xenobiotics is a global problem, and the development of phytoremediation technologies for the plant-based clean-up of contaminated soils is therefore of significant interest (Kramer 2005). Bacteria from the rhizosphere take part in the degradation of toxic compounds in a process called rhizodegradation (Kuiper et al. 2004). In polluted soils rhizodegradation may lead to a strong increase in plant yield (Lucy et al. 2004).

Many of the bacteria in the rhizosphere are unable to grow in the laboratory and thus culture-based methods are often inadequate for qualitative analysis of the rhizosphere bacterial populations. As a consequence, a number of culture-independent approaches have been applied to the study of microbial diversity (Kent and Triplett 2002; Artursson et al. 2005). Molecular approaches, such as 16S rRNA analyses and denaturing gradient gel electrophoresis, have been used to study the presence of un-culturable bacteria in the rhizosphere, such as the Archaea (Bomberg et al. 2003; Nicol et al. 2003). The first evidence for growth of the rhizosphere Archaea in culture was recently reported by Simon et al. (2005), and the interactions of Archaea with other micro-organisms in the rhizosphere is likely to become a focus of research in the future.

The multiple mechanisms as to how the plant beneficial bacteria promote plant fitness are only beginning to be resolved. It is obvious that the use of *Pseudomonas* and *Bacillus* spp. has yielded a mass of relevant results, but much remains to be learned from the bacteria in other taxa. The co-operation between bacteria and fungi in the rhizosphere adds much complexity to the interaction of these microbes with plant roots, and this will be further discussed in Sect. 1.2.1.

1.1.4 *Protozoa*

Protozoa are another very important group of rhizosphere micro-organisms. They have a diameter of 10–100 μm, and can transiently form large population numbers which, with regard to biomass, can be equal to most of all other animal biomass taken together. Significant effects on nutrient mineralization (e.g. nitrogen: Schröter et al. 2003) result from high rates of biomass turnover. The most important bacterial grazers are amoebae, which have access to bacterial biofilms as well as soil or root surface located colonies. Access is facilitated by the ability to form pseudopodia, which can exploit very remote locations of bacteria (see Bonkowski 2004). Grazing can be highly specific: Protozoa have been shown to prefer Gram-negative bacteria, while Gram-positive ones benefit (Griffith et al. 1999). They thus can considerably modify the bacterial soil composition. A shift to Gram-positive bacteria can alter plant resistance to pathogens (see Sect. 1.3.2).

In the presence of protozoa, plants show a stimulation of lateral root formation, which results in highly branched root systems. This is most probably due to the stimulation by amoebae of the release of auxin-like substances by bacteria (Bonkowski and Brandt 2002) and can result in a self-sustaining process, such as more branching – higher liberation of root exudates – better bacterial growth – better performance of protozoa – better nutrient supply for plants and so on. By the stimulation of nitrifying bacteria, protozoa can also increase local availability of nitrate, which in turn stimulates lateral root growth (Zhang and Forde 2000).

Both mycorrhiza-forming fungi and protozoa affect root architecture. While, e.g. ECM fungi induce shorter fine roots, protozoa stimulate growth of fine roots: they get longer and thinner. In experiments, where both types of micro-organisms were present, their individual effects were counterbalanced, obviously due to limiting carbon supply (Bonkowski et al. 2001). This was indicated by less protozoa and shorter fungal hyphae. The host plant, in contrast, could take advantage of this competition as shown by improved P (mycorrhizal fungus) and N nutrition (protozoa). The presence of both types of micro-organisms also reduced leaching losses, probably owing to the recovery of protozoa-mobilized nitrogen by mycorrhizal hyphae (Bonkowski et al. 2001). Restriction by symbiotic fungi of the allocation of plant carbon to the rhizosphere can also be the reason for reduced numbers of bacteria and protozoa (see Bonkowski 2004).

1.2   Organismic Interactions

1.2.1   Microbe-Microbe Interactions

Soil microbes sense, compete, and interact with each other in order to succeed within the complex microbial community. The exchange of signaling molecules, the production of antibiotics and siderophores and, maybe most important, the successful competition for nutrient sources lead to success in the microbial world.

1.2.1.1   Competition for Nutrient Supplies and Space

Fungi hold the monopoly on two important nutrient supply domains in the soil, namely mycorrhiza (Smith and Read 1997) and decomposition of lignocellulose (de Boer et al. 2005), which most probably exerted a strong evolutionary pressure on soil bacteria. The competition between bacteria and fungi for root exudates, cellulose, and lignin resulted in the dominance of fungal decomposers but also in the development of new niches for bacterial growth such as fungal exudates, living hyphal compartments and the walls of dead hyphae (de Boer et al. 2005). Fungi act as the dominant decomposers of complex recalcitrant organic material in the soil, aided by their mycelial growth habit (Griffin 1985) as hyphal growth allows the translocation of nutrients (Boddy and Watkinson 1995; Lindahl et al. 1999) and thus enables the fungus to bridge air filled voids in the soil and to cross nutrient-poor spots (Jennings 1987). Interestingly, one important bacterial group in soil that also effectively decomposes recalcitrant organic matter are the streptomycetes that also developed a mycelial growth (Griffin 1985).

Degradation of cellulose under natural conditions is mainly a domain of fungi. Even though certain filamentously growing bacteria, like streptomycetes, degrade cellulose (McCarthy and Williams 1992; Tuncer et al. 2004), which implies competition for cellulose resources, they seem to be unable to degrade crystalline cellulose as it is commonly encountered in plant cell walls (Wirth and Ulrich 2002). The cellulolytic potential of bacteria and fungi is comparable, but under natural conditions the activity seems to depend, e.g. on the pH value. Streptomycete (hemi-) cellulases are most effective at a neutral to alkaline pH, whereas fungal enzymes perform best at a more acid pH (McCarthy 1987), as it is encountered in wood (Rayner and Boddy 1988). The degradation of lignin is a largely fungal domain, even though some *Streptomyces* strains (Crawford 1978; Antai and Crawford 1981) and non-filamentous bacteria have been shown to grow on lignin or lignin-like compounds (Vicuña et al. 1993; Céspedes et al. 1997).

Fungal decomposition of lignin and cellulose releases huge amounts of water-soluble sugars that may serve as growth substrates for bacteria. A strong competition for these sugars with bacteria could deprive the fungus of its energy sources and could thus result in reduced fungal lignocellulose degradation, as has been shown with the white rot fungus *Dichomitus squalens* (Lang et al. 1997). This is apparently a species-specific interaction as another white rot fungus, *Pleurotus* sp., was not disturbed by the bacterial competition in its degradation, probably due to the production of bacteria-inhibiting compounds (Gramms et al. 1999; Andersson et al. 2003).

Although fungi are the main composers of recalcitrant organic matter, bacteria are equally effective with regard to simple organic substrates such as root exudates. To compete for root exudates, fungi and bacteria have evolved complex strategies. Interference competition through bacteria includes production of antagonistic metabolite such as antibiotics (Milner et al. 1996; Keel and Defago 1997; Thrane et al. 2000), lytic enzymes (Chernin et al. 1995, 1997; Nagarajkumar et al. 2004) and volatiles (Wheatley 2002), whereas competition for substrate is carried out by the production of nutrient sequestering compounds like siderophores (Loper and Henkels 1999; Whipps 2001). Antibiotics isolated from rhizosphere *Pseudomonas* spp. were shown to act against a range of fungi as well as bacteria (Raaijmakers et al. 2002), indicating that these strategies have not only evolved for the defence against these fungi but also against other bacteria (de Boer et al. 2005). Fungi, on the other hand, also developed strategies against antifungal substances of bacterial origin, including detoxification, transportation of antibiotics out of the cells or modification of bacterial gene expression (Duffy et al. 2003).

1.2.1.2 Synergism and Antagonism in the Soil

The dominating role of fungi with regard to the degradation of complex organic matter has not merely narrowed the niches available to bacterial occupation, but has also created new ones on and around fungal hyphae, on mycorrhizal roots, and within fungal fruiting bodies. Fungal exudates have been described as possibly the only source of nutrients for these bacteria (Linderman and Paulitz 1990; Andrade et al. 1997; Nurmiaho-Lassila et al. 1997). The selection of particular bacterial strains may be due to fungal storage sugars (trehalose, mannitol; Frey et al. 1997; Rangel-Castro et al. 2002), even though data on the exuded carbon compounds are limited (Finlay and Söderström 1992; Johansson et al. 2004). In fungus-bacterium interactions, the association of particular bacterial strains with a specific fungus might indicate specificity in co-operation, and not only a coincident (de Boer et al. 2005).

Fungal growth depends not only on soluble carbohydrates but also on growth factors produced by bacteria. Co-inoculation of wood chips with white rot fungi *Resinicium bicolor* or *Hypholoma fasciculare* and soil bacteria led to an increased wood degradation compared to inoculation with the fungus only. As wood degradation was not observed by bacteria alone (Murray and Woodward 2003), the question was raised as to whether this stimulation may have been caused by the bacterial production of growth factors, e.g. vitamins like thiamine (Henningston 1967), or by the stimulation of fungal enzyme activity due to removal of breakdown products through bacteria (de Boer et al. 2005). Furthermore, the production of cellulases and pectinases by bacteria might increase the accessibility of breakdown products for the fungus (Greaves 1971). Finally, detoxification of potentially harmful degradation products by the bacteria (Greaves 1971) or activities of nitrogen-fixing bacteria might promote fungal growth (Hendrickson 1991).

A streptomycete strain, *Streptomyces* AcH 505, that promotes the growth of the ectomycorrhizal fungus *Amanita muscaria* and suppresses growth of several pathogenic fungi (Maier et al. 2004; Schrey et al. 2005) was shown to produce both fungal growth-stimulating and -suppressing secondary metabolites (Riedlinger et al. 2006). Co-cultivation of the streptomycete with *A. muscaria* stimulated the production of the growth promoting substance, auxofuran, but reduced the production of the inhibitory secondary metabolites by the streptomycete. Furthermore, the high tolerance of *A. muscaria* against the growth suppressing compounds enabled the fungus to respond to auxofuran, thus creating an advantage over other microbial species that are more sensitive to these antibiotics (Riedlinger et al. 2006; see Sect. 1.2.3).

The main focus of research in fungus–bacterium interaction has traditionally been in the field of bacteria acting as biocontrol agents against pathogenic fungi. A wide range of bacteria such as *Agrobacterium, Bacillus* spp. (e.g. *B. cereus, B. pumilis*, and *B. subtilis*), *Streptomyces*, and *Burkholderia* are effective antagonists of soil-borne pathogens (Barea et al. 2005). The most widely studied bacteria in relation to biocontrol are, by far, *Pseudomonas* spp., such as *P. aeruginosa* and *P. fluorescens*, probably not only because they are very common and well adapted to life in the rhizosphere, but also due to the fact that they are fast growing, easy to culture and to manipulate genetically and thus amenable to experiment with (Whipps 2001). The production of antifungal metabolites in vitro has often been reported (Keel and Defago 1997; Kang et al. 1998; Nakayama et al. 1999). The use of mutants for antibiotic production or of reporter genes and probes has been applied to determine whether these antibiotics are also produced in the rhizosphere. Several antifungal substances have since been isolated from soil (Bonsall et al. 1997; Raaijmakers et al. 1999; Haas and Keel 2003).

Ectomycorrhizal fungi not only interact with plants but are part of a complex below-ground microbial community offering the opportunity to interact with pathogenic, saprotrophic and symbiotic organisms (Fitter and Garbaye 1994a). The extraradical mycelia of ectomycorrhizal fungi co-exist with saprotrophic fungi with which they interact in the organic layers of the forest soil. Ectomycorrhizal fungi may obtain at least a part of their carbon directly from soil organic matter (Chapela et al. 2001); thus they compete with saprotrophic fungi for nutrient resources. Gadgil and Gadgil (1975) demonstrated an increased degradation of pine litter in plots following removal of ectomycorrhizal fungi and it was concluded that possible antagonistic effects between the different fungi were eliminated. Lindahl et al. (1999, 2001) observed reduction of growth of the saprotrophic fungus *Hypholoma fasciculare* after encountering mycelia of the mycorrhizal fungus *Suillus variegatus* of which the growth was stimulated. Contact with a further ectomycorrhizal fungus, *Paxillus involutus*, in contrast did not result in comparable effects. Lindahl could also show that about 25% of labelled P that was captured in the mycelium of the saprophytic fungus *H. fasciculare* was transferred via the ectomycorrhizae to the host plant within 30 days, implying not only antagonistic effects of ectomycorrhizal fungi but also the ability to scavenge nutrients from the saprotroph (Lindahl et al. 1999). In contrast, merely a limited amount of labelled N was transferred from a *H. fasciculare* to an ectomycorrhizal fungus (*Tomentellus submollis*) when examined under more natural conditions. This transfer was suggested to be due to NH_4^+ released from *H. fasciculare* without any antagonistic interaction between the mycelia involved. The authors suggest that such interactions are highly species-specific and depend on environmental and experimental conditions (Wallander et al. 2006).

Clearly, the interactions between mycorrhizal and saprotrophic fungi are still poorly understood (Leake et al. 2002; Cairney and Merhag 2002; Hättenschwiler et al. 2005) making it difficult to draw general conclusions.

Antagonism of biocontrol fungi against plant pathogenic fungi remains an area of scientific interest, obviously due to the economic importance of plant pathogenic fungi. Best studied examples belong to the *Trichoderma* species and to non-pathogenic strains of *Fusarium* sp. (Whipps and Lumsden 2001). Biological control of *Fusarium* wilt in cereals, mediated through *Trichoderma*, as well as non-pathogenic *Fusarium* strains is associated with competition for carbon, nitrogen and iron (Mandeel and Baker 1991; Couteaudier and Alabouvette 1990; Larkin and Fravel 1999). Competition may also take place for infection sites on the root surfaces. Under the assumption that a root offers a certain number of possible infection sites (Mandeel and Baker 1991), increased amounts of inoculum of the non-pathogenic biocontrol strain might prevent the infection with pathogenic strains. Olivain and Alabouvette (1999) showed that both pathogenic and non-pathogenic strains of the soil-borne fungus *F. oxysporum* actively colonized the surface of tomato roots. Both penetrated the epidermal cells and colonized the upper layer of the cortical cells. Here, plant defence reactions were stronger towards the infection with the non-pathogenic strain as determined by formation of wall thickening and intracellular plugging (Olivain and Albouvette 1999), a result that was confirmed by Olivain et al. (2003) in flax (*Linum usitatissimum*) cell suspensions confronted with germinated microconidia of pathogenic and non-pathogenic *F. oxysporum* strains. The early physiological responses of the flax cells could thus be used to distinguish different strains of *F. oxysporum* with regard to their pathogenicity.

The question of direct competition within the host plant between non-pathogenic and pathogenic strains was addressed by Postma and Luttikholt (1996). Parallel and mixed inoculation of carnation roots with *F. oxysporum* f sp. *dianthi* and several non-pathogenic strains showed that some strains were able to reduce stem colonization by the pathogen thus reducing disease severity. It was hypothesized that locally induced resistance or direct competition between strains within the vessels results in the disease suppressive effect after mixed inoculation into the stem (Postma and Luttikholt 1996).

F. oxysporum Fo47 and other non-pathogenic *Fusarium* strains have been shown to exhibit the biocontrol effects through competition for nutrients in the soil as well as competition for root colonization and induced resistance (Mandeel and Baker 1991; Postma and Rattink 1992; Fravel et al. 2003). It may be expected that a single strain exhibiting all of these modes of action could constitute a more consistent biocontrol strain compared to a strain expressing only one of the described modes of action (Fravel et al. 2003).

During the competitive interaction between bacteria in the rhizosphere many peptide antibiotics called bacteriocins are distributed. These compounds often have an antimicrobial effect on closely related organisms (Rodelas et al. 1998), the bacterial membrane being the target for most bacteriocins (Klaenhammer 1993). Bacteria of the same genus or even species presumably share the same requirements regarding their environment (Sommers and Vanderleyden 2004). With respect to the population densities in soil habitat, the killing or inhibition of those neighbours that share the same environmental requirements enhances the possibility to survive and proliferate within the living space (Wilson et al. 1997; Sommers and Vanderleyden 2004). Acting as anti-competitors, bacteriocins could facilitate the conquest of established microbial communities. On the other hand they might be a defense strategy against invading strains or species (Riley and Wertz 2002).

The bacteriocin, trifolitoxin, is produced by *Rhizobium leguminosarum*. Gene transfer of the trifolitoxin genes to *R. etli* significantly increased nodule occupancy through this strain and suppressed the community of trifolitoxin-sensitive alpha-proteobacteria in the soil (Robleto et al. 1998). The authors show that a specific genetic alteration of a *Rhizobium* strain affects its efficacy under agricultural conditions. They could also show that the peptide antibiotic is produced in non-sterile soil conditions, in spite of its apparent instability in non-sterile soil (Robleto et al. 1998).

Antibiotics produced by *Pseudomonas* sp. are known to be important in competition with other soil inhabitants. The production of phenazine, a secondary metabolite that exhibits antimicrobial activity against a wide range of prokaryotic and eukaryotic microbes, is also important for the ecological adaptability of the producing strain (Mazzola et al. 1992; Cook et al. 1995). Population sizes of strains defective in phenazine production declined much faster than population sizes of phenazine producing strains. The authors suggest that that the decline of the non-producing strains is due to a reduced ability to compete with the resident micro-organisms. In the presence of the plant pathogenic fungus *Gaeumannomyces graminis var. tritici*, the target pathogen of the phenazine producing *Pseudomonas fluorescens* 2–79 and *Pseudomonas aureofaciens* 30–84, the ability to produce phenazine was less critical than without the pathogenic fungus. The authors suggest that the increased proliferation of bacteria on wheat roots infected with *G. graminis var. tritici* might be a result of increased leakage of nutrients from root lesions induced by this fungus which would override the importance of competition in the rhizosphere.

1.2.2 Plant Microbe Interactions

1.2.2.1 Plant Endophytes

Virtually all plants host a large variety of fungi and bacteria that cause no visible disease symptoms. To describe these organisms, the expression "endophyte" was introduced by De Bary (1866). Endophytes were defined as "microbes that colonize living internal tissues of plants without causing any immediate, overt negative effects" (Stone et al. 2000), thus organisms that spend at least part of their life cycle inter and/or intracellularly in healthy tissues of the host can be included (Petrini 1991; Sturz et al. 2000) as well as those that exhibit a more or less lengthy period of epiphytic growth, and even latent pathogens. Thus, organisms that may be described as saprobic or pathogenic under other circumstances are included (Boddy and Griffith 1989). Host-endophyte interactions may under certain circumstances result in disease formation (Schulz and Boyle 2005), which has been described as an unbalanced symbiosis (Kogel et al. 2006) or, vice versa, that the asymptomatic host-endophyte interaction is a balanced antagonism in which neither of the partners gains the upper hand (Schulz and Boyle 2005).

Common to endophytic interactions is the provision of nutrients for the endophyte by the plant and the protection from environmental stresses and competition with other microbes (Schulz and Boyle 2005). The infection of the plant with endophytic organisms may lead to improved ecological adaptability by enhancing the plant's tolerance to environmental stresses like drought (Ravel et al. 1997) or heat (Redman et al. 2002). Furthermore endophyte-infected plants often show improved growth compared to uninfected plants (Cheplick et al. 1989). This effect may be in part due to the production of phytohormones like indole-3-acetic acid (IAA), cytokines (Tan and Zou 2001) and/or by nitrogen fixation (Sevilla et al. 2001; Weidner et al. 2003; Gage 2004) as well as by the improvement of the plant's ability to take up other nutritional elements (Gasoni et al. 1997; Reis et al. 2000). Furthermore, fungal endophytes provide certain grasses with improved protection from herbivory (Preszler et al. 1996; Wilkinson et al. 2000), from some bacterial and fungal pathogens (Christensen 1996; Sturz et al. 1999), nematodes (Hallman and Sikora 1996) and mammals (Bacon et al. 1977).

The occurrence of endophytes has been recorded from almost all vascular plants examined thus far (Sturz et al. 2000; Arnold et al. 2000), as well as from marine algae (Smith et al. 1989), mosses and ferns (Petrini et al. 1992; Raviraja et al. 1996). Among the plant endophytic organisms that were isolated from roots, stem, leaves and seeds are fungi (Stone et al. 2000), bacteria (Kobayashi and Palumbo 2000), algae (Peters 1991), and insects (Feller 1995). Fungi and bacteria seem to represent the prevalent endophytic organisms due to their presence in almost all plant species studied so far (Stone et al. 2000; Strobel et al. 2004). Infection of roots with bacterial or fungal endophytes leads to an extensive and systemic spreading within the root tissue as has been shown for *Penicillium* sp. (Capellano et al. 1987), *Piriformospora indica* (Varma et al. 2000), and the dark septate endophytic fungi (Mandyam and Jumpponen 2005). Root endophytes may even spread into above-ground plant parts, as has been shown for certain rhizobia (Chi et al. 2005). Furthermore, root infection also often results in enhanced plant growth (Schulz et al. 2002). In contrast, infection of shoots or leaves with endophytic fungi seems to remain mainly localised and does not result in improved growth of the host plant (Stone et al. 1994; Carroll 1995).

Identification and characterization of endophytic fungi from the tree *Pinus monticola* showed that 90% of the 2019 fungal isolates belonged to the Rhytismataceae but none of the isolates showed matching sequences with the known species of that family (Glienke-Blanco et al. 2002). The authors concluded that if most of the fungal endophytes represent unknown taxa then the total amount of endophytic fungi may well exceed one million (Glienke-Blanco et al. 2002). Despite this potentially huge amount of unknown fungal species, little work has been conducted to isolate and characterize the species and their corresponding secondary metabolites. Merely the interaction between a specific group of fungal endophytes (Clavicipitales, Ascomycota) and grasses has received special attention. This interaction often results in toxic syndromes in animals feeding on infected grasses and increases the resistance towards insects. Due to the economic importance of this interaction, research has been carried out regarding grass endophyte systematics, genetics, chemistry, and ecology (Clay and Schardl 2002; Schardl et al. 2004; Spiering et al. 2006). The ecological influence such an endophytic fungal-plant co-operation might have was elucidated by Clay and Holah (1999) who demonstrated that tall fescue (*Festuca arundinacea*) infected with the fungal endophyte *Neotyphodium coenophialum* reduced the plant species diversity in its surroundings over the course of four years, resulting in the dominance of this single species. This example also shows a distinct difference between a mycorrhizal plant and an endophyte-inhabited grass. Mycorrhizal fungi colonise numerous host plants and may also transfer nutrients among the hosts (see Sects. 1.1.2 and 1.2.3). They are thus important for the plant community structure (Hartnett et al. 1993; Hartnett and Wilson 1999). In contrast, fungal endophyte symbiosis with tall fescue reduces plant species diversity and promotes the dominance of a single species, irrespective of the inconspicuous amount of biomass of the fungal partner (Clay and Holah 1999). The authors conclude that a "shared symbiosis (as in mycorrhiza) may equalize competitive abilities among plants and promote diversity, whereas private symbioses may increase competitive differentials and decrease diversity" (Clay and Holah 1999).

The best studied bacterial endophytic species, belonging to the rhizobia, are able to induce root nodules with legumes, which provide the plant with fixed nitrogen, leading to improved growth in nitrogen-limited soils (Weidner et al. 2003; Gage 2004). A study of the biodiversity of endophytic bacteria colonizing the nodules, roots, stems and foliage of clover showed that foliage tissue possessed the greatest number of different genera and species (Sturz et al. 1996). In addition to rhizobia, root nodules hosted a further 12 bacterial species including 8 species that were specific to this plant part (Sturz et al. 1996). A further observation of this plant microbe interaction by Tokala et al. (2002) showed a nodulation-helper-effect of a root-colonizing soil streptomycete, resulting in increased nodulation frequency during colonization of the root. The authors hypothezised that this effect is related to more vigorous and long-living rhizobia bacteroids resulting in a higher level of nitrogen fixation within nodules.

Species of the genus *Rhizobium* can frequently be found endophytically in the roots of certain cereal crops. Independently of root nodulation and nitrogen-fixation, plant growth and yields are improved (Biswass et al. 2000a,b; Lupway et al. 2004). Interestingly, following rice root infection with certain *Rhizobium* strains, the endophyte migrates from the roots to the leaves (Chi et al. 2005). These rice plants produced significantly higher biomass, showed enhanced growth, and accumulated higher levels of the growth-regulating phytohormones, indoleacetic acid and gibberellin. This emphasizes the potential value of these bacteria as biofertilizers for sustainable agriculture (Chi et al. 2005).

A novel technology using endophytes as biocontrol agents was described by Taghavi et al. (2005). The endophytic *Burkholderia cepacia* strain VM1468, which was originally isolated from yellow lupine, was transformed with a plasmid coding for constitutively expressed toluene degradation (Taghavi et al. 2005). Inoculation of poplar trees with this strain did not lead to establishment of the transformed bacterial strain but resulted in *in planta* horizontal gene transfer of the plasmid to different members of the endogenous endophytic community (Taghavi et al. 2005).

1.2.2.2 Soilborne Plant Pathogens

Severe plant diseases, including root and crown rots, wilts, and damping-off, are caused by soilborne bacteria, fungi, and oomycetes. These diseases are important yield-limiting factors in agriculture and forestry, and difficult to control by traditional means, such as through the use of resistant host cultivars or synthetic fungicides (Weller et al. 2002). Common and well investigated bacterial agents include Gram negative bacteria *Erwinia carotovora, Pseudomonas, Ralstonia* spp. and the Gram positive bacterium *Streptomyces scabies*. The fungal and oomycete phytopathogens include members of *Fusarium, Phytophthora, Pythium, Rhizopus, Rhizoctonia* and *Verticillium* (Tournas 2005). From the forest pathogens, among the most important are the filamentous fungi *Heterobasidion* and *Armillariella* (Asiegbu et al. 2005), and *Phytophthora* spp. (Rizzo et al. 2005).

The bacteria *Erwinia carotovora, Pseudomonas syringae* and *Ralstonia solanacearum* have become established as model soilborne phytopathogens (Setubal et al. 2005). For example, *P. syringae* pv. tomato causes bacterial speck, a disease that leads to plant stunting and reduced yields if young plants are severely infected. In addition to its natural host tomato, the bacterium causes bacterial speck in *Arabidopsis thaliana,* which eases its use as a model bacterium. *R. solanacearum* causes bacterial wilt in numerous solanaceous plants (e.g. eggplants, tomatoes, peppers), the symptoms of which include wilting, stunting and yellowing of the foliage. Both *R. solanacearum* and *E. carotovora* cause rots of underground organs, e.g. potato brown rot, where the potatoes become brownish and get filled with bacterial slime. The factors related to the pathogenicity of these three bacteria have gained much attention recently and will be covered in the following.

The interactions between the plant hosts and the microbes are initiated by the detection of host released chemical signals and growth towards these cues. Detection of these signals leads to altered patterns of gene expression in the target microbes causes a changed microbial physiology, which is required for successful disease development (Brencic and Winans 2005). Chemotactic behaviour has been indicated in *E. carotovora, R. solanacearum,* and *P. syringae* (McNamara and Wolfe 1997; Mazumder et al. 1999; Yao and Allen 2006). Recently, Yao and Allen (2006) showed that chemotaxis is required not only for virulence but also for competitive fitness in *R. solanacearum.*

Bacteria sense their local population density through quorum sensing (QS). In QS, the bacteria secrete and detect small, diffusible signal molecules. The pathogenic bacteria have incorporated QS mechanisms into the signalling cascades that control genes for pathogenicity and colonization of host surfaces (Henke and Bassler 2004). The studies with *E. carotovora, R. solanacearum,* and *P. syringae* have all yielded important information on quorum sensing and bacterial virulence (Henke and Bassler 2004). From the virulence factors of these phytopathogenic bacteria, quorum sensing systems regulate the production of diverse virulence factors including extracellular polysaccharides, degradative enzymes, antibiotics, and siderophores, as well as protein secretion systems. In addition, pathogen related plasmid transfer, motility, biofilm formation and fitness on and in plant surfaces are under QS controls (Von Bodman et al. 2003). The interference with QS signaling often leads to modulated disease development in plants, e.g. plant-derived metabolites may profoundly influence quorum sensing regulated bacterial processes (Newton and Fray 2004).

Suppression of host plant defenses is a key step for pathogenesis. *E. carotovora, R. solanacearum,* and *P. syringae* mediate this by delivering effector proteins into the plant cells using the so-called type III secretion system. In general, the type III effector proteins function to optimize the host cell environment for bacterial growth, and their main function is to manipulate and suppress host defenses. These molecules are essential for the virulence of *P. syringae, R. solanacearum* and *Erwinia* species (Whitehead et al. 2002; Nomura et al. 2005; Grant et al. 2006). In *Pseudomonas syringae* an additional form of defense suppression has been characterized. Coronatine, a bacterial toxin that structurally and functionally mimics methyl jasmonate, interferes with plant defense signaling during the infection (Toth and Birch 2005).

Among the hundreds of *Streptomyces* species described, only four species to date have been described as plant pathogens. These species, *Streptomyces scabies*, *S. acidiscabies*, *S. turgidiscabies* and *S. ipomoeae*, are agents of common scab disease in potato and other taproot crops (Loria et al. 2006). These diseases lead to reduction of root and shoot length, dramatic radial swelling of roots, tissue chlorosis and necrosis. The mechanisms of pathogenicity behind scab diseases are well documented, since plant disease symptoms are related to the production of a family of cyclic dipeptides, thaxtomins, by the streptomycete (Fig. 1.1). Two important findings have underlined the necessary role of thaxtomins in scab disease development. The application of purified thaxtomins was shown to lead to identical symptoms than the disease itself, i.e. cell hypertrophy and stunted growth (Lawrence et al. 1990; King et al. 1992). Second, when chemically mutagenized *S. scabies* strains were tested for their virulence, all of the mutants that produced lower levels of thaxtomin A relative to the parent strain showed reduced virulence in plant inoculation assays (Goyer et al. 1998). The plant pathogenic *Streptomyces* species possess a conserved biosynthetic pathway for the phytotoxin thaxtomin. The importance of the thaxtomin synthesis cluster was confirmed by an elegant genetic analysis (Kers et al. 2005). A large pathogenicity island, conserved among the plant pathogenic *Streptomyces* species, was transferred from *S. turgidiscabies* to the non-pathogen *S. diastatochromogenes*. As a result the latter bacterium conferred a plant pathogenic phenotype.

The vascular wilt diseases, often caused by the fungi from the genus *Fusarium* or genus *Verticillium*, lead to a particularly fast and effective killing of their plant hosts. *Fusarium* and *Verticillium* spp. enter the host roots directly through penetration hyphae and colonize the root cortex by intracellular and intercellular growth. Once reaching the vascular tissue, the pathogens spread rapidly upward through the xylem vessels, essentially blocking the transpiration stream and thus provoking the characteristic wilt symptoms.

Fusarium oxysporum is an important pathogen that causes vascular wilt, and leads to economically important losses in a wide variety of plants including

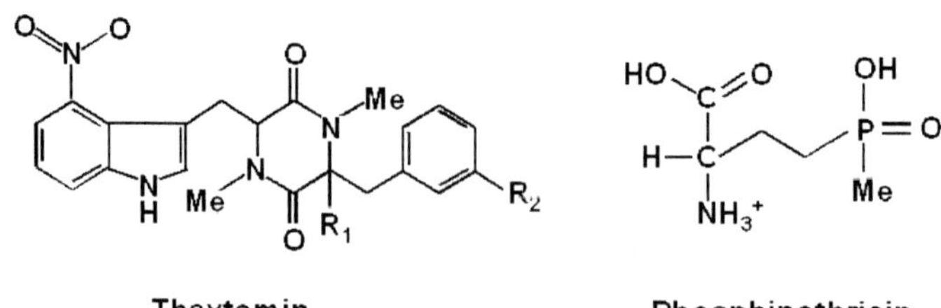

Fig. 1.1 Bioactive secondary metabolites from streptomycetes – the phytotoxins thaxtomin and phosphinothricin

vegetables, flowers and tree seedlings (Salerno et al. 2000; Bai and Shaner 2004). *F. oxysporum* is increasingly used for molecular studies. As an example the use of green fluorescent protein based visualisation techniques has enabled the *in situ* analysis of the plant infection process (Lagopodi et al. 2002), and the observation of *F. oxysporum* interacting with biocontrol organisms (Bolwerk et al. 2003, 2005). During plant colonization, *F. oxysporum* produces a necrosis- and ethylene-inducing peptide (Nep1) that inhibits both root and cotyledon growth and triggers cell death, thereby generating necrotic spots in the host plants. The majority of Nep1-induced plant genes are associated with general stress response and ethylene signal transduction (Bae et al. 2006), indicating that Nep1 is a facilitator of necrosis in *F. oxysporum*-plant interaction. The *Fusarium* spp. produce a variety of phytotoxins during plant colonization. These metabolites cause plant growth retardation, inhibition of seedling growth and suppressed plant regeneration (Rocha et al. 2005). *F. oxysporum* strains commonly produce toxins, and the search for the toxin biosynthesis genes in this fungus is in progress (Proctor et al. 2004; Llorens et al. 2006).

Members of the genus *Phytophthora* are among the most serious plant pathogens, causing devastating diseases in hundreds of plant hosts and immense losses in agriculture (Grunwald and Flier 2005) and forestry (Rizzo et al. 2005). These fungus-like eukaryotes generate asexual and sexual spores that are used for survival in adverse conditions, and as infectious propagules capable of actively locating host plants. *Phytophthora* spp. infect by motile zoospores that land on the plant leaves or infect the roots from the soil solution. The infection includes recognition of the host by sensing host-specific factors such as isoflavones, and host-nonspecific factors such as amino acids, calcium, and electrical fields. These not only influence the direction of spore movements, but also hyphal chemotropism in guiding the pathogen to potential infection sites. The infection is systemic, and leads mostly to the death of host plants.

Genetic analysis of *Phytophthora*-plant interaction has identified host resistance genes and pathogen avirulence genes that interact in a gene-for-gene manner (Huitema et al. 2004). Genomic resources developed over the past few years are now allowing detailed analysis of the *Phytophthora* life cycle (Huitema et al. 2004; Judelson and Blanco 2005). The speciation of *Phytophthora* has received increased interest, and central highlands of Mexico are considered to be a center of genetic diversity for these microbes. Not only the potato late blight pathogen *P. infestans*, but also several related *Phytophthora* species including *P. mirabilis, P. ipomoeae*, and possibly *P. phaseoli* originate from this area (Grunwald and Flier 2005).

The *Heterobasidion* root and butt rot is one of the most destructive diseases of forest trees. The ecology of the disease has been intensively studied, and recently, by the development of infection investigated by genetic tools (Li and Asiegbu 2004; Asiegbu et al. 2005; Adomas et al. 2006). Karlsson et al. (2003) identified the first fungal genes differentially expressed during contact with roots, including genes encoding mitochondrial proteins, a cytochrome P450 and a vacuolar ATP synthase. Samils et al. (2006) recently developed a rapid and simple *Agrobacterium*-mediated method of gene delivery into *H. annosum*, and Lind et al. (2006) the first genetic linkage map of the same fungus. *Heterobasidion annosum* may well develop to a model of a forest tree pathogen.

The extent of sequenced genomes of phytopathogens is ever increasing, and includes those of both bacteria and fungi. Examples of the bacterial species discussed here include the genomes of *Erwinia carotovora*, *Pseudomonas syringae* (pv. phaseolicola, syringae and tomato) and *Ralstonia solanacearum*. In these bacteria, genome analyses have already yielded significant information on their specific ways to cause plant disease (Setubal et al. 2005). From fungal soilborne pathogens, which do have considerably larger genomes than most bacterial species, the first to be finished was that of *Fusarium oxysporum*. However, several important plant pathogenic fungi are among those being currently sequenced (Xu et al. 2006).

1.2.2.3 Mycorrhiza

Defence Reactions During Colonization

Intruders of plant tissues generally induce plant defence responses. These include the formation of reactive oxygen species, such as superoxide and hydroxy radicals or hydrogen peroxide, which can prevent the spreading of pathogens (e.g., Gafur et al. 2004 and references therein). Another strategy is the induction of secondary metabolism (mainly phenylpropanoid metabolism; for references see, e.g. Hause and Fester 2005). The respective products precipitate proteins and can thus kill the respective (micro)organisms. Additionally, the production of proteins involved in defense reactions is increased (e.g. chitinases, peroxidases, glutathione *S*-transferase, pathogenesis-related proteins; Duplessis et al. 2005). Symbiotic fungi activate such mechanisms partly in a way similar to fungal pathogens (Dowkiw et al. 2003), but to a much lower extent and only transiently. They are largely repressed when the symbiosis develops (for ECM see Johansson et al. 2004). Dumas-Gaudot et al. (2000) suggested several mechanisms to explain these responses such as weak fungal elicitors, release of both fungal inhibitors of activation of plant secondary metabolism, or compounds that alleviate the efficiency of elicitors. Also the potential fungal symbiont can prevent damage due to plant defence reactions in the initial phase of contact. This results from an increased fungal expression / activity of detoxifying enzymes (Menotta et al. 2004; Morel et al. 2005).

Establishment of Mycorrhiza/Signalling Pathways

In order to form a functional mycorrhiza a genetic reprogramming of both partners has to take place. This way biochemical and morphological adaptations are introduced. For ECM it has been shown that the symbiotic fungus can support growth and development of its host plant already before physical contact is established

(Herrmann et al. 2004). It is suggested that the duration of such a presymbiotic phase largely depends on capacity of the host to deliver carbohydrates (Herrmann et al. 2004). Already in the presymbiotic phase, ECM-forming partners show changes in gene expression (Menotta et al. 2004), which are extended later on (Duplessis et al. 2005). In the early stages of interaction, fungi show an increased expression of cell wall proteins (hydrophobins, mannoproteins) while the host plant up-regulates defence-related gene expression (chitinases, peroxidases), together with respiration-related activities, especially of the fungus (Duplessis et al. 2005). Collections of symbiosis-related expressed sequence tags have been established for *Paxillus involutus*/*Betula pendula* (Johansson et al. 2004), *Hebeloma cylindrosporum* (Lambiliotte et al. 2004).

The establishment of an AM mycorrhiza and of root nodules (symbiosis of rhizobia with leguminous plants) has some signal transduction pathways in common (symbiosis receptor-like kinases; nodulation receptor kinase; Endre et al. 2002; Stracke et al. 2002).

Induced Systemic Resistance

AM can increase root resistance towards soil-borne pathogenic fungi (Dumas-Gaudot et al. 2000). They are thus able to induce a kind of bioprotection as reported for other non-pathogenic micro-organisms. This effect is not connected to changes in the host phenotype but increases plant resistance systemically to a wider range of pathogenic organisms (induced systemic resistance, ISR; Park et al. 1997; for a discussion with regard to AM see Hause and Fester 2005). The weak and transient induction of secondary metabolism in the host plant could be one of the reasons for the ISR response (see also Sect. 1.1.3).

1.2.2.4 Plants as Parasites on Fungi

A wide range of plants has succeeded in obtaining water and nutrients from fungi without delivering photoassimilates. Members of the *Monotropoidae*, e.g. are photosynthetically incompetent. Their seeds possess only marginal reserve products and depend on specific fungal support for germination (Bidartondo and Bruns 2005). Orchids are another example for fungus-dependent plants. About 200 species of the Orchidaceae are obligate myco-heterotrophs (Bidartondo et al. 2004). It is generally assumed, that these make use of fungi, which form ectomycorrhizas with forest trees. This way the respective orchids gain indirect access to external photoassimilates. In contrast, photoautotrophic orchids were reported to parasite on saprotrophic and pathogenic Basidiomycetes. Recent investigations also give evidence that photoautotrophic orchids, especially when they grow at low light sites (limited production of photoassimilates), also exploit ECM fungi (Bidartondo et al. 2004; Taylor et al. 2004; Selosse et al. 2004).

1.2.2.5 Influence of Invading Plants upon Soil Microorganismic Communities

As plants differ in their morphological (herbaceous, woody plants) and biosynthetic properties (e.g. secondary metabolites), the type of plant community will influence the microorganismic soil community (Wolfe and Klironomos 2005). Introduction of exotic plants into existing ecosystems can thus interfere with soil communities in many ways, exerting negative, positive or no effects (Wolfe and Klironomos 2005). For garlic mustard, e.g., it has been shown that the release of glucosinolates (antibiotic effects of the non-glucoside moiety) to the soil affects the abundance of endomycorrhiza-forming fungi (Roberts and Anderson 2001). Vice versa, the introduction of plants not naturally present in a given ecosystem can introduce new micro-organisms. Such an effect could be observed by reforestation of subtropical forests with pine (e.g., Brazil; own observations; tropical trees form endomycorrhizas, while trees originating from boreal zone are ectomycorrhizal). The ectomycorrhiza-forming fungi have spread in many regions of the Southern hemisphere this way (Richardson et al. 1994). Non-native plants can also be less sensitive to soil pathogens or even increase their abundance, with negative effects for native plants (Klironomos 2002).

1.2.3 Tri-Partite Interaction: Mycorrhizas and Helper Bacteria

The rhizosphere effect leads to the selection of distinct microbial communities (Smalla et al. 2001), where fungi play an important role (Linderman 1988; Andrade et al. 1997; Frey-Klett et al. 2005; de Boer et al. 2005). This is due to the fact that the fungal hyphae, which emanate from mycorrhizas, release a substantial amount of the acquired plant carbon to the soil. The energy-rich plant compounds promote bacterial growth and survival (Hobbie 1992), and the area rich in mycorrhizal fungal hyphae is therefore often depicted as the mycorrhizosphere (Foster and Marks 1967). Numerous reports exist on both ectomycorrhizal (Garbaye 1994) and endomycorrhizal (Artursson et al. 2006) fungi interacting with these mycorrhizosphere organisms, including bacteria, yeasts and filamentous fungi (Garbaye and Bowen 1989; Garcia-Romera et al. 1998; Fracchia et al. 2000; Sampedro et al. 2004; Duponnois et al. 2006).

The mycorrhizosphere effect leads to the enrichment of micro-organisms that improve plant fitness (Frey-Klett et al. 2005). Some of the mycorrhizosphere organisms directly influence the development and physiology of the plants through the production of plant growth regulators (Azcon et al. 1978) by increasing the root branching rate or root permeability. Others may interact in a more indirect way that supports plant growth. For example, the interacting organisms may improve nitrogen or phosphate availability, lead to better survival of the hyphae and the plants in a contaminated soil, assist the plant resistance against pathogens by biological control, or show direct effects on soil quality (Barea et al. 2005; Frey-Klett et al. 2005).

The most investigated group of micro-organisms interacting with mycorrhizal fungi and plants are the mycorrhizosphere bacteria (de Boer et al. 2005). These include intrahyphal bacteria in ectomycorrhizal (Bertaux et al. 2003) and intra-spore bacteria in arbuscular mycorrhizal (Bianciotto et al. 1996) fungi, and bacterial species that colonize the surfaces of fungal hyphae and mycorrhizal roots (Foster and Marks 1966; Nurmiaho-Lassila et al. 1997; de Boer et al. 2005; Artursson et al. 2006). Some of the mycorrhizosphere bacteria, classified as mycorrhiza helper bacteria (MHB, Garbaye 1994) can promote mycorrhiza formation. These include a variety of Gram-negative (Gryndler and Vosatka 1996; Founoune et al. 2002) and Gram-positive (Ames 1989; Abdel-Fattah and Mohamedin 2000; Maier et al. 2004; Schrey et al. 2005) species. In the following, we will give an overview on the relevance and the mechanisms of the mycorrhiza helper effect.

Numerous soil micro-organisms interact with arbuscular mycorrhizal (AM) fungi by producing substances that stimulate plant growth and mycorrhiza formation (Fitter and Garbaye 1994b). MHB of AM symbiosis include actinomycetes (Ames 1989; Abdel-Fattah and Mohamedin 2000), pseudomonads (Gryndler and Vosatka 1996; Gamalero et al. 2004) and bacilli (Vivas et al. 2003). The data with *Pseudomonas* species indicate that these mycorrhiza helper bacteria produce hyphal growth promoting substances. Gryndler and Vosatka (1996) observed that the mycorrhizal infection rate of the roots and the growth rate of soil substrate hyphae were significantly higher when plants were inoculated with the mycorrhizal fungus *Glomus fistulosum* together with *Pseudomonas putida* or with the cell extract of the bacterium than with the fungus alone. After fractionation of the liquid *Pseudomonas putida* culture, Vosatka and Gryndler (1999) observed, that the application of a fraction of low molecular weight increased the rate of mycorrhiza formation and the extension of extraradical hyphae. *Paenibacillus validus* stimulated the growth of the arbuscular mycorrhizal fungus *Glomus intraradices,* and has enabled the growth of this otherwise obligately symbiotic fungus *in vitro* without the plant partner (Hildebrand et al. 2002). Recently, Hildebrand et al. (2006) observed, that the hyphal growth of *G. intraradices* can be supported through the bacterial production of a specific C source, raffinose. This sugar alone was not able to support the production of fertile spores, indicating that other, additional bioactive substances are involved in the *Paenibacillus validus – G. intraradices* interaction.

The growth of ectomycorrhizal (ECM) fungi and mycorrhiza formation are similarly promoted by MHB. The MHB have been predominantly reported from fluorescent pseudomonads and sporulating bacilli (Garbaye and Bowen 1989; Founoune et al. 2002), and also from the genera *Burkholderia, Rhodococcus,* and *Streptomyces* (Poole et al. 2001; Maier et al. 2004; Schrey et al. 2005). Two main mechanisms behind the MHB function have been observed thus far: promoted fungal growth rate (faster colonization of root tips) and increased lateral root branching rate (more infection points). In other cases, a direct contact between MHBs and plant roots may be required for the promotion of the mycorrhizal symbiosis, as has been demonstrated for the MHB *Paenibacillus* sp. EJP73 (Aspray et al. 2006).

We have worked on a collection of actinomycetes that were isolated from the rhizosphere of a Norway spruce (*Picea abies*) stand. One of the isolates,

Streptomyces sp. AcH 505 significantly promoted the mycelial growth and mycorrhization rate of the symbiotic fungus *Amanita muscaria* but suppressed the mycelial extension of the plant pathogens *Armillariella obscura* and *Heterobasidion annosum* (Maier et al. 2004). This indicates an important application for the mycorrhiza helper bacterium AcH 505: the simultaneous promotion of mycorrhizal symbiosis as well as the suppression of pathogenic fungi. The bacterium influences the gene expression of the symbiotic fungus *A. muscaria*. The responsive fungal genes included members of signalling pathways, metabolism, cell structure, and cell growth-response (Schrey et al. 2005; Tarkka et al. 2006).

The interaction between mycorrhiza helper bacterium *Streptomyces* sp. AcH 505, symbiotic fungi and the plant host also involves the production of secondary metabolites by the streptomycete. The fungal growth is stimulated by the bacterial substance 5,6,7-trihydro-7-hydroxy-3-prolylbenzofuran-4-one, a novel compound classified as auxofuran (Fig. 1.2) due to its similar appearance to indole acetic acid (Riedlinger et al. 2006; Keller et al. 2006). Auxofuran is a non-selective growth promoter, showing its strongest effect at high nM to low µM concentrations (Riedlinger et al. 2006, N. Lehr and M. Tarkka unpublished). The different outcomes of the interaction between AcH 505 and the symbiotic and pathogenic

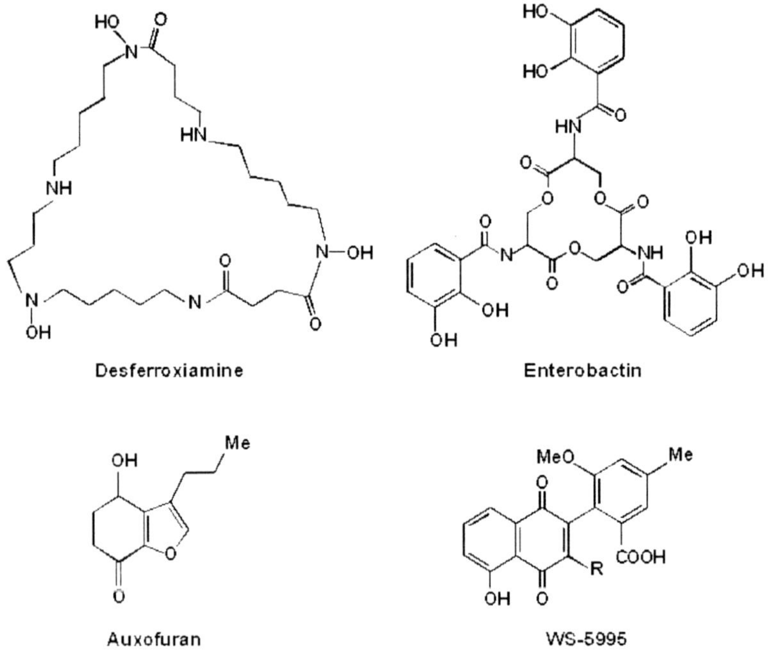

Desferroxiamine

Enterobactin

Auxofuran

WS-5995

Fig. 1.2 Secondary metabolites from streptomycetes – the siderophores desferroxiamine and enterobactin, fungal growth promoter auxofuran, and the antibiotic WS-5995. WS-5995 B, R=H; WS-5995 C, R=OH

fungal species could be explained by the bacterial production of the naphthoquinone antibiotics WS-5995 B and C. Since Schrey et al. (2005) observed an antifungal activity of strain AcH 505 against the ectomycorrhizal fungus *Hebeloma cylindrosporum*, the effect of WS-5995 B was tested on this fungus, and it was found that *H. cylindrosporum* is significantly more sensitive to WS-5995 B than *A. muscaria* (Riedlinger et al. 2006). These observations suggest that the outcome of the interaction between the bacterium and fungal species largely depends on the sensitivity of the fungi towards WS-5995 B and WS 5995 C: the fungi that are suppressed by AcH 505 in co-culture are sensitive against these antibiotics.

Mycelial morphology may be influenced during interaction with MHBs. Co-culture with *P. fluorescens* BBc6R8 leads to enhanced extension growth in *L. bicolor*, and in an increased branching angle and branching density in the fungal mycelium (Deveau et al. 2007). In the case of the MHB *Streptomyces* sp. AcH 505, although the bacterium promotes mycelial extension of *Amanita muscaria*, it sharply reduces hyphal biomass/colony area ratio due to a reduction in mycelial density (Schrey et al. 2007). Moreover, it reduces the diameter of the fungal hyphae (Maier 2003). We recently analysed the structural background of this hyphal thinning: bacterial inoculation leads to a changed organisation of the actin cytoskeleton in *A. muscaria* (Schrey et al. 2007).

Because of their selectivity, it has been suggested that mycorrhiza helper bacteria could become an alternative to soil fumigation, e.g. they could simultaneously be used for controlled mycorrhization and for antagonism against phytopathogenic fungi (Duponnois et al. 1993). However, our data on the interaction between the phytopathogen *H. annosum* and *Streptomyces* sp. AcH 505 are of concern (Lehr et al. 2007). We observed that while 11 *Heterobasidion annosum* strains tested were suppressed by AcH 505, root infection with one fungal isolate was promoted by the bacterium, involving a mechanism that leads to suppression of plant defence response. This suggests that some MHBs might behave as helpers of both symbiotic and pathogenic fungi.

The occurrence of MHB outside the traditionally investigated pseudomonads and bacilli (Poole et al. 2001; Schrey et al. 2005) suggests that the MHB activity could perhaps be found from all bacterial groups that exist in the rhizosphere. This argues for a taxonomically neutral approach while analysing potential MHB. In order to better investigate the MHB-fungus-plant interactions under natural conditions, it is necessary to isolate MHB strains that have the potential to exert their effects in the competitive soil substrate. For this, novel screening methodologies should be developed, where the endurance of the given bacterial species in the mycorrhizosphere should be addressed. As a recent technical breakthrough, Artursson et al. (2005) developed an effective bromodeoxyuridine incorporation method, which enables the identification of actively growing bacteria in the soil. This has proved to be a useful technique in the monitoring of mycorrhizosphere bacteria. The range of habitats that have been investigated in respect to MHB has recently broadened, for example through the work with *Glomus* MHB at heavy-metal-polluted sites (Vivas et al. 2003) and by the analyses of *Acacia* mycorrhizas in the sahelian areas (Duponnois et al. 2006).

1.2.4   Role of Micro-organisms in Weathering

Weathering is an important means to make essential elements available for plants. It is a result of physical disintegration and chemical decomposition of rocks and contributes substantially to soil fertility and ecosystem productivity (Hoffland et al. 2004). Micro-organisms such as bacteria and fungi play a pivotal role in this process, which depends on redox reactions and the liberation of organic acids. The latter can be released by fungal hyphae. Due to the high carbon cost, symbiotic fungi have an advantage over saprotrophic fungi, which are limited by carbon supply according to the availability of litter. Most efficient with regard to weathering are fungi, which form ecto and ericoid mycorrhizas (Landeweert et al. 2001). For a recent review on the topic see Hoffland et al. (2004).

1.3   Chemical Ecology of the Rhizosphere

1.3.1   Endophytic Organisms as Sources of Bioactive Secondary Metabolites

Secondary metabolites have been defined as "outwardly directed compounds produced during differentiation of a living organism" (Samson and Frisvad 2004), which are of low molecular weight and are not required for growth in pure culture. Instead, they fulfil essential functions during adaptation to natural surroundings (Demain 1981). Endophytes are especially interesting as a novel source of secondary metabolites, as these compounds can be biotope-specific (Schulz et al. 2002). Endophytes produce secondary metabolites in order to compete with other epiphytic organisms for space on the plant before entry, within the host tissue with pathogens, and presumably interfere with the fine-tuning of the host metabolism (Schulz et al. 2002). Thus, the isolation of metabolites from organisms with such distinct and unusual environments promises the detection of novel biologically active substances with the potential for medical, agricultural, and/or industrial exploitation. Plants from unique environmental settings promise to be interesting sources. Good candidates are those who have an unusual biology, exclusive survival strategies, a relevant ethnobotanical history, as well as plants with an unusual longevity, or plants growing in areas with large biodiversity (Strobel et al. 2004).

The diversity of endophyte secondary metabolites was reviewed by Tan and Zou (2001), by Strobel and Daisy (2003) and more recently by Strobel et al. (2004). The isolated secondary metabolites belong to several chemical categories like amines and amides, indole derivatives, pyrollizidines, steroids, terpenoids, sesquiterpenes and diterpenes, flavonoids and others (Tan and Zou 2001).

Endophytic fungi have been reported to produce a variety of secondary metabolites that are able to kill harmful micro-organisms like phytopathogens, bacteria, fungi, viruses, and protozoans (Tan and Zou 2001; Strobel and Daisy 2003; Strobel et al. 2004). A novel antimycotic substance, cryptocandin, was isolated from the endophytic fungus *Cryptosporiopsis* cf. *quercina* from the medicinal plant *Tripterigeum wilfordii*. This substance is a unique lipopeptide antibiotic with an inhibitory potential against, e.g. *Candida albicans* (human pathogenic fungus) and *Botrytis cinerea* (plant pathogenic fungus). However, the biological role of crypto-candin in nature has not been investigated. The authors suggest that cryptocandin might act in pathogen defense due to the general antimycotic potential of the sub-stance (Strobel et al. 1999). Another medicinal plant, *Erythrina crista-galli*, hosts species of the fungal endophyte *Phomopsis*. The secondary metabolite, phomol, extracted from fermentations of *Phomopsis* shows antifungal, antibacterial as well as weak cytotoxic activity (Weber et al. 2004). The plant harbouring the endophyte was traditionally used as an anti-inflammatory drug, a trait that could be repro-duced under laboratory conditions.

The isolation of bacteria belonging to the actinomycetes, or ideally to the strep-tomycetes, enhances the opportunity to find bioactive secondary substances, since these organisms are well known producers of antibiotics (Demain 1981). *Streptomyces* NRRL 3052 that was isolated from the Australian medicinal plant *Kennedia nigriscans,* used to dress bleeding wounds (Castillo et al. 2002), was shown to produce munumbicins A, B, C, D, E-4 and E-5. These constitute antibiot-ics with a broad activity against many human as well as plant pathogenic fungi and bacteria, and a malarial parasite, *Plasmodium* sp. (Castillo et al. 2002, 2006). Due to the occurrence of further bioactive compounds the authors claim that *Streptomyces* NRRL 3052 is the most biologically active endophytic *Streptomyces* spp. on record (Castillo et al. 2006).

Comparably, coronamycin is a novel antibiotic isolated from an endophytic streptomycete from *Monstera* sp., a plant growing in the upper Amazon region of Peru (Ezra et al. 2004). The peptide antibiotic shows activity against certain plant pathogenic fungi and the human fungal pathogen, *Cryptococcus neoformans*, but also has activity against *Plasmodium falciparum* (Ezra et al. 2004).

These examples show that endophytes are a promising source of bioactive secondary metabolites with great potential for use in agriculture, medicine and industry. They also show, however, that the ecological role of such metabolites in the interaction between endophyte and plant is largely unknown. One exception is the co-operation between endophytic fungi of the genus *Neophytodium* and *Festuca arundinacea* (tall fescue grass). Peramine, an alkaloid that was isolated in culture as well as *in planta* is toxic to insects without being harmful to mammals (Dew et al. 1990; Rowan and Latch 1994). In addition, other alkaloids have been described from the *Neophytodium–Festuca* interaction and their ecological roles have been addressed (Petroski et al. 1992). These compounds are toxic to mammals and insects, frequently producing toxicoses to feeding animals, and are thus also of economical interest.

Generally, very little is known about the metabolite spectrum of an endophytic organism in sterile culture compared to natural conditions. It is at least conceivable that certain metabolites are produced in vivo in order to occupy an ecological niche or to fine-tune the host metabolism, properties that might be lost during axenic cultivation. Very little is also known about both plant/endophyte and endophyte/ endophyte interactions (Strobel et al. 2004). Elucidation of the molecular and biochemical regulation of effector molecule production, as well as a deeper knowledge about their ecological functions is clearly needed.

1.3.2   The Roles of Streptomycetes in Plant Pathogenesis and Symbiosis

The streptomycetes are distinguished by their ability to produce diverse secondary metabolites (Goodfellow and Williams 1983). To date, approximately 17% of biologically active secondary metabolites (7600 out of 43,000; Berdy 2005) have been characterised from these filamentous bacteria. The streptomycetes commonly produce 2–3 dominant secondary metabolites, along with approximately 10 minor compounds. To ensure that these mixtures are effective, these bacteria often produce synergistically acting compounds, and the cost of their production is greatly decreased due to so-called combinatory biosynthesis. The acquisition and the retention of the substance diversity are enabled due to horizontal gene transfer between streptomycete isolates, together with the strong microbial competition (Firn and Jones 2000; Challis and Hopwood 2003; Davelos et al. 2004; Weissman and Leadlay 2005).

When applied as a single substance, most of the chemicals produced by streptomycetes do not possess any activity against specific target molecules unless tested at high concentrations (Firn and Jones 2000). In response to this, the streptomycetes simultaneously produce several bioactive secondary metabolites, which, in combination, possess a strong biological activity (Challis and Hopwood 2003). These mixtures, often consisting of antibiotics, metal chelators and growth regulators, may act in a synergistic way. For example, several reports suggest, that as single antibiotic production does not show any significant effect on the target organism, the mixture of many antibiotics kills the target (Cocito et al. 1997; Liras 1999). Other sorts of streptomycete chemicals may also act together. In their search for secondary metabolites from streptomycetes, Fiedler et al. (2001) detected an uncommon iron chelating substance from the culture extracts of two streptomycete strains. This was identified as enterobactin, a characteristic siderophore of *Enterobacteriaceae* sp. (Fig. 1.2). The production of enterobactin in addition to the common siderophores, desferri-ferrioxamine B and E, could be an important fitness factor in an iron-poor soil substrate (Challis and Hopwood 2003).

To lower the costs for secondary substance production, streptomycetes have evolved the combinatorial biosynthesis of secondary metabolites. This permits the production of a mixture of secondary metabolites by relatively small changes in the synthesis machinery. Polyketides represent an extremely rich source of biologically active compounds produced through combinatorial biosynthesis (Weissman and Leadlay 2005). Polyketide synthases contain subunits with differing functions, and small differences in the combination of these modules can cause an alteration in the molecular structure of the final product (Firn and Jones 2000).

The frequency of microbial encounters will increase at increased population densities. This has a strong effect on the extent to which streptomycetes produce chemicals in the soil. As an example, populations of high density have a stronger relative benefit from antibiotics production than low-density populations (Wiener 2000). Davelos et al. (2004) showed that the frequency and intensity of antibiotics production by streptomycetes are related to the location of the streptomycetes in the soil. Where high population densities were achieved, the isolates produced more frequently antibiotic substances (Davelos et al. 2004). In the rhizosphere, the microbial population densities are extremely high, and the data from Davelos et al. (2004) suggests that the rhizosphere could be a hotspot for antibiosis. Indeed, Frey-Klett et al. (2005) were able to show that the rhizosphere effect leads to the enrichment of bacteria that suppress plant pathogenic fungi.

The soil-borne plant pathogens – rots, wilts, and damping-off diseases – greatly reduce plant productivity in agriculture and forestry. Most of these pathogens are difficult to control by conventional methods such as the use of resistant plant cultivars or the controlled release of anti-microbial chemicals, since pathogens often become resistant against specific host defense mechanisms, and the spread of a single antibiotic species in the field rapidly leads to a selection for resistance genes in the target organisms (McManus et al. 2002; Fravel 2005). Biocontrol means the use of microbes for the amelioration of plant diseases. Streptomycetes have been characterised as an important group of biocontrol agents (Kerry 2000; Whipps 2001; Weller et al. 2002).

Streptomycetes can inhibit diverse plant pathogens, including Gram-positive and Gram-negative bacteria, fungi and nematodes (Crawford et al. 1993; Chamberlain and Crawford 1999; El-Abyad et al. 1993; Samac and Kinkel 2001). Biocontrol strains from *Streptomyces* spp. have often been isolated from so-called suppressive soils; soils whose microbial communities suppress plant disease development (Weller et al. 2002). For example, the common scab disease of potato, caused by *Streptomyces scabies* (Loria et al. 1997), is effectively suppressed by streptomycete isolates from *S. scabies* suppressive soils. The disease suppression effect can be long lasting. Liu et al. (1995) added streptomycete strains to a soil that was non-suppressive against potato scab. Even by the fourth year after the inoculation, the disease reduction due to these strains was at 63–73% in comparison to the mock-treated control. The biocontrol activity of these *Streptomyces* spp. can often be explained by direct inhibition of the growth of the pathogen (Eckwall and Schottel 1997; Liu et al. 1996), but it has been observed, that stronger biocontrol agents also have a capacity to resource competition (Neeno-Eckwall et al. 2001; Schottel at al. 2001).

The endophytic bacteria comprise several *Streptomyces* spp. (Hallmann et al. 1997). The streptomycete endophytes have been suggested as a promising source of biologically active natural products, since some streptomycetes associated with plants prevent the host from being attacked successfully by fungi and pests through the production of specific substances. These include antibiotics (Taechowisan et al. 2005), siderophores (Cao et al. 2005) and plant disease resistance inducing metabolites (Igarashi et al. 2000).

Streptomycetes are able to promote or suppress plant root symbioses. The outcome of the interaction is largely dependent on the level of resistance, which the symbiotic microbe possesses against the streptomycete-supplied antimicrobial substances. Gregor et al. (2003) showed, that wild type strains of *Bradyrhizobium japonicum* were unable to form root nodule symbiosis with soybean following co-inoculation with the antagonistic *Streptomyces kanamyceticus*. In contrast, *B. japonicum* mutants with increased antibiotic resistance formed significantly more root nodules in the presence of the streptomycetes than without it (Gregor et al. 2003), indicating that the antibiotic production by the bacteria may mask their nodulation promoting capabilities. *Streptomyces lydicus* WYEC108 suppresses root pathogenic fungi by mycoparasitism and by the secretion of antifungal metabolites (Crawford et al. 1993; Yuan and Crawford 1995). It also promotes plant growth, possibly through siderophore production. It was shown by Tokala et al. (2002) that *S. lydicus* colonizes the outer layers in pea root nodules and promotes root nodulation. The root nodule colonization by the streptomycete leads to a significantly increased rates of nitrogen fixation (Tokala et al. 2002), increasing the benefit of the symbiosis for the plant host. The promotion of mycorrhizal symbiosis establishment by *Streptomyces* sp. AcH 505 is dealt with in Sect. 1.2.3.

Several of the most potent natural phytotoxins have come from nonpathogenic microbes, including some produced by streptomycetes (Barazani and Friedman 2001). The ecological role for phytotoxin production has not gained much interest, except for the thaxtomins (Fig. 1.1). These are cyclic dipeptides produced by *Streptomyces scabies*, *S. acidiscabies*, *S. turgidiscabies* and *S. ipomoeae,* agents of common scab disease in potato and other taproot crops (Loria et al. 1997). The thaxtomins cause reduction of root and shoot length, dramatic radial swelling of roots, tissue chlorosis and necrosis. Thaxtomin production seems necessary for proper scab disease development, since all *S. scabies* mutants that produce lower levels of thaxtomin A relative to the parent strain show reduced virulence in plant inoculation assays (Goyer et al. 1998). Indeed, the application of purified thaxtomins was shown to lead to identical symptoms than the disease itself, i.e. cell hypertrophy and stunted growth (Lawrence et al. 1990; King et al. 1992).

The streptomycetes may have an impact on plant populations due to their ability to produce phytotoxins. As with antibiotic production, some of the streptomycete species may produce more than one phytotoxin; for example, three different phytotoxins, geldanamycin, nigericin, and hydanthocidin, have been isolated from *Streptomyces hygroscopicus* (Barazani and Friedman 2001). Many of these phytotoxic compounds have proved most useful as herbicides (Saxena and Pandey 2001). The glutamine synthase inhibitor phosphinothricin (PT) is the most successfully marketed natural product-based herbicide (glufosinate, Fig. 1.1). PT itself is produced by *Streptomyces viridochromogenes,* and its precursor bialaphos, requiring metabolic conversion in the target plant for herbicidal activity, is produced by some other *Streptomyces* spp (Kondo et al. 1973; Omura et al. 1984). PT is a relatively inexpensive herbicide, toxicologically and environmentally safe, and kills a wide range of target weeds. It has even bactericidal and fungicidal properties (Barazani and Friedman 2001). The complete PT biosynthetic gene cluster from *Streptomyces viridochromogenes* Tu494 was isolated by Schwartz et al. (2004), which makes it possible to investigate its biosynthesis in detail, and could later lead to the production of novel PT-based herbicides. In addition, the *Streptomyces* spp. produces a wide variety of other phytotoxic substances (Saxena and Pandey 2001).

In general, streptomycetes have received less attention than other micro-organisms as agents of biocontrol, as saprophytes or as modulators of plant symbioses. Expanded studies are surely warranted due to the abundance of these bacteria in soil, their ability to produce antibiotics, herbicides and other bioactive substances and to split down complex substrates. In addition to their dominant position as producers of pharmaceutical compounds (Berdy 2005), these bacteria could certainly become more important in agricultural use.

1.4 Concluding Remarks and Future Prospects

Soil health is an important parameter affecting plant productivity. Future work on rhizosphere interactions should thus address the functional importance of changes in plant and microbial communities and use this knowledge to, e.g. improve agricultural practices.

Many of the obstacles to increase our knowledge on plant-microbe interactions are of methodological origin. The development of molecular techniques has advanced our understanding of rhizosphere microbial communities and processes.

Ribosomal DNA based methods have provided some information on specific changes in microbial populations, but whether such populations are universally involved in specific processes, such as nutrient cycling, plant disease suppression etc., has to be specifically targeted.

Maybe the most important question to be asked concerns the functional consequence of the rhizosphere effect. The first metagenomic analyses with soil DNA and soil RNA have provided insights into to previously unknown gene clusters, and expressed sequences from uncultivable soil bacteria. Due to the continuing development of sequencing techniques and functional genomics platforms, we should, for example, be able to get access to ecosystem functions of unculturable rhizosphere micro-organisms in the future (see Sessitsch et al. 2006).

It can be argued that, for a wider application in agriculture, more knowledge is required on the genomics of agrobiotechnologically interesting bacteria. Due to this, the genomes of several PGPR, biocontrol bacteria and fungi have been sequenced. Careful analyses of these genomes have highlighted features that contribute to the knowledge of the properties of these plant-beneficial bacteria, analyses that are instrumental in developing biotechnological applications. For example, the analyses of complete genome sequences of microbial antagonists have identified genes with a role in biological control, rhizosphere colonization and transport capabilities for plant-derived compounds. One main result from these genomic analyses has been the detection of several previously unknown secondary metabolite gene clusters in biocontrol strains, predicting that the spectrum of antagonistic chemicals produced by the microbes may be wider than previously expected. In addition to the traditional biocontrol organisms, mycorrhizal fungi are also able to provide protection to the host plant against root and shoot pathogens. Genome sequencing projects with symbiotic fungi are underway and it will be most interesting to compare the genomes of symbiotic fungi to those of bacteria and fungi with biocontrol capacity.

There has been a considerable progress in techniques to study the mechanisms and patterns of root colonization by rhizosphere microbes. Advances in reporter gene technology could perhaps also be used to monitor C fluxes in the rhizosphere, production of antagonistic substances by biocontrol organisms, or colonization related changes in plant gene expression. Chemical communication is an essential part of the way in which microbial populations coordinate their behavior and interact with the plant roots. In recent years there has been a breakthrough in understanding what these signaling compounds are and how their perception is accomplished. Due to the ongoing development of methods in analytical chemistry and genetics, a continuing increase in knowledge in this area of rhizosphere research may be expected.

Research on two influential plant-microbe symbioses, mycorrhizas and root nodulation, has enabled us to unravel how plants are able to grow on nutrient deficient soils. Model legumes, *Lotus japonicus*, *Medicago truncatula*, and model angiosperm tree species from the genus *Populus* can be cultivated readily in the laboratory and are amenable to the techniques of modern molecular biology. The genomes of a number of rhizobial species, of *Lotus japonicus* and of *Populus trichocarpa* have been sequenced, and those of *Medicago truncatula* and selected mycorrhizal fungi will soon be available. This should allow progress in understanding the molecular mechanisms underlying the functioning of root symbioses. As an example, molecular data indicate that the evolution of mycorrhizal associations laid the foundation for rhizobial nodulation, and these symbiosis-related signaling pathways are under intensive investigation. The specificity of associations between plants and root symbionts, which range from the species-specific to the most promiscuous, has important consequences for species diversity in ecosystems. The interplay between plant and symbiont communities, and how this affects plant and microbial species richness, are some key questions to be answered in ongoing studies. The implications of changes in the symbiont community structure in response to biotic or abiotic interferences, and the resulting impact on plant productivity should be in focus in the future.

Carbon compounds lost from different plant species can vary markedly in quality and quantity. Conversely, the micro-organisms in the rhizosphere can influence plants in a variety of ways, and these interactions should have considerable potential in biotechnology. Selected strains of PGPR are being used as seed inoculants, and bacteria responsible for increasing the availability of phosphate and other nutrients in the soil have been described and tested as biofertilizers. Co-operation between taxonomically distant microbes shows the presence of physiological and genetic adaptation among the inhabitants of the rhizosphere (Fig. 1.3). Examples of these include the plant beneficial interactions between mycorrhizal fungi, phosphate solubilising bacteria and diazotrophs. *Pseudomonas-Trichoderma* interaction for efficient biocontrol provides another excellent example of a synergistic action

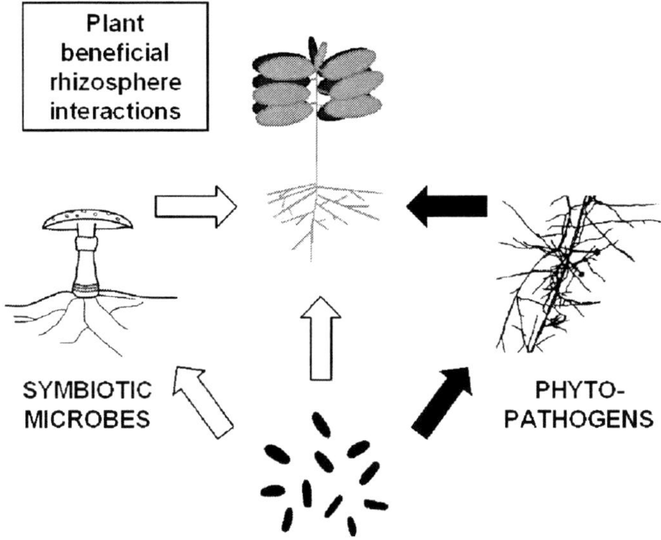

Fig. 1.3 Overview of plant microbe interactions in the rhizosphere. *White arrows* indicate a positive influence and *black arrows* indicates antagonism. Rhizosphere microbes (RM) may positively influence the formation of root symbioses through growth promotion of the symbiotic microbes or through the facilitation of the infection process. The RM may also have a positive influence on plant growth, e.g. through the exudation of plant hormones, acquisition and delivery of minerals, or the induction of plant disease resistance. Disease suppression may also result from a direct antagonism toward phytopathogens by the RM. The reader should note that the organisms often influence each other during the interaction process, which may enhance or suppress the plant beneficial effect

between microbes. In the light of these observations it becomes clear why a microbial inoculum consisting of a single bacterial or fungal strain often performs poorly in the field. Several reports show the potential of combining several biocontrol agents for a higher level of plant protection. Clearly, more research is needed for the successful application of microbial consortia in agricultural soils or in soil conservation. It should, however, be kept in mind that with increasing complexity, experimental access becomes more difficult. While it is relatively simple to investigate dual interactions, tripartite systems are already considerably more challenging whereas even larger system can not be properly handled any more.

Nevertheless, the availability of new and powerful technologies for studying microbial interactions in the field and *in vitro* guarantees further advances in the understanding of the complex interactive network connecting plants, microbes and their soil substrate. The future challenge is to combine field and *in vitro* data in such a way that gives us a coherent picture of the rhizosphere. This will be useful when considering environmental and agricultural applications with rhizosphere microbes.

Acknowledgements As far as our own results are presented we gratefully acknowledge financial support by the University of Tübingen (Strukturfonds) and the DFG-Graduiertenkolleg "Infection biology", which is the basis for the very efficient scientific collaboration with the microbiology group at the University of Tübingen (Prof. H.-P. Fiedler, Dr. J. Riedlinger, D. Schulz).

References

Abdel-Fattah GM, Mohamedin AH (2000) Interactions between a vesicular-arbuscular mycorrhizal fungus (*Glomus intraradices*) and Streptomyces coelicolor and their effects on sorghum plants grown in soil amended with chitin of brawn scales. Biol Fertil Soils 32:401–409

Adomas A, Eklund M, Johansson M, Asiegbu FO (2006) Identification and analysis of differentially expressed cDNAs during nonself-competitive interaction between *Phlebiopsis gigantea* and *Heterobasidion parviporum*. FEMS Microbiol Ecol 57:26–39

Ames BN (1989) Mycorrhiza development in onion in response to chitin-decomposing actinomycetes. New Phytol 112:423–427

Andersson BE, Lunderstadt S, Tornberg K, Schnurwr Y, Oberg LG, Mattiasson B (2003) Incomplete degradation of polycyclic aromatic hydrocarbons in soil inoculated with wood-rotting fungi and their effect on indigenous soil bacteria. Environ Toxicol Chem 22:1238–1243

Andrade G, Mihara KL, Lindermann RG, Bethlenfalvay GJ (1997) Bacteria from rhizosphere and hyphosphere soils of different arbuscular-mycorrhizal fungi. Plant Soil 192:71–79

Antai SP, Crawford DL (1981) Degradation of softwood, hardwood, and grass lignocelluloses by two streptomyces strains. Appl Environ Microbiol 42:378–380

Arnold AE, Maynard Z, Gilbert GS, Coley PD, Kursar TA (2000) Are tropical fungal endophytes hyperdiverse? Ecol Lett 3:267–274

Artursson V, Finlay R, Janssonet JK (2005) Combined bromodeoxyuridine immunocapture and terminal-restriction fragment length polymorphism analysis highlights differences in the active soil bacterial metagenome due to *Glomus mosseae* inoculation or plant species. Environ Microbiol 7:1952–1966

Artursson V, Finlay RD, Jansson JK (2006) Interactions between arbuscular mycorrhizal fungi and bacteria and their potential for stimulating plant growth. Environ Microbiol 8:1–10

Asiegbu F, Adomas A, Stenlid J (2005) Pathogen profile: conifer root and butt rot caused by *Heterobasidion annosum* (Fr.) Bref. s.l. Mol Plant Path 6:395–405

Aspray TJ, Jones E, Whipps JM, Bending GD (2006) Importance of mycorrhization helper bacteria cell density and metabolite localization for the *Pinus sylvestris–Lactarius rufus* symbiosis. FEMS Microbiol Ecol 56:25–33

Azcon R, Azcon G, de Aguilar C, Barea JM (1978) Effects of plant hormones present in bacterial cultures on the formation and responses to VA endomycorrhiza. New Phytol 80:359–364

Bacon CW, Porter JK, Robbins JD, Luttrell ES (1977) Epichloë typhina from toxic tall fescue grasses. Appl Environ Microbiol 34:576–581

Bae H, Kim MS, Sicher RC, Bae HJ, Bailey BA (2006) Necrosis- and ethylene-inducing peptide from *Fusarium oxysporum* induces a complex cascade of transcripts associated with signal transduction and cell death in *Arabidopsis*. Plant Physiol 141:1056–1067

Bai G, Shaner G (2004) Management and resistance in wheat and barley to fusarium head blight. Annu Rev Phytopathol 42:135–161

Barazani O, Friedman J (2001) Allelopathic bacteria and their impact on higher plants. Crit Rev Microbiol 27:41–55

Barea JM, Pozo MJ, Azcon R, Azcon-Aguilar C (2005) Microbial co-operation in the rhizosphere. J Exp Bot 56:1761–1778

Barka EA, Belarbi A, Hachet C, Nowak J, Audran J-C (2000) Enhancement of *in vitro* growth and resistance to gray mould of *Vitis vinifera* co-cultured with plant growth-promoting rhizobacteria. FEMS Microbiol Lett 186:91–95

Barrett B (1958) Synthesis of mycorrhiza with pure cultures of *Rhizophagus*. Phytopathology 48:391

Barriuso J, Pereyra MT, Lucas Garcia JA, Megias M, Gutierrez Manero FJ, Ramos B (2005) Screening for putative PGPR to improve establishment of the symbiosis *Lactarius deliciosus-Pinus* sp. Microb Ecol 50:82–89

Bashan Y, Holguin G, de-Bashan LE (2004) *Azospirillum*-plant relationships: physiological, molecular, agricultural, and environmental advances (1997–2003). Can J Microbiol 50:521–577

Berdy J (2005) Bioactive microbial metabolites: a personal view. J Antibiot 58:1–26

Bertaux J, Schmid M, Prevost-Boure NC, Churin JL, Hartmann A, Garbaye J, Frey-Klett P (2003) *In situ* identification of intracellular bacteria related to *Paenibacillus* spp. in the mycelium of the ectomycorrhizal fungus *Laccaria bicolor* S238N. Appl Environ Microbiol 69:4243–4248

Bevivino A, Sarrocco S, Dalmastri C, Tabacchioni S, Cantale C, Chiarini L (1998) Characterization of a free-living maize-rhizosphere population of *Burkholderia cepacia*: effect of seed treatment on disease suppression and growth promotion of maize. FEMS Microbiol Ecol 27:225–237

Bianciotto V, Bandi C, Minerdi D, Sironi M, Tichy HV, Bonfante P (1996) An obligately endosymbiotic mycorrhizal fungus itself harbors obligately intracellular bacteria. Appl Environ Microbiol 62:3005–3010

Bidartondo MI, Bruns TD (2005) On the origins of extreme mycorrhizal specificity in the Monotropoideae (Ericaceae): performance trade-offs during seed germination and seedling development. Mol Ecol 14:1549–1560

Bidartondo MI, Burghardt B, Gebauer G, Bruns TD, Read DJ (2004) Changing partners in the dark: isotopic and molecular evidence of ectomycorrhizal liaisons between forest orchids and trees. Proc R Soc Biol Sci 271:1799–1806

Biswass JC, Ladha JK, Dazzo FB (2000a) Rhizobia inoculation improves nutrient uptake and growth in lowland rice. Soil Sci Soc Am J 64:1644–1650

Biswass JC, Ladha JK, Dazzo FB, Yanni YG, Rolfe BG (2000b) Rhizobial inoculation influences seedling vigor and yield of rice. Agron J 92:880–886

Boddy L (1993) Saprotrophic cord-forming fungi: warfare strategies and other ecological aspects. Mycol Res 97:641–655

Boddy L (1999) Saprotrophic cord-forming fungi: meeting the challenge of heterogeneous environments. Mycologia 91:13–32

Boddy L (2000) Interspecific combative interactions between wood-decaying basidiomycetes. FEMS Microbiol Ecol 31:185–194

Boddy L, Griffith GS (1989) Role of endophytes and latent invasion in the development of decay communities in sapwood of angiospermous trees. Sydowia 41:41–73

Boddy L, Watkinson SC (1995) Wood decomposition, higher fungi, and their role in nutrient redistribution. Can J Bot 73:S1377–S1383

Bolwerk A, Lagopodi AL, Wijfjes AH, Lamers GE, Chin-A-Woeng TF, Lugtenberg BJ, Bloemberg GV (2003) Interactions in the tomato rhizosphere of two *Pseudomonas* biocontrol strains with the phytopathogenic fungus *Fusarium oxysporum* f. sp. radicis-lycopersici. Mol Plant Microbe Interact 16:983–993

Bolwerk A, Lagopodi AL, Lugtenberg BJ, Bloemberg GV (2005) Visualization of interactions between a pathogenic and a beneficial *Fusarium* strain during biocontrol of tomato foot and root rot. Mol Plant Microbe Interact 18:710–721

Bomberg M, Jurgens G, Saano A, Sen R, Timonen S (2003) Nested PCR detection of *Archaea* in defined compartments of pine mycorrhizospheres developed in boreal forest humus microcosms. FEMS Microbiol Ecol 43:163–171

Bonkowski M (2004) Protozoa and plant growth: the microbial loop in soil revisited. New Phytol 162:617–631

Bonkowski M, Brandt F (2002) Do soil protozoa enhance plant growth by hormonal effects? Soil Biol Biochem 34:1709–1715

Bonkowski M, Jentschke G, Scheu S (2001) Contrasting effects of microbial partners in the rhizosphere: interactions between Norway spruce seedlings (*Picea abies* Karst.), mycorrhiza (*Paxillus involutus* (Batsch) Fr.) and naked amoebae (protozoa) Appl Soil Ecol 18:193–204

Bonsall RF, Weller DM, Thomashow LS (1997) Quantification of 2,4-diacetylphloroglucinol produced by fluorescent *Pseudomonas* spp. in vitro and in the rhizosphere of wheat. Appl Environ Microbiol 63:951–955

Booth MG (2004) Mycorrhizal networks mediate overstorey-understorey competition in a temperate forest. Ecol Lett 7:538–546

Brencic A, Winans SC (2005) Detection of and response to signals involved in host-microbe interactions by plant-associated bacteria. Microbiol Mol Biol Rev 69:155–194

Brooks DS, Gonzales CF, Apple DN, Filer TH (1994) Evaluation of endophytic bacteria as potential biological control agents for oak wilt. Biol Control 4:373–381

Brundrett M, Kendrick B (1990) The roots and mycorhizas of herbaceous woodland plants. II. Structural aspects of morphology. New Phytologist 114:469–479

Cairney JWG, Merhag AA (2002) Interactions between ectomycorrhizal fungi and soil saprotrophs: implications for decomposition of organic matter in soils and degradation of organic pollutants in the rhizosphere. Can J Bot 80:803–809

Cao L, Qiu Z, You J, Tan H, Zhou S (2005) Isolation and characterization of endophytic streptomycete antagonists of *Fusarium* wilt pathogen from surface-sterilized banana roots. FEMS Microbiol Lett 247:147–152

Capellano A, Dequatre B, Valla G, Moiroud A (1987) Root nodule formation by *Penicillium* sp. on *Alnus glutinosa* and *Alnus incana*. Plant Soil 104:45–52

Carroll GC (1995) Forest endophytes: pattern and process. Cana J Bot 73:1316–1324

Castillo UF, Strobel GA, Ford EJ, Hess WM, Porter H, Jensen JB, Albert H, Robison R, Condron MAM, Teplow DB, Stevens D, Yaver D (2002) Munumbicins, wide-spectrum antibiotics produced by *Streptomyces* NRRL 30562, endophytic on *Kennedia nigriscans*. Microbiology 148:2675–2685

Castillo UF, Strobel GA, Mullenberg K, Condron MM, Teplow DB, Folgiano V, Gallo M, Ferracane R, Mannina L, VIel S, Codde M, Robison R, Porter H, Jensen J (2006) Munumbicins E-4 and E-5: novel broad-spectrum antibiotics from Streptomyces NRRL 3052. FEMS Microbiol Lett 255:296–300

Céspedes R, Gonzáles B, Vicuña R (1997) Characterisation of a bacterial consortium degrading the lignin model compound vanillyl-β-D-glucopyranoside. J Basic Microbiol 3:175–180

Challis GL, Hopwood DA (2003) Synergy and contingency as driving forces for the evolution of multiple secondary metabolite production by *Streptomyces* species. PNAS USA 100(2):14555–14561

Chamberlain K, Crawford DL (1999) *In vitro* and vivo antagonism of pathogenic turfgrass fungi by *Streptomyces hygroscopicu*s strains YCED9 and WYE53. J Ind Microbiol Biotechnol 23:641–646

Chapela IH, Osher LJ, Horton TR, Henn MR (2001) Ecomycorrhizal fungi introduced with exotic pine plantations induce soil carbon depletion. Soil Biol Biochem 33:1733–1740

Cheplick GP, Clay K, Marks S (1989) Interactions between infection by endophytic fungi and nutrient limitation in the grasses Lolium perenne and Festuca arundinacea New Phytol 111:89–97

Chernin L, Ismailov Z, Haran S, Chet I (1995) Chitinolytic *Enterobacter agglomerans* antagonistic to fungal plant pathogens. Appl Environ Microbiol 61:1720–1726

Chernin LS, de la Fuente L, Sobolev V, Haran S, Vorgias CE, Oppenheim AB, Chet I (1997) Molecular cloning, structural analysis and expression in *Escherichia coli* of a chitinase gene from *Enterobacter agglomerans*. Microbiology 63:834–839

Chi F, Shen S-H, Cheng H-P, Jing Y-X, Yanni YG, Dazzo FB (2005) Ascending migration of endophytic Rhizobia, from roots to leaves, inside rice plants and assessment of benefits to rice growth physiology. Appl Environ Microbiol 71:7271–7278

Chiarini L, Bevivino A, Dalmastri C, Tabacchioni S, Visca P (2006) Burkholderia cepacia complex species: health hazards and biotechnological potential. Trends Microbiol 14:277–286

Christensen MJ (1996) Antifungal activity in grasses infected with *Acremonium* and *Epichloë* endophytes. Australasian Plant Pathol 25:186–191

Clay K, Holah J (1999) Fungal endophyte symbiosis and plant diversity in successional fields. Science 285:1742–1744

Clay K, Schardl C (2002) Evolutionary origins and ecological consequences of endophyte symbiosis with grasses. Am Naturalist 160:99–127

Cocito C, Di Giambattista M, Nyssen E, Vannuffel P (1997) Inhibition of protein synthesis by streptogramins and related antibiotics. J Antimicrob Chemother 39:7–13

Compant S, Duffy B, Nowak J, Clement C, Barka EA (2005) Use of plant growth-promoting bacteria for biocontrol of plant diseases: principles, mechanisms of action, and future prospects. Appl Environ Microbiol 71:4951–4959

Cook RJ, Thomashow LS, Weller DM, Fuiimoto D, Mazzola, M, Bangera G, Kim D (1995) Molecular mechanisms of defense by rhizobacteria against root disease. PNAS USA 92:4197–4201

Cornelis P, Matthijs S (2002) Diversity of siderophore-mediated iron uptake systems in fluorescent pseudomonads: not only pyoverdines. Environ Microbiol 4:787–798

Couteaudier Y, Alabouvette C (1990) Quantitative comparison of *Fusarium oxysporum* competitiveness in relation to carbon utilization. FEMS Microbiol Let 74:261–267

Crawford DL (1978) Lignocellulose decomposition by selected streptomyces strains. Appl Environ Microbiol 35:1041–1045

Crawford DL, Lynch JM, Whipps JM, Ousley MA (1993) Isolation and characterization of actinomycete antagonists of a fungal root pathogen. Appl Environ Microbiol 59:3899–3905

Curl EA, Truelove B (1986) The Rhizosphere. Springer, Berlin Heidelberg New York

Davelos AL, Kinkel LL, Samac DA (2004) Spatial variation in frequency and intensity of antibiotic interactions among *Streptomycetes* from prairie soil. Appl Environ Microbiol 70:1051–1058

De Bary A (1866) Morphologie und Physiologie der Pilze, Flechten und Myxomyceten. Vol II. Hofmeister's handbook of physiological botany. Leipzig

de Boer W, Folman LB, Summerbell RC, Boddy L (2005) Living in a fungal world: impact of fungi on soil bacterial niche development. FEMS Microbiol Rev 29:795–811

Demain AL (1981) Industrial microbiology. Science 214:987–994

Deveau A, Palin B, Delaruelle C, Peter M, Kohler A, Pierrat JC, Sarniguet A, Garbaye J, Martin F, Frey-Klett P (2007) The mycorrhiza helper *Pseudomonas fluorescens* BBc6R8 has a specific priming effect on the growth, morphology and gene expression of the ectomycorrhizal fungus *Laccaria bicolor* S238N. New Phytol 175:743–755

Dew RK, Boissonneault GA, Gay N, Boling JA, Cross RI, Cohen DA (1990) The effect of the endophyte (*Acremonium coenophialum*) and associated toxin (s) of tall fescue on serum titer response to immunization and spleen cell flow cytometry analysis and response to mitogens. Vet Immunol Immunopathol 26:285–295

Dickie IA, Guza RC, Krazewski SE, Reich PB (2004) Shared ectomycorrhizal fungi between a herbaceous perennial (*Helianthemum bicknellii*) and oak (*Quercus*) seedlings. New Phytol 164:375–382

Dobbelaere S, Croonenborghs A, Thys A, Vande Broek A, Vanderleyden J (1999) Phytostimulatory effect of *Azospirillum brasilense* wild type and mutant strains altered in IAA production on wheat. Plant Soil 212:153–162

Dong Y-H, Xu J-L, Li X-Z, Zhang LH (2000) AiiA, a novel enzyme inactivates acyl homoserine-lactone quorum-sensing signal and attenuates the virulence of *Erwinia carotovora*. Proc Natl Acad Sci (USA) 97:3526–3531

Dowkiw A, Husson C, Frey P, Pinon J, Bastien C (2003) Partial resistance to *Melampsora larici-populina* leaf rust in hybrid poplars: genetic variability in inoculated excised leaf-disk bioassay and relationship with complete resistance. Phytopathology 93:421–427

Duffy B, Schouten A, Raaijmakers JM (2003) Pathogen self defense: mechanisms to counteract microbial antagonism. Annu Rev Phytopathol 41:501–538

Dumas-Gaudot E, Gollotte A, Cordier C, Gianinazzi S, Gianinazzi-Pearson V (2000) Modulation of plant defense systems. In: Douds D, Kapulnik Y (eds) Arbuscular mycorrhizas: molecular biology and physiology. Kluwer, Dordrecht, pp 173–200

Dunne C, Crowley JJ, Moënne-Loccoz Y, Dowling DN, de Bruijn FJ, O'Gara F (1998) Biological control of *Phytium ultimum* by *Stenotrophomonas maltophilia* W81 is mediated by an extracellular proteolytic activity. Microbiology 143:3921–3931

Dunne C, Moenne-Loccoz Y, de Bruijn FJ, O'Gara F (2000) Overproduction of an inducible extracellular serine protease improves biological control of *Pythium ultimum* by *Stenotrophomonas maltophilia* strain W81. Microbiology 146:2069–2078

Duplessis S, Courty PE, Tagu D, Martin F (2005) Transcript patterns associated with ectomycorrhiza development in *Eucalyptus globulus* and *Pisolithus microcarpus*. New Phytol 165:599–611

Duponnois R, Garbaye J, Bouchard D, Churin JL (1993) The fungus-specificity of mycorrhization helper bacteria (MHBs) used as an alternative to soil fumigation for ectomycorrhizal inoculation of bare-root Douglas-fir planting stocks with *Laccaria laccata*. Plant Soil 157:257–262

Duponnois R, Assikbetse K, Ramanankierana H, Kisa M, Thioulouse J, Lepage M (2006) Litter-forager termite mounds enhance the ectomycorrhizal symbiosis between *Acacia holosericea* A. Cunn. Ex G. Don and *Scleroderma dictyosporum* isolates. FEMS Microbiol Ecol 56:292–303

Eckwall EC, Schottel JL (1997) Isolation and characterization of an antibiotic produced by the scab disease-suppressive *Streptomyces diastatochromogenes* strain PonSSII J. Ind Microbiol Biotechnol 19:220–225

El-Abyad MS, El-Sayed MA, El-Shanshoury AR, El-Sabbagh SM (1993) Towards the biological control of fungal and bacterial diseases of tomato using antagonistic *Streptomyces* spp. Plant Soil 149:185–195

Endre G, Kereszt A, Kevel Z, Milhacea S, Kalo P, Kiss G (2002) A receptor kinase gene regulating symbiotic nodule development. Nature 417:962–966

Ezra D, Castillo UF, Strobel GA, Hess WM, Porter H, Jensen JB, Condron MAM, Teplow DB, Sears J, Maranta M, Hunter M, Weber B, Yaver D (2004). Coronamycins, peptide antibiotics produced by a verticillate *Streptomyces* sp. (MSU-2110) endophytic on *Monstera* sp. Microbiology 150:785–793

Feller IC (1995) Effects of nutrient enrichment on growth and herbivory of dwarf red mangrove (Rhizophora mangle). Ecol Monogr 65:477–505

Fiedler HP, Krastel P, Muller J, Gebhardt K, Zeeck A (2001) Enterobactin: the characteristic catecholate siderophore of *Enterobacteriaceae* is produced by *Streptomyces* species. FEMS Microbiol Lett 196:147–151

Finlay RD, Read DJ (1986a) The structure and function of vegetative mycelium of ectomycorrhizal plants. I. Translocation of "%C-labelled carbon between plants interconnected by a common mycelium. New Phytologist 103:143–156

Finlay RD, Read DJ (1986b) The structure and function of the vegetetive mycelium of ectomycorrhizal plants. II. The uptake and distribution of phosphorus by mycelial strands interconnecting host plants. New Phytologist 103:157–165

Finlay R, Söderström B (1992) Mycorrhiza and carbon flow in the soil. In: Allen MJ (ed) Mycorrhizal functioning. Chapman & Hall, New York, pp 134–160

Firn RD, Jones CG (2000) The evolution of secondary metabolism - a unifying model. Mol Microbiol 37:989–994

Fitter AH, Garbaye J (1994a) Interactions between mycorrhizal fungi and other soil organisms. Plant Soil 159:123–132

Fitter AH, Garbaye J (1994b) Interactions between the arbuscular mycorrhizal fungus *Glomus intraradices* and different rhizosphere micro-organisms. New Phytol 141:525–533

Foster RC, Marks GC (1966) The fine structure of the mycorrhizas of *Pinus radiata*. Australian J Biol Sci 19:1027–1038

Foster RC, Marks GC (1967) Observations on the mycorrhizas of forest trees. II. The rhizosphere of *Pinus radiata* D. Don. Austral J Biol Sci 20:915–926

Founoune H, Duponnois R, Ba AM, Sall S, Branget I, Lorquin J, Neyra M, Chotte JL (2002) Mycorrhiza helper bacteria stimulated ectomycorrhizal symbiosis of *Acacia holosericea* with *Pisolithus alba*. New Phytol 153:81–89

Fracchia S, Garcia-Romera I, Godeas A, Ocampo JA (2000) Effect of the saprophytic fungus Fusarium oxysporum on arbuscular mycorrhizal colonization and growth of plants in greenhouse and field trials. Plant Soil 223:175–184

Fravel DR (2005) Commercialization and implementation of biocontrol. Annu Rev Phytopathol 43:337–359

Fravel D, Olivain C, Alabouvette C (2003) *Fusarium oxysporum* and its biocontrol. New Phytologist 157:493–502

Frey P, Frey-Klett P, Garbaye J, Berge O, Heulin T (1997) Metabolic and genotypic fingerprinting of fluorescent pseudomands associated with the Douglas fir – *Laccaria bicolour* mycorrhizosphere. Appl Environ Microbiol 63:1852–1860

Frey-Klett P, Chavatte M, Clausse M-L, Courrier S, Le Roux C, Raaijmakers J, Martinotti MG, Pierrat J-C, Garbaye J (2005) Ectomycorrhizal symbiosis affects functional diversity of rhizosphere fluorescent pseudomonads. New Phytol 165:317–328

Gadgil RL, Gadgil PD (1975) Suppression of litter decomposition by mycorrhizal roots of *Pinus radiata*. New Zealand J Forest Sci 5:33–41

Gafur A, Schutzendubel A, Langenfeld-Heyser R, Fritz E, Polle A (2004) Compatible and incompetent *Paxillus involutus* isolates for ectomycorrhiza formation *in vitro* with poplar (*Populus x canescens*) differ in H2O2 production. Plant Biol 6:91–99

Gage DJ (2004) Infection and invasion of toots by symbiotic, nitrogen-fixing rhizobia during nodulation of temperate legumes. Microbiol Mol Biol Rev 68:280–300

Gallaud I (1905) Études sur les mycorrhizes endotrophs. Revue Générale de Botanique 17:5–48, 66–83, 123–135, 223–239, 313–325,425–433, 479–500

Gamalero E, Trotta A, Massa N, Copetta A, Martinotti MG, Berta G (2004) Impact of two fluorescent pseudomonads and an arbuscular mycorrhizal fungus on tomato plant growth, root architecture and P acquisition. Mycorrhiza 14:229–234

Garbaye J (1994) Mycorrhiza helper bacteria: a new dimension to the mycorrhizal symbiosis. New Phytol 128:197–210

Garbaye J, Bowen GD (1989) Stimulation of mycorrhizal infection of *Pinus radiata* by some micro-organisms associated with the mantle of ectomycorrhizas. New Phytol 112:383–388

Garcia de Salamone IE, Hynes RK, Nelson LM (2001) Cytokinin production by plant growth promoting rhizobacteria and selected mutants. Can J Microbiol 47:404–411

Garcia-Romera I, Garcia-Garrido JM, Martin J, Fracchia S, Mujica MT, Godeas A, Ocampo JA (1998) Interactions between saprotrophic *Fusarium strains* and arbuscular mycorrhizas of soybean plants. Symbiosis 24:235–246

Gasoni L, Stegman De Gurfinkel B (1997) The endophyte *Cladorrhinum foecundissimum* in cotton roots: phosphorus uptake and host growth. Mycological Res 101:867–870

Gerdemann JW (1965) Vesicular-arbuscular mycorrhizae of maize and tuliptree by *Endogone fasciculata*. Mycologia 57:562–575

Glienke-Blanco C, Aguilar-Vildoso CI, Vieira MLC, Barroso PAV, Azevedo JL (2002) Genetic variability in the endophytic fungus *Guignardia citricarpa* isolated from citrus plants. Genet Mol Biol 25:251–255

Goodfellow M, Williams ST (1983) Ecology of actinomycetes. Annu Rev Microbiol 37:189–216

Goyer C, Vachon J, Beaulieu C (1998) Pathogenicity of *Streptomyces scabies* mutants altered in Thaxtomin A production. Phytopathology 88:442–445

Graham JH (2001) What do root pathogens see in mycorrhizas? New Phytologist 149:357–359

Gramms G, Voigt K-D, Kirsche B (1999) Degradation of polycyclic aromatic hydrocarbons with three to seven aromatic rings by higher fungi in sterile and unsterile soil. Biodegradation 10:51–62

Grant SR, Fisher EJ, Chang JH, Mole BM, Dangl JL (2006) Subterfuge and manipulation: type III effector proteins of phytopathogenic bacteria. Annu Rev Microbiol 60:425–449

Greaves H (1971) The bacterial factor in wood decay. Wood Sci Technol 5:6–16

Gregor AK, Klubek B, Varsa EC (2003) Identification and use of actinomycetes for enhanced nodulation of soybean co-inoculated with *Bradyrhizobium japonicum*. Can J Microbiol 49:483–491

Griffin DM (1985) A comparison of the roles of bacteria and fungi. In: Leadbetter ER, Pointdexter JS (eds) Bacteria in nature . Bacterial activities in perspective, vol 1 Acdemic Press, London. pp 221–255

Griffith BS, Bonkowski M, Dobson G, Caul S (1999) Changes in soil microbial community structure in the presence of microbial-feeding nematodes and protozoa. Pedobiologia 43:297–304

Grunwald NJ, Flier WG (2005) The biology of *Phytophthora infestans* at its center of origin. Annu Rev Phytopathol 43:171–190

Gryndler M, Vosatka M (1996) The response of *Glomus fistulosum*-maize mycorrhiza to treatments with culture fractions from *Pseudomonas putida*. Mycorrhiza 6:207–211

Haahtela K, Laakso T, Korhonen TK (1986) Associative nitrogen fixation by *Klebsiella* spp.: adhesion sites and inoculation effects on grass roots. Appl Environ Microbiol 52:1074–1079

Haas D, Defago G (2005) Biological control of soil-borne pathogens by fluorescent pseudomonads. Nat Rev Microbiol 3:307–319

Haas D, Keel C (2003) Regulation of antibiotic production in root-colonizing *Pseudomonas* spp. and relevance for biological control of plant disease. Ann Rev Phytopathol 41:117–153

Hallman J, Sikora RA (1996) Toxicity of fungal endophyte secondary metabolites to plant parasitic nematodes and soil-borne plant pathogenic fungi. Eur J Plant Pathol 102:155–162

Hallmann J, Quadt Hallmann A, Mahaffee WF, Kloepper JW (1997) Bacterial endophytes in agricultural crops. Can J Microbiol 43:895–914

Hampp R, Schaeffer C (1998) Mycorrhiza – carbohydrate and energy metabolism. In: Varma A, Hock B (eds) Mycorrhiza – structure, function, molecular biology and biotechnology. Springer, Berlin Heidelberg New York, pp 273–303

Hartnett DC, Wilson GWT (1999) Mycorrhizae influence plant community structure and diversity in tallgrass prairie. Ecol 80:1187–1195

Hartnett DC, Hetrick BAD, Wilson GWT, Gibson DJ (1993) Mycorrhizal influence on intra- and interspecific neighbour interactions among co-occuring prairie grasses. J Ecol 81:787–795

Hättenschwiler S, Tiunov AV, Scheu S (2005) Biodiversity and litter decomposition in terrestrial ecosystems. Annu Rev Ecol Evol Syst 36:191–218

Hause B, Fester T (2005) Molecular and cell biology of arbuscular mycorrhizal symbiosis. Planta 221:184–196

Hendrickson OQ (1991) Abundance and activity of N2-fixing bacteria in decaying wood. Can J For Res 21:1299–1304

Henke JM, Bassler BL (2004) Bacterial social engagements. Trends Cell Biol 14:648–656

Henningston B (1967) Interactions between micro-organisms found in birch and aspen pulpwood. Stud For Suec 53:1–32

Herrmann S, Oelmüller R, Buscot F (2004) Manipulation of the onset of ectomycorrhiza formation by indole-3-acetic acid, activated charcoal or relative humidity in the association between oak microcuttings and *Piloderma croceum*: influence on plant development and photosynthesis. J Plant Phys 161:509–517

Hildebrandt U, Janetta K, Bothe H (2002) Towards growth of arbuscular mycorrhizal fungi independent of a plant host. Appl Environ Microbiol 68:1919–1924

Hildebrandt U, Ouziad F, Marner FJ, Bothe H (2006) The bacterium *Paenibacillus validus* stimulates growth of the arbuscular mycorrhizal fungus *Glomus intraradices* up to the formation of fertile spores. FEMS Microbiol Lett 254:258–267

Hiltner L (1904) Über neuere Erfahrungen und Probleme auf dem Gebiet der Bodenbakteriologie und unter besonderer Berücksichtigung der Gründüngung und Brache. Arbeiten der Deutschen Landwirtschaftlichen Gesellschaft 98:59–78

Hobbie SE (1992) Effects of plant species on nutrient cycling. Trends Ecol Evol 7:336–339

Hoffland E, Kuyper TW, Wallander H, Plassard C, Gorbushina AA, Haselwandter K, Holmström S, Landeweert R, Lundström US, Rosling A, Sen R, Smits MM, van Hees PAW, van Breemen N (2004) The role of fungi in weathering. Front Ecol Environ 2:258–264

Huitema E, Bos JI, Tian M, Win J, Waugh ME, Kamoun S (2004) Linking sequence to phenotype in *Phytophthora*-plant interactions. Trends Microbiol 12:193–200

Igarashi Y, Ogawa M, Sato Y, Saito N, Yoshida R, Kunoh H, Onaka H, Furumai T (2000) Fistupyrone, a novel inhibitor of the infection of Chinese cabbage by *Alternaria brassicicola*, from *Streptomyces* sp. TP-A0569. J Antibiot 53:1117–1122

Iniguez AL, Dong Y, Triplett EW (2004) Nitrogen fixation in wheat provided by *Klebsiella pneumoniae* 342. MPMI 17:1078–1085

Janos DP (1987) VA mycorrhizas in humid tropical ecosystems. In: Safir GR (ed) Ecophysiology of VA mycorrhizal plants. CRC Press, Boca Raton, pp 107–134

Jennings DH (1987) Translocation of solutes in fungi. Biol Rev 62:215–243

Johansson JF, Paul LR, Finlay RD (2004) Microbial inoculation in the mycorhizosphere and their significance for sustainable agriculture. FEMS Microbiol Ecol 48:1–13

Jonsson LM, Nilsson MC, Wardle DA, Zackrisson O (2001) Context dependent effects of ectomycorrhizal species richness on tree seedling productivity. Oikos 93:353–364

Judelson HS, Blanco FA (2005) The spores of *Phytophthora*: weapons of the plant destroyer. Nat Rev Microbiol 3:47–58

Kang Y, Carlson R, Trappe J, Schell MA (1998) Characterisation of genes involved in biosynthesis of a novel antibiotic from *Burkholderia cepacia* BC11 and their role in biological control of *Rhizoctonia solani*. Appl Environ Microbiol 64:3939–3947

Karlsson M, Olson A, Stenlid J (2003) Expressed sequences from the basidiomycetous tree pathogen *Heterobasidion annosum* during early infection of scots pine. Fungal Genet Biol 39:51–59

Keel C, Defago G (1997) Interactions between beneficial soil bacteria and root pathogens: mechanisms and ecological impact. In: Gange AC, Brown VK (eds) Multitrophic interactions in terrestrial system. Blackwell Science, Oxford, p 27

Keller S, Schneider K, Sussmuth RD (2006) Structure elucidation of auxofuran, a metabolite involved in stimulating growth of fly agaric, produced by the mycorrhiza helper bacterium *Streptomyces* AcH 505. J Antibiotics (Tokyo) 59:801–803

Kent AD, Triplett EW (2002) Microbial communities and their interactions in soil and rhizosphere ecosystems. Annu Rev Microbiol 56:211–236

Kerry BR (2000) Rhizospheric interactions and the exploitation of microbial agents for the biological control of plant parasitic nematodes. Annu Rev Phytopathol 38:423–441

Kers JA, Cameron KD, Joshi MV, Bukhalid RA, Morello JE, Wach MJ, Gibson DM, Loria R (2005) A large, mobile pathogenicity island confers plant pathogenicity on *Streptomyces* species. Mol Microbiol 55:1025–1033

King RR, Lawrence H, Calhount LA (1992) chemistry of phytotoxins associated with *Streptomyces scabies*, the causal organism of potato common scab. J Agric Food Chem 40:834–837

Klaenhammer TR (1993) Genetics of bacteriocins produced by lactic acid bacteria. FEMS Microbiol Rev 12:39–85

Klironomos JN (2002) Variation in plant responses to native and exotic arbuscular mycorrhizal fungi. Ecology 84:2292–2301

Kloepper JW, Leong J, Teintze M, Schroth MN (1978) *Pseudomonas* siderophores: a mechanism explaining disease-suppressive soils. Curr Microbiol 4:317–320

Kobayashi DY, Palumbo JD (2000) Bacterial endophytes and their effects on plants and uses in agriculture. In: Bacon CW, White JF (eds) Microbial endophytes. Marcel Dekker, New York, pp 199–236

Kogel K-H, Franken P, Hückelhoven R (2006) Endophyte or parasite – what decides? Curr Opin Plant Biol 9:358–363

Kondo Y, Shomura T, Ogawa Y, Tsuruoka T, Watanabe H, Totukawa K, Suzuki T, Moriyama C, Yoshida J, Inouye S, Niida T (1973) Studies on a new antibiotic SF-1293. 1. Isolation and physico-chemical and biological characterization of SF-1293 substances. Sci Rep Meiji Seika 13:34–41

Kramer U (2005) Phytoremediation: novel approaches to cleaning up polluted soils. Curr Opin Biotechnol. 16(2):133–141

Kuiper I, Lagendijk EL, Bloemberg GV, Lugtenberg BJ (2004) Rhizoremediation: a beneficial plant-microbe interaction. Mol Plant Microbe Interact 17:6–15

Lagopodi AL, Ram AF, Lamers GE, Punt PJ, Van den Hondel CA, Lugtenberg BJ, Bloemberg GV (2002) Novel aspects of tomato root colonization and infection by *Fusarium oxysporum* f. sp. radicis-lycopersici revealed by confocal laser scanning microscopic analysis using the green fluorescent protein as a marker. Mol Plant Microbe Interact 15:172–179

Lambilliotte R, Cooke R, Samson D, Fizames C, Gaymard F, Plassard C, Tatry MV, Berger C, Laudié M, Legeai F, Karsenty E, Delseny M, Zimmermann S, Sentenac H (2004) Large-scale identification of genes in the fungus *Hebeloma cylindrosporum* paves the way to molecular analyses of ectomycorrhizal symbiosis. New Phytologist 164:505–513

Landeweert R, Hoffland E, Finlay RD, Kuyper TW, Van Breemen N (2001) Linking plants to rocks: ectomycorrhizal fungi mobilize nutrients from minerals. Trends Ecol Evol 16:248–254

Lang E, Kleeberg I, Zadrazil F (1997) Competition of *Pleurotus* sp. and *Dichomitus squalens* with soil micro-organisms during lignocellulose decomposition. Biores Technol 60:95–99

Larkin RP, Fravel DR (1999) Mechanisms of action and dose-response relationships governing bio logical control of *Fusarium* wilt of tomato by non-pathogenic *Fusarium* spp. Phytopathology 89:1152–1161

Lawrence CH, Clark MC, King RR (1990) Induction of common scab symptoms in aseptically cultured potato tubers by the vivotoxin, thaxtomin. Phytopathology 80:606–608

Leake JR, Donnelly DP, Boddy L (2002) Interactions between ecto-mycorrhizal fungi and saprotrophic fungi. In: van de Heijden MGA, Sanders I (eds) Mycorrhizal ecology, ecological studies 157. Springer, Berlin Heidelberg New York, pp 345–372

Lehr NA, Schrey SD, Bauer R, Hampp R, Tarkka MT (2007) Suppression of plant defense response by a mycorrhiza helper bacterium. New Phytol 174:892–903

Li G, Asiegbu FO (2004) Use of Scots pine seedling roots as an experimental model to investigate gene expression during interaction with the conifer pathogen *Heterobasidion annosum* (P-type). J Plant Res 117:155–162

Lind M, Olson A, Stenlid J (2006) An AFLP-markers based genetic linkage map of *Heterobasidion annosum* locating intersterility genes. Fungal Genet Biol 42:519–527

Lindahl B, Stenlid J, Olsson S (1999) Translocation of ^{32}P between interacting mycelia of a wood-decomposing fungus and ectomycorrhizal fungi in microcosm systems. New Phytol 144:183–193

Lindahl B, Stenlid J, Finlay RD (2001) Effects of resource availability on mycelial interactions and ^{32}P-transfer between a saprotrophic and an ectomycorrhizal fungus in soil microcosms. FEMS Microbiol Ecol 38:43–52

Linderman RG (1988) Mycorrhizal interactions with the rhizosphere microflora – the mycorrhizosphere effect. Phytopathology 78:366–371

Linderman RG, Paulitz TC (1990) Mycorrhizal–rhizobacterial interactions In: Hornby D (ed) Biological control of soil-borne plant pathogens. CAB International, Wallingford, UK, pp 261–283

Liras P (1999) Biosynthesis and molecular genetics of cephamycins. Cephamycins produced by actinomycetes. Antonie Leeuwenhoek 75:109–124

Liu D, Anderson NA, Kinkel LL (1995) Biological control of potato scab in the field with antagonistic *Streptomyces scabies*. Phytopathology 85:827–831

Liu D, Anderson NA, Kinkel LL (1996) Selection and characterization of strains of *Streptomyces* suppressive to the potato scab pathogen Can J Microbiol 63:4516–4522

Llorens A, Hinojo MJ, Mateo R, Medina A, Valle-Algarra FM, Gonzalez-Jaen MT, Jimenez M (2006) Variability and characterization of mycotoxin-producing *Fusarium* spp isolates by PCR-RFLP analysis of the IGS-rDNA region. Antonie Van Leeuwenhoek 89:465–478

Loper JE, Henkels MD (1999) Utilization of heterologous siderophores enhances levels of iron available to *Pseudomonas putida* in the rhizosphere. Appl Environ Microbiol 65:5357–5363

Loria R, Kers J, Joshi M (2006) Evolution of plant pathogenicity in Streptomyces. Ann Rev Phytopathol 44:469–487

Lucy M, Reed E, Glick BR (2004) Applications of free living plant growth-promoting rhizobacteria. Antonie Van Leeuwenhoek 86:1–25

Lupway N, Clayton G, Hanson K, Rice W, Bierderbeck V (2004) Endophytic rhizobia in barley, wheat, and canola roots. Can J Plant Sci 84:37–45

Maier A (2003) Einfluss bakterieller Stoffwechselprodukte auf Wachstum und Proteom des Ektomykorrhizapilzes Amanita muscaria. PhD Thesis, University of Tübingen, Germany

Maier A, Riedlinger J, Fiedler H-P, Hampp R (2004) *Actinomycetales* bacteria from a spruce stand: characterisation and effects on growth of root symbiotic and plant parasitic soil fungi in dual culture. Mycol Prog 3:129–136

Mandeel Q, Baker R (1991) Mechanisms involved in biological control of *Fusarium* wilt of cucumber strains of non-pathogenic *Fusarium oxysporum*. Phytopathology 81:462–469

Mandyam K, Jumpponen A (2005) Seeking the elusive function of the root-colonising dark septate endophytic fungi. Stud Mycol 53:173–189

Marilley L, Aragno M (1999) Phylogenetic diversity of bacterial communities in degree of proximity of *Lolium perenne* and *Trifolium repens* roots. Appl Soil Ecol 13:127–136

Matsuda Y, Hijii N (2004) Ectomycorrhizal fungal communities in an *Abies firma* forest, with special reference to ectomycorrhizal associations between seedlings and mature trees. Can J Bot 82:822–829

Mazumder R, Phelps TJ, Krieg NR, Benoit RE (1999) Determining chemotactic responses by two subsurface microaerophiles using a simplified capillary assay method. J Microbiol Methods 37:255–263

Mazzola M, Cook RJ, Thomashow LS, Weller DM, Pierson LS III (1992) Contribution of phenazine antibiotic biosynthesis to the ecological competence of fluorescent pseudomonads in soil habitats. Appl Environ Microbiol 58:2616–2624

McCarthy AJ (1987) Lignocellulose-degrading actinomycetes. FEMS Microbiol Rev 46:145–163

McCarthy AJ, Williams ST (1992) Actinomycetes as agents of biodegradation in the environment – a review. Gene 115:189–192

McManus PS, Stockwell VO, Sundin GW, Jones AL (2002) Antibiotic use in agriculture. Ann Rev Phytopathol 40:443–465

McNamara BP, Wolfe AJ (1997) Coexpression of the long and short forms of CheA, the chemotaxis histidine kinase, by members of the family Enterobacteriaceae. J Bacteriol 179:1813–1818

Menotta M, Amicucci A, Sisti D, Gioacchini AM, Stocchi V (2004) Differential gene expression during pre-symbiotic interaction between *Tuber borchii* Vittad. and *Tilia americana* L. Curr Genet 46:158–165

Milner JL, Silo-Suh L, Lee JC, He H, Clardy J, Handelsmann J (1996) Production of kanosamine by *Bacillus cereus* UW85. Appl Environ Microbiol 62:3061–3065

Molina R, Massicotte H, Trappe JM (1992) Specifcity phenomena in mycorrhizal symbiosis: community-ecological consequences and practical implications. In: Allen MF (ed.) Mycorrhizal functioning: an integrated plant fungal process. Chapman and Hall, London, pp 357–423

Morel M, Jacob C, Kohler A, Johansson T, Martin F, Chalot M, Brun A (2005) Identification of genes differentially expressed in extraradical mycelium and ectomycorrhizal roots during *Paxillus involutus–Betula pendula* ectomycorrhizal symbiosis. Appl Environ Microbiol 71:382–391

Morgan JA, Bending GD, White PJ (2005) Biological costs and benefits to plant-microbe interactions in the rhizosphere. J Exp Bot 56:1729–1739

Murray AC, Woodward S (2003) In vitro interactions between bacteria isolated from sitka spruce stumps and *Heterobasidion annosum*. Forest Pathol 33:53–67

Nagarajkumar M, Bhaskaran R, Velzhahan R (2004) Involvement of secondary metabolites and extracellular lytic enzymes produced by *Pseudomonas fluorescens* in inhibition of Rhizoctonia solani, the rice sheath blight pathogen. Microbiol Res 159:73–81

Nakayama T, Homma Y, Hashidoko Y, Mizutani J, Tahara S (1999) Possible role of xanthobaccins produced by *Stenotrophomonas* sp. strain SB-K88 in suppression of sugar beet damping-off disease. Appl Environ Microbiol 65:4334–4339

Neeno-Eckwall EC, Kinkel LL, Schottel JL (2001) Competition and antibiosis in the biological control of potato scab Can J Microbiol 47:332–340

Newton JA, Fray RG (2004) Integration of environmental and host-derived signals with quorum sensing during plant-microbe interactions. Cell Microbiol 6:213–224

Nicol GW, Glover LA, Prosser JL (2003) The impact of grassland management on archaeal community structure in upland pasture rhizosphere soil. Environ Microbiol 5:152–162

Nomura K, Melotto M, He SY (2005) Suppression of host defense in compatible plant-*Pseudomonas syringae interactions*. Curr Opin Plant Biol 8:361–368

Nurmiaho-Lassila EL, Timonen S, Haahtela K, Sen R (1997) Bacterial colonization pattern of intact *Pinus sylvestris* mycorrhizospheres in dry pine forest soil: an electronmicroscopy study. Can J Microbiol 43:1017–1035

Olivain C, Alabouvette C (1999) Process of tomato root colonization by a pathogenic strain of *Fusarium oxysporum* f.sp. *lycopersici* in comparison with a non-pathogenic strain. New Phytol 141:497–510

Olivain C, Trouvelot S, Binet M-N, Cordier C, Pugin A, Alabouvette C (2003) Colonization of flax roots and early physiological responses of flax cells inoculated with pathogenic and non-pathogenic strains of *Fusarium oxysporum*. Appl Environ Microbiol 69:5453–5462

Omura RS, Murata M, Hanaki H, Hiotozawa K, Oiwa R, Tanaka H (1984) Phosalacine, a new herbicidal antibiotic containing phosphinothricin. Fermentation, isolation, biological activity and mechanism of action J Antibiot 37:829

Ongena M, Jourdan E, Schafer M, Kech C, Budzikiewicz H, Luxen A, Thonart P (2005) Isolation of an N-alkylated benzylamine derivative from *Pseudomonas putida* BTP1 as elicitor of induced systemic resistance in bean. Mol Plant Microbe Interact 18:562–569

Opelt K, Berg G (2004) Diversity and antagonistic potential of bacteria associated with bryophytes from nutrient-poor habitats of the Baltic Sea coast. Appl Environ Microbiol 70:6569–6579

Park KS, Moyne AL, Tuzun S, Kim CH, Kloepper JW (1997) Induction of PR-1a promoter in a transgenic tobacco reporter system by selected PGPR strains which induce resistance. In: Ogoshi A, Kobayashi K, Homma Y, Kodama F, Kondo N, Akino S (eds) Plant growth-promoting bacteria: present status and future prospects. Nakanishi Printing, Sapporo, Japan, pp 251–255

Pérez-Moreno J, Read DJ (2004) Ectomycorrhizal fungi, living ties that bind and nurture trees in nature. Interciencia 29:239–247

Persson T, Baath E, Clarholm M, Lundkvist H, Söderström BE, Sohlenius B, (1980) Trophic structure, biomass dynamics and carbon metabolism of soil organisms in a Scots pine forest. In: Persson T (ed) Structure and function of northern coniferous forests: an ecosystem study. Ecol Bull Stockholm 32:419–459

Peters AF (1991) Field and culture studies of *Streblonema-Macrocystis* new species *Ectocarpales Phaeophyceae* from Chile, a sexual endophyte of giant kelp. Phycologia 30:365–377

Petrini O (1991) Fungal endophytes of tree leaves. In: Andrews J, Hirano S (eds) Microbial ecology of leaves. Springer, Berlin Heidelberg New York, pp 179–197

Petrini O, Fisher PJ, Petrini LE (1992) Fungal endophytes of bracken (*Pteridium aquilinum*), with some reflections on their use in biological control. Sydowia 44:282–293

Petroski R, Powell RG, Clay K (1992) Alkaloids of *Stipa robusta* (sleepygrass) infected with an *Acremonium* endophyte. Nat Toxins 1:84–88

Poole EJ, Bending GD, Whipps JM, Read DJ (2001) Bacteria associated with *Pinus sylvestris-Lactarius rufus* ectomycorrhizas and their effects on mycorrhiza formation *in vitro*. New Phytol 151:743–751

Postma J, Luttikholt AJG (1996) Colonisation of carnation stems by a non-pathogenic isolate of *Fusarium oxysporum* and its effect on *Fusarium oxysporum* f.sp. *dianthi*. Can J Bot 74:1841–1851

Postma J, Rattink H (1992) Biological control of Fusarium wilt of carnation with a non-pathogenic isolate of Fusarium oxysporum. Can J Bot 70:1199–1205

Preszler RW, Gaylord ES, Boecklen WJ (1996) Reduced parasitism of a leaf-mining moth on trees with high infection frequencies of an endophytic fungus. Oecologia 108:159–166

Proctor RH, Plattner RD, Brown DW, Seo JA, Lee YW (2004) Discontinuous distribution of fumonisin biosynthetic genes in the *Gibberella fujikuroi* species complex. Mycol Res 108:815–822

Raaijmakers JM, Bonsall RF, Weller DM (1999) Effet of population density of *Pseudomonas fluorescens* on production of 2,4-diacetylphloroglucinol in the rhizosphere of wheat. Phytopathology 89:470–475

Raaijmakers JM, Vlami M, de Souza JT (2002) Antibiotic production by bacterial biocontrol agents. Antonie van Leeuwenhoek 81:537–547

Rangel-Castro JI, Danell E, Pfeffer PE (2002) A 13 C NMR study of exudation and storage of carbohydrates and amino acids in the ectomycorrhizal edible mushroom *Cantharellus cibarius*. Mycologia 94:190–199

Ravel C, Courty C, Coudret A, Charmet G (1997) Beneficial effects of *Neotyphodium lolii* on the growth and the water status in perennial ryegrass cultivated under nitrogen deficiency or drought stress. Agronomie 17:173–181

Raviraja NS, Sridhar KR, Bärlocher F (1996) Endophytic aquatic hyphomycetes of roots of plantation crops and ferms from India. Sydowia 48:152–160

Rayner ADM, Boddy L (1988) Wood decomposition: its biology and ecology. Wiley, Chichester, NY

Read DJ (1991) Mycorrhizas in ecosystems. Experientia 47:376–391

Read DJ (1993) Plant microbe mutualisms and community structure. In: Schulze ED, Mooney HA (eds) Biodiversity and ecosystem function (Ecological Studies 99). Springer, Berlin Heidelberg New York, pp 181–209

Redecker D, Morton JB, Bruns TD (2000) Ancestral lineages of arbuscular mycorrhizal fungi (Glomales). Mol Phylogen Evol 14:276–284

Redman RS, Sheehan KB, Stout RG, Rodriguez RJ, Henson JM (2002) Thermotolerance generated by plant/fungal symbiosis. Science 298:1581

Reis VM, Baldani JI, Baldani VLD, Dobereiner J (2000) Biological dinitrogen fixation in Gramineae and palm trees. Crit Rev Plant Sci 191:227–248

Richard F, Millot S, Gardes M, Selosse M-A (2005) Diversity and specificity of ectomycorrhizal fungi retrieved from an old-growth Mediterranean forest dominated by Quercus *ilex*. New Phytol 166:1011–1023

Richardson DM, Williams PA, Hobbs RJ (1994) Pine invasions in the southern hemisphere: determinants of spread and invadability. J Biogeo 21:511–527

Riedlinger JM, Schrey SD, Tarkka MT, Hampp R, Kapur M, Fielder HP (2006) Auxofuran, a novel metabolite that stimulates the growth of fly agaric, is produced by the mycorrhiza helper bacterium Streptomyces strain AcH 505. Appl Environ Microbiol 72:3550–3557

Riley MA, Wertz JE (2002) Bacteriocins: evolution, ecology, and application. Annu Rev Microbiol 56:117–137

Rizzo DM, Garbelotto M, Hansen EM (2005) *Phytophthora ramorum*: integrative research and management of an emerging pathogen in California and Oregon forests. Annu Rev Phytopathol 43:309–335

Roberts KJ, Anderson RC (2001) Effect of garlic mustard [Alliaria petiolata (Beib. Cavara & Grande)] extracts on plants and arbuscular mycorrhizal (AM) fungi. Am Midl Nat 146:146–152

Robleto EA, Kmiecik K, Oplinger ES, Nienhuis J, Triplett EW (1998) Trifolitoxin production increases nodulation competitivness of *Rhizobium etli* CE3 under agricultural comditions. Appl. Environ. Microbiol 64:2630–2633

Rocha O, Ansari K, Doohan FM (2005) Effects of trichothecene mycotoxins on eukaryotic cells: a review. Food Addit Contam 22:369–378

Rodelas B, Gonzales-Lozep J, Salmeron V, Martinez-Toledo MV, Pozo V (1998) Symbiotic effectiveness and bateriocin production by *rhizobium leguminosarum* bv *viceae* isolated from agricultural soils in Spain. Appl Soil Ecol 8:51–60

Rovira AD (1991) Rhizosphere research – 85 years of progress and frustration. In: Kleister DL, Cregan PB (eds) The rhizosphere and plant growth. Kluwer Academic Publishers, Amsterdam, pp 3–13

Rowan DD, Latch GCM (1994) Utilization of endophyte-infected perennial ryegrasses for increased insect resistance, In: Bacon CW, White JF Jr (eds) Biotechnology of endophytic fungi of grasses. CRC Press, Boca Raton, pp 169–183

Ryu CM, Farag MA, Hu CH, Reddy MS, Wie HX, Pare PW, Kloepper JW (2003) Bacterial volatiles promote growth in *Arabidopsis*. Proc Natl Acad Sci (USA) 100:4927–4932

Ryu CM, Farag MA, Hu CH, Reddy MS, Kloepper JW, Pare PW (2004) Bacterial volatiles induce systemic resistance in *Arabidopsis*. Plant Physiol 134:1017–1026

Saari SK, Campbell CD, Russell J, Alexander IJ, Anderson IC (2005) Pine microsatellite markers allow roots and ectomycorrhizas to be linked to individual trees. New Phytol 165:295–304

Salerno MI, Gianinazzi S, Gianinazzi-Pearson V (2000) Effects on growth and comparison of root tissue colonization patterns of *Eucalyptus viminalis* by pathogenic and nonpathogenic strains of *Fusarium oxysporum*. New Phytol 146:317–324

Samac DA, Kinkel LL (2001) Suppression of the root-lesion nematode (*Pratylenchus penetrans*) in alfalfa (*Medicago sativa*) by *Streptomyces* spp. Plant Soil 235:35–44

Samils N, Elfstrand M, Czederpiltz DL, Fahleson J, Olson A, Dixelius C, Stenlid J (2006) Development of a rapid and simple *Agrobacterium tumefaciens*-mediated transformation system for the fungal pathogen *Heterobasidion annosum*. FEMS Microbiol Lett 255:82–88

Sampedro I, Aranda E, Scervino JM, Fracchia S, Garcia-Romera I, Ocampo JA, Godeas A (2004) Improvement by soil yeasts of arbuscular mycorrhizal symbiosis of soybean (*Glycine max*) colonized by *Glomus mosseae*. Mycorrhiza 14:229–234

Samson RA, Frisvad JC (2004) Penicillium: new taxonomic schemes and mycotoxins and other extrolites. Stud Mycol 49:1–266

Saxena S, Pandey AK (2001) Microbial metabolites as eco-friendly agrochemicals for the next millennium. Appl Microbiol Biotechnol 55:395–403

Schardl CL, Leuchtmann A, Spiering MJ (2004) Symbioses of grasses with seedborne fungal endophytes Annu Rev Plant Biol 55:315–340

Schottel JL, Shimizu K, Kinkel LL (2001) Relationships of in vitro pathogen inhibition and soil colonization to potato scab biocontrol by antagonistic *Streptomyces* spp. Biol Control 20:101–112

Schrey SD, Schellhammer M, Ecke M, Hampp R, Tarkka MT (2005) Mycorrhiza helper bacterium *Streptomyces* AcH 505 induces differential gene expression in the ectomycorrhizal fungus *Amanita muscaria*. New Phytol 168:205–216

Schrey SD, Salo V, Raudaskoski M, Hampp R, Nehls U, Tarkka MT (2007) Interaction with mycorrhiza helper bacterium *Streptomyces* sp. AcH 505 modifies organisation of actin cytoskeleton in the ectomycorrhizal fungus *Amanita muscaria* (fly agaric). Curr Genet, DOI: 10.1007/s00294-007-0138-x (in press)

Schröter D, Wolters V, De Ruiter PC (2003) C and N mineralisation in the decomposer food webs of a European forest transect. Oikos 102:294–308

Schulz B, Boyle C (2005) The endophytic continuum. Mycol Res 109:661–686

Schulz B, Boyle C, Draeger S, Römmert A-K, Krohn K (2002) Endophytic fungi: a source of novel biologically active secondary metabolites. Mycol Res 106:996–1004

Schwartz D, Berger S, Heinzelmann E, Muschko K, Welzel K, Wohlleben W (2004) Biosynthetic gene cluster of the herbicide phosphinothricin tripeptide from *Streptomyces viridochromogenes* Tu494. Appl Environ Microbiol 70:7093–7102

Selosse M-A, Faccio A, Scappaticci G, Bonfante P (2004) Chlorophyllous and achlorophyllous specimens of *Epipactis microphylla* (Neottieae, Orchidaceae) are associated with ectomycorrhizal septomycetes, including truffles. Microb Ecol 47:416–426

Sessitsch A, Hackl E, Wenzl P, Kilian A, Kostic A, Stralis-Pavese N, Sandjong BT, Bodrossy L (2006) Diagnostic microbial microarrays in soil ecology. New Phytol 171:719–736

Setubal JC, Moreira LM, da Silva AC (2005) Bacterial phytopathogens and genome science. Curr Opin Microbiol 8:595–600

Sevilla M, Burris RH, Gunapala N, Kennedy C (2001) Comparison of benefit to sugarcane plant growth and 15N2 incorporation following inoculation of sterile plants with Acetobacter diazotrophicus wild-type and Nif- mutants strains. Mol Plant Microbe Interact 14:358–366

Simon HM, Jahn CE, Bergerud LT, Sliwinski MK, Weimer PJ, Willis DK, Goodman RM (2005) Cultivation of mesophilic soil crenarchaeotes in enrichment cultures from pPlant roots. Appl Environ Microbiol 71:4751–4760

Smalla K, Wieland G, Buchner A, Zock A, Parzy J, Kaiser S, Roskot N, Heuer H, Berg G (2001) Bulk and rhizosphere soil bacterial communities studied by denaturing gradient gel electrophoresis: plant-dependent enrichment and seasonal shifts revealed. Appl Environ Microbiol 67:4742–4251

Smith CS, Chand T, Harris RF, Andrews JH (1989) Colonization of a submersed aquatic plant, Eurasian water milfoil (*Myriophyllum spicatum*) by fungi under controlled conditions. Appl Environ Microbiol 55:2326–2332

Smith SE, Read DJ (1997) *Mycorrhizal Symbiosis*. 2nd edn. Academic Press, London

Smith SE, Smith FA (1990) Structure and function on the interfaces in biotrophic symbioses as they relate to nutrient transport. New Phytol 114:1–38

Sommers E, Vanderleyden J (2004) Rhizosphere bacterial signalling: a love parade beneath our feet. Cri Rev Microbiol 30:205–240

Spiering MJ, Greer DH, Schmid J (2006) Effects of the fungal endophyte, *Neotyphodium lolii*, on net photosynthesis and growth rates of perennial Ryegrass (*Lolium perenne*) are independent of in planta endophyte concentration. Ann Bot 98:379–387

Steenhoudt O, Vanderleyden J (2000) *Azospirillum*, a free-living nitrogen-fixing bacterium closely associated with grasses: genetic, biochemical and ecological aspects. FEMS Microbiol Rev 24(4):487–506

Stone JK, Viret O, Petrini O, Chapela I (1994) Histological studies of host penetration and colonization by endophytic fungi. In: Petrini O, Ouellette GB (eds) Host wall alterations by parasitic fungi. American Phytopathological Society Press, St Paul, MN, pp 115–128

Stone JK, Bacon CW, White JF (2000) An overview of endophytic microbes: endophytism defined. In: Bacon CW, White F (eds) Microbial endophytes. Marcel Dekker, New York

Stracke S, Kistner C, Yoshida S (2002) A plant receptorlike kinase required for both bacterial and fungal symbiosis. Nature 417:959–962

Strobel G, Daisy B (2003) Bioprospecting for microbial endophytes and their natural products. Microbiol Mol Biol Rev 67:491–502

Strobel GA, Miller RV, Martinez-Miller C, Condron MM, Teplow DB, Hess WM (1999) Cryptocandin, a potent antimycotic from the endophytic fungus *Cryptosporiopsis* cf. *quercina*. Microbiology 145:1919–1926

Strobel G, Daisy B, Castillo U, Harper J (2004) Natural products from endophytic micro-organisms. J Nat Prod 67:257–268

Sturz AV, Christie BR, Matheson BG, Nowak J (1996) Biodiversity of endophytic bacteria which colonize red clover nodules, roots, stems and foliage and their influence on host growth. Biol Fertil Soils 25:13–19

Sturz AV, Christie BR, Matheson BG, Arsenault WJ, Buchanan NA (1999) Endophytic bacterial communities in the periderm of potato tubers and their potential to improve resistance to soil-borne plant pathogens. Plant Pathol 48:360–369

Sturz AV, Christie BR, Nowak J (2000) Bacterial endophytes: potential role in developing sustainable systems of crop production. Crit Rev Plant Sci 19:1–30

Taechowisan T, Lu C, Shen Y, Lumyong S (2005) Secondary metabolites from endophytic *Streptomyces aureofaciens* CMUAc130 and their antifungal activity. Microbiology 151:1691–1695

Taghavi S, Barac T, Greenberg B, Borremans B, Vangronsveld J, van der Lelie D (2005) Horizontal gene transfer to endogenous endophytic bacteria from poplar improves phytoremediation of toluene. Appl Environ Microbiol 71:8500–8505

Tan RX, Zou WX (2001) Endophytes: a rich source of functional metabolites. Nat Prod Rep 18:448–459

Tarkka MT, Schrey S, Nehls U (2006) The alpha-tubulin gene *AmTuba1*: a marker for rapid mycelial growth in the ectomycorrhizal basidiomycete *Amanita muscaria*. Curr Genet 49:294–301

Taylor DL, Bruns TD, Hodges SA (2004) Evidence for mycorrhizal races in a cheating orchid Proc R Soc B Biol Sci 271:35–43

Thrane C, Harder Nielsen T, Neiendam Nielsen M, Sørensen J, Olsson S (2000) Viscosinamide-producing *Pseudomonas fluorescens* DR54 exerts a biocontrol effect on *Pythium ultimum* in sugar beet rhizosphere. FEMS Microbiol Ecol 33:139–146

Tokala RK, Strap JL, Jung CM, Crawford DL, Salove MH, Deobald LA, Bailey JF, Morra MJ (2002) Novel plant-microbe rhizosphere interaction involving *Streptomyces* lydicus WYEC108 and the pea plant (*Pisum sativum*). Appl Environ Microbiol 68:2161–2171

Toth IK, Birch PR (2005) Rotting softly and stealthily. Curr Opin Plant Biol 8:424–429

Tournas VH (2005) Spoilage of vegetable crops by bacteria and fungi and related health hazards. Crit Rev Microbiol 31:33–44

Tuncer M, Kuru A, Isikli M, Sahin N, Çelenk FG (2004) Optimization of extracellular endoxylanase, endoglucanase and peroxidase production by *Streptomyces* sp. F2621 isolated in Turkey. J Appl Microbiol 79:783–791

van Loon LC, Bakker PA, Pieterse CM (1998) Systemic resistance induced by rhizosphere bacteria. Annu Rev Phytopathol 36:453–483

Varma A, Singh A, Sahay NS, Sharma J, Roy A, Kumari M, Raha D, Thakran S, Deka D, Bharti K, Hurek T, Blechert O, Rexer K-H, Kost G, Hahn A, Maier W, Walter M, Strack D, Kranner I (2000) Piriformospora indica: an axenically culturable mycorrhiza-like endosymbiotic fungus. In: Hock B (ed) The mycota. Vol IX. Fungal associations. Springer, Berlin Heidelberg New York, pp 125–150

Vessey JK (2003) Plant growth promoting rhizobacteria as biofertilizers. Plant Soil 255:571–586

Vicuña R, Gonzáles B, Seelenfreund D, Ruttimann C, Salas L (1993) Ability of natural bacterial isolates to metabolize high and low-molecular-weight lignin-derived molecules. J Biotechnol 30:9–13

Vivas A, Voros I, Biro B, Campos E, Barea JM, Azcon R (2003) Symbiotic efficiency of autochthonous arbuscular mycorrhizal fungus (*G. mosseae*) and *Brevibacillus* sp. isolated from cadmium polluted soil under increasing cadmium levels. Environ Pollut 126:179–189

Von Bodman SB, Bauer WD, Coplin DL (2003) Quorum sensing in plant-pathogenic bacteria. Annu Rev Phytopathol 41:455–482

Vosatka M, Gryndler M (1999) Treatment with culture fractions from *Pseudomonas putida* modifies the development of *Glomus fistulosum* mycorrhiza and the response of potato and maize plants to inoculation. Appl Soil Ecol 11:245–251

Wallander H, Lindahl BD, Nilsson LO (2006) Limited transfer of nitrogen between wood decomposing and ectomycorrhizal mycelia when studied in the field. Mycorrhiza 16:213–217

Weber D, Strerner O, Anke T Gorzalczancy S, Martino V, Acevedo C (2004) Phomol, a new anti-inflammatory metabolite from an endophyte of the medicinal plant Erythrina crista-galli. J. Antibiot (Tokyo) 57:559–563

Weidner S, Pühler A, Küster H (2003) Genomics insights into symbiotic nitrogen fixation. Curr Opin Biotechnol 14:200–205

Weissman KJ, Leadlay PF (2005) Combinatorial biosynthesis of reduced polyketides. Nat Rev Microbiol 3:925–936

Weller DM (1988) Biological control of soilborne plant pathogens in the rhizosphere with bacteria. Annu Rev Phytopahol 26:379–407

Weller DM, Raaijmakers JM, Gardener BB, Thomashow LS (2002) Microbial populations responsible for specific soil suppressiveness to plant pathogens. Annu Rev Phytopathol 40:309–348

Wheatley RE (2002) The consequences of volatile organic compound mediated bacterial and fungal interactions. Antonie van Leeuwenhock 81:357–364

Whipps JM (2001) Microbial interactions and biocontrol in the rhizosphere. J Exp Bot 52:487–511

Whipps, JM, Lumsden RD (2001) Commercial use of fungi as plant disease biological control agents: status and prospects. In: Butt T, Jackson C, Magan N (eds) Fungal biocontrol agents: progress, problems and potential, CABI Publishing, Wallingford, pp 9–22

Whipps JM, Lynch JM (1986) The influence of the rhizosphere on crop productivity

Whitehead NA, Byers JT, Commander P, Corbett MJ, Coulthurst SJ, Everson L, Harris AK, Pemberton CL, Simpson NJ, Slater H, Smith DS, Welch M, Williamson N, Salmond GP (2002) The regulation of virulence in phytopathogenic *Erwinia* species: quorum sensing, antibiotics and ecological considerations. Antonie Van Leeuwenhoek 81:223–231

Wiener P (2000) Antibiotic production in a spatially structured environment. Ecol Lett 3:122–133

Wilkinson H, Siegel MR, Blankenship JD, Mallory AC, Bush LP, Schardl CL (2000) Contribution of fungal loline alkaloids to protection from aphids in a grass-endophyte mutualism. MPMI 13:1027–1033

Wilson RA, Handley BA, Beringer JE (1997) Bacteriocin production and resistance in a field population of *Rhizobium leguminosarum* biovar *viciae*. Soil Biol Biochem 30:413–417

Wirth S, Ulrich A (2002) Cellulose-degrading potentials and phylogenetic classification of carboxymethyl-cellulose decomposing bacteria isolated from soil. System Appl Microbiol 25:584–591

Wolfe BE, Klironomos JN (2005) Breaking new ground: soil communities and exotic plant invasion. BioSci 55:477–487

Woo S, Fogliano V, Scala F, Lorito M (2002) Synergism between fungal enzymes and bacterial antibiotics may enhance biocontrol. Antonie Van Leeuwenhoek 81:353–356

Xu JR, Peng YL, Dickman MB, Sharon A (2006) The dawn of fungal pathogen genomics. Annu Rev Phytopathol 44:337–366

Yao J, Allen C (2006) Chemotaxis is required for virulence and competitive fitness of the bacterial wilt pathogen *Ralstonia solanacearum*. J Bacteriol 188:3697–3708

Yuan WM, Crawford DL (1995) Charaterization of Streptomyces lydicus WYEC108 as a potential biocontrol agent against fungal root and seed rots. Appl Environ Microbiol 61:3119–3125

Zhang L, Birch RG (1997) The gene for albicidin detoxification from *Pantoea dispersa* encodes an esterase and attenuates pathogenicity of *Xanthomonas albilineans* to sugarcane. Proc Natl Acad Sci (USA) 94:9984–9989

Zhang H, Forde BG (2000) Regulation of *Arabidopsis* root development by nitrate availability. J Exp Bot 51:51–59

Zhang L, Xu J, Birch RG (1998) High affinity binding of albicidin phytotoxins by the AlbA protein from *Klebsiella oxytoca*. Microbiology 144:555–559

Chapter 2
Role of Microbial Diversity for Soil, Health and Plant Nutrition

C.R. Bhatia

2.1 Introduction

Soil provides the medium for root development, and with the exception of carbon, hydrogen, oxygen and some nitrogen, plants depend on soil for all other nutrients and water. Soils develop by the disintegration of rocks, and minerals therein, through biotic actions of the microbes and the fauna sustained by them. Earlier, only the physical and chemical properties of soil were considered important. However, the role of soil biodiversity in maintaining fertility, and the interdependence of soil biological activities with physical and chemical characteristics is well recognized now (Abbott and Murphy 2003; Fitter 2005; Suzuki et al. 2005; Madsen 2005; Manlay et al. 2007). Physical properties and the amount of soil organic matter (SOM) determine the microbial diversity that varies with depth, and soil health. SOM adds to soil fertility, water retention and has a great influence on the growth of the above ground vegetation. Biological indicators such as microbial biomass, soil respiration, enzyme activities and microbial diversity indicate soil health. Significance of soil biodiversity for sustainability of the farming systems has been discussed at length (Brussard et al. 2007). Microbial diversity is an excellent indicator of soil health (Nielsen and Winding 2002). They report that variation in microbial population or activities precede changes that can be noticed in some cases as early signs of soil degradation or amelioration. Water and nutrient supply from soil, particularly N and P, determine the plant growth both in natural and agro-ecosystems. The above ground vegetation is the ultimate source of C for the microbes in the rhizosphere that, in turn, support the macro-fauna. Thus, the above ground vegetation influences the below ground microbial community structure and soil properties (Orwin and Wardale 2005).

This chapter gives an overview of the role of soil microbial diversity and processes controlled by them to enhance C sequestration into soil and meet the growing needs of increasing crop productivity. The effects of genetically engineered (GE)

C.R. Bhatia
17, Rohini, Plot No. 29–30, Sector 9-A, Vashi, Navi Mumbai 400703, India
e-mail: neil@bom7.vsnl.net.in

C.S. Nautiyal, P. Dion (eds.) *Molecular Mechanisms of Plant and Microbe Coexistence.* Soil Biology 15, DOI: 10.1007/978-3-540-75575-3

crops and microbes used for biocontrol on soil microbial diversity are examined. Possibilities of introducing peptides and enzymes that interact with the pathogenic soil microbes or enhance nutrient availability are considered. Two major challenges are to effectively manipulate soil microbial and genetic diversity and its interactions with the crop plants to (i) minimize the chemical fertilizer inputs, and enhance their use efficiency without reduction in productivity and (ii) enhance carbon sequestration into soil to reduce atmospheric CO_2 increasing at an alarming rate due to the anthropogenic factors.

2.2   Soil Microbial Diversity

In natural ecosystems C, N, P, K, Ca, Mg, S and all other mineral nutrients are cycled back into soil through litter fall and decay of the organic matter. The soil microbes that include bacteria, fungi, actinomycetes, protozoa and algae play a significant role in the nutrient cycling. Though it is widely accepted that soil biodiversity is vital for maintaining productivity in natural and managed agro ecosystems, the understanding of the microbial communities, soil fauna and their diversity is extremely limited (Buckley and Schmidt 2003; Nannipieri et al. 2003; Lynch et al. 2004; Fitter 2005; Fitter et al. 2005; Nannipieri and Smalla 2006). Of the soil microbes, 99% cannot be cultured; identification, characterization and finding their role are particularly difficult for such organisms. Estimates on the number of microbial species present in the soil vary from few thousands to millions. Lately, the nucleic acid based techniques including analysis of DNA and rRNA molecules from soil samples have revealed enormous diversity (Buckley and Schmidt 2003; Suzuki et al. 2005). The molecular methods used for soil microbial diversity are covered in the reviews by Nannipieri et al. (2003) and Lynch et al. (2004). High throughput DNA sequencing techniques developed for the human genome project are now being used to determine the soil genomes and diversity of sequences (Deutschbauer et al. 2006; Gewin 2006). Soil genome sequencing is reported to be highly complex. Hopefully, as a result of these investigations, in future it may be possible to characterize and quantify the microbial diversity in soil samples, and to follow the effect of plant communities and agronomic practices. The tools of metagenomics where the isolated genes are expressed in *E. coli* are providing new information on specific genes isolated from the DNA extracted from soil samples (Rondon et al. 2000; Gewin 2006).

Torsvik et al. (2002) compared the genome complexity among three terrestrial niches, where the number of prokaryotic cells per cubic centimeter of soil was similar (about 10 billion). The pristine pasture and forest soils contained over 10 times the genome complexity (equivalent to 3500–8800 *E. coli* genomes) compared to that of the agricultural field soils (equivalent to 140–350 *E. coli* genomes). Using improved analytical methods (Gans et al. 2005) suggests that more than 1 million distinct genomes might exist in the pristine soil, exceeding the previous estimates by two orders of magnitude. Further, it was shown that metal pollution could reduce the genomic diversity of pristine soil by more than 99.9%, revealing the highly toxic effect of metal contamination, especially for rare microbial taxa (Gans et al. 2005).

Interactions amongst soil microbes and between plants and microbes in the rhizosphere are emerging as fascinating areas of molecular biology. Hopefully elucidation of the mechanisms and specific role of the recognized species would lead to better farm management practices in future. These new approaches for investigating soil microbial diversity based on nucleic acids and proteins (soil proteome) are discussed by Nannipieri and Smalla (2006). Besides the species diversity and functional redundancy, there is enormous genetic diversity within the same species which remains largely unexplored, except for the species producing the antibiotics. The new evidences also show that soil bacteria often swap genes among themselves (Chandler 2006).

Using the molecular techniques significant changes in soil microbial communities have been shown in recent years both in the natural and cultivated areas. Microbial activities using rRNA abundance in soil samples showed that the differences between conventionally managed and never cultivated fields were significant but not between the latter and those abandoned for nine years (Buckley and Schmidt 2003). Fields abandoned for more than 45 years showed microbial activities similar to those that were never cultivated. In wooded mountain pastures, cattle grazing which involves repeated mowing, trampling and addition of urine and dung had more profound effects on microbial communities in sunny areas compared to the shaded (Kohler et al. 2005a,b). These studies led the authors to infer that in natural pastures the microbial communities below ground also change similar to the shifting mosaic of the plant species above ground. Plant species composition and richness determined the soil microbial community resistance and resilience to experimentally imposed drying on pasture plant species (Orwin and Wardle 2005). Cluster roots of white lupine (*Lupinus albus* L.) plants that secrete organic acids into soil were used to monitor the community structure in microcosms (Weisskopf et al. 2005). Frequencies of auxin producers were higher in juvenile and mature cluster roots and significantly decreased with their senescence. Proportion of the active population was higher in proximity to the roots. Comparative analyses of prokaryotic genomic sequences suggest the importance of ecology in determining microbial genome size and gene content. The significant variability in genome size and gene content among strains and species of prokaryotes indicate the highly fluid nature of prokaryotic genomes, a result consistent with those from multilocus sequence typing and representational difference analyses. The integration of various levels of ecological analyses coupled to the application and further development of high throughput technologies are accelerating the pace of discovery in microbial ecology (Xu 2006). The evidences based on molecular techniques cited above point to large diversity. However, identification of microbes responsible for the key processes remains an immense challenge (Madsen 2005).

2.3 Evolution of Farming

A brief consideration of the evolution of farming systems is necessary for a correct perception of the change from natural to agro-ecosystems. Some 10–12,000 years back, humans started clearing the natural vegetation, and planting seeds of the crops such as wheat and barley which earlier they had been collecting from natural stands. Cultivation of crops was possible only after clearing the existing natural vegetation, and often burning the same at site, similar to slash and burn agriculture still practiced in many places by tribal populations. Farming practices continue to evolve with developments in other areas of technology. Early farmers soon realized that repeatedly growing cereals in the same land reduced the soil fertility, which could be restored by keeping the plots fallow or through cultivation of leguminous crops. Practices of growing and turning in green manure crops were evolved. With the domestication of animals and the use of farmyard manures, various methods of composting were developed to maintain soil productivity. This type of subsistence agriculture continued in most parts of the world till the development of synthetic chemical fertilizers – N, P and K. The use of chemical fertilizers enhanced productivity several fold, and with the resultant increased harvests the micronutrient deficiencies started showing up. The need to control insect pests, pathogens and weeds led to the development of chemical pesticides which ultimately end up in the soil. As the pesticide residues and their degradation products increased, soil biological activity and crop productivity were affected in contaminated fields. At the same time various physical processes that cause soil degradation such as excessive use of irrigation, mining of ground water and nutrients, loss of top soil, unbalanced use of fertilizers etc. also affect microbial biodiversity resulting in deterioration of soil quality.

Soil health and productivity management have gained immense importance in recent years, especially in the intensively cropped regions in the tropics and sub-tropics as soils in these areas are often poor in SOM. The concepts of Integrated Soil Management (ISM) and Integrated Nutrient Management (INM) have been developed for more sustainable production systems. Practices such as no till farming that cause minimal disturbance to the top soil are increasing (Kirchmann and Thorvaldsson 2000).

A strong lobby against the use of chemical fertilizers and pesticides has emerged, especially in the European Economic Community. The opponents of modern farming practices would like to revert back to organic or ecological (also biodynamic) farming without the use of synthetic fertilizers and pesticides. The ability to feed the present and projected population in year 2050 based on organic farming has been questioned by those supporting modern farming practices. The soils, their SOM content as well the climatic conditions vary in different parts, and therefore the ISM and INM practices cannot be universal. They necessarily need location specific development.

2.3.1   Present and the Future Scenario

Globally more food and other farm products are needed to meet the growing requirements of increasing, economically ascendant world population. Predictions based on the trend of the past 50 years indicate that by 2050, N, P and pesticide use would increase more than 2.5 times the amount used in the year 2000 (Tilman et al. 2001). In general, it is not possible to increase the crop land area in most parts of the world and hence, the increased production must come from the presently cropped area by enhancing land productivity. Further, the additional production must use less of water, chemical fertilizers, pesticides, energy, and manual labor to bring about reduction in the adverse impact on the different components of the environment – soil, water and atmosphere.

Firewood obtained from trees was the main source of energy for cooking and heating homes till the coal, electricity and oil emerged as the alternative and more convenient energy sources. With the realization that the fossil fuels are not unlimited and their increasing cost has brought focus on crops as a renewable source of liquid fuels – ethanol and bio-diesel. This implies that the same land resources should meet the growing demand of food as well as part of the energy. The human civilization developed entirely depending on renewable resources for food and energy but the population was limited at that time. Moreover, virgin land was available for cultivation and grazing of animals. In order to meet the food, and a substantial part of the energy needs of the present 6 billion plus, expected to reach 9 billion by 2050, using only renewable resources does not appear feasible with the existing technologies and knowledge. At the same time, pursuing the path followed in the last century would further exacerbate the environmental problems. The challenge therefore is to increase productivity in an environmentally sustainable manner. This may involve exploitation of soil microbe–microbe and microbe–plant interactions (Morrissey et al. 2004) particularly in low input cropping systems (Johansson et al. 2004). Considering the present level of crop productivity, population growth rates and other factors, countries can be broadly put into four groups:

1. Abundant food production, declining population and stable demand for food, but not major exporters of food items, stabilized soils with high SOM, adequate water resources and forest cover; mainly in the temperate regions.
2. Abundant food production, stabilized or slow population growth, major exporters of food grains and other agri-products, cultivable land set aside, large forest cover, very limited soil degradation. Such regions can easily divert access food grain production for bio-fuels, and or land area to energy crops.
3. Regions with high population growth where food self sufficiency has been achieved in recent years with high inputs of chemical fertilizers and pesticides. Water resources are limited, poor soils with high degradation, low SOM, and limited forest cover.
4. High food insecurity, subsistence farming, very limited use of irrigation, fertilizer and pesticides. High population growth rate; poor soils low in SOM and inadequate water resources.

The developed countries of Europe and North America fall in 1 or 2 of the above, while most of the developing countries with poor economic growth come under 3 and 4.

2.4 Carbon Flow into Agro- and Natural Ecosystems

Cropping systems aim to maximize the fixation of solar energy, a free resource into phytomass within the constraints of temperature, water and plant nutrients through human intervention. Soils provide water and nutrients for enlarging the leaf canopy for intercepting solar radiation. Carbon, hydrogen, oxygen, nitrogen and sulfur, along with the other macro and micronutrients, are incorporated into organic molecules through light dependent reactions. These are utilized for construction, maintenance and turnover of different macromolecules and plant organs. Improved farming practices such as irrigation, fertilizers, pesticides and management amplify the fixation of solar energy, and energy flow into the cropping systems. A part of the resources are used to protect plants from insects and pathogens (Mitra and Bhatia 1982). The above ground plant cover provides C, N and other nutrients to the soil microbes through root exudates, and decomposition of aerial and root phytomass litter (Fig. 2.1). It is decomposed by the soil microbes and partly mineralized; CO_2 and methane are released into the atmosphere. The respiratory activities of plant roots, associative mycorrhizal fungi and free living heterotrophs in soil determine the CO_2 evolution from soil as a component of global carbon cycle. Plant root systems are the main pathway for C inputs into soil. Benefits of plant–microbial interactions also has a cost in terms of C inputs (Morgan et al. 2005). Dynamic simulation models have been developed (Wu et al. 2007). Current photosynthesis in boreal forests in the northern latitudes has been shown to drive soil respiration (Högberg et al. 2001). Girdling of the pine trees through removal of the bark to prevent the supply of current photosynthates to the roots reduced the number of ectomycorhizal fungi from 11 in the control plots to 1. The same was also true for wet tropical forest where phytomass litter inputs were positively correlated to fungal and bacterial biomass, and the latter to soil CO_2 efflux (Li et al. 2005).

Soil is considered as an important net sink for carbon, estimated to contain around 1500×109 tons globally (Copley 2000). This amount is estimated to be 300 times the amount of carbon released currently through burning of fossil fuels. It was assumed that the carbon locked in the soil is inert and stays there. However, recently it has been shown that SOC is more vulnerable to land use and changing climate (Bellamy et al. 2005; Schulze and Freibauer 2005). Soils could therefore also be a source rather than the sink for atmospheric C (Chen et al. 2005; Janzen 2006) largely due to the activities of the soil microbes.

2.4.1 Increase of Greenhouse Gases and Climatic Change

There is increasing evidence that the world's climate is changing and that the rate of change since the onset of the industrial revolution is greater than would be expected from natural variability alone. Clearing of forests for agriculture and burning of fossil fuels are considered the two main cause of elevated CO_2 concentration in the atmosphere (Schimel et al. 2001; Oren et al. 2001). Besides CO_2 increase in methane and nitrous oxide (N_2O) in the atmosphere are linked to agriculture. These gases trap the earth's outgoing infra red part of the solar radiation leading to increase in temperature as in greenhouses, and hence referred as greenhouse gases. They are the cause of global warming and predicted climate changes. CO_2 is the major component of greenhouse gases and the cause of global rise in temperature. It is foreseen that the increase

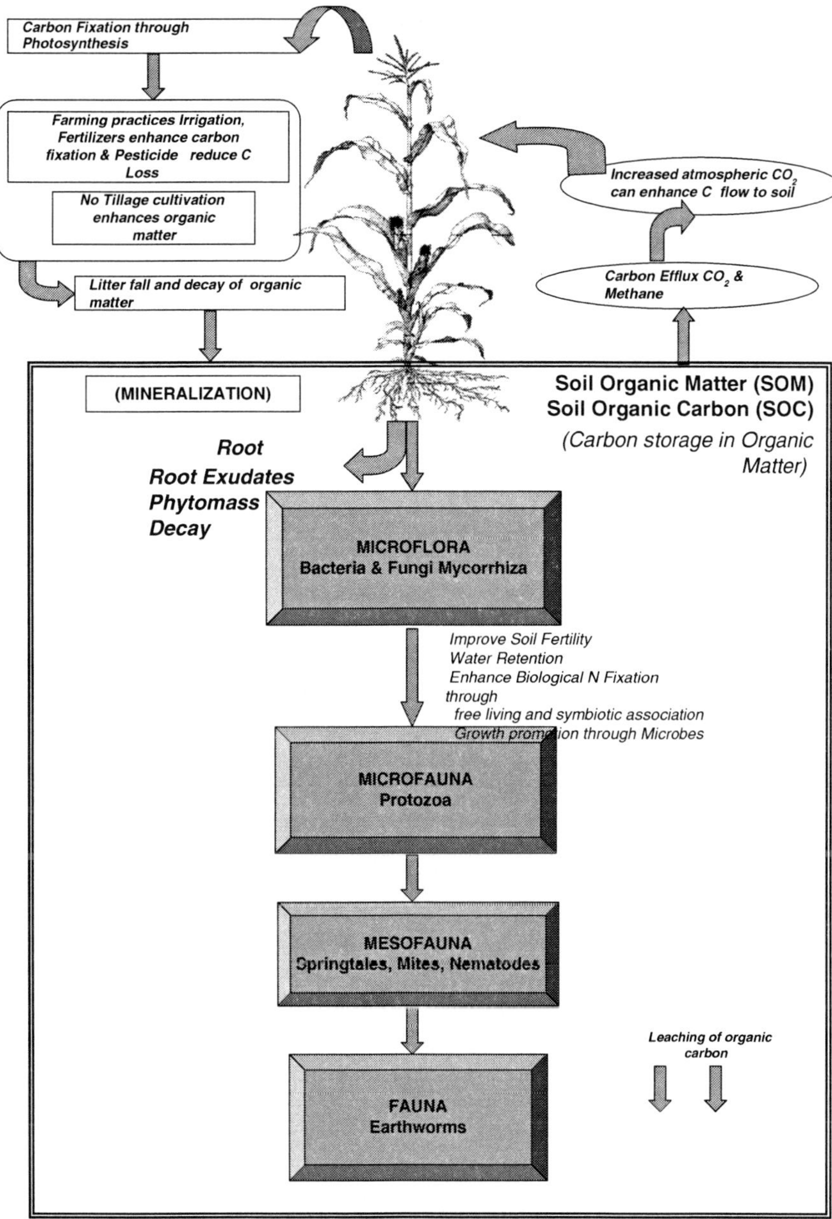

Fig. 2.1 Carbon flow and efflux from soil

in temperature would lead to melting of the snow in the polar regions, raising the sea level causing flooding, and submergence of the low lying coastal areas. All these events would impact plant growth, and alter soil biodiversity that are difficult to model. Climate change associated with greenhouse gas (GHG) emissions was recognized as a global concern in 1979, which led to the adoption of Kyoto Protocol as a first step to achieve stabilization of greenhouse gases. Projected scenarios indicate that increased temperatures and CO_2 concentrations have the potential to enhance herbage growth, but changes in seasonal precipitation would reduce these benefits particularly in areas with low rainfall. Increased frequency of droughts, storms and other extreme events may have further implications for grasslands. Potential farm-scale adaptive responses to climate change have been identified; at the same time grassland agriculture also contributes to GHG emissions, particularly methane and nitrous oxide. The net carbon balance and carbon sequestration depends on the management practices. Hopkins and Prado (2007) have recently reviewed management for mitigating grassland's contribution to GHG emissions which need to be further developed in a holistic way considering other location specific constraints.

2.4.2 *Increasing Carbon Sequestration in Soils*

The implications of global warming and elevated CO_2 on man made agro- and natural ecosystems with plants as the primary producers are enormous and difficult to predict. Increased atmospheric CO_2 would enhance photosynthesis leading to increased flow of carbon assimilates into soil. Higher temperatures and associated alterations in precipitation would lead to considerable changes in cropping patterns, and may have profound effects on soil biodiversity. However, soil microbes due to their short life cycles, and extensive exchange of genetic material are better endowed to cope with the environmental changes than the plants.

It is advocated that sequestration of CO_2 in SOM could contribute to its reduction in the atmosphere (Schlesinger 1999). Adoption of conservation tillage, including no till cultivation practices, could sequester all CO_2 released from agricultural activities. Application of N fertilizers and CO_2 enrichment are the other alternatives for increased C sequestration into soil. The Royal Society of UK disagrees with the proposition of soil as C sink (Adam 2001) stating that improved management could enhance C sequestration on short term to bring about 25% reduction in atmospheric CO_2 by 2050, but thereafter the potential would be limited. Uncertainties were pointed out that some soils could even release their C. This has been shown by Bellamy et al. (2005) mentioned earlier. Use of N fertilizers to enhance C sequestration could also increase release of other greenhouse gases – methane and NO.

CO_2 enrichment experiments provide data on the possible effects of increased CO_2 concentration in the atmosphere. Annual grass land exposed to enhanced CO_2 levels increased N uptake by the plants resulting in reduction of extractable N in soil (Hu et al. 2001). Microbial biomass C increased while N remained unchanged resulting in a higher C:N ratio, suggesting increased fungal/bacterial ratio. Bacterial (*Pseudomonas fluorescence* and *Pantoea agglomerans*) inoculation of maize increased carbon flow into the system, with higher net CO_2 assimilation rate, larger C allocation to the roots with increased amino acid exudation, and higher root/rhizosphere respiration indicating intensive C turnover (Schulze and Pöschel 2005). Other studies also indicate that in the short term CO_2 enrichment enhances C gain of the ecosystem through stimulation of photosynthesis (Diaz et al. 1993; Zak et al. 1993; DeLucia et al. 1999). Microbial processes in the soil gain with the increased C availability, but when N becomes limiting, plants and microbes compete for N (Hu et al. 2001, 2005). Enhanced C input and limitation of N favors fungi over bacteria as fungal biomass has a lower C:N ratio. In northern mid-latitude pine forests N fertilization and C enrichment increased growth by 74% (Oren et al. 2001). Under poor soil fertility C enrichment had no effect. C and N metabolisms are interdependent in plants (Swank et al. 1982) and microbes. The effect of CO_2 enrichment in different soil types and defined microbial population can provide new knowledge to enhance C sequestration and limiting the loss of C and N from farm and forest land (Fontaine and Barot 2005). In temperate and boreal forests in the Northern hemisphere it has been shown that C sequestration is largely driven by N deposition (Magnani et al. 2007). Legume based cropping systems have reduced C and N losses (Drinkwater et al. 1998). The future of the terrestrial carbon sink seems to be uncertain. The elevated global temperatures and changes in precipitation predicted for 2050 could induce a switch from soil as sink to source of carbon since warmer soils release more carbon, and forests suffer increasing periods of drought. Humans have been managing terrestrial ecosystems for their own ends for millennia – from deforestation and clearing of natural vegetation for cultivation of crops to increasing fertilizer and pesticide use. Of the 50 billion tonnes of carbon currently locked up in terrestrial biomass and vulnerable to release in the next 20 years, 40 billion tonnes is put at risk not by changes in climate but by changes in land use. As pointed out by Reay et al. (2007) climate change may be the greatest threat to this huge carbon stock towards the end of the century. In the shorter term, it is chainsaws and ploughs, not drought and extreme temperatures that we must address.

Human interventions could exploit enhanced plant biomass accumulation in elevated atmospheric CO_2 concentration to reduce the future rate of increase in CO_2 levels, and associated global warming. However, N availability constraints could limit CO_2-induced stimulation of plant growth and biomass accumulation. Reich et al. (2006) have recently indicated that variation in both the availability of soil N and deposition of atmospheric N are likely to influence plant biomass accumulation under elevated atmospheric CO_2. Considering that productivity of both natural and managed vegetations are limited by the availability of N, soil N would be a major constraint on global terrestrial responses to elevated CO_2.

2.5 Nitrogen Fixation Through Soil Microbes

The role of soil microbes in fixation of atmospheric N is the best known and agronomically exploited area of soil microbiology (Sprent and Sprent 1990). Free living bacterial species such as *Azospirullum*, *Azotobacter* and several photosynthetic cyanobacteria fix atmospheric N. However, the biological nitrogen fixation (BNF) by the leguminous crops and tree species is an important component of the nitrogen cycle in agricultural and natural ecosystems. Legumes form an important component in natural pastureland vegetation. They have been traditionally used as a component of cereal–legume rotations or specific legumes as green manure crops to restore soil fertility. Such farming systems were sustained over a long period though cereal productivity was much lower in comparison to those currently harvested using chemical fertilizers. Ley farming and cultivation of green manure crops has become uneconomical in densely populated regions (Ali 1999). Atmospheric N fixation by leguminous plants is a complex process that has been extensively investigated, yet its understanding remains inadequate to realize the full potential. It involves interactions between the legume plant and the *Rhizobium, Bradyrhizobium, Sinorhizobium* and *Azorhizobium* species of soil bacteria collectively referred as rhizobia.

In response to a variety of substances exuded by the host plant roots, specific soil bacteria are attracted to a particular legume species, leading to the formation of specialized root organ - the nodules. The different steps involved in the establishment of successful symbiotic association are processes controlled by the host and bacterial genes in the following sequence: multiplication and colonization at the root surface (*Roc* **ro**ot **c**olonization), adhesion of bacteria to root hair surface (*Roa* **ro**ot **a**dhesion), curling or branching of root **hair** (*Hab* and *Hac* root **hair b**ranching and **c**urling), formation of **inf**ection thread (*Inf*), induction of meristem in the host roots for the nodule initiation and differentiation, **ba**cteroid **r**elease from the infection thread (*Bar*), **ba**cterial **d**ifferentiation (*Bad*), onset of **ni**trogen **f**ixation (*Nif*), **no**dule function **p**ersistence/maintenance (*Nop*), development of **co**mplimentary **f**unctions associated with N fixation (*Cof*) and its transport (Caetano-Anolles and Gresshoff 1991). Nodulation mutants have been isolated in many of the legume crops (Gresshoff 1993) that broadly classified into four classes (Sagan et al. 1994): nod − (no nodule formation), nod +/− (few nodule formation), fix − (ineffective nodules), nod ++ (super or hyper nodulation) and nts (**n**itrate **t**olerant **s**ymbiosis that fix N even in the presence of high nitrate in soil). In field experiments having native microbes a more descriptive classification – non-nodulating with native root nodulating bacteria (RNB); non-nodulating with a specific strain, low nodulating with native RNB; low nodulating at low N; high nodulating at low N but low nodulating at high N and high nodulating at high N have been used (Wani et al. 1995). How the different steps in the nodulation process are affected after inoculation with specific *Rhizobium* strains, in presence of abundant native populations, is not known except for the fact that often the introduced strains lose in competition to the native strains.

Genetic manipulation of the legume host as well as *Rhizobium* species provide enormous opportunities to enhance BNF, and reduce fertilizer N requirement for the following cereal or other non-legume crops (Bhatia et al. 2001). The available non-nodulation and hyper-nodulation mutants can be used for selection of the best host and bacterial combinations. The characterization of soybean as well as the *Rhizobium* genome would further facilitate development of such combinations. At the same time, there is a need to develop user friendly methods for the delivery of the desired strains to minimize competition from the native rhizobia already present in soil.

The possibilities of making cereals, particularly rice, fix nitrogen like the grain legumes have been explored (Dey and Datta 2002). Cereals lack the large number of genes involved in successful symbiotic partnership with rhizobia. Some homologues of nodulin genes have been identified in rice, formation of nodular structures on rice seedling roots were reported (Al-Mallah et al. 1989). Occurrence of early nodulin genes have been reported in rice (Reddy et al. 1999) and two soybean genes related to nodulation have been transferred into rice (Day et al. 2000). However, rice or other cereals having root nodules with nitrogen fixing microbes are still a long way off if at all feasible.

2.6 Effect of GE Crops on Microbial Diversity

A large number of plants expressing alien genes have been developed using recombinant DNA techniques in the last 20 years, and many of them are under commercial cultivation after mandatory clearances from regulatory agencies. The first generation GE plants were aimed to control pest and diseases (Christou et al. 2006), and now more plants with altered metabolic pathways are being developed. Ever since GE crops were developed in the 1980s, supporters and those opposing cultivation of such crops have made conflicting claims on their biosafety and environmental effects. Many questions have been raised regarding the unintended environmental effects of such plants (Bhatia and Mitra 1998; Conner et al. 2003; Ammann 2005). In the present context, the potential short- and long term effects of GE crops on the rhizosphere are a matter of concern as brought out by Bruinsma et al. 2003. Model systems to monitor the effect on non target soil microorganisms have been developed (Turrini et al. 2005). The regulatory aspects related to the soil systems for GE crops, and for the GE microbial pesticides have been discussed by Sayre and Seidler (2005). The possible risks associated with such plants are evaluated in relation to the overall benefits. The results obtained have been reviewed by Lynch et al. (2004) and Liu et al. (2005). Large scale cultivation of GE crops can alter the soil microbial communities in two ways: release of transgene product into the rhizosphere through root exudates or through phytomass degradation, and transfer and integration of plant DNA into resident microbes through horizontal gene transfer. Both these are also possible from the cultivation of crop cultivars developed conventional plant breeding methods. The need for comparing the environmental perturbations caused by GE crops with conventionally bred crop cultivars grown has been advocated (Conner et al. 2003; Ammann 2005). A major limitation in following such effects is the absence of base line data on microbial diversity and the methods to quantify soil microbial diversity. Maize plants expressing *Bacillus thuringiensis* (Bt) *cry1Ab* gene released the toxin into the rhizosphere through root exudates and litter decay (Saxena et al. 1999; Saxena and Stotzky 2001). It remained bound to soil and active for 234 days – the duration of the experiment.

In vitro experiments using the toxin from three different strains of Bt showed no microbiocidal or microstatic activity against selected bacteria, fungi and algae (Koskella 2002). In another study the two maize lines expressing Bt genes, the root exudates of Bt 176 significantly reduced pre-symbiotic hyphal growth of arbuscular mycorrhizal fungi compared to Bt 11 and non-transgenic corn lines. However, no differences were found between control and defensin expressing lines. Bt toxin as well as defensin did not affect the fungal-host recognition mechanism. Bt 176 affected appressoria development; 36% of them failed to produce infection pegs (Turrini et al. 2005). Higher numbers of fungi were recovered from the roots of potato lines expressing a synthetic anti-microbial peptide magainin II, in comparison to the control plants (Callaghan et al. 2005). Rhizosphere bacterial isolates showed different susceptibility to magainin analogues in in vitro experiments. Effect of transgenic potato line, modified for its starch composition characteristics by RNA antisense, on soil and rhizosphere bacterial and fungal diversity were investigated using molecular techniques based on bacterial and fungal rDNA (Milling et al. 2005). No significant differences between the transgenic, unmodified parental line and another cultivar on microbial community structure were observed. When *Pseudomonas* specific primers were used differences in the rhizosphere patterns of transgenic and the parental cultivar were observed. However, similar differences in *Pseudomonas* community were observed in comparisons between two standard, non transgenic, cultivars.

In one of the most comprehensive studies Heuer et al. (2002) investigated the effects of T4-lysozyme release from transgenic potato roots which enhances bactericidal activity against *Bacillus subtilis*. Genetically engineered T4-lysozyme producing, control without T-4 lysozyme gene, and the parental lines were compared at two different field sites for three years. Soil bacterial communities were analyzed using three different complimentary techniques – fatty acid analysis of cultured organisms, Biolog GN microplates, and 16S rRNA gene fragments using DDGE or by cloning and sequencing. No significant effects of the T4-lysozyme expression on rhizosphere communities, over the other environmental variables were observed.

Horizontal transfer of genes from GE crops to the soil bacteria has been reviewed by Lynch et al. (2004) and Mercier et al. (2006). Persistence of plant DNA in soil for a long period has been shown. Soil colloids adsorb biological molecules retarding their microbial degradation. The other variable is the presence of bacteria in competent state to take up the exogenous DNA in the soil. Lastly, the frequency of such events will determine the gene transfer into soil bacteria. The significance of such events should be considered in view of the normal high gene exchange among the different bacteria (Chandler 2006).

Effect of Bt-corn on soil macroflora has been investigated. Finely ground *Bt*-corn leaves, expressing Cry1Ab protein, when added to soil had no deleterious effects on survival, growth, development and reproduction of earthworm population (Vercesi et al. 2006). Juvenile earthworms in pots with *Bt*-corn plants had no effect. A slight negative effect was observed on cocoon hatchability. It is apparent that the GE crops approved for commercial cultivation cause minor changes in microbial community structure and function. However, it is widely accepted now that all transgenic events need to be evaluated on case by case basis for different biosafety concerns before approval for commercial cultivation.

Cultivation of herbicide glyphosate resistant (GR) wheat and canola rotations had little effect on soil microorganisms (Lupwayi et al. 2007). Soil microbial biomass, bacterial functional diversity, community structure and dehydrogenase enzyme activity were monitored in the rhizosphere and bulk soil samples at six sites in Canada. These included fields that followed direct seeding or conventional tillage and different wheat – canola rotations. Out of 22–40 plots significant differences were observed in only in 2–3 plots; however, the observed differences were not consistent. The authors conclude that "Overall, GR crop frequency effects on soil microorganisms were minor and inconsistent over a wide range of growing conditions and crop management."

The above discussion brings out that some genetically engineered crops may affect soil ecosystems, but the long-term significance of any of these changes was not clear. Alterations of soil ecosystems could decrease plant decomposition rates and hence soil C and N levels and soil fertility. Similarly, declining species diversity of soil microorganisms, in some cases, can cause lower community diversity and productivity above ground (van der Heijden et al. 1998; Wolfenbarger and Phifer 2000).

2.6.1 GE Crops Exuding Specific Molecules into Rhizosphere and Possibilities of Enhancing Microbial Cooperation

Plant roots are known to secrete as much as 20% of their stored assimilates into the rhizosphere (Whipps 1990; Uren 2001; Walker et al. 2003). These exudates contain many different compounds including sugars, amino acids, proteins and signal peptides (Uren 2001), and are reported to change the rhizosphere biology and increase the availability of micronutrients (Mraschner and Römheld 2001). Root exudates increase the growth of soil bacteria and their predators enhancing SOM degradation and N mineralization. Model for the same has been developed by Raynaud et al. (2006). Genetic engineering provides unique opportunities for introducing the desired recombinant proteins into the rhizosphere by growing such modified crops. This opens up possibilities for enhancing microbial cooperation in the rhizosphere (Barea et al. 2005) or what Brussard et al. (2007) has referred as "planned microbial diversity".

Tesfaye et al. (2005) reported transfer of fungal endochitinase gene from *Trichoderma harzianum* Riafi into alfalfa (*Medicago sativa* L.) plants. Root exudates of these transgenic plants showed the presence of endochitinase with antifungal activity as demonstrated by inhibition of spore germination of two fungal pathogens. These experiments have shown the potential for introducing the desired proteins into the rhizosphere that could be exploited as a biocontrol method for protecting the plants from soil borne pathogens or for accelerating bio-remediation processes. The recombinant proteins/enzymes can also be used to influence growth promoting microbes, and microbial symbiosis (Austin et al. 1995; Tesfaye et al. 2005).

Transgenic *Arabidopsis thaliana* plants expressing phytase gene from *Aspergillus niger* were first shown to release extra cellular phytase and utilize P supplied as phytate in the medium (Richardson et al. 2001). Later using a root hair specific promoter increased acquisition of P from phytate in the medium in *Arabidopsis* (Mudge et al. 2003) and potato (Zimmerman et al. 2003). Expression of the fungal phytase gene in *Nicotiana tabaccum* improved P nutrition in amended soils (George et al. 2005a). Limitations to the potential of transgenic plants that exude phytase in different soil types were brought out by George et al. (2005b). *Trifolium subterraneum* L. plants constitutively expressing a chimerical phytase gene, and showing a 77-fold increase in exuded phytase activity, were grown in a range of soils differing in organic P content.

Transgenic plants that exuded phytase showed better growth and P nutrition only in a soil containing large concentration of organic P amenable to hydrolysis by plant derived phytase, and also total organic P. In the transgenic line the root growth was shorter than that of the control plants and the longer root system of the latter may have given greater access to soil P. The natural microbial diversity in the soil had no significant influence on P availability. The authors point out the inherent limitations of single trait alterations and infer that "such approaches can be successful under certain edaphic conditions". With better understanding of the processes involved it should be possible to develop plants with improved P nutrition by incorporating more than one gene. Transfer of nine genes into *Brassica juncea* for metabolic engineering for production of long chain fatty acids have been reported (Wu et al. 2005).

Spaepen et al. (2007) have recently reviewed the role of bacterial IAA in different microorganism–plant interaction and highlight the fact that bacteria use this phytohormone to interact with plants as part of their colonization strategy, including phytostimulation and circumvention of basal plant defense mechanisms.

2.7 Genetically Engineered Bio-control Agents

As a result of increased concern over the use of chemical pesticides in agriculture, the biological control of pests and disease-causing organisms using antagonistic micro organisms remains a viable option for sustainable agriculture. The interactions within microbial populations and between microbes and higher organisms are the basis for biocontrol. Use of antagonistic microorganisms is thus emerging as an environment friendly means to control pest and disease causing organisms in integrated pest management (IPM) programs. New and improved strains produced by genetic manipulations provide opportunities for developing more effective biocontrol organisms. Information on the effects of released wild-type or GE organisms on resident communities is important to assess the potential risks associated with the introduction of such organisms into agro ecosystems. Rhizocompetent *Pseudomonas* species have been used for suppression of crop diseases (Nautiyal et al. 2002). They have also been identified as ideal candidate for strain improvement using the rDNA techniques. In view of the biosafety considerations, there is considerable interest in the impact of released GE bio-control agents (GE-BCA) on non target species and microbial diversity (Glandorf et al. 2001; Moenne-Loccoz et al. 2001). Impact of field releases of GE *Pseudomonas fluorescence* on indigenous microbial populations of wheat (De Leij et al. 1995, Glandorf et al. 2001) and sugar beet (Thompson et al. 1995; Moenne-Loccoz et al. 2001) has been reported. Timms-Wilson et al. 2004 improved a *Pseudomonas fluorescens* strain SBW25 by chromosomal insertion of constitutively expressed phzABCDEFG genes for the biosynthesis of antifungal compound phenazine-1-carboxylic acid (PCA). The most effective GE strain 23.10 was used on pea, wheat and sugar beet grown on soil infected with *Phythium ultimum* that causes damping-off in several crops. Colony isolation and different molecular methods were used to assess the impact on microbial diversity. Root-mycorrhiza associations were followed using microscopic examination. The results indicate that plant type, age and disease have a greater effect on the abundance, diversity and succession patterns of rhizosphere bacterial and fungal communities than the inocula of GE-BCA. Further, the presence of inocula reduced the impact of disease on the microbial diversity and function. A transient, small decrease in mycorrhizal association was observed four days after inoculation. The authors infer that no major groups were excluded or enriched as a result of GE-BCA.

Pseudomonas fluorescens CHA0-Rif and its derivative CHA0 Rif/pME3424, which has improved biocontrol activity and enhanced production of the antibiotics 2,4-diacetylphloroglucinol (Phl) and pyoluteorin (Plt), were introduced into soil microcosms and the culturable bacterial community developing on cucumber roots was investigated (Natsch et al. 1998). The introduction of either of the two strains led to a transiently enhanced metabolic activity of the bacterial community on glucose dimers and polymers as measured with BIOLOG GN plates. The introduced strains did not significantly affect the abundance of dominant groups of culturable bacteria discriminated by restriction analysis of amplified 16S rDNA of 2500 individual isolates. About 30–50% of the resident bacteria were very sensitive to Phl and Plt, but neither the wild-type nor CHA0-Rif/pME3424 changed the proportion of sensitive and resistant bacteria in situ. In microcosms with a synthetic bacterial community, both biocontrol strains reduced the population of a strain of *Pseudomonas* but did not affect the abundance of four other bacterial strains including two highly antibiotic-sensitive isolates. The authors concluded that detectable perturbations in the metabolic activity of the resident bacterial community caused by the biocontrol strain CHA0-Rif are (i) transient, (ii) similar for the genetically improved derivative CHA0 Rif/pME3424 and (iii) less pronounced than changes in the community structure during plant growth. From a biological safety assessment point of view, the data therefore suggested that a genetically improved biocontrol strain may have specific interactions with fungal pathogens rather than general effects on bacterial communities. However, the possibility that specific interactions could occur between introduced strains and non-culturable resident bacteria can not be ruled out.

The antibiotic 2,4-diacetylphloroglucinol (Phl) is produced by a range of naturally occurring fluorescent pseudomonads. One isolate, *P. fluorescens* F113, protects pea plants from the pathogenic fungus *Pythium ultimum* by reducing the number of pathogenic lesions on plant roots, but with a concurrent reduction in the emergence of pea plants. The genes from F113 were isolated and a 6.7-kb gene cluster was inserted into the chromosome of the non-Phl-producing *P. fluorescens* strain SBW25 *EeZY6KX* (Bainton et al. 2004). Pea roots inoculated with SBW25 *EeZY6KX* have significantly lower indigenous populations than with F113 and the control. The authors achieved the integration of the Phl antibiotic and competitive exclusion mechanisms into a single strain Pa21. Impact of Pa21 on survival and plant emergence was investigated following inoculation of pea seedlings where it provided protection against *P. ultimum* but did not cause lower seed emergence. Thus, strain Pa21 possesses the necessary qualities to provide effective integrated biocontrol, through maintaining its wt trait of competitive exclusion on the plant roots and expressing the antibiotic genes. Such modified candidate strains may be of some agronomic benefit.

2.8 Conclusions and Outlook for the Future

The recent reports on the soil microbial diversity using the molecular techniques show that the microbial communities are highly diverse, and change with minor perturbations. This may turn out to be a universal phenomenon as more information is gathered from different soils and climatic conditions. As knowledge of the so-called "black box" of the soil microbial diversity increases, an enormous source of knowledge of species and genes would be available. These can be utilized with the application of conventional and new molecular techniques to evolve more productive and sustainable cropping patterns.

There would be increased demands for food, feed, fiber, timber and bio-fuels. The choice of the crops grown for bio-fuels would vary in different regions depending upon the climatic conditions. The legume oil yielding plants are the likely crops of choice for bio-fuels in many areas as they meet their N requirement through fixation, in comparison to sugarcane and sweet sorghum that require application of N fertilizers for increased productivity.

Tremendous increase in the productivity of crops has been brought about with the application of contemporary genetic knowledge and tools since 1900, when Mendel's Laws of Inheritance were rediscovered. All the improvements made so far have aimed at the alterations in the above ground parts of the plants. The genetic diversity in the "hidden part" below ground, in the root characteristics, the soil microbes in the rhizosphere, and their positive interactions have hardly been exploited. The increasing amount of genomic data will further broaden insight into the role of specific key molecules in the rhizosphere to enhance cooperation among the interacting organisms for improved productivity. These along with the tools and techniques of no till or limited tillage, precision farming using slow release forms of agro-chemicals, built in disease and pest resistance, beneficial microbial inoculants for growth promotion or the control of soil borne pests and pathogens should play a significant role in future. With the genetic tools now available it may be relatively easy to introduce the desired traits in the soil microbes of choice, but to make them successfully compete with the native strains already present, and environmental safety of such modified organisms would be a bigger challenge.

Discussions of the environmental risks and benefits of adopting GE organisms are highly polarized between pro- and anti-biotechnology groups. The current state of available knowledge is frequently overlooked in this debate. A review of existing scientific literature reveals that key experiments on the long term environmental risks are lacking. The complexity of ecological systems presents considerable challenges for experiments to assess the long term risks. The existing studies emphasize that these can vary depending upon the trait and organism modified.

References

Abbott LK, Murphy DV (eds) (2003) Soil biological fertility: a key to sustainable land use in agriculture. Springer, Berlin Heidelberg New York, p 276

Adam D (2001) Royal Society disputes value of carbon sinks. Nature 412:108

Ali M (1999) Evaluation of green manure technology in tropical low land rice system. Field Crop Res 61:61–78

Al-Mallah MK, Davey MR, Cocking EC (1989) Formation of nodular structures on rice seedlings by rhizobia. J Expt Bot 40:473–478

Ammann K (2005) Effects of biotechnology on biodiversity: herbicide-tolerant and insect-resistant GM crops. Trends Biotechnol 23:388–394

Austin S, Bingham ET, Mathew DE, Shahan MN, Will J, Burgess RR (1995) Production and field performance of transgenic alfalfa (*Medicago sativa* L) expressing α-amylase and manganese dependent peroxidase. Euphytica 85:381–393

Bainton NJ, Lynch JM, Naseby D, Way JA (2004) Survival and ecological fitness of *Pseudomonas fluorescens* genetically engineered with dual Biocontrol mechanisms. Microbial Ecol 48:349–357

Barea JM, Pozo MJ, Azcon R, Concepcion AA, Azcón-Aguilar C (2005) Microbial co-operation in the rhizosphere. J Exp Bot 56:1761–1778

Bellamy PH, Loveland PJ, Bradley RI, Lark RM, Kirk JD (2005) Carbon losses from all soils across England and Wales 1978–2003. Nature 437:245–248

Bhatia CR, Mitra R (1998) Biosafety of transgenic crop plants. Proc Indian Natl Sci Acad B64:293–318

Bhatia CR, Nichterlein K, Maluszynski M (2001) Mutations affecting nodulation in grain legumes and their potential in sustainable cropping systems. Euphytica 120:415–432

Bruinsma M, Kowalchuk GA, van Veen JA (2003) Effects of genetically modified plants on microbial communities and processes in soil. Biol Fertil Soils 37:329–337

Brussard L, de Ruiter PC, Brown GC (2007) Soil biodiversity for agricultural sustainability. Agric Ecosyst Environ 121:233–244

Buckley DH, Schmidt TM (2003) Diversity and dynamics of microbial communities in soils from agro-ecosystems. Environ Microbiol 5:441–452

Caetano-Annoles G, Gresshoff PM (1991) Plant genetic control of nodulation. Annu Rev Microbiol 45:345–382

Callaghan MO, Gerard EM, Waipara NW, Young SD, Glare TR, Barell PJ, Conner AJ (2005) Microbial communities of *Solanum tuberosum* and magainin-producing transgenic lines. Plant Soil 266:47–56

Chandler M (2006) Molecular biology: singled out for integration. Nature 440:1121–1122

Chen X, Hutley LB, Eamus D (2005) Soil organic carbon content at a range of north Australian tropical savannas with contrasting site histories. Plant Soil 268:161–171

Christou P, Capell T, Kohli A, Gatehouse JA, Gatehouse AMR (2006) Recent developments and future prospects in insect pest control in transgenic crops. Trends Plant Sci 11: 302–308

Conner AJ, Glare TR, Nap JP (2003) The release of genetically modified crops into the environment: overview of ecological risk assessment. Plant J 33:19–46

Copley J (2000) Ecology goes underground. Nature 406:452–454

Day BR, McAlwin CB, Loh JT, Denny RL, Wood TC, Young ND, Stacey G (2000) Differential expression of two soybean apyrases, one of which is an early nodulin. Mol Plant Microbe Interact 13:1053–1070

Dey M, Datta SK (2002) Promiscuity of hosting nitrogen fixation in rice: an overview from the legume perspective. Critical Rev Biotechnol 23:281–314

De Leij FAAM, Sutton EJ, Whipps JM, Fenlon JS, Lynch JM (1995) Impact of field release of genetically modified *Pseudomonas fluorescens* on indigenous microbial population of wheat. Appl Environ Microbiol 61:3443–3453

DeLucia EH, Hamilton JG, Naidu SL, Thomas RB, Andrews JA, Finzi A, Lavine M, Matamala R, Mohan JE, Hendrey GR, Schlesinger WH (1999) Net primary production of a forest ecosystem with experimental CO_2 enrichment. Science 284:1177–1179

Deutschbauer AM, Dylan Chivian D, Arkin AP (2006) Genomics for environmental microbiology. Curr Opin Biotechnol 17:229–235

Diaz S, Grime JP, Harris J, McPherson E (1993) Evidence of a feed back mechanism limiting plant response to elevated carbon dioxide. Nature 364:616–617

Drinkwater LE, Wagoner P, Sarrantonio M (1998) Legume base cropping systems have reduced carbon and nitrogen losses. Nature 396:262–265

Fitter AH (2005) Darkness visible: reflections on underground ecology. J Ecol 93:231–243

Fitter AH, Gilligan CA, Hollingworth K, Kleczkowski A, Twyman RM, Pitchford JW (2005) Biodiversity and ecosystem function in soil. Funct Ecol 19:369–377

Fontaine SB, Barot SB (2005) Size and functional diversity of microbe populations control plant persistence and long-term soil carbon accumulation. Ecol Lett 8:1075–1087

Gans J, Wolinsky M, Dunbar J (2005) Computational improvements reveal great bacterial diversity and high metal toxicity in soil. Science 309:1387–1390

George TS, Richardson AE, Smith JB, Hadobas PA, Simpson RJ (2005a) Expression of a fungal phytase gene in *Ncotiana tabaccum* improves phosphorus nutrition of plants grown in amended soils. Plant Biotech J 3:129–140

George TS, Richardson AE, Smith JB, Hadobas PA, Simpson RJ (2005b) Limitations to the potential of transgenic *Trifolium subterraneum* L. plants that exude phytase when grown in soils with a range of organic P content. Plant Soil 258:263–274

Gewin V (2006) Genomics: discovery in the dirt. Nature 439:384–386

Glandorf DCM, Verheggen P, Jansen T, Jorritsma LS, Thomashow LS, Leeflang P, Smit E, Wernars K, Lauerijs E, Thomas-Oates JE, Bakker PAHM, van Loon LC (2001) Effect of genetically modified *Psedomonas putida* WCS358r on the fungal rhizosphere microfllora of field grown wheat. Appl Environ Microbiol 67:3371–3378

Gresshoff PM (1993) Molecular genetic analysis of nodulation genes in soybean. Plant Breed Rev 11:275–318

Heuer H, Kroppenstedt RM, Lottmann J, Berg G, Smalla K (2002) Effects of T4-lysozyme release from transgenic potato roots on bacterial rhizosphere communities are negligible relative to natural factors. Appl Environ Microbiol 68:1325–1335

Högberg, P, Nordgren A, Buchmann N, Taylor AFS, Ekblad A, Hogberg MN, Nyberg G, Ottosson-Löfvenius M, Read DJ (2001) Large-scale forest girdling shows that current photosynthesis drives soil respiration. Nature 411:789–792

Hopkins A, Prado D (2007) Implications of climate change for grassland in Europe: impacts, adaptations and mitigation options: a review. Grass Forage Sci 62:118–126

Hu S, Chapin FS, Firestone MK, Fields CB, Chiariello NR (2001) Nitrogen limitation of microbial decomposition in grassland under elevated CO_2. Nature 409:188–190

Hu S, Wu J, Burkey KO, Firestone MK (2005) Plant and microbial N acquisition under elevated atmospheric CO_2 in two mesocosm experiments with annual grasses. Global Change Biol 11:213–223

Janzen HH (2006) The soil carbon dilemma: shall we hoard it or use it? Soil Biol Biochem 38:419–424

Johansson JF, Paul LR, Roger D, Finlay RD (2004) Microbial interactions in the mycorrhizosphere and their significance for sustainable agriculture. FEMS Microbiol Ecol 48:1–13

Kirchmann H, Thorvaldsson G (2000) Challenging targets for future agriculture. Eur J Agron 12:145–161

Kohler F, Hamelin J, Gillet F, Gobat JM, Butler A (2005a) Plant species composition effects on belowground properties and the resistance and resilience of the soil microflora to a drying disturbance. Plant Soil, 278:205–221

Kohler F, Hamelin J, Gillet F, Gobat JM, Butler A (2005b) Soil microbial community changes in wooded mountain pastures due to simulated effects of cattle grazing. Plant Soil 278: 327–340

Koskella JS (2002) Larvicidal toxins from *Bacillus thuringiensis* subspp krustaki, morisoni (strain tenebrionis) and israelensis have no microbicidal or microstatic activity against selected bacteria, fungi and algae in vitro. Can J Microbiol 48:262–267

Li Y, Xu M, Zou X, Xia Y (2005) Soil CO_2 efflux and fungal and bacterial biomass in a plantation and a secondary forest in wet tropics in Puerto Rico. Plant Soil 268:151–160

Liu B, Zeng Q, Yan F, Xu H, Xu C (2005) Effects of transgenic plants on soil microorganisms. Plant Soil 271:1–13

Lupwayi NJ, Hanson KG, Harker KN, Clayton GW, Blackshaw RE, O'Donovan JT, Johnson EN, Gan Y, Irvine RB, Monreal MA (2007) Soil microbial biomass, functional diversity and enzyme activity in glyphosate-resistant wheat–canola rotations under low-disturbance direct seeding and conventional tillage. Soil Biol Biochem 39:1418–1427

Lynch JM, Benedetti A, Insam H, Nuti MP, Smalla K, Torsvik V, Nannipieri P (2004). Microbial divesity in soil: ecological theories, the contribution of molecular techniques and the impact of transgenic plants and transgenic organisms. Biol Fertil Soils 40:363–385

Madsen EL (2005) Identifying microorganisms responsible for ecologically significant biogeochemical processes. Nat Rev Microbiol 3:439–446

Magnani F, Mencuccini M, Borghetti M, Berbigier P, Berninger F, Delzon S, Grelle A, Hari P, Jarvis PG, Kolari P, Kowalski AS, Lankreijer H, Law BE, Lindroth A, Loustau D, Manca G, Moncrieff JB, Rayment M, Tedeschi V, Valentini R, Grace J (2007) The human footprint in the carbon cycle of temperate and boreal forests. Nature 447:849–851

Manlay RJ, Feller C, Swift MJ (2007) Historical evolution of soil organic matter concepts and their relationships with the fertility and sustainability of cropping systems. Agric Ecosyst Environ 119:217–233

Mercier A, Kay E, Simonet P (2006) Horizontal gene transfer by natural transformation. In: Nannipieri P, Small K (eds) Nucleic acids and proteins in soil. Springer, Berlin Heidelberg New York

Milling A, Smalla K, Maidl FX, Schloter M, Munch JC (2005) Effect of transgenic potatoes with an altered starch composition on the diversity of soil and rhizosphere bacteria and fungi. Plant Soil 266:23–39

Mitra R, Bhatia CR (1982) Bioenergetic considerations in breeding for insect resistance in plants. Euphytica 31:429–437

Moenne-Locoz Y, Tichy HV, O'Donnell A, Simson R, O'Gara F (2001) Impact of 2,4-diacetylphloroglucinol producing biocontrol strain *Pseudomonas fluoresces* F113 on intraspecific diversity of resident culturable fluorescent pseudomonads associated with the roots of field grown sugar beet seedlings. Appl Environ Microbiol 67:3418–3425

Morgan JAW, Bending GD, White PJ (2005) Biological costs and benefits to plant–microbe interactions in the rhizosphere. J Exp Bot 56:1729–1739

Morrissey JP, Dow JM, G. Louise Mark GL, O'Gara F (2004) Are microbes at the root of a solution to world food production? EMBO Rep 5:922–926

Mraschner H, Römheld V (2001) Root induced changes in the availability of micronutrients in the rhizosphere. In: Waise Y, Eshel A, Kafkafi U (eds) Plant roots: the hidden half. Marcel Dekker, New York, pp 557–581

Mudge SR, Smith FW, Richardson AE (2003) Root specific and phosphate regulated expression of phytase under the control of a phosphate transporter promoter enables *Arabidopsis* to grow on phytate as a sole P source. Plant Sci 165:871–878

Nannipieri P, Smalla K (eds) (2006) Nucleic acids and proteins in soil. Springer, Berlin Heidelberg New York, p 458

Nannipieri P, Ascher J, Ceccherini MT, Landi L, Pietramellara G, Renella G (2003) Microbial diversity and soil functions. Eur J Soil Sci 54:655–670

Natsch A, Keel C, Hebecker N, Laasik E, Défago G (1998) Impact of *Pseudomonas fluorescens* strain CHA0 and a derivative with improved biocontrol activity on the culturable resident bacterial community on cucumber roots. FEMS Microbiol Ecol 27(4):365–380

Nautiyal CS, Johri JK, Singh HB (2002) Survival of the rhizosphere – competent biocontrol strain *Pseudomonas fluorescens* NBRI2650 in the soil and phytosphere. Can J Microbiol 48: 588–601

Nielsen MN, Winding A (2002) Microorganisms as indicators of soil health. NERI Technical Report No. 388. National Environmental Research Institute, Ministry of the Environment, Denmark URL: http://www.dmu.dk

Oren R, Ellsworth DS, Johnsen KH, Phillips N, Ewers BE, Maier C, Schafer KVR, McCarthy H, Hendrey G, McNulty SG, Katul GS (2001) Soil fertility limits carbon sequestration by forest ecosystem in a CO_2 enriched atmosphere. Nature 411:469–472

Orwin KH, Wardle DA (2005) Plant species composition effects on belowground properties and the resistance and resilience of the soil microflora to a drying disturbance. Plant Soil 278:205–221

Raynaud X, Lata JC, Leadley PW (2006) Soil microbial loop and nutrient uptake by plants: a test using a coupled C:N model of plant–microbial interactions. Plant Soil 287:95–116

Reay D, Sabine C, Smith P, Hymus G (2007) Climate change spring-time for sinks. Nature 446:727–728

Reddy PM, Aggarwal RK, Ramos MC, Ladha JK, Brar DS, Kouchi H (1999) Widespread occurrence of the homologs of the early nodulin (ENOD) genes in *Oryza* species and related grasses. Biochem Biophys Res Commun 258:148–154

Reich PB, Hobbie SE, Lee T, Ellsworth DS, West JB, Tilman D, Knops JMH, Naeem S, Trost J (2006) Nitrogen limitation constrains sustainability of ecosystem response to CO_2. Nature 440:922–925

Richardson AE, Hadobas PA, Hayes JE (2001) Extracellular secretion of *Aspergillus* phytase from *Arabidopsis* roots enables plants to obtain phosphorus from phytate. Plant J 25:1–10

Rondon MR, August PR, Bettermann AD, Brady SF, Grossman TH, Liles MR, Loiacono KA, Lynch BA, MacNeil IA, Minor C, Tiong CL, Gilman M, Osburne MS, Clardy J, Handelsman J, Goodman RM (2000) Cloning the soil metagenome: a strategy for accessing the genetic and functional diversity of uncultured microorganisms. Appl Environ Microbiol 66:2541–2547

Sagan MT, Huguet TG, Duc G (1994) Phenotypic characterization and classification of nodulation mutants of pea (*Pisum sativum* L.). Plant Sci 100:59–70

Saxena D, Stotzky G (2001) *Bacillus thuringiensis* (Bt) toxin released from root exudates and biomass of Bt corn has no apparent effect on earthworm, nematodes, protozoa, bacteria, fungi in soil. Soil Biol Biochem 33:1225–1230

Saxena D, Flores S, Stotzky G (1999) Transgenic plants – insecticidal toxin in root exudates from Bt. corn. Nature 402:480

Sayre P, Seidler RJ (2005) Application of GMO in the US EPA research and regulatory considerations related to soil systems. Plant Soil 275:77–91

Schimel DS, House JI, Hibbard KA, Bousquet P, Ciais P, Peylin P, Braswell BH, Apps MJ, Baker D, Bondeau A, Canadell J, Churkina G, Cramer W, Denning AS, Field CB, Friedlingstein P, Goodale C, Heimann M, Houghton RA, Melillo JM, Moore B, Murdiyarso

D, Noble I, Pacala SW, Prentice IC, Raupach MR, Rayner PJ, Scholes RJ, Steffen WL, Wirth C (2001) Recent patterns and mechanism of carbon exchange by terrestrial ecosystems. Nature 414:169–172

Schlesinger WH (1999) Carbon sequestration in soils. Science 284:209

Schulze ED, Freibauer A (2005) Carbon unlocked from soils. Nature 437:205–206

Schulze ED, Pöschel G (2005) Bacterial inoculation of maize affects carbon allocation to roots and carbon turnover in the rhizosphere. Plant Soil 267:235–241

Spaepen S, Vanderleyden J, Remans R (2007) Indole-3-acetic acid in microbial and microorganism-plant signaling. FEMS Microbiol Rev 31:425–448

Sprent JI, Sprent P (1990) Nitrogen fixing organisms: pure and applied aspects. Chapman and Hall, London

Suzuki C, Kunito T, Aono T, Liu CT, Oyaizu H (2005) Microbial indices of soil fertility. J Appl Microbiol 98:1062–1074

Swank JC, Below FE, Lambert RJ, Hageman RJ (1982) Interaction of carbon and nitrogen metabolism in productivity of maize. Plant Physiol 70:1185–1190

Tesfaye M, Denton MD, Samac D, Vance CP (2005) transgenic alfalfa secretes endochitinase protein to the rhizosphere. Plant Soil 269:233–243

Thompson IP, Lilley AK, Ellis RJ, Bramwell PA, Bailey MJ (1995) Survival colonization and dispersal of genetically modified *Pseudomonas fluorescens* SBW25 in the phytosphere of field grown sugar beet. Nature Bio/Technol 13:1493–1497

Tilman D, Fargione J, Wolff B, D'Antonio C, Dobson A, Howarth R, Schindler D, Schlesinger WH, Simberloff D, Swackhamer D (2001) Forecasting agriculturally driven global environment change. Science 292:281–284

Timms-Wilson TM, Kilshaw K, Bailey MJ (2004) Risk assessment for engineered bacteria used in biocontrol of fungal diseases in agricultural crops. Plant Soil 266:57–67

Torsvik V, Ovreas L, Thingstad TF (2002) Prokaryotic diversity-magnitude, dynamics, and controlling factors. Science 296:1064–1066

Turrini A, Sbrana C, Nuti MP, Pietrangeli BM, Giovannetti M (2005) Development of model system to assess the impact of genetically modified corn and aubergine plants on arbuscular mycorrhizal fungi. Plant Soil 266:69–75

Uren NC (2001) Types, amounts and possible functions of compounds released into rhizosphere by soil grown plants. In: Pinton R, Varanini Z, Nannipieri P (eds) The rhizosphere. Marcel Dekker, New York, pp19–40

van der Heijden MGA, Klironomos JN, Ursic M, Moutoglis P, Streitwolf-Engel R, Boller T, Wiemken A, Sanders IR (1998) Mycorrhizal fungal diversity determines plant biodiversity ecosystem variability and productivity. Nature 396:69–72

Vercesi ML, Krogh PH, Holmstup M (2006) Can Bacillus thuringiensis (Bt) corn residues and Bt-corn plants affect life history traits in the earthworm *Aporrectodea calliginosa*? Appl Soil Ecol 32:180–187

Walker TS, Bais HP, Grotewold E, Vivanco JM (2003) Root exudation and rhizosphere biology. Plant Physiol 132:44–51

Wani SP, Rupella OP, Lee KK (1995) Sustainable agriculture in the semi-arid tropics through biological nitrogen fixation in grain legumes. Plant Soil 174:29–49

Weisskopf L, Fromin N, Tomasi N, Aragno M, Matinoia E (2005) Secretion activity of white lupines cluster roots influences bacterial abundance, function and community structure. Plant Soil 268:181–194

Whipps JM 1990 Carbon economy. In: Lynch JM (ed) The rhizosphere. John Wiley, Chichester, pp 59–99

Wolfenbarger LL, Phifer PR (2000) The ecological risks and benefits of genetically engineered plants. Science 290:2088–2093

Wu G, Truska M, Datla N, Vrinten P, Bauer J, Zabk T, Cirpus P, Heinz E, Qiu X (2005) Stepwise engineering to produce high yields of very long-chain fatty acids in plants. Nature Biotechnol 23:1013–1017

Wu L, McGechan MB, McRoberts N, Baddeley JA, Watson CA (2007) SPACSYS: integration of a 3D root architecture component to carbon, nitrogen and water cycling—model description. Ecol Model 200:343–359

Xu J (2006) Microbial ecology in the age of genomics and metagenomics: concepts, tools, and recent advances. Mol Ecol 15:1713–1731

Zak DR, Pregitzer K, Curtis PS, Teeri JA, Fogel R, Randlett DL (1993) Elevated atmospheric CO_2 and feed back between carbon and nitrogen cycles. Plant Soil 151:105–117

Zimmermann P, Zardi G, Lehmann M, Zelder C, Amrhein N, Frossard E, Bucher M (2003) Engineering the root–soil interface via targeted expression of a synthetic phytase gene in trichoblasts. Plant Biotechnol J 1:353–360

Chapter 3
Reconstructing Soil Biology

Patrice Dion

All words would fall far short of what it was.

– Dante

3.1   Introduction

Soil bacteria and archaea first evolved as principal inhabitants of primordial environments, and then in association with fungi, plants and animals as new environments emerged. Thus, transitions occurred, which did not correspond to a replacement of ancient processes by new ones, but to a superposition or accretion of processes. The maintenance of primordial environments depends on the development of complexity to compensate for the effects of increasing diversity. Much of the persistence that is observed along the biological history of soils depends on symbiogenesis, which might be understood in terms, not of natural selection among biological systems, but of natural maintenance.

Soil microbial diversity has no equivalent in other environments carrying heavy microbial loads, such as aquatic or human gut systems. Variability and complexity of soil microbial communities originate in part from heterogeneous distribution of water, oxygen and nutrients within the structured soil matrix (Young and Ritz 2000). Evolutionary change also has the potential to contribute to soil microbial diversity, insofar as it occurs without replacement. In the present review, the biological history of soil will be presented as a cumulative succession of colonizations. Under the combined influence of biological variation and global environmental changes, new groups of microbes and multicellular organisms have colonized land without necessarily replacing previously established soil inhabitants. New soil organisms might have appeared suddenly, in a punctuated equilibrium pattern (Eldredge and Gould 1972). Such sudden releases from species homeostasis might have generated waves of new soil colonists, organisms from one such wave collectively

P. Dion
Département de phytologie, Pavillon Charles-Eugène Marchand, 1030,
avenue de la Médecine, Université Laval. Québec (Québec) G1V 0A6, Canada
e-mail: patrice.dion@fsaa.ulaval.ca

C.S. Nautiyal, P. Dion (eds.) *Molecular Mechanisms of Plant and Microbe Coexistence.* Soil Biology 15, DOI: 10.1007/978-3-540-75575-3
© Springer-Verlag Berlin Heidelberg 2008

forming what will be referred to here as an evolutionary layer. As new evolutionary layers did not eliminate previous ones, the persistence of ancient and novel cellular mechanisms, organisms, and whole ecosystems created opportunities and pressure for the assembly of ever more intricate biological networks.

The persistence of evolutionary layers and of their components both promotes and depends on symbiogenesis, in a self-reinforcing mode. It contributes to shape selection and may also directly trigger genome rearrangements (Moran and Plague 2004; Pitman et al. 2005). Symbiotic processes taking place as a result of the persistence of evolutionary layers may eventually lead to the fusion of components of different evolutionary layers in a single, common entity.

Here, it will be attempted to determine how the famous saying of Theodosius Dobzhansky, to the effect that "Nothing in biology makes sense except in the light of evolution", might apply to the case of soil microbial diversity. Whereas discussions on soil biological relationships often offer an evolutionary perspective (Phillips et al. 2003; Karandashov and Bucher 2005; Lloret and Martínez-Romero 2005), this review will specifically focus on the evolutionary structure of soil communities. It is

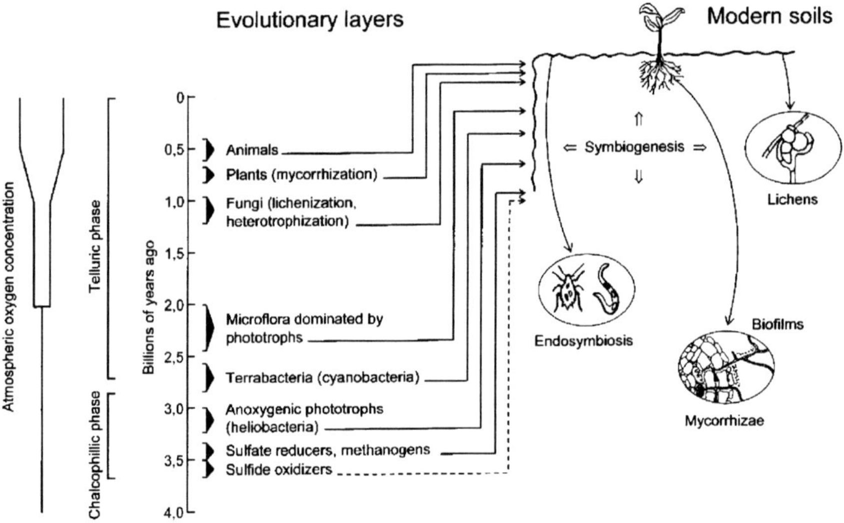

Fig. 3.1 Evolutionary assembly of the soil biological horizons. The figure includes a time scale providing approximate dates for the emergence of individual evolutionary layers. These are identified *to the right* of the time scale. *To the left* they are assigned either to a chalcophilic phase, for those evolutionary layers that depended on the activity of sulfur-loving microbes in subaerial springs, or else to a telluric phase during which new evolutionary layers appeared directly into the soils. *To the far left* of the figure is a schematic rendering of the variations in atmospheric oxygen concentration along the time scale provided. Modern soils are represented *to the far right* of the figure, and as such this *right section* of the figure relates strictly to the point "0" of the time scale. Evolutionary layers are linked to modern soils by *full or dashed lines*, the latter indicating that only extant functional equivalents to the postulated ancient members of the considered evolutionary layer are discussed in this review. Accretion of the various evolutionary layers and persistence of environments depends at least in part on symbiogenesis. Some of the symbioses and modes of interactions outlined in this review are presented *below* the modern soils depiction

envisioned that deciphering biological functions of ancient soils (Retallack 2001, 2003) will help categorize current soil biology.

A case has been made for the initial progression of early organisms towards a "Darwinian threshold" before which no fixed genealogical lineages would have existed (Woese 2002). Perhaps soils were colonized early, and before genetically stable cellular templates had emerged? Indeed, the possibility that life originated on shorelines has been explored (Bywater and Conde-Frieboes 2005). However, the current review will focus on the colonization of soils by DNA-containing cells definable by classical biology (Fig. 3.1).

3.2 Postulated Origin of Land Colonization in Sulfide-rich Springs

Convergent evidence (Retallack 2001) indicates that land microbial colonization might have begun in the mid-Archean, or 3500 Ma (million years ago). It is proposed here that the first land organisms originated from biofilm communities arranged around subaerian sulfide-rich springs. Given the virtual absence of atmospheric oxygen at that time (Canfield et al. 2000), possible electron acceptors for anaerobic sulfide oxidation are nitrate (Kühl and Jørgensen 1992) and arsenate (Hollibaugh et al. 2006), these possibilities not being mutually exclusive. Nitrate could have been produced abiotically from atmospheric dinitrogen by lightning or comet impacts (Mancinelli and McKay 1988), while arsenate would have formed from arsenite through reactions involving stronger oxidants such as nitrate (Oremland and Stolz 2003). Both dissimilatory nitrate (Castresana and Moreira 1999; Cabello et al. 2004) and arsenate (Oremland and Stolz 2003) reductases are widely shared among prokaryotes and postulated to be ancient. A cytoplasmic arsenate reductase, involved in arsenic resistance, has also been presented as an ancient trait shared by a wide variety of organisms, suggesting abundant contact of early life with arsenate (Jackson and Dugas 2003).

It has been suggested that the complete sulfur cycle may be of great antiquity (Grassineau et al. 2001), but controversy exists on this matter (Nisbet and Fowler 1999). Extant obligatory chemoautotrophs capable of anaerobic sulfide oxidation with nitrate (Beller et al. 2006), or arsenate (Hoeft et al. 2004) as electron acceptors are members of the Proteobacteria, and no deep-branching examples are known. The pathways for oxidation of sulfur compounds are complex (Friedrich et al. 2001) and bear the trace of convergent evolution (Kelly et al. 1997). Most anaerobic sulfur oxidizers have been studied in marine settings; however a complete anaerobic sulfur cycle might be operating in subterranean environments (Gevertz et al. 2000). While the phylogeny, physiology and biochemistry of sulfide oxidizers are under revision (Megonigal et al. 2005), additional studies specifically addressing the question of anaerobic sulfur oxidation in freshwater or soil systems, such as the recent work by Haaijer et al. (2006), might help elucidate the origin of soil colonization.

The sulfide-oxidizing primary producers would have nurtured a secondary community composed of sulfate reducers, methanogens and methanotrophs (Nisbet and Sleep 2001; Retallack 2003). Isotopic evidence for mesophilic microbial sulfate reduction as early as 3470 Ma has been found (Shen et al. 2001). Methanogenesis and anaerobic methanotrophy are also ancient metabolisms (Teske et al. 2003).

Modern soils exhibit a continuum from those that are flooded and continuously anaerobic to well-drained soils that may have limited areas and periods of anaerobiosis. The extent and intensity of anaerobic processes, including nitrate and sulfate reduction and methanogenesis, vary accordingly in these soils (Tiedje et al. 1984). Even oxic soils such as savanna and desert soils today harbor presumably dormant populations of methane-producing and sulfate-reducing bacteria (Peters and Conrad 1995). It would be of interest to know more about the distribution of arsenate respirers.

Early microbial mats that colonized pools formed at the vicinity of sulfide-rich springs evolved by accretion of evolutionary layers, which also remained physically stratified. In this model, prephotosynthetic biofilms would have had bacterial sulfide processers on top, underlain by sulfate respirers as well as methanogenic and methanotrophic archaea. Then, anoxygenic photosynthetizers would have added an upper layer to these original mats, introducing a new source of reduction power (Nisbet and Sleep 2001). Phylogenetic evidence suggests that anoxygenic photosynthesis appeared 3190 Ma (Battistuzzi et al. 2004).

Phylogenetic analysis of photosynthesis genes show that purple bacteria represent the basal lineage for photosynthetic bacteria, with green sulfur and non-sulfur bacteria being each other's closest relatives and being rooted intermediately between heliobacteria and purple bacteria. Heliobacteria are closest to the last common ancestor of the oxygenic photoautotrophs (Guest 1994; Xiong et al. 2000). Heliobacteria still occur today as photoheterotrophic, endospore-forming bacteria colonizing terrestrial habitats (Ormerod et al. 1996; Stevenson et al. 1997), and it may be that the common ancestor of heliobacteria and cyanobacteria played an important role in early, anaerobic colonization of sulfide-rich springs and neighboring habitats.

3.3 The Advent of the Cyanobacteria

In many cyanobacteria, oxygenic photosynthesis is inhibited at low hydrogen sulfide concentrations, whereas anoxygenic photosynthesis is induced at the high sulfide concentration of 3 mM. The bacteria then rely on photosystem I only and sulfide as an electron donor (Padan 1979). Sulfide-rich springs have concentrations of up to 10 mM hydrogen sulfide, and are home to a diverse anaerobic microbial flora comprising cyanobacteria (Elshahed et al. 2003). Hence, cyanobacteria have the ability to blend in anaerobic communities based on sulfur metabolism, while switching to oxygenic photosynthesis when spreading to surrounding land. Resolution of deep-branching phylogenetic relationships suggests that the earliest cyanobacterial lineages were unicellular rods that lived in terrestrial or freshwater environments (Sánchez-Baracaldo et al. 2005).

Primordial life on other planets might be endolithic (Walker et al. 2005). Endolithic methanogens might have existed on Earth 3500 Ma, colonizing silica dykes down to 1 km below the surface (Canfield 2006; Ueno et al. 2006). Similarly, early telluric cyanobacterial expansion from sulfide-rich springs might have occurred endolithically at first, the overlaying rock shielding the photosynthetic microbes from desiccation and harsh solar radiation. Such an endolithic microbial community would have contributed to initial weathering of rock minerals, through production of carbon dioxide and the excretion of a variety of organic acids (Gorbushina and Krumbein 2005).

Although disagreements occur (Brown et al. 2001; Ciccarelli et al. 2006), various phylogenomic analyses suggest that cyanobacteria, actinomycetes and *Deinococcus* form a high-level phylogenetic clade (Wolf et al. 2001; Battistuzzi et al. 2004; Bern and Goldberg 2005). Members of the putative cyanobacteria-actinomycetes-Deinococcales clade were collectively referred to as "Terrabacteria", on the basis that their common ancestor may have pioneered land colonization between 2800 and 3100 Ma. The Terrabacteria are characterized by desiccation resistance and light protection mediated by carotenoid and other pigments. These properties, associated to phototrophism, would have permitted efficient soil colonization (Battistuzzi et al. 2004). Reduced carbon measurements and other geochemical evidence suggest cyanobacterial mat formation at the surface of thick and clayey paleosols 2600 Ma (Watanabe et al. 2000).

Cyanobacteria are thought to have carried out oxygenic photosynthesis before 2700 Ma (Brocks et al. 1999). However, photosynthetic oxygen production would not have resulted in an immediate increase in atmospheric oxygen level. Thus, massive methanogenesis would have remained contemporary to early oxygenic photosynthesis, resulting in atmospheric methane accumulation to 50–500 times present day levels (Catling et al. 2001). The mechanism involved in the eventual rise in atmospheric oxygen is a matter of some debate (Towe et al. 2002). Various hypotheses suggest that gradual loss of atmospheric hydrogen originating from volcanic gas emissions (Holland 2002), or methanogenesis coupled to CH_4-induced hydrogen escape (Catling et al. 2001), would have triggered oxidation of the atmosphere 2200–2400 Ma, from less than 0.002 atm (or a fraction of 0.01 of the present atmospheric level, PAL, which is 0.21 atm), to 0.03 atm, or 0.14 PAL (Rye and Holland 1998). During the 400-million year period separating the advent of oxygenic photosynthesis and the massive rise of atmospheric oxygen, oxygen accumulation occurred locally and transiently (Kurland and Andersson 2000; Catling et al. 2001). Because of their heterogeneous structure and their capacity for organic matter immobilization, soils might have been particularly prone to generation of these "oxygen oases" (Catling et al. 2001). Thus, early, localized accumulation of oxygen in soils might have favored the extensive development of microbial diversity that seems to be a characteristic of certain paleosols (Gutzmer and Beukes 1998; Beukes et al. 2002).

Cyanobacteria were instrumental in early soil colonization, and, given their capacity to use abundant substances such as water and atmospheric nitrogen, might well have represented a major source of biomass (Sánchez-Baracaldo et al. 2005). More specifically, cyanobacteria illustrate some of the basic phenomena underlying soil microbial colonization, particularly the accretion of evolutionary layers and the persistence of environments. Both phenomena are connected, and occur at least in part through symbiogenesis.

With respect to evolutionary layer accretion, extant cyanobacteria occur in a wide variety of environments, including various types of fertile soils (Rippka 1988), desert biological soil crusts (Belnap 2003) and Arctic soils (Liengen 1999). In these various environments, their presence as a component of vastly different microbial communities reflects evolutionary history of the soils and gradual assembly of layers as influenced by ambient conditions. As regards environmental persistence, it is proposed here that a modern equivalent to early soil formation and initial weathering of rock can be found in the cryptoendolithic activity of cyanobacteria. Endolithic phototrophs, primarily cyanobacteria but also algae, occur at depths varying from 0.5 to 5 mm in various types of rocks, and are responsible for primary productivity sustaining an heterotrophic microbial community (Hughes and Lawley 2003). Today, as is postulated to have occurred in ancient soils (see above), cyanobacteria still play a major role in rock weathering (Büdel et al. 2004).

Evolutionary layers are interconnected and mutually dependent, and, again cyanobacteria are a case in point. The development of interactions between cyanobacteria and organisms belonging to different evolutionary layers is exemplified by the observation of numerous extant symbioses involving cyanobacterial participation. Such associations occur today with a wide range of photosynthetic and nonphotosynthetic eukaryotic hosts, including fungi, liverworts, mosses, pteridophytes, higher plants and echiuroid worms. The eukaryotic host generally benefits from the photosynthetic, and in many cases also from the nitrogen-fixing abilities of cyanobacteria (Raven 2002). Soil symbioses will be considered in more detail below (see Sects. 3.5–3.8).

3.4   Biogeochemical Cycling Based on Microbial Phototrophic Inputs

As well as natural rocks (see above section), stone buildings are home to microbial communities, the development and transitions of which have been studied in some detail. Again, the microbial colonization of stone buildings is initiated by phototrophic organisms which build up a visible biofilm on the nutrient-depleted stone surface (Warscheid and Braams 2000). These phototrophs may be cyanobacteria or green algae (Crispim et al. 2003). Among the cyanobacteria, unicellular forms are often preponderant (Crispim et al. 2003), whereas filamentous forms occur as short filaments with thick pigmented sheaths (Gaylarde and Gaylarde 2005). Phototrophic microorganisms may grow on the stone surface or may penetrate some millimeters into the rock pore system. Phototrophs excrete carbohydrates and growth factors, thus facilitating the establishment of a complex microbial community. Fungi are especially concentrated in stone crusts. They are able to penetrate into the rock material by hyphal growth and by biocorrosive activity, due to the excretion of organic acids or the oxidation of mineral-forming cations, preferably iron and manganese. Heterotrophic bacteria are also present, especially Gram-positive bacteria of the Terrabacteria group, such as actinomycetes and related coryneform bacteria. These organisms predominate over the more sensitive Gram-negative bacteria. Finally, colonization of building stones by chemoautotrophic bacteria also occurs, as the result of the accumulation of reduced sulfur and nitrogen compounds (Warscheid and Braams 2000).

Because of their filamentous growth mode, actinomycetes are particularly well suited for growth in the heterogeneous rock and soil environments, and in ancient soils they were perhaps closely associated to the cyanobacteria as actinolichens (Retallack 2003). Associations between algae and streptomycetes have been constructed in the laboratory, and similar associations have been found in nature (Hawksworth 1988).

Colonization of early soils by chemoautotrophs allowed the functioning of complete biogeochemical cycles. A mixed aerobic and anaerobic soil colonization process is postulated here, which might have been characterized initially by the establishment of aerobic micropockets, that would have formed in daytime around photosynthesizing cyanobacteria. Outside these pockets or at the chemocline, the heterogeneous distribution of oxygen in the ancient soils might have supported transitional geochemical cycling, similar to what is now observed in aquatic consortia and biofilms (Paerl and Pinckney 1996). In these systems characterized by a heterogeneous oxygen distribution, anoxygenic photosynthetic bacteria and aerobic chemoautotrophs cooperate in sulfur oxidation, whereas hydrogen from fermentation of fixed carbon is used by sulfate-reducing bacteria and methanogens.

Following the partial rise in atmospheric oxygen occurring 2200 Ma (see above), the soil aerobic pockets expanded. Without being completely eliminated, the anaerobes would have migrated to more profound horizons. By the time a second increase in atmospheric oxygen concentration occurred by the late Precambrian, or 1100–550 Ma, the eukaryotic lineage was diversifying (Douzery et al. 2004), hence providing anaerobic niches and allowing persistence of an evolutionary layer that was originally associated with primordial anoxic or partly anoxic conditions.

3.5 Factors Involved in the Establishment of Mutualistic Symbioses

Early fossil evidence suggests that colonization of Precambrian paleosols involved free-living organisms resembling cyanobacteria (Horodyski and Knauth 1994; Selosse and Le Tacon 1998). Today, however, mutualistic symbioses represent an essential aspect of soil biology (Buscot 2005). Symbiogenesis was stimulated by the emergence of organisms belonging to different evolutionary layers, as well as by stepwise changes in environmental conditions such as atmospheric oxygen concentrations. For example, lichenization has been proposed to have contributed to the persistence and diversification of algae and cyanobacteria following the onset of competition by multicellular plants (Selosse and Le Tacon 1998).

The establishment of novel symbioses depends on organismal diversification, and in turn associated organisms are subjected to novel selection pressures (Saffo 2002). This interdependence, as applied to soil biology, is in line with the conception of the soil as a self-organizing system (Young and Crawford 2004). It should also be considered in the light of the notion that soil organisms are subjected to interactions from pore to landscape scales (Crawford et al. 2005). Very local interactions facilitate close physical and metabolic contact, which trigger competition and consequent selection. Wide-scale interactions, as arise from long-distance dispersal through air or water flows, establish confrontations as a general outcome of evolutionary innovations. In the long term, however, stably maintained coexistence promotes the development of cooperation strategies. For example, bacteria and fungi tend to compete for utilization of simple and rapidly metabolized plant-derived substrates, whereas mutualistic strategies have evolved for utilization of more recalcitrant substrates such as cellulose and lignin (de Boer et al. 2005).

The establishment of mutalistic symbioses has different consequences for the microsymbiont and the macrosymbiont. In the case of the larger, eukaryotic, partner, symbiosis might result in exposure to a more stressful environment (Lutzoni and Pagel 1997), as well as to increased aggressiveness and expansion range (Rudgers et al. 2005), leading to enhanced diversification. On the contrary, observations on *Buchnera*, an endosymbiont of aphids, suggest that selection on the microsymbiont is ineffective to eliminate slightly deleterious mutations (Wernegreen and Moran 1999), at least for those characters that are not directly subjected to host pressure (Canbäck et al. 2004). Endosymbiosis favors the loss of apparently beneficial genes and the accumulation of deleterious mutations. The inefficiency of natural selection during endosymbiosis may be due to the partitioning of populations among hosts and large fluctuations in population sizes (Ochman and Moran 2001). Since the efficiency of natural selection tends to decrease with increasing organism size (Lynch 2006), the microsymbiont might escape the effect of selection through its association with a larger organism. The results of long-term evolution without a stringent selection surveillance include increased stability of genome structure and the development of an irreversible host dependency (Andersson and Kurland 1998; Klasson and Andersson 2004). Whereas smaller genomes are noted to evolve faster (Ciccarelli et al. 2006), such accelerated reductive evolution, occurring in the initial stages of symbiogenesis, might be followed by structural genomic stasis resulting from the loss of genetic elements that mediate recombination events (Tamas et al. 2002).

Obligate mutual dependency may conduct to organismal fusion, the endosymbiotic origin of mitochondria and chloroplasts (Margulis 1970) being the most dramatic illustration of this process. Accretion of evolutionary layers, which occurs largely as a result of symbiotic interactions between representatives of distinct layers, promotes the fusion and blurring of these layers to generate a hybrid evolutionary state. In this sense, evolution can be envisioned as functioning through persistence rather than purifying selection.

3.6 The Symbiotic Continuum

The recently proposed notion of an endophytic continuum (Schulz and Boyle 2005) can be expanded, in the sense that the various soil symbioses that exist today may be regarded as a continuum of interactions. These are related to each other by one or various factors such as time, space, mechanisms and evolutionary consequences. Indications on the existence, nature and implications of the symbiotic continuum are provided by the cyanobacteria and the fungi. These two types of organisms are well known for their capacity to establish a wide range of interactions with different partners (Selosse and Le Tacon 1998; Raven 2002).

The Glomeromycota, to which the arbuscular mycorrhizal fungi belong, originated early in fungal evolution, about 1200–1400 Ma. This was after Chytrid divergence, but before Ascomyceta split from Basidiomyceta, 1200 Ma. As the current members of the Glomeromycota are terrestrial (Schüßler et al. 2001), such an early occurrence suggests that colonization of land by fungi occurred deep in the Proterozoic, more than 900 Ma (Heckman et al. 2001). These early terrestrial fungi might have established extracellular or intracellular associations with phototrophic algae or cyanobacteria.

Extracellular interactions between fungi and cyanobacteria evolved independently on several occasions (Gargas et al. 1995), and the fossil record dates the origin of lichens to at least 600 Ma (Yuan et al. 2005). However, considering the probable antiquity of terrestrial colonization by fungi and cyanobacteria (see above) it was suggested that non-septate lichens formed earlier (Selosse and Le Tacon 1998). A high level of stress on the fungus is associated with a transition to the lichen symbiosis, resulting in accelerated evolution that has an impact, not only on lichen-forming fungi, but also on non-lichen-forming fungi through delichenization (Lutzoni et al. 2001).

Early intracellular interactions between fungal and phototrophic partners may have resembled the present-day *Geosiphon pyriformis*, which is the result of endo-cyanosis, or an intracellular colonization of a fungus by a cyanobacterium (Kluge 2002). The observation that the fungal partner in *Geosiphon* is a member of an ancestral lineage within the Glomeromycota (Schüßler 2002) suggests that endomycorrhizal fungi first established interactions with unicellular phototrophs before the advent of land plants. Hence, the symbiotic continuum was initiated very early in the history of soil colonization, as fungi interacted in two alternative ways with cyanobacteria and other unicellular phototrophs, to generate the lichens by exocyanosis, and the arbuscular mycorrhizae by endocyanosis and later association with land plants.

Although there are no undisputed eukaryotic land fossils before 460–480 Ma (Wellman et al. 2003), green algae appeared 1000 Ma, while bryophytes and higher plants diverged 703 Ma (Heckman et al. 2001). This suggests that land colonization by plants dates back to at least 700 Ma. The first terrestrial plants were bryophytes (Hedges 2002), and primitive arbuscular mycorrhizal associations with these rootless phototrophs might have been instrumental in early land colonization. The glomalean fungus *Glomus claroideum*, known to form arbuscular mycorrhizae with vascular plants, is also able to form a mycorrhiza-like symbiosis with the hornwort *Anthoceros punctatus* (Schüßler 2000). Mycorrhiza-like associations are also observed with other lower land plants, including hepatics and lycopods (Read et al. 2000). Whereas the functional significance of these associations might not be clear, their occurrence and morphological similarities with true mycorrhizae call for their interpretation within the framework of the symbiotic continuum hypothesis. By the time roots evolved in the Early Devonian, 410–395 Ma, the original axis of early vascular plants was already host to abundant below-ground mycorrhizal fungi (Raven and Edwards 2001).

A further extension of the symbiotic continuum is revealed by comparative analyses of mycorrhizal and nitrogen-fixing rhizobial associations (Hirsch et al. 2001; Szczyglowski and Amyot 2003). It has been proposed that elements of the signaling pathway between legume roots and infecting rhizobia may have derived from a pre-existing pathway that regulates the more ancient arbuscular mycorrhizal symbiosis (Albrecht et al. 1999). Plant response components to both the rhizobial and the mycorrhizal stimuli include a receptor kinase (Endre et al. 2002; Stracke et al. 2002), a protein that shares similarities with ligand-gated cation channels (Ané et al. 2004) and a protein resembling calcium and calmodulin-dependent protein kinases (Lévy et al. 2004). Plant genes that are activated in response to infection by both the rhizobial and the fungal partner have been termed symbiosin genes (Kistner et al. 2005). Such commonalities have led to predictions and initial hints that mycorrhizal fungi would produce Myc signals analogous to the rhizobial Nod factors (Albrecht et al. 1999; Harrison 2005). In particular, the *Medicago trunculata* plant cell responds to contact with fungal hypha by undergoing cytoskeletal reorganization. This reaction depends on gene products required for rhizobial and mycorrhizal infections (Genre et al. 2005).

Both the Nod factor and the Myc factor signaling pathways might be derived from the more ancient chitin signaling pathway, that triggers a plant response to fungal pathogenic attack (Stacey et al. 2006). A set of rice genes was similarly expressed in response to both mycorrhizal interaction and infection by the fungal pathogens *Magnaporthe grisea* and *Fusarium moniliforme* (Güimil et al. 2005).

Strigolactones are sesquiterpene signals produced by plant roots that induce hyphal branching, which is an early reaction of the infecting mycorrhizal fungus in the vicinity of a susceptible root, and may also induce fungal synthesis of the Myc factor. In yet a further striking illustration of the symbiotic continuum, strigolactones also induce seed germination of the parasitic weeds *Striga* and *Orobanche* (Akiyama et al. 2005; Akiyama and Hayashi 2006). Strigolactone synthesis by plant roots is stimulated by soil phosphorus deficiency, when *Striga* infestation is most prevalent (Yoneyama et al. 2001) and also when potential benefits to the plant of the mycorrhizal association are greatest. Thus, preservation of mechanisms along the symbiotic continuum involves a continuity from molecular up to environmental relationships.

Root knot nematodes act at a distance to induce cytoskeletal, nuclear, and morphological effects in *Lotus japonicus* root hairs similar to those caused by the rhizobial nodulation factors. Specific components of the host perception machinery are involved in both interactions, implying a degree of common function between these prokaryotic and eukaryotic symbionts. The nematode also uses its stylet to inject peptide signals that trigger plant developmental pathways common to both nodule and gall formation (Davis and Mitchum 2005).These observations suggest that evolution of parasitism in root knot nematodes was accompanied by conscription of older symbiotic pathways (Weerasinghe et al. 2005). Root knot nematodes have acquired rhizobial genes by horizontal gene transfer, including a gene for Nod factor synthesis, which suggests routes for generating similarities in interaction mechanisms with the host plant (Scholl et al. 2003). In spite of these commonalities between different symbioses, each particular interaction system retains specific developmental and physiological mechanisms (Parniske 2000).

One mechanism for establishment of the symbiotic continuum may be described as "abuse", whereby a parasite takes advantage of a signal which is produced by its host to attract benevolent symbionts. Strigolactone recognition by parasitic plants (see above) was so acquired (Bouwmeester et al. 2007).

The recent observation that some photosynthetic bradyrhizobia did not harbor *nod* genes (Giraud et al. 2007) is intriguing as regards the range of the symbiotic continuum. For example, the possibility of similarities between infection mechanisms of photosynthetic bradyrhizobia and the actinorhizal *Frankia* has been evoked (Giraud et al. 2007).

3.7 Microsymbiotic and Macrosymbiotic Factors Shaping Interactions

Metabolic networks in bacteria, archaea and eukaryotes are arranged into four discrete groups of increasing size and connectivity, the latest and largest of these involving oxygen. The networks in smaller groups are nested within those in larger groups (Raymond and Segrè 2006). Hence, persistence of metabolic networks may be imprinted in evolutionary relationships and life itself. The evolution of life in soils might only represent an extension of this persistence principle to the hierarchical levels of organisms and environments.

Different bacteria have deployed common molecular mechanisms to initiate and maintain pathogenesis or mutualism with hosts (Hentschel et al. 2000). A case in point is the sharing, by various plant and animal pathogens, of similar type III and type IV secretion systems for translocation of DNA and protein substrates to target cells (Wren 2000). Phylogenetic analyses of the various type III secretion systems suggest these to be of an ancient origin and to have diverged independently in unrelated bacteria (He et al. 2004b).

Acquisition of the capacity to establish mutualisms and diseases is facilitated by horizontal transfer of DNA fragments. The transferred DNA elements may be of large size, and their dissemination among bacterial hosts may be facilitated by conjugative transfer and phage infections. However, acquisition of symbiotic capacity through horizontal gene transfer is possible only in bacteria that already have a long history of interaction with their host (Ochman and Moran 2001).

The convergence observed when different bacteria share symbiotic mechanisms may also reflect phylogenetic relationships. Parallels exist in the infection strategies of *Sinorhizobium meliloti* and the related animal pathogen, *Brucella abortus*, which adopt similar lifestyles inside the host. Both rhizobia and brucellae are endocytosed by host cells, where they then undergo adaptive changes and ultimately live for prolonged periods in intracellular, acidic, host-membrane-bound compartments. Mutants altered in *S. meliloti bacA* or its *B. abortus* ortholog are unable of long-term residence in their respective host cell (LeVier et al. 2000). BacA from the two bacteria affects the fatty acid content of lipid A, and as such may be involved in the capacity of the plant and mammalian pathogens to avoid recognition by the host (Ferguson et al. 2004).

Alternatively, the conservation of interactive mechanisms may be driven by the macrosymbiont, which imposes a common selective pressure on different associated microbes. This homogenizing effect is exerted by an individual host interacting with different microbes, and also by different hosts sharing defining symbiotic characters. Some macrosymbiotic idiosyncrasies persist across large evolutionary distances. Plants and animals recognize similar, albeit not identical, pathogen-associated molecular patterns. Recognition is mediated by Toll and Toll-like receptors in animals, and in plants by proteins resembling Toll-like receptors and sharing with these similar structural modules, in particular leucine-rich repeats (Nürnberger et al. 2004). Whether these similarities are the result of conservation of ancient defence mechanisms or convergent evolution, they suggest that the conservation or reproduction of recognition mechanisms in the macrosymbiont encouraged the conservation or reproduction of interactive mechanisms in the microbial symbiont. The result of evolution or coevolution of symbiotic partners is maintenance, in a process that sustains and strengthens itself.

3.8 The Evolution of Symbionts and Hypersymbionts

Following symbiogenesis, pressure is exerted to preserve the interaction and expand its applicability. To preserve the interaction, microsymbionts increase and diversify benefits to the host and develop powerful mechanisms for excluding competitors. For example, the endophytic fungus *Piriformospora indica* induces a variety of beneficial responses in its host, including stress tolerance and disease resistance (Waller et al. 2005). Other endophytes synthesize inhibitory compounds active against a wide variety of fungi, bacteria and other organisms (Strobel 2006).

The nematode mutualist and entomopathogenic Proteobacterium *Photorhabdus* also produces a range of biocidal compounds whilst growing in its insect host, so as to out-compete or repel potential competitors that would come from the insect gut microflora or the soil (ffrench-Constant et al. 2003). In the case of this ternary interaction, niche revendication benefits the nematode macrosymbiont, which is also multiplying in the dead insect. As such, it preserves the accretion of three distinct evolutionary layers, represented by the bacterial, nematode and insect partners.

Niche revendication may also depend on trophic relationships, such as those established by nitrogen-fixing rhizobia (Lodwig et al. 2003) and plant cell-transforming agrobacteria (Zhu et al. 2000). For rhizobia (Denison and Kiers 2004) as for agrobacteria (Brencic et al. 2005), such relationships exert a continuum of effects on the host plant, ranging from parasitism to commensalism and mutualism. The development of specific mechanisms for syntrophic exchange of nutrients and metabolites (Kooijman and Hengeveld 2005) represents a fundamental process in evolutionary layer networking.

To expand the applicability of the interaction, some symbionts evolve towards promiscuity, and become "hypersymbionts". Certain strains of *Pseudomonas aeruginosa* are promiscuous pathogens, with an ability to infect a wide variety of plants and animals. The genome of one of these strains harbors pathogenicity islands derived from a wide array of bacterial species, including plant and animal pathogens (He et al. 2004a).

Amoebae serve as training grounds for acclimatization of symbionts to an intracellular lifestyle. The capacity of bacteria to colonize amoebae has evolved several times (Molmeret et al. 2005), and today both environmental and clinical isolates of amoebae harbor bacterial endosymbionts. For example, approximately 25% of environmental and clinical *Acanthamoeba* isolates are infected. A wide variety of Gram-negative endosymbionts have been described in amoebae. Some are highly adapted to this intracellular milieu and cannot be cultured without amoebae (Winiecka Krusnella and Linder 2001). These are phylogenetically related to obligate endosymbionts of insects or intracellular parasites of animals, including Rickettsiae and Chlamydiae. Interaction with unicellular eukaryotes might facilitate evolution of mechanisms for intracellular survival that are then useful in other eukaryotic cells (Harb et al. 2000; Molmeret et al. 2005). Various bacteria, including *Burkholderia*, *Chlamydia*, *Legionella* and Rickettsiae, are facultative hosts of the amoebae and are also intracellular animal pathogens. Growth inside amoebae renders legionellae more invasive to macrophages and epithelial cells and turns them into more virulent pathogens of experimental animals (Winiecka-Krusnella and Linder 2001). Amoebae secrete some factors that enhance replication of various *Legionella* species in human monocytes (Neumeister et al. 2000).

Upon superimposition of a new evolutionary layer, more ancient organisms tend to persist through the establishment of interactions. Thus, symbiogenesis appears as a powerful mechanism for evolutionary layer accretion, acting by promoting concomitant evolution of partners and niche definition. The environments favorable to ancient organisms turned into microsymbionts become dispersed, or emulsified, within new environments shaped by more recent life forms.

3.9 Evolution and the Persistence of Environments

The persistence of ancient organisms and of the niches that these organisms inhabit allows environments to be maintained on a micro scale despite global changes and the appearance of new and more complex life forms. The heterogeneity that is required for environmental maintenance is generated along evolutionary time by the increasing complexity of biological systems. The recreation of ancient environments within a new biological order occurs, for example, upon replacement of physical with biological heterogeneity.

By the time atmospheric oxygen accumulated, 1500–2000 Ma, primitive eukaryotes or proto-eukaryotes already existed which had the capacity to engulf bacteria (Kurland and Andersson 2000; Kurland et al. 2006). Anaerobic bacteria so enclosed might have persisted in continuing oxygen-free environments found in protists. These relationships were reconstructed experimentally through co-cultivation of *Mobiluncus curtisii*, an obligate non-sporeforming anaerobe, with free living *Acanthamoeba* spp. under aerobic conditions. In these experiments, internalization, multiplication and persistence of bacterial cells were established for several weeks (Tomov et al. 1999). Similarly, obligately anaerobic clostridia exist as endophytes of graminaceous plants. They are found in substantial numbers in the subterraneous and aerial parts of the plant, and some clostridial types were preferentially isolated from plants, as compared to the surrounding soil (Miyamoto et al. 2004).

Land colonization by arthropods is believed to have occurred rather late, after the origin of the myriapod lineage 642 Ma and before the millipede-centipede divergence, which occurred 442 Ma. Arthropods are likely to have been preceded by nematodes and other animals leaving little fossil records (Pisani et al. 2004). Amphibians derived from fish in the Late Devonian, 360 Ma, and adapted to life in fully terrestrial environments in the succeeding 20–30 million years (Carroll 2001). The intestines of soil animals are well documented niches favoring colonization by strict anaerobes. Termites have been particularly well studied in this regard, and are host to a complex and diverse microflora, including anaerobic protozoa, *Desulfovibrio*, *Bacteroides*, methanogenic archaea (Brune and Friedrich 2000), and clostridia (Tokuda et al. 2000). Cockroaches (Sprenger et al. 2000) and humus-feeding coleopteran larvae (Egert et al. 2003) are also hosts to strictly anaerobic bacteria or archaea. Such interactions across evolutionary layers lead to further evolutionary changes, as is suggested by the observation that the clostridia colonizing the gut of termites are related to vertebrate pathogens that demonstrate an affinity with gut tissue (Tokuda et al. 2000). In some cases, the interactions between evolutionary layers are only transient, as they do not involve permanent colonization of an animal host. This is the case, for example, in the earthworm gut, where the activity of denitrifyers is enhanced during transit (Horn et al. 2006).

Biofilms may be thought of as minimal ecosystems (Guerrero et al. 2002). They developed in the early Archean (around 3500 Ma) or perhaps even earlier, in various settings including shallow waters in the vicinity of hydrothermal vents (Nisbet and Fowler 1999). Biofilm formation is a current behavior of bacteria interacting with soil fungi (Bianciotto et al. 2001; Bianciotto and Bonfante 2002) or plant roots (Morris and Monier 2003; Ramey et al. 2004), and as such illustrates environmental persistence, with the continued expression of associated organizational traits (Davey and O'Toole 2000).

3.10   The Heterotrophization of Soil Microbial Life

It has been hypothesized that the appearance of large animals about 575 Ma, which required that atmospheric oxygen be at present-day levels, was facilitated by the construction of an elaborate soil microflora. The activity of this abundant soil microbial population resulted in enhanced weathering and the formation of biogenic clay, which in turn plays a major role in organic carbon sequestration. Enhanced organic carbon immobilization in clay particles would then have resulted in the rise of atmospheric oxygen from 0.03 atm (see above) to levels sufficient for the evolution of large animals (Kennedy et al. 2006). This postulated intensification of soil microbial activity may have corresponded to a transition from an autotroph-dominated to an heterotroph-dominated microflora such as that which occurs in modern soils (Janssen 2006). The gradual superposition of evolutionary layers, occurring with the advent of lichens, then of bryophytes and vascular plants, led to a partial heterotrophization of the microbial soil metabolism, realized through extensive interactions of microbes with eukaryotic phototrophs.

The association of additional microorganisms with the primary symbiotic partners enhances the predominantly heterotrophic character of the soil microflora and the environmental impact of the symbioses. For example, secondary microbial colonization of the lichen zone affects the rate and mechanisms of mineral weathering reactions (Banfield et al. 1999). On the other hand, various aspects of the impact of intracellular (Bonfante 2003) and extracellular (Artursson and Jansson 2003) bacterial colonization of arbuscular mycorrhizal fungi remain to be identified (Artursson et al. 2006). The reduced genome size of the obligate endosymbiont "*Candidatus* Glomeribacter gigasporarum*" suggests that this organism maintains an ancient relationship with its host, the arbuscular mycorrhizal fungus *Gigaspora margarita* (Jargeat et al. 2004).

3.11   Conclusions

In soil systems, evolutionary layer accretion both stimulates and depends upon the interactive coexistence of mechanisms, organisms and environments. A layered mode of evolution has also been proposed for the immune system (Herzenberg and Herzenberg 1989), the function of which, as does that of the soil, depends on extensive interactions between its components.

Soil biology is a result of the interplay of symbioses, which in turn drive coevolution of the symbiotic partners. The coevolutionary process produces local maladaptation to symbiotic existence, which may be transient or permanent, this deviation from adaptation driving further selection (Thompson et al. 2002). As the geographical component of complex landscape is a crucial factor contributing to symbiotic maladaptation (Thompson et al. 2002), the spatially and temporally heterogeneous soil might be considered as a privileged milieu for evolution of microbes and other species interacting with them. Heterogeneities in space, immediate time and resource distribution do not fully account for the maintenance of microbial soil diversity. Evolutionary heterogeneity also matters, and evolutionary distance may exert a perceptible effect on soil organisms, as do spatial and temporal distances.

Following the Darwinian presentation of competitive selection as an essential component of evolutionary processes, an alternative model of evolution through symbiotic relationships has gained support (Roossinck 2005). Mutualism might in fact represent a general strategy for life (Müller-Hill 2006): originally applied to function and evolution of macromolecules, this notion might retain some truth up to the soil organism or ecosystem levels. At these higher levels of complexity, mutualism may be based on mutual exploitation, rather than reciprocal cooperation (Saikkonen et al. 2004). In this case, natural selection might be considered as being shaped, at least in part, by association between organisms belonging to different evolutionary layers. The emergence of cheaters, which have the potential to disrupt mutualistic systems by obtaining a disproportionate share of their resources, skews the dynamics of interactions (Velicer 2003), and their influence within the framework outlined here calls for examination. High relatedness between members of a social population contributes to cheater control (Gilbert et al. 2007), pointing to an additional factor for persistence of interactions.

As suggested earlier in this review, mutualism, evolutionary layer accretion and environmental persistence might be the expression, at superior hierarchical levels of life, of a duplicative and conservative mode of evolution. It is a recurrent theme in modern biology that two apparently distinct biomolecular systems found in the same or different cells show derivation relationships or share common ancestry. It would be of interest to investigate whether or not aggregative evolution might be established as a founding principle of life, and, eventually, how this principle imprinted the changes of biomolecules, cells, organisms and environments.

Given the absence of a tangible experimental object, the reconstruction proposed here remains a literary exercise, only with less evocative value than Dante's cosmology. Current models of planet formation suggest that relatively small, rocky planets should be much more common than giant gas planets such as Jupiter (Beaulieu et al. 2006). Furthermore, liquid-water habitable zones are relatively wide around main-sequence stars not too different from our sun (Kasting and Catling 2003). So the words may eventually come to tell the soil story.

Acknowledgements The author thanks Cécile Gauthier and Bernard Pouliot for their help in preparation of Fig. 3.1.

References

Akiyama K, Hayashi H (2006) Strigolactones: chemical signals for fungal symbionts and parasitic weeds in plant roots. Ann Bot 97:925–931

Akiyama K, Matsuzaki K, Hayashi H (2005) Plant sesquiterpenes induce hyphal branching in arbuscular mycorrhizal fungi. Nature 435:824–827

Albrecht C, Geurts R, Bisseling T (1999) Legume nodulation and mycorrhizae formation; two extremes in host specificity meet. EMBO J 18:281–288

Andersson SGE, Kurland CG (1998) Reductive evolution of resident genomes. Trends Microbiol 6:263–268

Ané J-M, Kiss, GB, Riely, BK et al. (2004) *Medicago truncatula DMI1* required for bacterial and fungal symbioses in legumes. Science 303:1364–1367

Artursson V, Jansson JK (2003) Use of bromodeoxyuridine immunocapture to identify active bacteria associated with arbuscular mycorrhizal hyphae. Appl Environ Microbiol 69:6208–6215

Artursson V, Finlay RD, Jansson JK (2006) Interactions between arbuscular mycorrhizal fungi and bacteria and their potential for stimulating plant growth. Environ Microbiol 8:1–10

Banfield JF, Barker WW, Welch SA, Taunton A (1999) Biological impact on mineral dissolution: Application of the lichen model to understanding mineral weathering in the rhizosphere. Proc Natl Acad Sci USA 96:3404–3411

Battistuzzi F, Feijao A, Hedges SB (2004) A genomic timescale of prokaryote evolution: insights into the origin of methanogenesis, phototrophy, and the colonization of land. BMC Evol Biol 4:44

Beaulieu J-P, Bennett DP, Fouqué P et al. (2006) Discovery of a cool planet of 5.5 Earth masses through gravitational microlensing. Nature 439:437–440

Beller HR, Chain PSG, Letain TE et al. (2006) The genome sequence of the obligately chemolithoautotrophic, facultatively anaerobic bacterium *Thiobacillus denitrificans*. J Bacteriol 188:1473–1488

Belnap J (2003) The world at your feet: desert biological soil crusts. Front Ecol Environ 1:181–189

Bern M, Goldberg D (2005) Automatic selection of representative proteins for bacterial phylogeny. BMC Evol Biol 5:34

Beukes NJ, Dorland H, Gutzmer J, Nedachi M, Ohmoto H (2002) Tropical laterites, life on land, and the history of atmospheric oxygen in the Paleoproterozoic. Geology 30:491–494

Bianciotto V, Bonfante P (2002) Arbuscular mycorrhizal fungi: a specialised niche for rhizospheric and endocellular bacteria. Antonie van Leeuwenhoek 81:365–371

Bianciotto V, Andreotti S, Balestrini R, Bonfante P, Perotto S (2001) Mucoid mutants of the biocontrol strain *Pseudomonas fluorescens* CHA0 show increased ability in biofilm formation on mycorrhizal and nonmycorrhizal carrot roots. Mol Plant Microbe Interact 14:255–260

Bonfante P (2003) Plants, mycorrhizal fungi and endobacteria: a dialog among cells and genomes. Biol Bull 204:215–220

Bouwmeester HJ, Roux C, Lopez-Raez JA, Bécard G (2007) Rhizosphere communication of plants, parasitic plants and AM fungi. Trends Plant Sci 12:224–230

Brencic A, Angert ER, Winans SC (2005) Unwounded plants elicit *Agrobacterium vir* gene induction and T-DNA transfer: transformed plant cells produce opines yet are tumour free. Mol Microbiol 57:1522–1531

Brocks JJ, Logan GA, Buick R, Summons RE (1999) Archean molecular fossils and the early rise of eukaryotes. Science 285:1033–1036

Brown JR, Douady CJ, Italia MJ, Marshall WE, Stanhope MJ (2001) Universal trees based on large combined protein sequence data sets. Nat Genet 28:281–285

Brune A, Friedrich M (2000) Microecology of the termite gut: structure and function on a microscale. Curr Opin Microbiol 3:263–269

Büdel B, Weber B, Kühl M, Pfanz H, Sültemeyer D, Wessels D (2004) Reshaping of sandstone surfaces by cryptoendolithic cyanobacteria: bioalkalization causes chemical weathering in arid landscapes. Geobiology 2:261–268

Buscot F (2005) What are soils? In: Buscot F, Varma A (eds) Microorganisms in soils: roles in genesis and functions. Springer, Berlin Heidelberg New York, pp 3–17

Bywater RP, Conde-Frieboes K (2005) Did life begin on the beach? Astrobiology 5:568–574

Cabello P, Roldán MD, Moreno-Vivián C (2004) Nitrate reduction and the nitrogen cycle in archaea. Microbiology 150:3527–3546

Canbäck B, Tamas I, Andersson SGE (2004) A phylogenomic study of endosymbiotic bacteria. Mol Biol Evol 21:1110–1122

Canfield DE (2006) Gas with an ancient history. Nature 440:426–427

Canfield DE, Habicht KS, Thamdrup B (2000) The Archean sulfur cycle and the early history of atmospheric oxygen. Science 288:658–661

Carroll RL (2001) The origin and early radiation of terrestrial vertebrates. J Paleont 75:1202–1213

Castresana J, Moreira D (1999) Respiratory chains in the last common ancestor of living organisms. J Mol Evol 49:453–460

Catling DC, Zahnle KJ, McKay C (2001) Biogenic methane, hydrogen escape, and the irreversible oxidation of early Earth. Science 293:839–843

Ciccarelli FD, Doerks T, von Mering C, Creevey CJ, Snel B, Bork P (2006) Toward automatic reconstruction of a highly resolved tree of life. Science 311:1283–1287

Crawford JW, Harris JA, Ritz K, Young IM (2005) Towards an evolutionary ecology of life in soil. Trends Ecol Evol 20:81–87

Crispim CA, Gaylarde PM, Gaylarde CC (2003) Algal and cyanobacterial biofilms on calcareous historic buildings. Curr Microbiol 46:79–82

Davey ME, O'Toole GA (2000) Microbial biofilms: from ecology to molecular genetics. Microbiol Mol Biol Rev 64:847–867

Davis EL, Mitchum MG (2005) Nematodes. Sophisticated parasites of legumes. Plant Physiol 137:1182–1188

de Boer W, Folman LB, Summerbell RC, Boddy L (2005) Living in a fungal world: Impact of fungi on soil bacterial niche development. FEMS Microbiol Rev 29:795–811

Denison RF, Kiers ET (2004) Lifestyle alternatives for rhizobia: mutualism, parasitism, and forgoing symbiosis. FEMS Microbiol Lett 237:187–193

Douzery EJP, Snell EA, Bapteste E, Delsuc F, Philippe H (2004) The timing of eukaryotic evolution: does a relaxed molecular clock reconcile proteins and fossils? Proc Natl Acad Sci USA 101:15386–15391

Egert M, Wagner B, Lemke T, Brune A, Friedrich MW (2003) Microbial community structure in midgut and hindgut of the humus-feeding larva of *Pachnoda ephippiata* (Coleoptera: Scarabaeidae). Appl Environ Microbiol 69:6659–6668

Eldredge N, Gould SJ (1972) Punctuated equilibria: an alternative to phyletic gradualism. In: Schopf TJM (ed) Models in paleobiology. Freeman, Cooper, San Francisco, pp 82–115

Elshahed MS, Senko JM, Najar FZ et al. (2003) Bacterial diversity and sulfur cycling in a mesophilic sulfide-rich spring. Appl Environ Microbiol 69:5609–5621

Endre G, Kereszt A, Kevei Z, Mihacea S, Kaló P, Kiss GB (2002) A receptor kinase gene regulating symbiotic nodule development. Nature 417:962–966

Ferguson GP, Datta A, Baumgartner J, Roop RM II, Carlson RW, Walker GC (2004) Similarity to peroxisomal-membrane protein family reveals that *Sinorhizobium* and *Brucella* BacA affect lipid-A fatty acids. Proc Natl Acad Sci USA 101:5012–5017

ffrench-Constant R, Waterfield N, Daborn P et al. (2003) *Photorhabdus*: towards a functional genomic analysis of a symbiont and pathogen. FEMS Microbiol Rev 26:433–456

Friedrich CG, Rother D, Bardischewsky F, Quentmeier A, Fischer J (2001) Oxidation of reduced inorganic sulfur compounds by bacteria: emergence of a common mechanism? Appl Environ Microbiol 67:2873–2882

Gargas A, DePriest PT, Grube M, Tehler A (1995) Multiple origins of lichen symbioses in fungi suggested by SSU rDNA phylogeny. Science 268:1492–1495

Gaylarde CC, Gaylarde PM (2005) A comparative study of the major microbial biomass of biofilms on exteriors of buildings in Europe and Latin America. Int Biodeterior Biodegrad 55:131–139

Genre A, Chabaud M, Timmers T, Bonfante P, Barker DG (2005) Arbuscular mycorrhizal fungi elicit a novel intracellular apparatus in *Medicago truncatula* root epidermal cells before infection. Plant Cell 17:3489–3499

Gevertz D, Telang AJ, Voordouw G, Jenneman GE (2000) Isolation and characterization of strains CVO and FWKO B, two novel nitrate-reducing, sulfide-oxidizing bacteria isolated from oil field brine. Appl Environ Microbiol 66:2491–2501

Gilbert OM, Foster KR, Mehdiabadi NJ, Strassmann JE, Queller DC (2007) High relatedness maintains multicellular cooperation in a social amoeba by controlling cheater mutants. Proc Natl Acad Sci USA 104:8913–8917

Giraud E, Moulin L, Vallenet D et al. (2007) Legumes symbioses: absence of *nod* genes in photosynthetic bradyrhizobia. Science 316:1307–1312

Gorbushina AA, Krumbein WE (2005) Role of microorganisms in wear down of rocks and minerals. In: Buscot F, Varma A (eds) Microorganisms in soils: roles in genesis and functions. Springer, Berlin Heidelberg New York, pp 59–84

Grassineau NV, Nisbet EG, Bickle MJ et al. (2001) Antiquity of the biological sulphur cycle: evidence from sulphur and carbon isotopes in 2700 million-year-old rocks of the Belingwe Belt, Zimbabwe. Proc R Soc Lond B 268:113–119

Guerrero R, Piqueras M, Berlanga M (2002) Microbial mats and the search for minimal ecosystems. Int Microbiol 5:177–188

Guest H (1994) Discovery of the heliobacteria. Photosynth Res 41:17–21

Güimil S, Chang H-S, Zhu T et al. (2005) Comparative transcriptomics of rice reveals an ancient pattern of response to microbial colonization. Proc Natl Acad Sci USA 102:8066–8070

Gutzmer J, Beukes NJ (1998) Earliest laterites and possible evidence for terrestrial vegetation in the Early Proterozoic. Geology 26:263–266

Haaijer SCM, Van der Welle MEW, Schmid MC, Lamers LPM, Jetten MSM, Op den Camp HJM (2006) Evidence for the involvement of betaproteobacterial *Thiobacilli* in the nitrate-dependent oxidation of iron sulfide minerals. FEMS Microbiol Ecol 58:439–448

Harb OS, Gao L-Y, Kwaik YA (2000) From protozoa to mammalian cells: a new paradigm in the life cycle of intracellular bacterial pathogens. Environ Microbiol 2:251–265

Harrison MJ (2005) Signaling in the arbuscular mycorrhizal symbiosis. Annu Rev Microbiol 59:19–42

Hawksworth DL (1988) The fungal partner. In: Galun M (ed) CRC handbook of lichenology. CRC Press, Boca Raton, FL., pp 35–38

He J, Baldini RL, Deziel E et al. (2004a) The broad host range pathogen *Pseudomonas aeruginosa* strain PA14 carries two pathogenicity islands harboring plant and animal virulence genes. Proc Natl Acad Sci USA 101:2530–2535

He SY, Nomura K, Whittam TS (2004b) Type III protein secretion mechanism in mammalian and plant pathogens. Biochim Biophys Acta 1694:181–206

Heckman DS, Geiser DM, Eidell BR, Stauffer RL, Kardos NL, Hedges SB (2001) Molecular evidence for the early colonization of land by fungi and plants. Science 293:1129–1133

Hedges SB (2002) The origin and evolution of model organisms. Nat Rev Genet 3:838–849

Hentschel U, Steinert M, Hacker J (2000) Common molecular mechanisms of symbiosis and pathogenesis. Trends Microbiol 8:226–231

Herzenberg LA, Herzenberg LA (1989) Toward a layered immune system. Cell 59:953–954

Hirsch AM, Lum MR, Downie JA (2001) What makes the rhizobia-legume symbiosis so special? Plant Physiol 127:1484–1492

Hoeft SE, Kulp TR, Stolz JF, Hollibaugh JT, Oremland RS (2004) Dissimilatory arsenate reduction with sulfide as electron donor: experiments with Mono Lake water and isolation of strain MLMS-1, a chemoautotrophic arsenate respirer. Appl Environ Microbiol 70:2741–2747

Holland HD (2002) Volcanic gases, black smokers, and the great oxidation event. Geochim Cosmochim Acta 66:3811–3826

Hollibaugh JT, Budinoff C, Hollibaugh RA, Ransom B, Bano N (2006) Sulfide oxidation coupled to arsenate reduction by a diverse microbial community in a soda lake. Appl Environ Microbiol 72:2043–2049

Horn MA, Drake HL, Schramm A (2006) Nitrous oxide reductase genes (*nosZ*) of denitrifying microbial populations in soil and the earthworm gut are phylogenetically similar. Appl Environ Microbiol 72:1019–1026

Horodyski RJ, Knauth LPS (1994) Life on land in the Precambrian. Science 263:494–498

Hughes KA, Lawley B (2003) A novel Antarctic microbial endolithic community within gypsum crusts. Environ Microbiol 5:555–565

Jackson CR, Dugas SL (2003) Phylogenetic analysis of bacterial and archaeal *arsC* gene sequences suggests an ancient, common origin for arsenate reductase. BMC Evol Biol 3:18

Janssen PH (2006) Identifying the dominant soil bacterial taxa in libraries of 16S rRNA and 16S rRNA genes. Appl Environ Microbiol 72:1719–1728

Jargeat P, Cosseau C, Ola'h B et al. (2004) Isolation, free-living capacities, and genome structure of "*Candidatus* Glomeribacter gigasporarum," the endocellular bacterium of the mycorrhizal fungus *Gigaspora margarita*. J Bacteriol 186:6876–6884

Karandashov V, Bucher M (2005) Symbiotic phosphate transport in arbuscular mycorrhizas. Trends Plant Sci 10:22–29

Kasting JF, Catling D (2003) Evolution of a habitable planet. Annu Rev Astron Astrophys 41:429–463

Kelly DP, Shergill JK, Lu W-P, Wood AP (1997) Oxidative metabolism of inorganic sulfur compounds by bacteria. Antonie van Leeuwenhoek 71:95–107

Kennedy M, Droser M, Mayer LM, Pevear D, Mrofka D (2006) Late Precambrian oxygenation; inception of the clay mineral factory. Science 311:1446–1449

Kistner C, Winzer T, Pitzschke A et al. (2005) Seven *Lotus japonicus* genes required for transcriptional reprogramming of the root during fungal and bacterial symbiosis. Plant Cell 17:2217–2229

Klasson L, Andersson SGE (2004) Evolution of minimal-gene-sets in host-dependent bacteria. Trends Microbiol 12:37–43

Kluge M (2002) A fungus eats a cyanobacterium: the story of the *Geosiphon pyriformis* endocyanosis. Biol Environ 102B:11–14

Kooijman SALM, Hengeveld R (2005) The symbiontic nature of metabolic evolution. In: Reydon TAC, Hemerik L (eds) Current themes in theoretical biology. Springer, Berlin Heidelberg New York, pp 159–202

Kühl M, Jørgensen BB (1992) Microsensor measurements of sulfate reduction and sulfide oxidation in compact microbial communities of aerobic biofilms. Appl Environ Microbiol 58:1164–1174

Kurland CG, Andersson SGE (2000) Origin and evolution of the mitochondrial proteome. Microbiol Mol Biol Rev 64:786–820

Kurland CG, Collins LJ, Penny D (2006) Genomics and the irreducible nature of eukaryote cells. Science 312:1011–1014

LeVier K, Phillips RW, Grippe VK, Roop RM II, Walker GC (2000) Similar requirements of a plant symbiont and a mammalian pathogen for prolonged intracellular survival. Science 287:2492–2493

Lévy J, Bres C, Geurts R et al. (2004) A putative Ca^{2+} and calmodulin-dependent protein kinase required for bacterial and fungal symbioses. Science 303:1361–1364

Liengen T (1999) Environmental factors influencing the nitrogen fixation activity of free-living terrestrial cyanobacteria from a high arctic area, Spitsbergen. Can J Microbiol 45:573–581

Lloret L, Martínez-Romero E (2005) Evolución y filogenia de *Rhizobium*. Rev Latinoameric Microbiol 47:43–60

Lodwig EM, Hosie AHF, Bourdes A et al. (2003) Amino-acid cycling drives nitrogen fixation in the legume-*Rhizobium* symbiosis. 422:722–126

Lutzoni F, Pagel M (1997) Accelerated evolution as a consequence of transitions to mutualism. Proc Natl Acad Sci USA 94:11422–11427

Lutzoni F, Pagel M, Reeb V (2001) Major fungal lineages are derived from lichen symbiotic ancestors. Nature 411:937–940

Lynch M (2006) The origins of eukaryotic gene structure. Mol Biol Evol 23:450–468

Mancinelli RL, McKay CP (1988) The evolution of nitrogen cycling. Origins Life Evol Biosphere 18:311–325

Margulis L (1970) Origin of eukaryotic cells. Yale University Press, New Haven

Megonigal JP, Hines ME, Visscher PT (2005) Anaerobic metabolism: linkages to trace gases and aerobic processes. In: Schlesinger WH (ed) Biogeochemistry, vol 8, Treatise on geochemistry (eds Holland HD, Turekian KK). Elsevier-Pergamon, Oxford

Miyamoto T, Kawahara M, Minamisawa K (2004) Novel endophytic nitrogen-fixing clostridia from the grass *Miscanthus sinensis* as revealed by terminal restriction fragment length polymorphism analysis. Appl Environ Microbiol 70:6580–6586

Molmeret M, Horn M, Wagner M, Santic M, Abu Kwaik Y (2005) Amoebae as training grounds for intracellular bacterial pathogens. Appl Environ Microbiol 71:20–28

Moran NA, Plague GR (2004) Genomic changes following host restriction in bacteria. Curr Opin Genet Dev 14:627–633

Morris CE, Monier J-M (2003) The ecological significance of biofilm formation by plant-associated bacteria. Annu Rev Phytopathol 41:429–453

Müller-Hill B (2006) What is life? The paradigm of DNA and protein cooperation at high local concentrations. Mol Microbiol 60:253–255

Neumeister B, Reiff G, Faigle M, Dietz K, Northoff H, Lang F (2000) Influence of *Acanthamoeba castellanii* on intracellular growth of different *Legionella* species in human monocytes. Appl Environ Microbiol 66:914–919

Nisbet EG, Fowler CMR (1999) Archaean metabolic evolution of microbial mats. Proc R Soc Lond B 266:2375–2382

Nisbet EG, Sleep NH (2001) The habitat and nature of early life. Nature 409:1083–1091

Nürnberger T, Brunner F, Kemmerling B, Piater L (2004) Innate immunity in plants and animals: Striking similarities and obvious differences. Immunol Rev 198:249–266

Ochman H, Moran NA (2001) Genes lost and genes found: evolution of bacterial pathogenesis and symbiosis. Science 292:1096–1099

Oremland RS, Stolz JF (2003) The ecology of arsenic. Science 300:939–944

Ormerod JG, Kimble LK, Nesbakken T, Torgersen YA, Woese CR, Madigan MT (1996) *Heliophilum fasciatum* gen. nov. sp. nov. and *Heliobacterium gestii* sp. nov.: Endospore-forming heliobacteria from rice field soils. Arch Microbiol 165:226–234

Padan E (1979) Facultative anoxygenic photosynthesis in cyanobacteria. Annu Rev Plant Physiol 30:27–40

Paerl HW, Pinckney JL (1996) A mini-review of microbial consortia: their roles in aquatic production and biogeochemical cycling. Microb Ecol 31:225–247

Parniske M (2000) Intracellular accommodation of microbes by plants: a common developmental program for symbiosis and disease? Curr Opin Plant Biol 3:320–328

Peters V, Conrad R (1995) Methanogenic and other strictly anaerobic bacteria in desert soil and other oxic soils. Appl Environ Microbiol 61:1673 1676

Phillips DA, Ferris H, Cook DR, Strong DR (2003) Molecular control points in rhizosphere food webs. Ecology 84:816–826

Pisani D, Poling L, Lyons-Weiler M, Hedges S (2004) The colonization of land by animals: molecular phylogeny and divergence times among arthropods. BMC Biol 2:1. DOI: 10.1186/1741-7007-2-1

Pitman AR, Jackson RW, Mansfield JW, Kaitell V, Thwaites R, Arnold DL (2005) Exposure to host resistance mechanisms drives evolution of bacterial virulence in plants. Curr Biol 15:2230–2235

Ramey BE, Koutsoudis M, von Bodman SB, Fuqua C (2004) Biofilm formation in plant-microbe associations. Curr Opin Microbiol 7:602–609

Raven JA (2002) The evolution of cyanobacterial symbioses. Biol Environ 102B:3–6

Raven JA, Edwards D (2001) Roots: evolutionary origins and biogeochemical significance. J Exp Bot 52:381–401

Raymond J, Segrè D (2006) The effect of oxygen on biochemical networks and the evolution of complex life. Science 311:1764–1767

Read DJ, Duckett JG, Francis R, Ligrone R, Russell A (2000) Symbiotic fungal associations in "lower" land plants. Phil Trans R Soc Lond B 355:815–831

Retallack GJ (2001) Soils of the past. An introduction to paleopedology, 2nd edn. Blackwell Science, Oxford

Retallack GJ (2003) Soils and global change in the carbon cycle over geological time. In: Drever JI (ed) Treatise on geochemistry, vol 5. Elsevier, Amsterdam, pp 581–605

Rippka R (1988) Recognition and identification of cyanobacteria. Method Enzymol 167:28–67

Roossinck MJ (2005) Symbiosis versus competition in plant virus evolution. Nat Rev Micro 3:917–924

Rudgers JA, Mattingly WB, Koslow JM (2005) Mutualistic fungus promotes plant invasion into diverse communities. Oecologia 144:463–471

Rye R, Holland HD (1998) Paleosols and the evolution of atmospheric oxygen: a critical review. Am J Sci 298:621–672

Saffo MB (2002) Themes from variation: probing the commonalities of symbiotic associations. Integr Comp Biol 42:291–294

Saikkonen K, Wäli P, Helander M, Faeth SH (2004) Evolution of endophyte-plant symbioses. Trends Plant Sci 9:275–280

Sánchez-Baracaldo P, Hayes PK, Blank CE (2005) Morphological and habitat evolution in the Cyanobacteria using a compartmentalization approach. Geobiology 3:145–165

Scholl E, Thorne J, McCarter J, Bird DMcK (2003) Horizontally transferred genes in plant-parasitic nematodes: a high-throughput genomic approach. Genome Biol 4:R39

Schulz B, Boyle C (2005) The endophytic continuum. Mycol Res 109:661–686

Schüßler A (2000) *Glomus claroideum* forms an arbuscular mycorrhiza-like symbiosis with the hornwort *Anthoceros punctatus*. Mycorrhiza 10:15–21

Schüßler A (2002) Molecular phylogeny, taxonomy, and evolution of *Geosiphon pyriformis* and arbuscular mycorrhizal fungi. Plant Soil 244:75–83

Schüßler A, Schwarzott D, Walker C (2001) A new fungal phylum, the *Glomeromycota*: phylogeny and evolution. Mycol Res 105:1413–1421

Selosse M-A, Le Tacon F (1998) The land flora: a phototroph-fungus partnership? Trends Ecol Evolut 13:15–20

Shen Y, Buick R, Canfield DE (2001) Isotopic evidence for microbial sulphate reduction in the early Archaean era. Nature 410:77–81

Sprenger W, van Belzen M, Rosenberg J, Hackstein J, Keltjens J (2000) *Methanomicrococcus blatticola* gen. nov., sp. nov., a methanol- and methylamine-reducing methanogen from the hindgut of the cockroach *Periplaneta americana*. Int J Syst Evol Microbiol 50:1989–1999

Stacey G, Libault M, Brechenmacher L, Wan J, May GD (2006) Genetics and functional genomics of legume nodulation. Curr Opin Plant Biol 9:110–121

Stevenson AK, Kimble LK, Woese CR, Madigan MT (1997) Characterization of new phototrophic heliobacteria and their habitats. Photosynth Res 53:1–12

Stracke S, Kistner C, Yoshida S et al. (2002) A plant receptor-like kinase required for both bacterial and fungal symbiosis. Nature 417:959–962

Strobel G (2006) *Muscodor albus* and its biological promise. J Indust Microbiol Biotechnol 33:514–522

Szczyglowski K, Amyot L (2003) Symbiosis, inventiveness by recruitment? Plant Physiol 131:935–940

Tamas I, Klasson L, Canback B et al. (2002) 50 million years of genomic stasis in endosymbiotic bacteria. Science 296:2376–2379

Teske A, Dhillon A, Sogin ML (2003) Genomic markers of ancient anaerobic microbial pathways: sulfate reduction, methanogenesis, and methane oxidation. Biol Bull 204:186–191

Thompson JN, Nuismer SL, Gomulkiewicz R (2002) Coevolution and maladaptation. Integr Comp Biol 42:381–387

Tiedje J, Sexstone A, Parkin T, Revsbech N (1984) Anaerobic processes in soil. Plant Soil 76:197–212

Tokuda G, Yamaoka I, Noda H (2000) Localization of symbiotic clostridia in the mixed segment of the termite *Nasutitermes takasagoensis* (Shiraki). Appl Environ Microbiol 66:2199–2207

Tomov AT, Tsvetkova ED, Tomova IA, Michailova LI, Kassovski VK (1999) Persistence and multiplication of obligate anaerobe bacteria in amoebae under aerobic conditions. Anaerobe 5:19–23

Towe KM, Catling D, Zahnle K, McKay C (2002) The problematic rise of Archean oxygen. Science 295:1419–1421

Ueno Y, Yamada K, Yoshida N, Maruyama S, Isozaki Y (2006) Evidence from fluid inclusions for microbial methanogenesis in the early Archaean era. Nature 440:516–519

Velicer GJ (2003) Social strife in the microbial world. Trends Microbiol 11:330–337

Walker JJ, Spear JR, Pace NR (2005) Geobiology of a microbial endolithic community in the Yellowstone geothermal environment. Nature 434:1011–1014

Waller F, Achatz B, Baltruschat H et al. (2005) The endophytic fungus *Piriformospora indica* reprograms barley to salt-stress tolerance, disease resistance, and higher yield. Proc Natl Acad Sci USA 102:13386–13391

Warscheid T, Braams J (2000) Biodeterioration of stone: a review. Int Biodeterior Biodegrad 46:343–368

Watanabe Y, Martini JEJ, Ohmoto H (2000) Geochemical evidence for terrestrial ecosystems 2.6 billion years ago. Nature 408:574–578

Weerasinghe RR, Bird DMcK, Allen NS (2005) Root-knot nematodes and bacterial Nod factors elicit common signal transduction events in *Lotus japonicus*. Proc Natl Acad Sci USA 102:3147–3152

Wellman CH, Osterloff PL, Mohiuddin U (2003) Fragments of the earliest land plants. Nature 425:282–285

Wernegreen J, Moran N (1999) Evidence for genetic drift in endosymbionts (*Buchnera*): analyses of protein-coding genes. Mol Biol Evol 16:83–97

Winiecka-Krusnella J, Linder E (2001) Bacterial infections of free-living amoebae. Res Microbiol 152:613–619

Woese CR (2002) On the evolution of cells. Proc Natl Acad Sci USA 99:8742–8747

Wolf YI, Rogozin IB, Grishin NV, Tatusov RL, Koonin EV (2001) Genome trees constructed using five different approaches suggest new major bacterial clades. BMC Evol Biol 1:8

Wren BW (2000) Microbial genome analysis: insights into virulence, host adaptation and evolution. Nat Rev Genet 1:30–39

Xiong J, Fischer WM, Inoue K, Nakahara M, Bauer CE (2000) Molecular evidence for the early evolution of photosynthesis. Science 289:1724–1730

Yoneyama K, Takeuchi Y, Yokota T (2001) Production of clover broomrape seed germination stimulants by red clover root requires nitrate but is inhibited by phosphate and ammonium. Physiol Planta 112:25–30

Young IM, Crawford JW (2004) Interactions and self-organization in the soil-microbe complex. Science 304:1634–1637

Young IM, Ritz K (2000) Tillage, habitat space and function of soil microbes. Soil Tillage Res 53:201–213

Yuan X, Xiao S, Taylor TN (2005) Lichen-like symbiosis 600 million years ago. Science 308:1017–1020

Zhu J, Oger PM, Schrammeijer B, Hooykaas PJJ, Farrand SK, Winans SC (2000) The bases of crown gall tumorigenesis. J Bacteriol 182:3885–3895

Chapter 4
Rhizosphere Colonization: Molecular Determinants from Plant-Microbe Coexistence Perspective

Chandra Shekhar Nautiyal(✉), Suchi Srivastava, and Puneet Singh Chauhan

4.1 Introduction

Inoculation of plants with beneficial microbes can be traced back for centuries. Although bacteria were not proven to exist until in 1683 von Leeuwenhoek discovered microscopic 'animals' under the lens of his microscope, their utilisation to stimulate plant growth in agriculture has been exploited since ancient times. Theophrastus (372–287 BC) suggested the mixing of different soils as a means of 'remedying defects and adding heart to the soil' (Vessey 2003). Colonisation of plant root system is the very first step in nearly all interactions between plants and soil borne microbes. The importance of this compartment for plant growth and soil microbiology had already been realised in the very pioneer times of microbiology in the late 19th century. In 1888, Hellriegel and Wilfarth proved the special case of nitrogen nutrition of legumes through their root nodule bacteria, which Beijerinck finally isolated in 1889. In 1887–1888 Hiltner was certainly fascinated by the findings of Hellriegel and Wilfarth about the special case of nitrogen nutrition in legumes. Together with Professor Nobbe, intensive studies were conducted about the nature of the symbiotic interaction of nodule bacteria and legume roots (Nobbe and Hiltner 1896). Apparently during these studies, Hiltner became aware of the importance of the ecological interactions in the root zone. Together with his teacher he developed the first inoculant based on root nodule bacteria for agricultural practice which they called 'Nitragin' in 1890. As agricultural practice was in close contact to the new achievements of basic sciences, the challenge quickly arose whether at all, and how, these discoveries of the – at that time – very young science 'soil bacteriology' could be applied in the field. Region of contact between root and soil where soil is affected by roots was designated as "rhizosphere" by Hiltner (1904).

The rhizosphere of plants is one of the most fascinating microbial habitats for basic and applied studies in the field of microbiology, as it is shaped by the soil, the

---

C.S. Nautiyal

National Botanical Research Institute, Rana Pratap Marg, Lucknow 226001, India

e-mail: csn@nbri.res.in, nautiyalnbri@lycos.com

C.S. Nautiyal, P. Dion (eds.) *Molecular Mechanisms of Plant and Microbe Coexistence.* Soil Biology 15, DOI: 10.1007/978-3-540-75575-3

© Springer-Verlag Berlin Heidelberg 2008

plant and the microorganisms. Given the tremendous diversity of soil microbes, soil fauna, and plants, it is virtually impossible to investigate the intricacies of every potential rhizosphere interaction in every environmental circumstance. However, an understanding of controls over belowground function is becoming increasingly important as natural and agro-ecosystems around the globe are exposed to anthropogenic pressures. In addition, the chemistry and development of soil present today have been strongly affected by the actions of rhizospheres over evolutionary time frames and the evolution of true plant roots and their extension deep into substrate is hypothesised to have led to a revolution in planetary carbon and water cycling during the Devonian period (Beerling and Berner 2005; Cardon and Gage 2006). What is the biogeochemical function of the rhizosphere on Earth today? In what major ways is rhizosphere function belowground similar across terrestrial ecosystems, and in what fundamental ways can it differ? However, since the microbial inoculations would mainly be performed in soils before the plant is grown up, the strains should also be able to survive in the soil and show a good saprophytic ability. To fulfill these requirements, progress must be made in our knowledge of which bacterial traits affect the soil- and rhizosphere-colonising ability of microbes (Nautiyal 2006; Nautiyal et al. 2006a,b).

This review focuses on methods to improve root colonisation by introduced bacteria using molecular biology strategy and tools to enhance plant-microbe coexistence. The plausible mechanisms adopted by these rhizobacteria in root colonisation, though abundantly documented but still remains to be fully explored. It is thus presumed that better understanding of the molecular basis of plant-rhizobacteria coexistence will enable researchers to make more informed decisions in designing inoculants that will successfully colonise host rhizosphere and consistently promote the growth of host plants, biological control and bioremediation.

4.2 Genetic Regulation of Plant–Microbe Association

For most of the past century, microbiology has been devoted to studying the physiology and genetics of bacteria in vitro. This means that we now understand a great deal about the lives of bacteria on agar plates, but have little understanding of the determinants of ecological success in the wild. Without an understanding of the causes of ecological success our understanding of the biology of bacteria will remain incomplete and the full potential of these organisms in biotechnology will remain unrealised. Bottom-up (genes to population) and top-down (population to genes) approaches are both useful. The bottom-up approach is commonly employed for studies of bacteria, although is rarely pursued to the population level. Bottom-up positive selection approaches open the door to understand ecological performance in bacteria at a mechanistic level in the wild. In vivo expression technology (IVET) strategies, despite their tremendous power, have been little used, except in studies of animal pathogenicity. Rainey and Preston (2000) have reviewed IVET strategies, their development, and potential in biotechnology. Using 22,800 genechip probe array of FPT9601-T5-treated *Arabidopsis* plants in comparison to control plants, Tosa et al. (2005) reported on up- or down-regulation of 95 and 105 genes respectively. Up-regulated ones include genes involved in metabolism, signal transduction, and stress response, putative auxin-regulated genes and nodulin-like genes and some ethylene-responsive genes were down-regulated. Isabel et al. (2005) identified 28 *rap* genes (*root-activated promoters*) preferentially expressed in the maize rhizosphere during rhizosphere colonisation by *P. putida* KT2440. IVET is a powerful tool to identify promoters and genes that are expressed specifically under certain conditions of interaction. Among modern techniques for the study of rhizosphere colonisation in inoculated and uninoculated controls microarray analysis besides IVET, differential display (DD) technique is also very useful. Differential display of genes is mRNA based technique where differentially expressed genes of plants infected with rhizobacteria resulted in enhanced expression of PR proteins (Hassan et al. 2003). In another cDNA-amplified fragment length polymorphism technique, 81 transcript derived fragments showing differential expression during exponential growth phase in an aggregation-inducing medium containing high C:N ratio. RT-PCR analyses confirmed the differential patterns observed by cDNA-AFLP of *nodQ*, involved in sulfation; *narK*, involved in nitrite/nitrate transport, and flp, involved in auto aggregation; as well as genes encoding a biopolymer transport protein, and the signal recognition particle (Valverde et al. 2006).

Correlations have been reported between rhizosphere competence and growth rate of a plant root adhered bacteria and it has a prerequisite to multiply using organic compounds and other physiological traits which may further contribute to their rhizosphere competence and these rhizosphere competent bacteria helps in increasing the plant productivity by multiple changing in the gene expression.

Auxin produced by bacteria in the rhizosphere can stimulate the activity of the 1-aminocyclopropane-1-carboxylate (ACC) synthase, an enzyme normally used by plants to form ethylene (Xie et al. 1996) and transcription of *ipdC*, an *Erwinia* IAA biosynthetic gene, is induced in response to bean and tobacco compounds (Brandl and Lindow 1997) and the bacterial auxin synthesis is dependent upon plant-exuded tryptophan. Analyses of the more than 4000 ORFs of *Bacillus subtilis* revealed that *yqkF* are growth regulatory gene related to auxin which may manipulate hormonal processes in plants (Andrews and Harris 2000). To control the ethylene production, increased amount of ACC is hydrolysed by an ACC deaminase and ACC deaminase-producing rhizobacteria upregulate genes involved in cell division and proliferation but down-regulate stress genes thus reducing plant stress and induce root elongation and proliferation in plants, largely by lowering ethylene levels (Glick et al. 1998; Penrose and Glick 2003) which bring about ISR by fortifying the physical and mechanical strength of the cell wall (Knoester et al. 1999). Ahn et al. 2002 – by showing augmented, rapid transcript accumulation of defense-related genes including PR-1a, phenylalanine ammonia-lyase (PAL), and 3-hydroxy-3-methylglutaryl CoA reductase (HMGR) following inoculation with PGPRs – concluded that PGPRs by a typical phenomenon of potentiation using the jasmonate pathway of defence provide ISR to the plants. Phenotypically ISR in plants is similar to pathogen-induced systemic acquired resistance (SAR) known defense-related genes, i.e. the SA-responsive genes PR-1, PR-2, and PR-5, the ethylene-inducible gene (Karen et al. 2005).

Plant-microbe coexistence is strongly influenced by abiotic stress conditions induced by drought, high salt and temperature. Stress is an unhealthy condition and prolonged stress weakens the immune system, opening the door for a variety of ailments that an otherwise healthy plant could overcome. The major consequences of drought stress is the loss of water from the protoplasm and leads to the concentration of ions in the protoplasm, higher concentration of which are toxic to plants. However, a lack of water results in the overproduction of free radicals, leading to the damage of cellular membranes and ultimately loss of solutes from the cell and organelles (Apel and Hirt 2004). Damage to DNA under these conditions severely hinders the ability of the plant to recover, as DNA stores the genetic information that is ultimately used for the synthesis of new proteins (Fig. 4.1).

Drought is one of the most serious world-wide problems for agriculture. Four-tenths of the world's agricultural land lies in arid or semi-arid regions and many variables play a part in reaching drought conditions; these include lack of natural rainfall, soil type, air temperature, humidity, wind conditions, sun exposure and also the plant type (root depth). Plants respond to these conditions with an array of biochemical and physiological adaptations and being sessile they develop tolerance mechanisms to cope with the detrimental effects of environmental stress. They evolved some of the genes coding for an antioxidant enzyme to protect DNA against free radicals, cell membrane stabilizing enzymes and a protein which is thought to stabilize osmotic imbalances by actively transporting solutes across the cell membranes, thereby minimizing water loss during drought.

Timmusk and Wagner (1999) reported that the *Arabidopsis thaliana* plants inoculated with *Paenibacillus polymyxa* plants were more resistant to drought stress and revealed that ERD 15 and RAB 18 is a drought responsive genes and gets differentially expressed in case of *P. polymyxa* treated plants in drought and its mechanism of action of the plant growth promotion was studied by Timmusk et al. (2003). Using plasmid-borne *gfp* gene tagging of *P. polymyxa,* Timmusk et al. (2005) concluded that it colonises on root and form biofilm on plant root tips and enters intercellular spaces but does not spread throughout the plant.

Adaptation to environmental stresses is dependent upon the activation of cascades of molecular networks involved in stress perception, signal transduction, activation and regulation of specific stress tolerant genes. During stress conditions dehydration responsive transcription factor after binding with the cis acting elements promotes both, the stress inducible genes codes for osmoprotectants, scavengers or stress proteins such as cold responsive proteins (COR) or late embryogenesis abundant (LEA) with an undefined mechanism of action (Vinocur and Altman 2005) and regulatory proteins such as transcription factors or components of signal cascades and regulate the expression of a set of genes involved in stress. Both categories of genes have been shown to impart tolerance when overexpressed in plants. Significant improvement of stress tolerance in case of *A. thaliana* has been noticed by the interaction of a single transcription factor, which activates

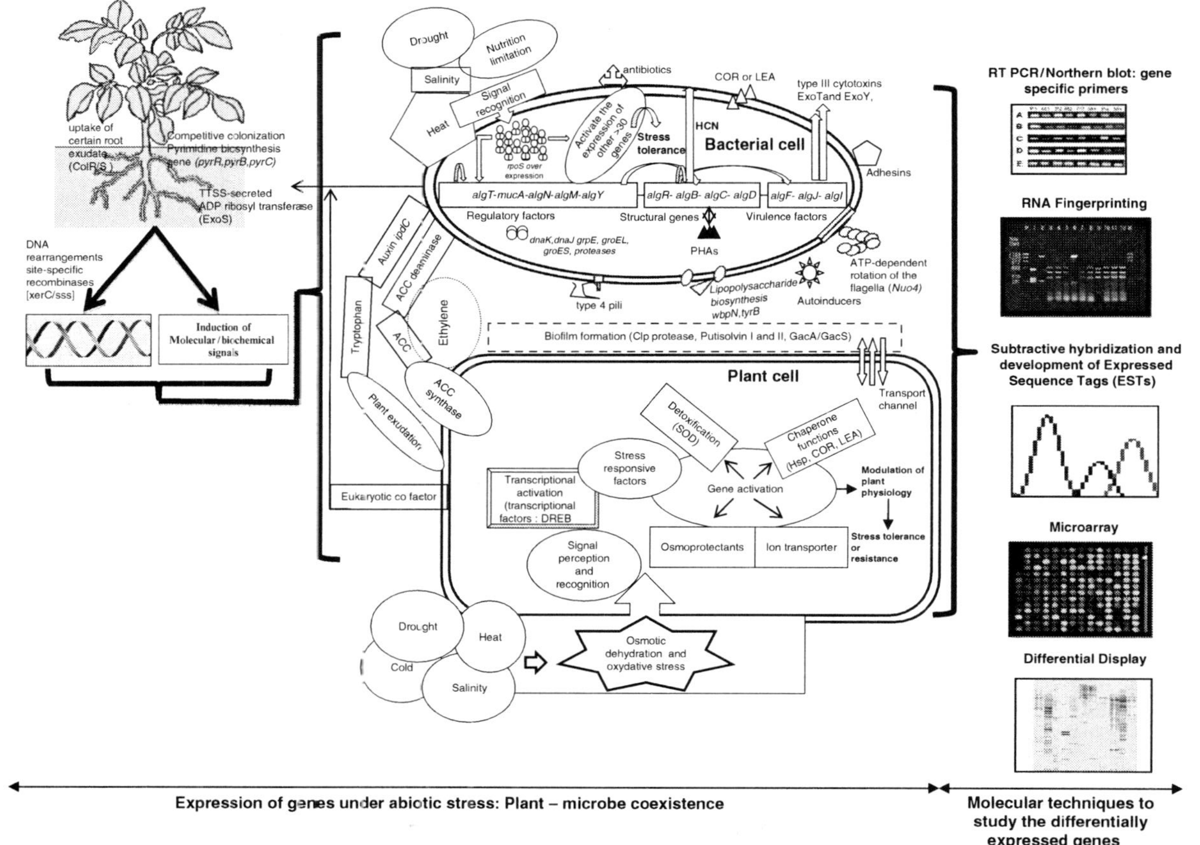

Fig. 4.1 Abiotic stress inducible gene expression in rhizobacteria

expression of downstream genes involved in drought- and salt-stress (Sakuma et al. 2006). Huang and Liu (2006) reported a novel cDNA encoding DRE-binding transcription factor, designated GhDBP3 from *Gossypium hirsutum*, showed enhanced expression by drought, NaCl, low temperature and ABA treatment. During different environmental stresses, basic cellular processes such as DNA replication, repair, recombination, transcription, ribosome biogenesis and translation initiation play essential role and all these functions are genetically controlled by helicases of DEAD-box protein superfamily. Thus, helicases might playing an important role in regulating plant growth and development under stress conditions by regulating some stress-induced pathways (Vashisht and Tuteja 2006). Kwak et al. (2007) reported that *A. thaliana* genome contains eight genes of high mobility group B (HMGB) proteins, and gets upregulated by cold salt and drought stress and down-regulated by drought or salt stress. Various strategies have been used to produce transgenic plants with increased tolerance to dehydration stress. These include the overproduction of enzymes responsible for biosynthesis of osmolytes, late-embryogenesis-abundant proteins and detoxification enzymes. However, in case of crop improvement regulatory gene would have potential to play a broader role in stress tolerance and still a careful appraisal for the selection of the genes is expected (Chinnusamy et al. 2004; Sreenivasulu et al. 2007).

4.3 Genomics and Proteomics of Plant–Microbe Coexistence

The ability of microbes to exploit water resources that are less available to plants helps in buffering the soil microbial community from stress as during stressed physiological conditions these rhizomicrobes enter a state of reduced metabolic activity and go for different morphological changes (Kulkarni and Nautiyal 1999). At the onset of starvation, a regulatory response leads to the enhanced expression of particular metabolic genes and almost all virulence gene of functions such as plant cell-wall-degrading enzymes, phytotoxins, ice nucleation activity, exopoly-saccharide production, and the type III protein secretion machinery of plant pathogenic bacteria exhibit increased transcription at temperatures well below the respective growth optima (Angela et al. 2001).

Several genes have been reported to be up- or down-regulated in response to different stresses in PGPR (Fig. 4.1). These genes might generate products either directly involved in protection against environmental stress or that play a role in stress regulation. Polyols such as glycerol, mannitol, sorbitol, sucrose and quaternary ammonium compounds such as glycine betaine, proline, betaine, β-alanine betaine, choline *O*-sulphate and the tertiary sulphonium compound dimethylsulphoniopropionate are effective osmoprotectants widely distributed in bacteria, marine algae and many plant families (Rathinasabapathi 2000). Lee et al. (2005) has isolated a novel strain *P. stutzeri* CJ38, that enabled direct transformation of maltose to trehalose and by the introduction of the yeast trehalose-6-phosphate synthase (TPS1) gene in tomato showed improved tolerance under drought, salt and oxidative stress suggested that carbohydrate alterations produced by trehalose biosynthesis is linked to the stress response and increased tolerance of abiotic stress, without decreasing productivity, under both stress and nonstress conditions through trehalose biosynthesis (Cortina and Culiáñez-Macià 2005).

Under certain stress conditions local changes in the superhelicity of DNA also induce or repress genes both at the level of transcription and translation (Dorman 1991). Transcriptional profiling of *P. aeruginosa* grown under steady-state hyperosmotic stress conditions showed an up-regulation of 116 and 81 genes at least threefold in cells grown in the presence of 0.3 M NaCl and 0.7 M sucrose, respectively. However, 66 genes showed a change in expression of at least threefold in response to both stressors 40 of which are associated with virulence factors, encoding proteins of a type III secretion system (TTSS), the type III cytotoxins ExoT and ExoY, and two ancillary chaperones (Aspedon et al. 2006) (Fig. 4.1).

Heat shock response, a universally conserved stress response is governed by the positive transcriptional control of the $\sigma32$ (*rpoH*) polypeptide. Functional genomic study of stress tolerance in *P. putida* by the global mRNA expression studies in 392 regulated genes showed that 36 genes were differentially expressed more than twofold and 32 genes of 23 operons are indispensable in response to abiotic conditions (Reva et al. 2006).

Laville et al. (1992) demonstrated that *P. fluorescens* CHA0 colonizes plant roots, produces several secondary metabolites in stationary growth phase, and suppresses a number of plant diseases, including *Thielaviopsis basicola*-induced black root rot of tobacco. Mutations in a *P. fluorescens* gene named *gacA* (for global antibiotic and cyanide control) pleiotropically block the production of the secondary metabolites 2,4-diacetylphloroglucinol (Phl), HCN, and pyoluteorin. The *gacA* mutants of strain CHA0 drastically reduced ability to suppress black root rot under gnotobiotic conditions, supporting the previous observations that the antibiotic Phl and HCN individually contribute to the suppression of black root rot. The *gacA* gene is directly followed by a *uvrC* gene. Double *gacA-uvrC* mutations render *P. fluorescens* sensitive to UV irradiation. The *gacA-uvrC* cluster is homologous to the orf-2 (= *uvrY*)-*uvrC* operon of *E. coli*. The *gacA* gene specifies a trans-active 24-kDa protein. Sequence data indicate that the GacA protein is a response regulator in the FixJ/DegU family of two-component regulatory systems. Expression of the *gacA* gene itself was increased in stationary phase. Thus Laville et al. (1992) proposed that GacA, perhaps activated by conditions of restricted growth, functions as a global regulator of secondary metabolism in *P. fluorescens*. Later Schnider et al. (1995) reported that CHA0 produces a variety of secondary metabolites, in particular the antibiotics pyoluteorin and 2,4-diacetylphloroglucinol, and protects various plants from diseases caused by soilborne pathogenic fungi. The *rpoD* gene encoding the housekeeping sigma factor sigma 70 of *P. fluorescens* was sequenced. The deduced RpoD protein showed 83% identity with RpoD of *P. aeruginosa* and 67% identity with RpoD of *E. coli*. Attempts to inactivate the single chromosomal *rpoD* gene of strain CHA0 were unsuccessful, indicating an essential role of this gene. When *rpoD* was carried by an IncP vector in strain CHA0, the production of both antibiotics was increased severalfold and, in parallel, protection of cucumber against disease caused by *P. ultimum* was improved, in comparison with strain CHA0.

At the onset of any stressed conditions bacteria ensure its survival in changing environment by its regulatory mechanisms. These regulatory mechanisms which make essential contributions to bacterial survival under stressed conditions are the alternative sigma factors – RpoS (σ^S) and RpoE (σ^{22}). In response to starvation for carbon, nitrogen, phosphate and amino acids cells leads to enter into stationary phase and the global regulator sigma S (*rpoS*) level induced and resulting in a partial reduction of the growth rate. Additional inducing conditions for *rpoS* are hyperosmolarity, high or low temperature, acidic pH and high cell density and probably all generate a common intracellular signal. Initially it seemed that a reduction or cessation of growth might be a signal. Using an *rpoS* mutant of a cosmopolitan strain *P. aeruginosa,* Jorgenson et al. (1999) proved that during exposure to different stressors such as hydrogen peroxide, high temperature, hyperosmolarity, low pH and ethanol, disruption of the *rpoS* gene resulted in a two- to threefold increase in the rate of kill of stationary-phase cells, thus alternative sigma factor encoded by the *rpoS* gene is a general stress response regulator playing important role for survival under stressful conditions by controlling the expression of genes which confer increased resistance to various stresses. A knockout mutant study during exposure to different stressors (hydrogen peroxide, high temperature, hyperosmolarity, low pH and ethanol) has proven that *rpoS* along with some other factors regulate the desiccation tolerance in *P. aeruginosa* (Jorgensen et al. 1999). By using *rpoS-lacZ* fusions studies it has been reported that *rpoS* transcription induced in late exponential phase and reach a peak upon entry into stationary phase (Benchamas et al. 2003) and during osmotic upshock *rpoS* mRNA expression remains steady, however the protein levels increase (Henge-Aronis 1996). Hulsmann et al. (2003) also reported that *rpoS* in *V. vulnificus* is important for adaptation to environmental changes and may have a role in virulence. The expression of *rpoS*, which encodes an RNA polymerase subunit results in a 30-fold induction at the onset of stressors/stationary phase and regulate the expression of more than 30 genes simultaneously (Fig. 4.1). Kazmierczak et al. (2005) reported that sigma factors provide promoter recognition specificity to the polymerase and contribute to DNA strand separation and then they dissociate from RNA polymerase core enzyme following transcription initiation of many genes as the regulon of a single sigma factor can be comprised of hundreds of genes, sigma factors provide effective mechanisms for simultaneously regulating large numbers of prokaryotic genes (Weber et al. 2005). Arden et al. (2006) – by microarray analysis of osmotically stressed *P. aeruginosa* – have also showed differential gene expression.

Expression of *rpoS* also regulates the production of several exoproducts to alter their environment for the survival in adverse conditions and to impart a role in virulence (Roberson and Firestone 1992; Sledjeski et al. 2001). Bacteria in many environments including soil generally live in colonies within a matrix largely composed of extracellular polysaccharides (EPS). During desiccation, maintenance of functional cell envelop is also a prerequisite for matric stress adaptation as it is exposed to the external environment and is host to various essential metabolic and structural features thus EPS envelope may protect bacteria from drying and fluctuations in water potential. Pseudomonads used to produce alginate as a main constituent of EPS towards desiccation and osmotic tolerance and which is controlled by the sigma factor AlgU (AlgT) (Keel et al. 2001). Induction of alginate biosynthesis genes and the production of alginate only under water-limited conditions, indicate that EPS production is a fitness trait for survival in low-water-content habitats. Alternatively, dehydration leads to improper folding of the transporters, which in turn activates expression of genes for their biosynthesis. Twenty-six genes homologues of *P. putida* KT2400 genome sequence have been identified for bacterial growth and survival in water-limited environments; these are involved in protein fate, nutrient or solute acquisition, energy generation, motility, alginate biosynthesis or cell envelope structure (Nelson et al. 2002).

P. aeruginosa is capable of synthesising polyhydroxyalkanoic acids (PHAs) and rhamnolipids, both of which are composed of 3-hydroxydecanoic acids connected by ester bonds and synthesising the biofilm matrix polymer alginate. Alginate is a biopolymer of D-glucouronic acid and mannouronic acid and is responsible for adherence, barrier to phagocytosis and neutralizes the oxygen radical. At normal and stress conditions comparison of mRNA concentrations shows difference in the expression of *algD*, the key gene leading to overproduction of alginate. After growth on 3% ethanol but not after heat-shock, an increase in *algD* mRNA levels and a corresponding decrease in *mucB* (a regulatory gene) mRNA levels were detected (Edwards and Saunders 2001). An isogenic knock-out mutants study on *P. aeruginosa* PAO1 and the alginate-overproducing *P. aeruginosa* FRD1 suggested that PHA biosynthesis and alginate biosynthesis are in competition with respect to a common precursor and the PAO1 PHA-negative mutant form a stable biofilm with large, distinct and differentiated microcolonies characteristic of alginate-overproducing strains of *P. aeruginosa* (Pham et al. 2004).

Alginate gene expression is transcriptionally controlled by a gene cluster at 68 min on the chromosome: *algT* (*algU*)-*mucA*-*mucB* (*algN*)-*mucC* (*algM*)-*mucD* (*algY*). The *algT* gene encodes a 22-kDa alternative sigma factor (σ22) that autoregulates its own as well as the promoters of *algR*, *algB*, and *algD* and two AlgR-binding sites, RB1 and RB2, located far upstream from the *algD* mRNA start site, are essential for the high-level activity of *algD* (Mohr et al. 1992; Mathee et al. 1997). The *rpoS* mutant impaired in transcription of *algD* showed that GacA play a role in controlling alginate production and gene expression during the stationary phase in *A. vinelandii* (Castanjeda et al. 2001). Whitchurch et al. (2002b) also stated that AlgR govern the fimbrial biogenesis, twitching motility, and biofilm formation in *P. aeruginosa*. The presence of *O*-acetyl groups plays an important role in the ability of the polymer to act as a virulence factor, and the *algF*, *algJ*, and *algI* genes are essential for the addition of *O*-acetyl groups to alginate. Random fusion study with *phoA* (alkaline phosphatase [AP] gene) to *algF*, *algJ*, and *algI* showed alkaline phosphatase activity, indicating that both AlgF and AlgJ were exported to the periplasm and AlgI, is an integral membrane protein with seven transmembrane domains and they form a complex (AlgI-AlgJ-AlgF) in the membrane which acts as a reaction center for *O*-acetylation of alginate (Franklin and Ohman 2002) (Fig. 4.1).

In *P. aeruginosa*, the response regulator AlgR is required for transcription of *algC* and *algD*, which encode key enzymes in the alginate biosynthetic pathway. In *P. syringae* FF5, however, *algR* is not required for the activation of *algD*. Interestingly, *algR* mutants of *P. syringae* remain nonmucoid, indicating an undefined role for this response regulator in alginate biosynthesis. AlgC promoter has two potential binding sites for AlgR and $\sigma54$, the alternative sigma factor encoded by *rpo*N thus AlgR has a positive role in the activation of *algC* in *P. syringae* and contributes to both virulence, epiphytic fitness and systemic movement (Alejandro et al. 2004). Expression analysis revealed that, during mid-logarithmic growth, AlgR activated the expression of 58 genes while it repressed the expression of 37 others, while during stationary phase, it activated expression of 45 genes and repression of 14 genes. Thus the new roles for AlgR are that it can repress gene transcription, can activate *fimTU-pilVWXY1Y2E* operon, regulate HCN production, and controls the transcription of the putative *cbb*3-type cytochrome PA1557 (Lizewski et al. 2004) (Fig. 4.1).

Previous work has shown that functional genomics, in particular DNA microarrays and proteomics, provides a powerful tool to study global gene expression in bacterial biofilms (Sauer and Camper 2001). Successful colonisation of rhizomicrobes includes weakening or destruction of target and competing organisms, efficient uptake of low concentrations of nutrients and competition with other cells of the same or unrelated species for nutrients and the accumulation of extracellular metabolic products, which have a regulatory role on bacterial colonisation. During these interactions microbes entering the rhizospheric environment utilize minor, non-nutrient components of root exudates as signals to guide their movement towards the root surface and elicit changes in gene expression appropriate to the complex beneficial associations and maintain themselves in a competitive manner on the root system (Kiely et al. 2006). The exact composition of the exudates is determined by many factors, including species and nutritional status of the plant, soil structure, micronutrient status and plant developmental stage and the exudation from plant root is also stimulated by microorganisms and these compounds which are mainly exuded by the plant roots promote the bacterial populations from 3- to 100-fold (Bais et al. 2006). Depending on the exact nature of the compound in the root exudates, they may play a role in activation of microbial genes responsible for recognition and initiation of symbiotic association, act as an antimicrobial plant defense, activate or disrupt key microbial genes responsible for biofilm formation, or they may simply act as an easy source of moisture, nutrients, and energy. Rhizodeposition of the roots are playing important role in plant growth promotion was confirmed by the addition of mucilage to the soil which increased the microbial C up to 23% and the number of cultivable bacteria was enhanced by 450%. Catabolic (Biolog GN2) and 16S–23S intergenic spacer fingerprints exhibited significant differences between control and mucilage treatments indicates that mucilage can affect both the metabolic and genetic structure of the bacterial community (Benizri et al. 2007).

Through transposon mutagenesis, Roberts et al. (2007) reported that mutants showing insertion at *aceF* [encodes the dihydrolipoamide acetyltransferase subunit of the pyruvate dehydrogenase complex (PDHC, enzyme responsible for the conversion of pyruvate to acetyl-CoA)] are impaired of rhizosphere colonisation and maintenance of *Enterobacter cloacae* population density in diverse crop plants.

Many bacteria have evolved sensory transduction systems that utilize diffusible signals or "pheromones" to sense and respond to their biotic environment, including their own population density (Kaiser and Losick 1993; Salmond et al. 1995; Swift et al. 1996; Gray 1997). A two component sensor kinase [ColR/S] is involved in the competitive root colonisation ability of *P. fluorescens* WCS365 in response to an environmental stimulus and maintains the cells internal pH, enhance the uptake of certain root exudate due to the proton motive force generated by the NADH dehydrogenases (Dekkers et al. 1998c) and also involved in chemotaxis towards exudate compounds to initiate a high growth rate in the rhizosphere (Dekkers et al. 1998a). Lambda integrase family of site-specific recombinases [xerC/sss] also seems to be essential for colonisation. In certain rhizomicrobes efficient root colonisation is linked to their ability to secrete a site-specific recombinase (Dekkers et al. 1998b).

Margarita et al. (2002) provided evidence that shows that the lack of NADH dehydrogenase I, an enzyme of the aerobic respiratory chain encoded by the *nuo* operon, is responsible for the impaired root-colonisation ability of PCL1201. The complete sequence of the *nuo* operon (ranging from *nuoA* to *nuoN*) of *P. fluorescens* WCS365 was identified, including the promoter region and a transcriptional terminator consensus sequence downstream of *nuoN*. It was shown biochemically that PCL1201 is lacking NADH dehydrogenase I activity. Since it was assumed that low-oxygen conditions were present in the rhizosphere, the activity of the *nuo* and the *ndh* promoters at different oxygen tensions were analysed. The results showed that both promoters are up-regulated by low concentrations of oxygen and that their levels of expression vary during growth. By using *lacZ* as a marker, it was shown that both the *nuo* operon and the *ndh* gene are expressed in the tomato rhizosphere. In contrast to the *nuo* mutant PCL1201, an *ndh* mutant of WCS365 appeared not to be impaired in competitive root tip colonisation.

Martínez-Granero et al. (2006) has demonstrated that in the biocontrol agent *P. fluorescens* F113, phenotypic variation is mediated by the activity of two site-specific recombinases, Sss and XerD. By overexpressing the genes encoding either of the recombinases, large numbers of variants (mutants) after selection either by prolonged laboratory cultivation or by rhizosphere passage were generated. All the isolated variants were more motile than the wild-type strain and appear to contain mutations in the *gacA* and/or *gacS* gene. By disrupting these genes and complementation analysis, it was observed that the Gac system regulates swimming motility by a repression pathway. Variants isolated after selection by prolonged cultivation formed a single population with a swimming motility that was equal to the motility of *gac* mutants, being 150% more motile than the wild type. The motility phenotype of these variants was complemented by the cloned *gac* genes. Variants isolated after rhizosphere selection belonged to two different populations: one identical to the population isolated after prolonged cultivation and the other comprising variants that besides a *gac* mutation harbored additional mutations conferring higher motility. Results showed that *gac* mutations are selected both in the stationary phase and during rhizosphere colonisation. The enhanced motility phenotype is in turn selected during rhizosphere colonisation. Several of these highly motile variants were more competitive than the wild-type strain, displacing it from the root tip within two weeks.

Certain well-conserved genes in fluorescent *Pseudomonas* spp. are involved in pathogenic interactions between the bacteria and evolutionarily diverse hosts including plants, insects and vertebrate animals. One such gene, *dsbA*, encodes a periplasmic disulfide-bond-forming enzyme implicated in the biogenesis of exported proteins and cell surface structures. Role of *dsbA* in *P. fluorescens* Q8r1-96, a biological control strain that produces the antibiotic 2,4-diacetylphloroglucinol (2,4-DAPG) and is known for its exceptional ability to colonise the roots of wheat and pea was elucidated (Mavrodi et al. 2006a). The deduced DsbA protein from Q8r1-96 is similar to other predicted thiol:disulfide interchange proteins and contains a conserved DsbA catalytic site, a pattern associated with the thioredoxin family active site, and a signal peptide and cleavage site. A *dsbA* mutant of Q8r1-96 exhibited decreased motility and fluorescence, and altered colony morphology; however, it produced more 2,4-DAPG and total phloroglucinol-related compounds and was more inhibitory in vitro to the fungal root pathogen *Gaeumannomyces graminis* var. *tritici* than was the parental strain. When introduced separately into a natural soil, Q8r1-96 and the *dsbA* mutant did not differ in their ability to colonise the rhizosphere of wheat in greenhouse experiments lasting 12 weeks. However, when the two strains were co-inoculated, the parental strain consistently out-competed the *dsbA* mutant. It was concluded that *dsbA* does not contribute to the exceptional rhizosphere competence of Q8r1-96, although the *dsbA* mutation reduces competitiveness when the mutant competes with the parental strain in the same niche in the rhizosphere. The results also suggest that exoenzymes and mul-timeric cell surface structures are unlikely to have a critical role in root colonisation by this strain (Mavrodi et al. 2006a). Recently, interaction of *Pseudomonas* spp. with various hosts were investigated to determine their contributions to the unusual colonisation properties of strain Q8r1-96 (Mavrodi et al. 2006b). Mutants were characterized to determine their 2,4-DAPG production, motility, fluorescence, colony morphology, exoprotease and hydrogen cyanide (HCN) production, carbon and nitrogen utilization, and ability to colonise the rhizosphere of wheat grown in natural soil. The data suggested novel functions for two genes, *ptsP* and *orfT*, that previously were linked with pathogenesis in *P. aeruginosa*. The *ptsP* and *orfT* mutants of Q8r1-96 did not have nonspecific growth defects in vitro, and the effects of the mutations became apparent only when the mutants were tested in the rhizo-sphere, either individually or in competition with the parental strain. Both genes fulfill the criteria for "true" rhizosphere colonisation determinants, and is the first report to provide evidence that *ptsP* is involved in rhizosphere colonisation by fluorescent pseudomonads.

Environmental signals through quorum sensing trigger many plant-associated bacteria to compete with a diverse community of microorganisms including multi-cellular differentiation, fruiting body development, and sporulation by a process of signal exchange, which enables bacterial populations to coordinate gene expression in rhizosphere competence (Mazzola et al. 1992). At the onset of effective root colonisation bacteria goes to phase variation, a regulatory process for DNA rearrangements which is orchestrated by site-specific recombinase (Dybvig 1993). Urgel et al. (2000) studied some transposon mutants defective in attachment to corn seeds and sequence analysis of these mutants showed similarity with genes of known functions such as putative surface and membrane proteins, including a calcium-binding protein, a hemolysin, a peptide transporter, and a potential multidrug efflux pump. The second attachment step requires the synthesis of bacterial cellulose fibrils that cause a tight and irreversible binding of the bacteria to the roots and overexpression of a colonisation gene (with functions in the cell envelope, chemotaxis and motility, transport, secretion, DNA metabolism and defense mechanism, regulation, energy metabolism, stress, detoxification, and protein synthesis) of a rhizobacterium of tomato plant caused an increase in the extent of colonisation. Transposon insertion in two overlapping genes with different orientations, *wbpN* and *tyrB* (aromatic amino acid amino transferase) genes, plays a role in root colonisation, whose function in lipooligosaccharide synthesis is still unknown; a regulatory protein (*pyrR*) gene and biosynthetic genes (*pyrB* and *pyrC*) of pyrimidine biosynthetic pathway were also defective mutants (Tn5*lac*) of competitive colonisation (Lugtenberg et al. 2001) and *dsbA*, encodes a periplasmic disulfide-bond-forming enzyme implicated in the biogenesis of exported proteins; cell surface structures were also reported as a gene for competitive root colonisation (Mavrodi et al. 2006a). Recently Rodriguez et al. (2007) also confirmed that cyclic glucans, capsular polysaccharide, and cellulose fibrils are involved in the phenomenon of root colonisation.

More than 80% of known bacterial species shows motility by means of flagella and flagellar motility among bacteria is found as a prerequisite for the movement towards favourable conditions and avoiding detrimental factors and it also allows the bacteria to compete with other rhizobacteria present in the environment. Weger et al. (1987) reported that the flagella are involved in the colonisation of the deeper root parts in potato. During flagellar movement bacteria has to expend too much of energy and a *nuo4* genes coding for the subunits of NADH:ubiquinone oxidoreductase, generate a proton motive force to drive ATP synthesis, active transport and ATP-dependent rotation of the flagella play an important role in root colonisation (Anraku and Gennis 1987; Moens and Vanderleyden 1996). Howie et al. (1987) reported through transposon study that the most severely impaired colonisation mutants are non-motile mutants and mutants impaired in *O*-antigen synthesis (Dekkers et al. 1998c). In addition to motility and attachment, *O*-antigenic side chain of the outer membrane lipopolysaccharides also enhances survival of *P. fluorescens* within tomato root (Duijff et al. 1997). In a steep nutrient concentration gradient environment bacterial chemotaxis is thought to play a critical role in structuring microbial communities, characterisation of the function of two chemotaxis gene clusters (*che1* and *che2*) in *R. leguminoserum* in controlling motility behaviour for effective nodulation also supporting the previous evidences (Miller et al. 2007).

Scher et al. (1988) reported that in wheat and soybean nonmotile mutants of *Pseudomonas* are not impaired in root colonisation and type 4 pili are involved in colonisation of both plants and these pili perform competitive root colonisation through twitching motility. In *P. aeruginosa*, type 4 pili mediate the initial contact between the bacteria and the epithelial cell surface by driving the locomotion through twitching motility and attaching to abiotic surfaces to form biofilms. Once rhizobacteria reach a root and are tightly bound to it they colonise the root by their ability to maintain and grow on the root system to initiate biofilm formation and cytoplasmic Clp protease protein, participates in biofilm formation (O'Toole and Kolter 1998). The matrix, which holds bacterial biofilms together, has been presumed to be derived from lysed cells and is not a important component of biofilm structure including exopolysaccharides, proteins, and DNA which bacteria produce in substantial quantities through a mechanism independent of cellular lysis and it has been confirmed by the treatment of bacterial biofilm to DNaseI (Whitchurch 2002a). The bacterial cells on the surface of the biofilm are different from the cells within the biofilm matrix. The embedded cells behavior can change the thickness of the biofilm. The surface cells are metabolically active and large. These surface cells divide and increase the thickness of the biofilm. Little oxygen is available to the embedded cells, therefore they are smaller and grow slower. The bacteria exist in a somewhat dormant state, becoming active when cells in the outer layers are killed. Recently, a (*mvaT*) gene, a negative regulator of *cupA* (chaperone-usher pathway) was shown to be required for biofilm formation on abiotic surfaces (Isabelle et al. 2004). Among Gram-positives, *B. subtilis* is a ubiquitous soil bacterium that forms biofilms in a process that is negatively controlled by the transcription factor AbrB and different AbrB-regulated genes. YoaW, a putative secreted protein, and SipW, a signal peptidase had a role in biofilm formation (Hamon et al. 2004). The lipopolysaccharide/exopolysaccharide-coupled biosynthetic genes *rmlA*, *rmlC*, and *xanB* are necessary for biofilm formation and twitching motility (Huang et al. 2006). A transposon (Tn5*luxAB*) study of PCL1627 (biosurfactant mutant) has shown that heat shock genes (*dnaK*, *dnaJ* and *grpE*) regulate Putisolvin biosynthesis at the transcriptional level and is dependent on the GacA/GacS two-component signaling and the lipopeptides Putisolvin I and II with surfactant activity, which is induced by the quorum-sensing signals and consequently control biofilm formation by *P. putida* PCL1445 (Kitagawa et al. 1991; Dubern et al. 2006). In response to these quorum sensing signals, bacteria interact with plant tissues through adhesins including polysaccharides, surface proteins, and the recognition of plant produced lectins and their cognate carbohydrates is a common means of specificity; the whole process of biofilm development and intimate interactions is under the control of cell-cell communication between colonising bacteria. These plant-associated biofilms undergo chromosomal rearrangements and are hot spots for conjugative plasmid transfer, favored by the close proximity between cells and the constant supply of nutrients coming from the plant in the form of exudates or leachates (Danhorn and Fuqua 2007).

In the comparison of the protein or transcription profiles of *Pseudomonas* spp. employing comparative proteomics, Sauer and Camper (2001) detected

45 differentially expressed proteins in biofilms of *P. putida* ATCC 39168 following 6 h of attachment, indicating that this strain undergoes a global change in gene expression after surface-adherence. Furthermore, the expression rate of 16 proteins was changed when planktonic cells were grown in a medium supplemented with 3-oxo-C12-HSL. Only one protein, the periplasmatic putrescine binding protein PotF, was found to be downregulated in biofilm cells as well as in the presence of AHL signal-molecules. On the basis of this result it was suggested that QS does not play an important role in the initial attachment process. Sauer and Camper (2001) studied global changes in the expression profile of intracellular proteins during the initial stage of biofilm formation of an AHL-negative *P. putida* strain. At present there is an ongoing debate on whether biofilm differentiation into mature biofilms necessarily requires a defined genetic programme or merely constitutes the sum of cellular adaptations and growth cycles influenced by the nutrient diffusion conditions in individual communities (Kjelleberg and Molin 2002). As a consequence of the latter assumption even global methodologies employing DNA-array technology or proteome analyses will hardly be able to define a "universal biofilm regulon". Moreover, as bacterial cells are distributed in a structured biofilm in numerous microenvironments with different physiological activities, results can only be considered as an overlay of gene expression alterations of all occurring sub-populations. A dissection and consequent proteome analysis of these sub-populations would certainly improve our understanding of the interplay of quorum sensing and sessile lifestyle to regulate gene expression in *P. putida* but is technically very challenging as conventional 2-DE requires high amounts of biomass.

Differentially expressed spots were identified by matrix-assisted time of flight mass spectrometry (MALDI-TOF MS) and database search in the recently completed genome sequence of *P. putida* KT2440 (Nelson et al. 2002). Arevalo-Ferro et al. (2005) investigated the impact of QS and biofilm formation on the protein profile of surface-associated proteins of *P. putida* IsoF. This was accomplished by comparative proteome analyses of the *P. putida* wild type IsoF and the QS-deficient mutant F117 grown either in planktonic cultures or in 60 h old mature biofilms. Differentially expressed proteins were identified by peptide mass fingerprinting and database search in the completed *P. putida* KT2440 genome sequence. The sessile life style affected 129 out of 496 surface proteins, suggesting that a significant fraction of the bacterial genome is involved in biofilm physiology. In surface-attached cells 53 out of 484 protein spots were controlled by the QS system, emphasizing its importance as global regulator of gene expression in *P. putida* IsoF. Most interestingly, the impact of QS was dependent on whether cells were grown on a surface or in suspension; about 50% of the QS controlled proteins identified in planktonic cultures were found to be oppositely regulated when the cells were grown as biofilms. Of all identified surface-controlled proteins, 57% were also regulated by the *ppu* QS system. The data provide strong evidence that the set of QS-regulated proteins overlaps substantially with the set of proteins differentially expressed in sessile cells. In fact, the most striking result of comparative proteome analysis was the finding that expression of QS-regulated proteins in *P. putida* IsoF is strongly dependent on the life style of the organism.

4.4 Strategies to Enhance Plant–Microbe Coexistence

Weller's group has focused on the role of the antifungal metabolite DAPG in biological control of soil-borne pathogens by fluorescent *Pseudomonas* spp. (Thomashow and Weller 1996). Genetic studies, modeled after Koch's postulates, demonstrated unequivocally that DAPG plays a major role in the suppression of a variety of soil-borne plant pathogens by fluorescent *Pseudomonas* strains (Raaijmakers et al. 1999). They have demonstrated that genotypic diversity within a group of antagonistic microorganisms that share a common biocontrol trait has great potential for improving biological control. This approach capitalizes on existing knowledge concerning mechanisms, while exploiting the differences among strains to face the challenges of diverse soil and rhizosphere environments. By matching rhizobacterial genotypes with crops or varieties for which they have a colonisation preference, root colonisation and biocontrol can be increased without increasing the amount of inoculum (Raaijmakers and Weller 2001). The biosynthetic locus for 2,4-DAPG includes five genes, *phlACBDE. phlD*, a key gene in the biosynthesis of the antibiotic and highly conserved in nature (De La Fuente et al. 2006), is widely used for identification of 2,4-DAPG producers. Using molecular fingerprinting by BOX-PCR and *phlD*-RFLP, Weller et al. (2007) have described 22 genotypes. A distinction was made between the superior ("premier") root colonisation of strains of certain genotypes, which reach and maintain large population sizes for long periods of time, and the "average" colonisation of most rhizobacteria, whose rhizosphere populations decline within days or weeks after introduction into the soil. For example, D-genotype strains (i.e. Q8r1-96) are premier colonists of wheat and pea roots, whereas genotype B and L strains (i.e. Q2-87 and 1M1-96, respectively) are average colonists of these crops. Superior rhizosphere competence is a trait that permits a 2,4-DAPG producer to consistently protect roots against soilborne pathogens (Weller et al. 2007). Thus the factor(s) responsible for superior rhizosphere competence of 2,4-DAPG-producing *P. fluorescens* belonging to certain genotypes on some crops remain(s) elusive and variable colonisation of introduced strains remains a major impediment to the widespread use of biocontrol agents.

Genomic subtraction is among the best methods currently available for exploring structural differences between the genomes of closely related bacteria (Straus and Ausubel 1990; Lan and Reeves 2000), including fluorescent pseudomonads (Schmidt et al. 1998). Thomashow's group described the application of genomic suppressive subtractive hybridization (SSH) (Akopyants et al. 1998; Westbrock-Wadman et al. 1999) as one approach to identify genes that contribute to the exceptional rhizosphere competence of D-genotype strains. DNA sequences present in the superior root coloniser *P. fluorescens* Q8r1-96 but not in the less rhizosphere-competent strain Q2-87 were cloned, their sequences determined and analyzed, and their expression in the rhizosphere and distribution among 29 other 2,4-DAPG-producing strains representative of 17 different genotypes were assessed. Several subtracted fragments distributed primarily among isolates of the D genotype or expressed in the rhizosphere were identified as candidates for further analysis

(Mavrodi et al. 2002; Landa et al. 2002, 2003). The development of bacterial artificial chromosome (BAC) systems has allowed the construction of large insert-sized DNA libraries (Shizuya et al. 1992). Compared with YAC and cosmid cloning, BAC has a lower rate of chimerism and higher efficiency of cloning. It is also easier to handle and is more stably maintained. To date, BAC libraries have been constructed in many kinds of organisms, e.g. human (Kim et al. 1996), plants (Woo et al. 1994), fungi (Nishimura et al. 1998) and bacteria (Dewar et al. 1998) have become a powerful tool for genome analysis (Blomberg and Lugtenberg 2001). Because it employs large-sized DNA inserts, the BAC system offers the following significant advantages for cloning and analysis of bacterial genomes: (i) it requires only a relatively small number of clones to provide complete coverage of the bacterial genome and (ii) it facilitates cloning of clustered genes, such as those for certain metabolic processes, for secretion, or for pathogenicity (e.g. *hrp* genes). Such libraries are powerful tools for genome analysis, physical mapping, map-based cloning, and simple screening for specific genomic sequences because of their low chimeric clone formation rates and high cloning efficiency. Moreover, they are easy to handle and can be stably maintained. The physical organisation of phytobeneficial genes was investigated in the plant growth promoting rhizobacterium *A. lipoferum* 4VI by hybridization screening of a BAC library (Blaha et al. 2005).

Ryu et al. (2005) have investigated the mechanisms by which PGPR, elicit plant growth promotion from the viewpoint of signal transduction pathways within plants. Results suggest that elicitation of growth promotion by PGPR in *Arabidopsis* is associated with several different signal transduction pathways and that such signaling may be different for plants grown in vitro vs in vivo.

Recently Zuo et al. (2007) reported that an ERF transcription factor gene (GbERF2) was cloned by suppression subtraction hybridization from sea-island cotton after *Verticillium dahliae* attack. These results show that GbERF2 plays an important role in response to ethylene stress and fungal attack in cotton.

Developments in high-throughput DNA sequencing have resulted in elucidation of the whole-genome sequences. Current data (as of August 2, 2007 on BLAST with microbial genomes 912 bacterial/46 archaeal/139 eukaryotic genomes tree) can be obtained from URL NCBI data base (http://www.ncbi.nlm.nih.gov/sutils/genome table.cgi) data base and references quoted therein). *P. fluorescens* SBW25 (PfSBW25) is a Gram-negative bacterium that grows in close association with plants. In common with a broad range of functionally similar bacteria it plays an important role in the turnover of organic matter and certain isolates can promote plant growth. Despite its environmental significance, the causes of its ecological success are poorly understood. Gal et al. (2003) described the development and application of a simple promoter trapping strategy IVET to identify PfSBW25 genes showing elevated levels of expression in the sugar beet rhizosphere. A total of 25 rhizosphere-induced (*rhi*) fusions are reported with predicted roles in nutrient acquisition, stress responses, biosynthesis of phytohormones and antibiotics.

One of the promises in the genomics era is an improved ability to identify causal relationships among genotype, phenotype and the environment, and to do this on a genome wide scale. In particular, the advent and dissemination of genomics information and technologies has resulted in the development of several powerful new approaches that allow one to simultaneously analyze both the phenotype and the genotype of thousands of mutants. Collectively, these tools improve the ability to map phenotypic landscapes and develop integrated models connecting genetic alterations and their resultant phenotypes. As a result of advances in genomics technologies, several techniques now exist that substantially improve researchers ability to identify such genes. Metabolic engineers now have the ability using DNA micro-arrays to map phenotypic landscapes of considerable genetic diversity, which should improve understanding of the relationships that exist among phenotype, genotype, and environment. It has became apparent that bacteria of a certain species living in close association with different plants either as associated rhizosphere bacteria or as plant pathogens or symbiotic organisms typically reflect this relationship in their genetic relatedness and is markedly influenced by soil management and soil features. Further studies must address the consequences of the co-operation between microbes in the rhizosphere under field conditions to assess their ecological impacts and biotechnological potential.

4.5   Concluding Remarks and Future Prospects

There are more opportunities available for microbiologists today than at any time in the history of the field. Although the microbiological advancements of the last century have been profound, a great deal of biology remains to be discovered and described through study of the microbial world. Knowledge of microbial diversity and function in soils is limited because of the taxonomic and methodological limitations associated with studying these organisms. Although methods to study diversity (numerical, taxonomic, and structural) are improving for both bacteria and fungi, there is still not a clear association between diversity and function. Even if an organism is functionally redundant in one function, chances are it is not redundant in all functions and will have different susceptibilities and tolerances to abiotic and biotic stresses. It is generally thought that a diverse population of organisms will be more resilient to stress and more capable of adapting with environmental changes. Our understanding of plant–soil interactions can be greatly refined through the development of "smart" field technology, where real-time, computer-controlled electronic diagnostic devices can be used to monitor rhizosphere and plant health. The maximization of production efficiencies will also involve the development of crop cultivars that are bred specifically to capitalize on beneficial plant–microbial associations.

The application of molecular tools is enhancing our ability to understand and manage the rhizosphere and will lead to new products with improved effectiveness. Thus, future research in rhizosphere biology will rely on the development of molecular and biotechnological approaches to increase our knowledge of rhizosphere biology and to achieve an integrated management of soil microbial populations. The new tools of recombinant DNA technology, mathematical modeling, and computer technology combined with a continuation of the more classical approaches such as crop rotation, various tilling strategies, addition of organic amendments such as compost, mulch or manures, should quickly open up new ways to harness the power of microbes to improve soil, plant, human and the environment health.

Acknowledgements Thanks are due to the Director, National Botanical Research Institute, Lucknow for providing the necessary facilities. The work was supported by Task Force grant SMM-002 and New Millennium Indian Technology Leadership Initiative (NMITLI) programme from Council of Scientific & Industrial Research (CSIR), New Delhi, India awarded to CSN.

References

Ahn IP, Park K, Kim CH (2002) Rhizobacteria-induced resistance perturbs viral disease progress and triggers defense-related gene expression. Mol Cell 13:302–308

Akopyants NS, Fradkov A, Diatchenko L, Hill JE, Siebert PD, Lukyanov SA, Sverdlov ED, Berg DE (1998) PCR-based subtractive hybridization and differences in gene content among strains of *Helicobacter pylori*. Proc Natl Acad Sci USA 95:13108–13113

Alejandro P, Aloza V, Mohamed KF, Ana MB, Carol LB (2004) AlgR functions in *algC* expression and virulence in *Pseudomonas syringae* pv. *Syringe*. Plant-Microbe Interact 150:2727–2737

Andrews JH, Harris RF (2000) The ecology and biogeography of microorganisms on plant surfaces. Annu Rev Phytopathol 38:145–180

Angela S, Hongqiao L, Helge W, Stephan A, Antje B, Katharina F, Alexander S, Matthias SU (2001) Thermoregulated expression of virulence factors in plant-associated bacteria. Arch Microbiol 176:393–399

Anraku Y, Gennis R (1987) The aerobic respiratory chain of *Escherichia coli*. Trend Biochem Sci 12:262–266

Apel K, Hirt H (2004) Reactive oxygen species: metabolism, oxidative stress, and signal transduction. Ann Rev Plant Biol 55:373–399

Arden A, Kelli P, Marvin W (2006) Microarray analysis of the osmotic stress response in *Pseudomonas aeruginosa*. J Bacteriol 188:2721–2725

Arevalo-Ferro C, Reil C, Görg A, Eberl L, Riedel K (2005) Biofilm formation of *Pseudomonas putida* IsoF: the role of quorum sensing as assessed by proteomics. Syst Appl Microbiol 28:87–114

Aspedon A, Palmer K, Whiteley M (2006) Microarray analysis of the osmotic stress response in *Pseudomonas aeruginosa*. J Bacteriol 188:2721–2725

Bais HP, Tiffany LW, Laura GP, Simon G, Jorge MV (2006) The role of root exudates in rhizosphere interactions with plants and other organisms. Ann Rev Plant Biol 57:233–266

Beerling DJ, Berner RA (2005) Feedbacks and the coevolution of plants and atmospheric CO_2. Proc Natl Acad Sci USA 102:1302–1305

Benchamas S, Thomas Mark S, Katzenmeier G, Jonathan GS, Tungpradabkul S, Kunakorn M (2003) Role of the stationary growth phase sigma factor RpoS of *Burkholderia pseudomallei* in response to physiological stress conditions. J Bacteriol 185:7008–7014

Benizri E, Nguyen C, Piutti S, Slezack-Deschaumes S, Philippot L (2007) Additions of maize root mucilage to soil changed the structure of the bacterial community. Soil Biol Biochem 39:1230–1233

Blaha D, Sanguin H, Robe P, Nalin R, Bally R, Moënne-Loccoz R (2005) Physical organization of phytobeneficial genes nifH and ipdC in the plant growth-promoting rhizobacterium *Azospirillum lipoferum* $4V_L$. FEMS Microbiol Lett 244:157–163

Blomberg GV, Lugtenberg BJJ (2001) Molecular basis of plant growth promotion and biocontrol by rhizobacteria. Curr Opin Plant Biol 4:343–350

Brandl MT, Lindow SE (1997) Environmental signals modulate the expression of an indole-3-acetic acid biosynthetic gene in *Erwinia herbicola*. Mol Plant Microb Interact 10:499–505

Cardon ZG, Gage DJ (2006) Resource exchange in the rhizosphere: molecular yools and the microbial perspective. Annu Rev Ecol Evol Syst 37:459–488

Castanjeda M, Sanchez J, Moreno S, Nunez C, Espi NG (2001) The global regulators GacA and GacAS form part of a cascade that controls alginate production in *Azotobacter Vinelandii*. J Bacteriol 183:787–793

Chinnusamy V, Schumaker K, Zhu JK (2004) Molecular genetic perspectives on cross-talk and specificity in abiotic stress signaling in plants. J Exp Bot 55:225–236

Cortina C, Culiáñez-Macià FA (2005) Tomato abiotic stress enhanced tolerance by trehalose biosynthesis. Plant Sci 169:75–82

Danhorn T, Fuqua C (2007) Biofilm formation by plant-associated bacteria. Ann Rev Microbiol (in press)

De La Fuente L, Mavrodi DV, Landa BB, Thomashow LS, Weller DM (2006) *phlD*-based genetic diversity and detection of genotypes of 2,4-diacetylphloroglucinol-producing *Pseudomonas fluorescens*. FEMS Microbio Ecol 56:64–78

Dekkers LC, Bloemendaal CP, de Weger LA, WijffelmanCA, Spaink HP, Lugtenberg BJJ (1998a) A two-component system plays an important role in the root-colonizing ability of *Pseudomonas fluorescens* strain WCS365. Mol Plant Microbe Interact 11:45–56

Dekkers LC, Phoelich CC, van der Fits L, Lugtenberg BJJ (1998b) A site-specific recombinase is required for competitive root colonization by *Pseudomonas fluorescens* WCS365. Proc Natl Acad Sci 95:7051–7056

Dekkers LC, van der Bij AJ, Mulders IHM, Phoelich CC, Wentwoord RAR, Glandorf DC, Wijffelman CA, Lugtenberg BJJ (1998c) Role of the O-antigen of lipopolysaccharide, and possible roles of growth rate and NADH: ubiquinone oxidoreductase in competitive tomato root-tip colonization by *Pseudomonas fluorescens* WCS365. Mol Plant Microb Interact 11:763–771

Dewar K, Sabbagh L, Cardinal G, Veilleux F, Sanschagrin F, Birren B, Levesque RC (1998) *Pseudomonas aeruginosa* PAO1 bacterial artificial chromosomes: strategies for mapping, screening, and sequencing 100 kb loci of the 5.9 Mb genome. Microb Comp Genomic 2:105–117

Dorman CJ (1991) DNA supercoiling and environmental regulation of gene expression in pathogenic bacteria. Infect Immun 59:745–749

Dubern JF, Lugtenberg BJJ, Bloemberg GV (2006) The *ppuI-rsaL-ppuR* quorum-sensing system regulates biofilm formation of *Pseudomonas putida* PCL1445 by controlling biosynthesis of the cyclic lipopeptides putisolvins I and II. J Bacteriol 188:2898–2906

Duijff BJ, Gianinazzi-Pearson V, Lemanceau P (1997) Involvement of the outer membrane lipopolysaccharides in the endophytic colonisation of tomato roots by biocontrol *Pseudomonas fluorescens* strain WCS417r. New Phytol 135:325–334

Dyhvig K (1993) DNA rcarrangements and phenotypic switching in prokaryotes. Mol Microbiol 10:465–471

Edwards KJ, Saunders NA (2001) Real-time PCR used to measure stress-induced changes in the expression of the genes of the alginate pathway of *Pseudomonas aeruginosa*. J Appl Microbiol 91:29–37

Franklin MJ, Ohman DE (2002) Mutant analysis and cellular localization of the AlgI, AlgJ, and AlgF Proteins required for O acetylation of alginate in *Pseudomonas aeruginosa*. J Bacteriol 184:3000–3007

Gal M, Preston GM, Massey RC, Spiers AJ, Rainey PB (2003) Genes encoding a cellulosic polymer contribute toward the ecological success of *Pseudomonas fluorescens* SBW25 on plant surfaces. Mol Ecol 12:3109–3121

Glick BR, Penrose DM, Jiping LA (1998) Model for the lowering of plant ethylene concentrations by plant growth promoting bacteria. J Theor Biol 190:63–68

Gray KM (1997) Intercellular communication and group behavior in bacteria. Trends Microbiol 5:184–188

Hamon MA, Stanley NR, Britton RA, Grossman AD, Lazazzera BA (2004) Identification of AbrB-regulated genes involved in biofilm formation by *Bacillus subtilis*. Mol Microbiol 52(3):847–860

Hassan NE, Vereecke D, Goethals K, Jaziri M, Baucher M (2003) Screening for differential gene expression in *Atropa belladonna* leafy gall induced following *Rhodococcus fascians* Infection. Euro J Plant Pathol 109:327–330

Henge-Aronis R (1996) Regulation of gene expression during entry into stationary phase. In: Neidhardt FC, Curtiss R III, Ingraham JL, Lin ECC, Low KB, Magasanik B, Reznikoff WS, Riley M, Schaechter M, Umbarger HE (eds) *Escherichia coli* and *Salmonella*: cellular and molecular biology, 2nd edn, vol 1. ASM Press, Washington, pp 1497–1512

Hiltner L (1904) Uber neuere Erfahrungen und Probleme auf dem Gebiete der Bodenbakteriologie unter bessonderer Ber&$$$;¨ ucksichtigung der Gr&$$$;¨ undung und Brache. Arb Dtsch Landwirtsch Ges Berl 98:59–78

Howie WJ, Cook RJ, Weller DM (1987) Effects of soil matrix potential and cell motility on wheat root colonization by fluorescent pseudomonads suppressive to takeall. Phytopathol 77:286–292

Huang B, Liu JY (2006) Cloning and functional analysis of the novel gene GhDBP3 encoding a DRE-binding transcription factor from *Gossypium hirsutum*. Biochim Biophys Acta 1759:263–269

Huang TP, Eileen BS, Wong ACL (2006) Differential biofilm formation and motility associated with lipopolysaccharide/exopolysaccharide-coupled biosynthetic genes in *Stenotrophomonas maltophilia*. J Bacteriol 188:3116–3120

Hulsmann A, Rosche TM, Kong IS, Hassan HM, Beam DM, Oliver JD (2003) RpoS-dependent stress response and exoenzyme production in *Vibrio vulnificus*. Appl Environ Microbiol 69:6114–6120

Isabel MRG, Campos MJ, Ramos JL (2005) Analysis of *Pseudomonas putida* KT2440 gene expression in the maize rhizosphere: in vitro expression technology capture and identification of root-activated promoters. J Bacteriol 187:4033–4041

Isabelle V, Stephen PD, Rachael ES, Miguel C, Isabelle V, Stephen L, Lazdunski A, Williams P, Filloux A (2004) Biofilm formation in *Pseudomonas aeruginosa*: fimbrial *cup* gene clusters are controlled by the transcriptional regulator MvaT. J Bacteriol 186:2880–2890

Jorgensen F, Bally M, Chapon HV, Michel G, Lazdunski A, Williams P, Stewart GS (1999) RpoS-dependent stress tolerance in *Pseudomonas aeruginosa*. Microbiol 145:835–844

Kaiser D, Losick R (1993) How and why bacteria talk to each other. Cell 73:873–885

Karen MLK, Verhagen BWM, Keurentjes Joost JB, Johan AVP, Martijn R, Loon LCV, Pieterse CMJ (2005) Colonization of the *Arabidopsis* rhizosphere by fluorescent *Pseudomonas* spp. activates a root-specific, ethylene-responsive *PR-5* gene in the vascular bundle. Plant Mol Biol 57:731–748

Kazmierczak MJ, Wiedmann M, Kathryn JB (2005) Alternative sigma factors and their roles in bacterial virulence. Microbiol Mol Biol Rev 69:527–543

Keel US, Lezbolle KB, Behler E, Haas D, Keel C (2001) The sigma factor AlgU(AlgT) controls exopolysaccharide production and tolerance towards dessication and osmotic stress in the biocontrol agent *P. flourescens* CHAO. Appl Environ Microbiol 67:5683–5693

Kiely PD, Haynes JM, Higgins CH, Franks A, Mark GL, Morrissey JP, O'Gara F (2006) Exploiting new systems-based strategies to elucidate plant-bacterial interactions in the rhizosphere. Microb Ecol 51:257–266

Kim U, Birren B, Slepak T, Mancino V, Boysen C, Kang H, Simon M, Shizuya H (1996). Construction and characterization of a human bacterial artificial chromosome library. Genomic 34:213–218

Kitagawa M, Chieko W, Yoshioka S, Takashi Y (1991) Expression of ClpB, an analog of the ATP-dependent protease regulatory subunit in *Escherichia coli* is controlled by a heat shock sigma factor. J Bacteriol 173:4247–4253

Kjelleberg S, Molin S (2002) Is there a role for quorum sensing signals in bacterial biofilm? Curr Opin Microbiol 5:254–258

Knoester M, Pieterse CM, Bol JF, Van Loon LC (1999) Systemic resistance in *Arabidopsis* induced by rhizobacteria requires ethylene-dependent signaling at the site of application. Mol Plant Microbe Interact 12:720–727

Kulkarni S, Nautiyal CS (1999) Effects of salt and pH stress on temperature tolerant *Rhizobium* sp. NBRI330 nodulating *Prosopis juliflora*. Curr Microbiol 40:221–226

Kwak KJ, Kim JY, Kim YO, Kang H (2007) Characterization of transgenic *Arabidopsis* plants overexpressing high mobility group B proteins under high salinity, drought or cold stress. Plant Cell Physiol 48:221–231

Lan R, Reeves PR (2000) Intraspecies variation in bacterial genomes: the need for a species genome concept. Trends Microbiol 8:396–401

Landa BB, Mavrodi OV, Raaijmakers JM, McSpadden Gardener BB, Thomashow LS, Weller DM (2002) Differential ability of genotypes of 2,4-diacetylphloroglucinol-producing *Pseudomonas fluorescens* strains to colonize the roots of pea plants. Appl Environ Microbiol 68:3226–3237

Landa BB, Mavrodi DM, Thomashow LS, Weller DM (2003) Interactions between strains of 2,4-diacetylphloroglucinol-producing *Pseudomonas fluorescens* in the rhizosphere of wheat. Phytopathology 93:982–994

Laville J, Voisard C, Keel C, Maurhofer M, Défago G, Haas D (1992) Global control in *Pseudomonas fluorescens* mediating antibiotic synthesis and suppression of black root rot of tobacco. Proc Natl Acad Sci USA 189:1562–1566

Lee JH, Lee KH, Gyeom KC, Young LS, Joong KG, Young HP, Oh CS (2005) Appl Microbiol Biotechnol 68:213–219

Lizewski SE, Jill RS, Debra WJ, Anders F, Alexander JC, Michael JS (2004) Identification of AlgR-regulated genes in *Pseudomonas aeruginosa* by use of microarray analysis. J Bacteriol 186:5672–5684

Lugtenberg BJJ, Dekkers L, Bloemberg GV (2001) Molecular determinants of rhizosphere colonization by *Pseudomonas*. Ann Rev Phytopathol 39:461–490

Margarita M, Carvajal C, André C, Wijfjes HM, Mulders IHM, Lugtenberg BJJ, Bloemberg GV (2002) Characterization of NADH dehydrogenases of *Pseudomonas fluorescens* WCS365 and their role in competitive root colonization. Mol Plant Microbiol Inter 15:662–671

Martínez-Granero F, Rivilla R, Martín M (2006) Rhizosphere selection of highly motile phenotypic variants of *Pseudomonas fluorescens* with enhanced competitive colonization ability. Appl Environ Microbiol 72:3429–3434

Mathee K, Craig MJ, Ohman DE (1997) Posttranslational control of the *algT* (*algU*)-encoded s22 for expression of the alginate regulon in *Pseudomonas aeruginosa* and localization of its antagonist proteins MucA and MucB (AlgN). J Bacteriol 179:3711–3720

Mavrodi DV, Mavrodi OV, McSpadden-Gardener BB, Landa BB, Weller DM, Thomashow LS (2002) Identification of differences in genome content among *phlD*-positive *Pseudomonas fluorescens* strains by using PCR-based subtractive hybridization. Appl Envir Microbiol 68:5170–5176

Mavrodi OV, Mavrodi DV, Park AA, Weller DM, Thomashow LS (2006a) The role of *dsbA* in colonization of the wheat rhizosphere by *Pseudomonas fluorescens* Q8r1-96 Microbiology 152:863–872

Mavrodi OV, Mavrodi DV, Weller DM, Thomashow LS (2006b) Role of ptsP, orfT, and sss recombinase genes in root colonization by Pseudomonas fluorescens Q8r1-96. Appl Environ Microbiol 72:7111–7122

Mazzola M, Cook RJ, Thomashow LS, Weller DM, Pierson LS (1992) Contribution of phenazine antibiotic biosynthesis to the ecological competence of fluorescent pseudomonads in soil habitats. Appl Environ Microbiol 58:2616–2624

Miller LD, Yost CK, Hynes MF, Alexandre G (2007) The major chemotaxis gene cluster of *Rhizobium leguminosarum* bv. *viciae* is essential for competitive nodulation. Mol Microbiol 63:348–362

Moens S, Vanderleyden J (1996) Functions of bacterial flagella. Crit Rev Microbiol 22:67–100

Mohr CD, Leveau JHJ, Krieg DP, Hibler NS, Deretic V (1992) AlgR-binding sites within the algD promoter make up a set of inverted repeats separated by a large intervening segment of DNA. J Bacteriol 174:6624–6633

Nautiyal CS (2006) Biological control of plant diseases by natural and genetically engineered fluorescent *Pseudomonas* spp. In: Ray RC, Owen PW (eds) Microbial biotechnology in horticulture. Science Publishers, Enfield (NH) USA, pp 125–162

Nautiyal CS, Mehta S, Singh HB (2006a) Biological control and plant growth-promoting *Bacillus* strains from milk. J Microbiol Biotechnol 16:184–192

Nautiyal CS, Mehta S, Singh HB, Pushpangadan P (2006b) Synergistic bioinoculant composition comprising bacterial strains of accession nos. NRRL B-30486, NRRL B-30487 and NRRL B-30488 and method of producing said composition thereof. US Patent 7097830

Nelson KE, Weinel C, Paulsen IT, Dodson RJ, Hilbert H, Martins VA dos Santos, Fouts DE, Gill SR, Pop M, Holmes M, Brinkac L, Beanan M, DeBoy RT, Daugherty S, Kolonay J, Madupu R, Nelson W, White O, Peterson J, Khouri H, Hance I, Chris Lee P, Holtzapple E, Scanlan D, Tran K, Moazzez A, Utterback T, Rizzo M, Lee K, Kosack D, Moestl D, Wedler H, Lauber J, Stjepandic D, Hoheisel J, Straetz M, Heim S, Kiewitz C, Eisen JA, Timmis KN, Dusterhoft A, Tummler B, Fraser CM (2002) Complete genome sequence and comparative analysis of the metabolically versatile *Pseudomonas putida* KT2440. Environ Microbiol 4:799–808

Nishimura M, Nakamura S, Hayashi N, Asakawa S, Shimizu N, Kaku H, Hasebe A, Kawasaki S (1998) Construction of a BAC library of the rice blast fungus *Magnaporthe grisea* and finding specific genome regions in which its transposons tend to cluster. Biosci Biotechnol Biochem 62:1515–1521

Nobbe F, Hiltner L (1896) Inoculation of the soil for cultivating leguminous plants. US Patent 570 813

O'Toole GA, Kolter R (1998) Flagellar and twitching motility are necessary for *Pseudomonas aeruginosa* biofilm development. Mol Microbiol 30:295–304

Penrose DM, Glick BR (2003) Methods for isolating and characterizing ACC deaminase-containing plant growth-promoting rhizobacteria. Physiol Plantar 118:10–15

Pham TH, Web JS, Rehm BHA (2004) The role of polyhydroxyalkanoate biosynthesis *Pseudomonas aeruginosa* in rhamnolipid and alginate production as well as stress tolerance and biofilm formation. Microbiology 150:3405–3413

Raaijmakers JM, Weller DM (2001) Exploiting genotypic diversity of 2,4-diacetylphloroglucinol-producing *Pseudomonas* spp.: characterization of superior root-colonizing *P. fluorescens* strain Q8r1-96. Appl Environ Microbiol 67:2545–2554

Raaijmakers JM, Bonsall RF, Weller DM (1999) Effect of population density of *Pseudomonas fluorescens* on production of 2,4-diacetylphloroglucinol in the rhizosphere of wheat. Phytopathology 89:470–475

Rainey PB, Preston GM (2000) In vivo expression technology strategies: valuable tools for biotechnology. Curr Opin Biotechnol 11:440–444

Rathinasabapathi B (2000) Metabolic engineering for stress tolerance: installing osmoprotectant synthesis pathways. Ann Bot 86:709–716

Reva ON, Weine C, Weine M, Bohm K, Stjepandic D, Hoheisel JD, Tummler B (2006) Functional genomics of stress response in *P. putida* KT2440. J Bacteriol 188:4079–4092

Roberson EB, Firestone MK (1992) Relationship between desiccation and exopolysaccharide production in a soil *Pseudomonas* sp. Appl Environ Microbiol 1284–1291

Roberts DP, McKenna LF, Lohrke SM, RehnerS, de Souza JT (2007) Pyruvate dehydrogenase activity is important for colonization of seeds and roots by *Enterobacter cloacae*. Soil Biol Biochem 39:2150–2159

Rodriguez NDN, Dardanelli MS, Ruíz-Saínz JE (2007). Attachment of bacteria to the roots of higher plants. FEMS Microbiol Lett 1–10

Ryu CM, Chia-Hui Hu CH, Locy RD, Kloepper JW (2005) Study of mechanisms for plant growth promotion elicited by rhizobacteria in *Arabidopsis thaliana*. Plant Soil 268:285–292

Sakuma Y, Maruyama K, Qin F, Osakabe Y, Shinozaki K, Yamaguchi-Shinozaki K (2006) Dual function of an *Arabidopsis* transcription factor DREB2A in water-stress-responsive and heat-stress-responsive gene expression. Proc Natl Acad Sci USA 103:18822–18827

Salmond GPC, Bycroft BW, Stewart GSAB, Williams P (1995) The bacterial 'enigma': cracking the code of cell-cell communication. Mol Microbiol 16:615–624

Sauer K, Camper AK (2001) Characterization of phenotypic changes in *Pseudomonas putida* in response to surfaceassociated growth. J Bacteriol 183:6579–6589

Scher FM, Kloepper JW, Singleton C, Zaleski I, Laliberte M (1988) Colonization of soybean roots by *Pseudomonas* and *Serratia* species: relationship to bacterial motility, chemotaxis and generation time. Phytopathology 78:1055–1059

Schmidt KD, Schmidt-Rose T, Romling U, Tummler B (1998) Differential genome analysis of bacteria by genomic subtractive hybridization and pulsed field gel electrophoresis. Electrophoresis 19:509–514

Schnider U, Keel C, Blumer C, Troxler J, Defago G, Haas D (1995) Amplification of the housekeeping sigma factor in *Pseudomonas fluorescens* CHA0 enhances antibiotic production and improves biocontrol abilities. J Bacteriol 177:5387–5392

Shizuya H, Birren B, Kim U, Mancino V, Slepak T, Tachiiri Y, Simon M (1992) Cloning and stable maintenance of 300-kilobase-pair fragment of human DNA in *Escherichia coli* using an F-factor-based vector. Proc Natl Acad Sci USA 89:8794–8797

Sledjeski DD, Whitman C, Zhang A (2001) Hfq is necessary for regulation by the untranslated RNA DsrA. J Bacteriol 183:1997–2005

Sreenivasulu NA, Sopory SK, Kavi Kishor PB (2007) Deciphering the regulatory mechanisms of abiotic stress tolerance in plants by genomic approaches. Gene 388:1–13

Straus D, Ausubel FM (1990) Genomic subtraction for cloning DNA corresponding to deletion mutations. Proc Natl Acad Sci USA 87:1889–1893

Swift S, Throup JP, Williams P, Salmond GP, Stewart GS (1996) Quorum sensing: a population-density component in the determination of bacterial phenotype. Trends Biochem Sci 21:214–219

Thomashow LS, Weller DM (1996) Current concepts in the use of introduced bacteria for biological disease control: mechanisms and antifungal metabolites. In: Stacey G, Keen NT (eds) Plant microbe interactions. Chapman & Hall, New York, NY, vol 1, pp 187–236

Timmusk S, Wagner EGH (1999) The plant growth promoting Rhizobacterium *Pennibacillus polymyxa* induces changes in *Arabidopsis thaliana* gene expression: a possible connection between biotic and abiotic stress responses. Mol Plant Microbe Interact 12:951–959

Timmusk S, van West P, Gow NAR, Wagner EGH (2003) Antagonistic effects of *Paenibacillus polymyxa* towards the oomycete plant pathogens *Phytophthora palmivora* and *Pythium aphanidermatum*. In: Mechanism of action of the plant growth promoting bacterium *Paenibacillus polymyxa*. Uppsala University, Uppsala, Sweden, pp 1–28

Timmusk S, Grantcharova N, Gerhart E, Wagner H (2005) *Paenibacillus polymyxa* Invades plant roots and forms biofilms. Appl Environ Microbiol 71:7292–7300

Tosa Y, Mayama S, Nakayashiki H, Ohara Y, Wang Yan Qing (2005) Microarray analysis of the gene expression profile induced by the endophytic plant growth-promoting rhizobacteria, *Pseudomonas fluorescens* FPT9601-T5 in *Arabidopsis*. Mol Plant Microb Interact 18:385–396

Urgel ME, Salido A, Ramos JL (2000) Genetic analysis of functions involved in adhesion of *Pseudomonas putida* to seeds. J Bacteriol 182:2363–2369

Valverde A, Yaacov O, Burdman S (2006) How rhizobacteria promote plant growth: direct impact on plant development. FEMS Microbiol 265:186–194

Vashisht AA, Tuteja N (2006) Stress responsive DEAD-box helicases: a new pathway to engineer plant stress tolerance. J Photochem Photobiol 84:150–160

Vessey JK (2003) Plant growth promoting rhizobacteria as biofertilizers. Plant Soil 255:571–586

Vinocur B, Altman A (2005) Recent advances in engineering plant tolerance to abiotic stress: achievements and limitations. Curr Opin Biotechnol 16:123–132

Weber H, Polen T, Heuveling J, Wendisch VF, Hengge R (2005) Genome-wide analysis of the general stress response network in *Escherichia coli*: σ^S-dependent genes, promoters, and sigma factor selectivity. J Bacteriol 187:1591–1603

Weger de LA, van der Vlugt CIM, Wijfjes AHM, Bakker PAHM, Schippers B, Lugtenberg BJJ (1987) Flagella of a plant growth stimulating *Pseudomonas fluorescens* strain are required for colonization of potato roots. J Bacteriol 169:2769–2773

Weller DM, Landa BB, Mavrodi OV, Schroeder KL, De La Fuente L, Bankhead LB, Allende MS, Bonsall RRF, Mavrodi DV, Thomashow LS (2007) Role of 2,4-diacetylphloroglucinol- producing fluorescent *Pseudomonas* spp. in the defense of plant roots. Plant Biol 9:4–20

Westbrock-Wadman S, Sherman DR, Hickey MJ, Coulter SN, Zhu YQ, Warrener P, Nguyen LY, Shawar RM, Folger KR, Stover CK (1999) Characterization of a *Pseudomonas aeruginosa* efflux pump contributing to aminoglycoside impermeability. Antimicrob Agents Chemother 43:2975–2983

Whitchurch CB, Nielsen TT, Ragas PC, Mattick JS (2002a) Extracellular DNA required for bacterial biofilm formation. Science 295:1487

Whitchurch CB, Erova TE, Emery JA, Sargent JL, Harris JM, Semmler ABT, Young MD, Mattick JS, Wozniak DJ (2002b) Phosphorylation of the *Pseudomonas aeruginosa* response regulator AlgR is essential for type IV fimbria-mediated twitching motility. J Bacteriol 184:4544–4554

Woo SS, Jiang J, Gill B, Paterson AH, Wing RA (1994) Construction and characterization of a bacterial artificial chromosome library of *Sorghum bicolor*. Nucleic Acids Res 22:4922–4931

Xie H, Pasternak JJ, Glick BR (1996) Isolation and characterization of mutants of the plant growth-promoting rhizobacterium *Pseudomonas putida* GR12-2 that overproduce indoleacetic acid. Curr Microbiol 32:67–71

Zuo KJ, Qin J, Zhao JY, Ling H, Zhang LD, Cao YF, Tang KS (2007) Over-expression GbERF2 transcription factor in tobacco enhances brown spots disease resistance by activating expression of downstream genes. Gene 391:80–90

Chapter 5
Belowground Mycorrhizal Endosymbiosis and Aboveground Insects: Can Multilevel Interactions be Exploited for a Sustainable Control of Pests?

Emilio Guerrieri(✉) and Maria Cristina Digilio

5.1 Introduction

Terrestrial plants interact with an incredible variety of organisms. Some of these interactions are beneficial, some are detrimental; some develop in the aerial part of the plant, some at root level. The study of these interactions is a precious source of information that could be used to increase plant fitness, especially plant defence against insect pests and microbial pathogens.

Until the end of the last century, there had been a dramatic separation between research on belowground and aboveground interactions, although it has been possible to understand the basic rules of plant responses to beneficial and harmful organisms in both of the two "areas".

In 1980 it was suggested for the first time to investigate plant-insect interactions following a multitrophic approach, that is by considering each species as being an element of a food chain, having at its base the plant, at the intermediate level the complex of herbivore species (consumers) and at the top level the complex of entomophagous species (carnivores) (Price et al. 1980). This milestone paper opened a completely new field of investigation, and led, in a relatively short period of time, to the identification of unexpected mechanisms that regulate plant defences against insects (Fig. 5.1). In fact, along with the well known physical (e.g. thorns) and chemical (e.g. anti-feedant and toxic compounds) defences that directly affect the development and the reproduction of an invading insect, it has been demonstrated that plants can *indirectly* reduce the populations of herbivore insects by recruiting or enhancing the efficiency of natural enemies, either predators (e.g. ladybirds) and/or parasitoids (e.g. Ichneumonidea and Chalcidoidea wasps). This is accomplished through the production/release of attracting volatile organic compounds (VOC), by supplying food (such as extra floral nectars) or by providing shelter (e.g. domatia) to entomophagous species (Agrawal and Karban 1997; Dicke et al. 2003, Wäckers et al. 2005).

E. Guerrieri
Istituto per la Protezione delle Piante CNR Via Università 133 - 80055 Portici (NA) Italy
e-mail: guerrieri@ipp.cnr.it

C.S. Nautiyal, P. Dion (eds.) *Molecular Mechanisms of Plant and Microbe Coexistence.* Soil Biology 15, DOI: 10.1007/978-3-540-75575-3

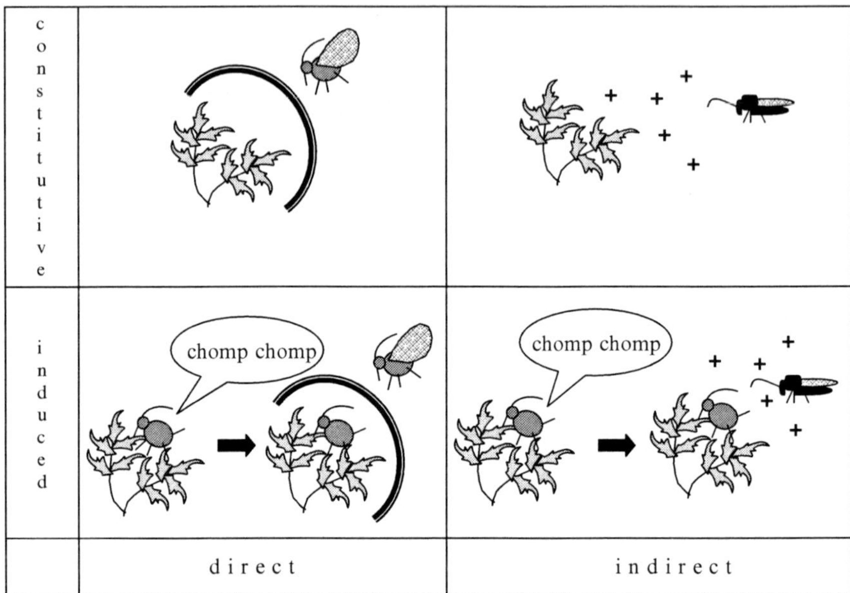

Fig. 5.1 Plant-defences against herbivore insects. Indirect defences are represented as the emission of VOC (+) attractive towards a parasitoid wasp

In many cases, plant defences, either direct or indirect, are activated only following herbivore damage, and for this reason they are referred to as *induced*, to distinguish them from *constitutive defences* that are always expressed, independently of herbivore attack (Agrawal et al. 1999 and references therein). It is documented that induced defences have a lower metabolic cost for the plant (Zangerl 2003) compared to constitutive ones, but they always need some plant damage to be activated; hence sometimes they can be economically inefficient. Nonetheless, the metabolic cost associated with induced defences is exacted only if pest attack occurs and can thus be less than that involved in constitutive defences (Simms and Fritz 1990). However, little is known about the ecological costs of induced defences that may include an increased susceptibility to untargeted herbivores (Cipollini et al. 2003).

The study of VOC that are attractive towards the natural enemies of insect pests has been one of the main research topics of agricultural entomology and biological control from the beginning of the 1990s (Vet and Dicke 1992). The theory predicted that for natural enemies of herbivore insects, the use of herbivore-induced VOC to locate their hosts/preys represents a winning strategy because they are both highly detectable and reliable (Vet and Dicke 1992). In any agricultural and forestry ecosystem the plant biomass is dominant; hence plant VOC are produced in large quantities that are easily detected by insect antennae. The release of such compounds in response to herbivore feeding activity makes them highly reliable for a natural enemy, given that there is a selective pressure on herbivore populations towards a strong reduction of the emission of individual (colony) odours (Vet and Dicke 1992).

So far, volatile compounds involved in these multitrophic interactions have been characterized in several herbaceous (e.g. Turlings et al. 1991; Birkett et al. 2003) and perennial systems (e.g. Scutareanu et al. 1997). This has led to the identification of insect elicitors (Mattiacci et al. 1995; Alborn et al. 1997), of metabolic pathways induced by plant damage (Walling 2000 and references therein; Schaller et al. 2005 and references therein), and most recently of the genes regulating the production/release of these semiochemicals (Schnee et al. 2006).

The time that the plant needs to activate the induced defences depends on the feeding habit of the invading herbivore. In this view, insects are usually divided into two main groups: chewers and suckers. Feeding larvae of Lepidoptera and Coleoptera are typical chewers, whose activity is always associated to a massive mechanical damage to plant tissues. Conversely, sap feeders like aphids and whiteflies, or cell-content feeders like thrips and spider mites, all belong to suckers, which cause low or null mechanical damage to infested tissues. As a consequence, evidence of plant response to chewers in terms of semiochemical production can be recorded in hours from the beginning of the attack (Turlings et al. 1998), whilst days are needed in the case of sap feeders (Guerrieri et al. 1999).

Interestingly, common patterns of plant responses to herbivore insects have been found regardless of the site of interaction. For example, maize root exudates released in response to the attack of a beetle (*Diabrotica virgifera*) selectively guide a parasitic nematode of this pest (*Heterorhabditis megidis*) to its host larvae (Rasmann et al. 2005). One of the compounds involved in this attractiveness was the terpene E-ß-caryophyllene that, in a different multitrophic system, is released by the aerial part of the plant in response to the attack/oviposition by a bug pest (*Nezara viridula*), and showed a similar attractive function towards the egg-parasitoid (*Trissolcus basalis*) of this pest (Colazza et al. 2004). More recently, this compound has been proved to regulate the flight behaviour of the aphid parasitoid *Aphidius ervi* towards tomato plants infested by aphids (Sasso et al. 2007).

Nonetheless, it has been demonstrated that plant responses to herbivores, regardless of the site of interaction, are usually systemic and thus their effects can also be recorded in the undamaged parts of the plant (Turlings and Tumlinson 1992; Rose et al. 1996; Guerrieri et al. 1999; Soler et al. 2007). These findings, along with the consideration that plants often suffer multiple attacks by different organisms, prompted a series of investigations where the plant was considered as a living connection between the two separated environments. Hence, the hypothesis was formulated that there could be a mutual influence between organisms living belowground and those living aboveground (Rillig 2004; Wardle et al. 2004; Bezemer et al. 2005).

Arbuscular mycorrhizal fungi (AMF, phylum Glomeromycota) are among the most common microbial organisms in the soil, constituting endotrophic symbiotic associations with plant roots (Arbuscular Mycorrhizae, AM) reported for approximately 80% of vascular land plants. This symbiosis is considered a crucial factor for most terrestrial ecosystems and has a high potential of application to plant production and defence (Smith and Read 1997).

There are several advantages that the plant experiences from AM symbiosis, including phosphate and other nutrients supply (Marschner and Dell 1994; Harrison and van Buuren 1995, Joner et al. 2000; Hodge et al. 2001) especially in phosphorus-deficient soils, a better resistance to drought (Augè 2001) and a significant higher degree of bioprotection against various pathogens, including nematodes (Pinochet et al. 1996; Borowicz 2006 and references therein), fungi (Azcón-Aguilar and Barea 1997; Borowicz 2001; Fritz et al. 2006) and even insect pests (Guerrieri et al. 2004). A positive effect of AM fungi on soil structure has been indicated, making them a key component of sustainable agriculture (Johanson et al. 2004; van der Heijden et al. 2006). There is also evidence that defence responses induced by AM are systemic (see Liu et al. 2007 and references therein), thus indicating these symbioses as potential candidates for corroborating the hypothesis of mutual belowground-aboveground interactions.

Many parameters affect the final outcome of belowground AM symbiosis, one being the species-specificity of several plant-fungal associations (Kendrick 1992; Klironomos 2000). For example, by using a molecular approach it has been demonstrated that co-occurring grass species associate with a non-random set of AMF (Bever et al. 1996; Vandenkoornhuyse et al. 2003). Even on the same plant species, interesting differences in a number of physiological traits emerged by comparing different species/isolates of AMF (Hart and Reader 2002), leading to differences in the effects on plant physiology and growth (Klironomos 2003). Similarly, it has been recently demonstrated that the presence and the identity of AMF has a direct influence on the competitiveness between legume crops and weeds (Scheublin et al. 2007). On the other hand, the presence of plant growth promoting rhizobacteria (PGPR) seems to play a role in the outcome of plant-AM interaction (Filion et al. 1999; Gamalero et al. 2004 and references therein).

For all these reasons, it is not surprising that biological, genetic and chemical aspects of AM symbiosis have been thoroughly investigated (Franken and Requena 2001; Strack et al. 2003; Rillig 2004; Balestrini and Lanfranco 2006).

In this chapter we will focus on the mutual influence between aboveground insects and belowground AM fungi as mediated by the plant, indicating the outcome and the main parameters that regulate the top-down and bottom-up effects. We will also examine how modern techniques can be used to characterize these multilevel interactions as well as the signal-transduction pathways that are involved, indicating the possible cross talk between them. Finally, we will discuss how the thorough characterization of these multilevel interactions among AM symbiosis, herbivore insects and their natural enemies, can be used in the postulation of novel and sustainable strategies to control insect pests.

5.2   The Effect of Aboveground Herbivory on AM Symbiosis

It is not always simple to separate clearly the top-down from the bottom-up effects during a contemporary presence of herbivore insects and AMF on the same plant, especially in long-term interactions. It is probably for this reason that so far, only two mycorrhizal systems have been deeply investigated to assess the influence of a herbivore insect on the development of the fungal symbiont (Gange et al. 2002; Wamberg et al. 2003), whilst several papers have been published in which aboveground damage was performed by vertebrate species or simulated by artificial clipping (reviewed by Ghering and Whitham 2002; Klironomos et al. 2004).

Following initial observations carried out in 1997, Gange and collaborators examined the influence of the leaf-chewing caterpillar *Arctia caja* (Lepidoptera) on root colonization of *Plantago lanceolata* by *Glomus intraradices* through both laboratory and field experiments (Gange and Bower 1997; Gange et al. 2002). In the laboratory, the feeding activity of the lepidopteran larvae nearly halved the levels of AM colonization, although this effect was reached after five events of defoliation. Similarly, in manipulative field experiments that included selective applications of insecticide and fungicide, a negative interaction between *A. caja* attack and *G. intraradices* colonization was recorded. The authors hypothesised that in this system the mycorrhizal symbiont suffered of the reduction of nutritive compounds following severe herbivory by insect chewers (Gange et al. 2002).

More recently, in a detailed study on pea plants, it was demonstrated that the effect of aboveground insect herbivory on root colonization by the same AMF species (*G. intraradices*) changes in relation to the physiological status of the plant (Wamberg et al. 2003). More precisely, the feeding activity by adult weevils of *Sitona lineatus* induced a marked increase of root colonization by *G. intraradices* during the nutrient acquisition phase, whilst the reverse was recorded during the reproductive phase of the plant (Wamberg et al. 2003). It was theorized that, during the vegetative phase (days 0–25), the plant compensates the loss of nutrients due to herbivory by investing on roots, i.e. by transferring more carbon belowground, that is exploited by AM symbiosis. During a later reproductive phase (from day 30 onwards), the plant transfers more resources to flowers and seeds; hence, there is a lack of carbon to be sent to roots and this leads to a progressive reduction of AM colonization (Wamberg et al. 2003).

A further piece of knowledge to the puzzle of top-down effects of aboveground herbivory on AM colonization has been recently added by Klironomos et al. (2004), although no living organism but artificial clipping was used to cause foliar damage to *Bromus* plants. In this study, it was demonstrated that the extent of the clipping effect was dependent on which fungal species was associated with the plant (Klironomos et al. 2004). The authors concluded that it is extremely important to know the composition of fungal inoculum because the response of individual AMF monocultures cannot be used to predict the response of multi-species AMF assemblages (van der Heijden et al. 1998a,b). In other words, the latter is neither a linear function of single AMF species responses nor a mirror of the most responsive AM fungal species. In this study, it was also demonstrated that there could be qualitative effects of aboveground stresses on AM development other than, or along with, quantitative ones. In this view, many other parameters (i.e. vesicular colonization, arbuscular colonization, extraradical hyphal length) must be measured in addition to the most commonly used "root length colonised" (percent AMF colonization × root length), because phenological changes in AMF do not necessarily occur in all mycorrhizal structures at the same time (Klironomos et al. 2004).

In a wider perspective, given the general negative influence of insect herbivory on the AM symbiosis, it cannot be excluded that in agricultural ecosystems at least a part of the "product losses" referred to insects are in fact caused by a reduction of AM symbiosis.

Finally, it must be noted that all the studies about the effect of aboveground herbivory on AM colonization have considered plant damage as caused exclusively by chewers, either insects or vertebrates (including artificial clipping). As a consequence, it remains completely unexplored whether feeding activity by aboveground insect suckers, e.g. aphids and whiteflies, has or has not consequences on the development/colonization by AMF.

5.3 The Effect of AM Symbiosis on Plant Direct Defences Against Herbivore Insects

Plant induced defences can turn into resistance against herbivores through either a compensating replacement of damaged tissues (tolerance) or a reduction of the herbivore's fitness (true resistance). In other words, the metabolic pathways involved in plant resistance belong to either primary or secondary metabolism of the plant.

The larger availability of soil nutrients, in particular of P, N and K, delivered to plant roots by AM fungi (Marschner and Dell 1994; Joner et al. 2000; Hodge et al. 2001), could be the ideal pre-requisite for an induced tolerance response by mycorrhizal plants towards herbivore insects, especially chewers (McNaughton and Chapin 1985).

So far, only two detailed studies, both involving plant chewers, have investigated this possibility with different results. Borowicz (1997) indicated that the presence of the AM fungus *Glomus etunicatum* did not change the tolerance level of soybean plants towards the Mexican bean beetle *Epilachna varivestis*. More recently, a different outcome of mycorrhizal influence on the tolerance of prairie plants towards grasshoppers was reported (Kula et al. 2005). In this study, root colonization of eight prairie plant species by a complex mix of AM fungi (including several species of *Glomus*) resulted in a compensatory regrowth after defoliation by the grasshopper *Melanoplus bivittatus* (Kula et al. 2005). In particular, the total aboveground plant biomass of mycorrhizal plants was nearly double the biomass production of non-mycorrhizal plants after they have been grazed by grasshoppers. This result was due principally to the response of two dominant C_4 grass species (Kula et al. 2005). It is not possible to compare the results of these two studies because they have been recorded in two completely different multitrophic systems. However, in the prairie experiments, the higher number of possible combinations between plant and mycorrhizal fungi species or a synergistic effect of "multiple partners" could have played a role in the final tolerance response towards the herbivore insect (van der Heijden 1998a,b).

In contrast, quite a few studies have investigated the effect of belowground AM symbiosis on plant true resistance towards aboveground herbivores, with results that pointed out the extreme complexity of these interactions. However, a general trend of negative impact of AM on chewer insect performances has been indicated.

In a pioneering paper, Rabin and Pacovsky (1985) reported a slower development and a lower weight at pupation in both *Spodoptera fungiperda* and *Helicoverpa zea* fed on excised leaves of soybean plants colonized by the AM fungus *Glomus fasciculatum*, in respect of larvae fed with leaves coming from P-fertilized plants. The authors then concluded that a true resistance response was induced by AM symbiosis in soybean plants (Rabin and Pacovsky 1985). Surprisingly, these results were not confirmed later on by Borowicz (1997), who investigated the effect of AM symbiosis on soybean resistance towards the Mexican beetle *Epilachna varivestis*. In fact, the presence of *G. etunicatum*, associated with low-P-fertilizer, resulted in higher larval mass and pupation rate of the herbivore (Borowicz 1997). In order to understand whether it is the species specificity of AM-plant interaction that plays a key role in the soybean response to chewers (see Sanders 2002 and references therein), it would be extremely interesting to cross test the development of *E. varivestis* on excised leaves of soybean plants colonised by *G. fasciculatum*, as well as the response of both *S. fungiperda* and *H. zea* when reared on soybean plants colonized by *G. etunicatum* and grown at low level of P-fertilizer. However, it must be noted that *Epilachna* feeds in a completely different way from *Spodoptera* (and *Helicoverpa*), and this can influence the plant response. The former usually scratches the leaf surface to feed on plant juices, more or less like thrips (Thripidae) whilst *Spodoptera* ingests plant tissues by using its powerful mandibles.

The same negative effect on the resistance of *Lotus corniculatus* towards a chewer insect, the common blue butterfly *Polymmatus icarus*, was reported by Goverde et al. (2000). However, a species-specific effect of AM fungus on herbivore fitness was demonstrated in this system (Goverde et al. 2000), as it was for the plant *Leucanthemum vulgare* and the leaf-miner *Chromatomyia syngensiae* (Diptera: Agromyzidac) in laboratory tests (Gange et al. 2003).

In detail, the percentage of plant leaves that were mined by the fly significantly varied with the AM fungal species considered, with the association *Glomus caledonium+G. fasciculatum* inducing the highest rate of attack and *G. caledonium+G. fasciculatum+G. mosseae* the lowest (Gange et al. 2003). Conversely, the theory of species specificity of plant-fungal associations did not apply to strawberry, where both root-feeding larvae and shoot-feeding adults of the black vine weevil (*Otiorrhyncus sulcatus*) were negatively affected by the presence of either *Glomus mosseae* or *G. fasciculatum* (Gange 2001). Similarly, the thistle gall fly was reported to reduce its performances, in terms of number of galls/plant and average weight of larvae, on mycorrhizal *Cirsium arvense* plants (Gange and Nice 1997).

An interesting link between the physiological state of the plant and the effect of AM plant on leaf-chewing insects has been reported for pea plants colonised by *Glomus intraradices* and attacked by adult weevils (*Sitona lineatus*) (Wamberg et al. 2003). During the vegetative phase (days 0–25) there was no difference in the range of leaf damage by chewing insects whilst a drastic reduction of *S. lineatus* attack was recorded during the reproductive phase (from day 30 onward). The authors hypothesised that the translocation of nutrients and carbon to the reproductive organs leads to a decrease of leaf quality, which in turn hampers herbivore attack (Wamberg et al. 2003).

Finally, field experiments with natural occurring populations of AM, shaped by means of application of fungicides, confirmed the trend of a negative effect of endomycorrhizal symbiosis on chewing insects, an outcome that resulted permanent on *Plantago lanceolata* attacked by *Arctia caja* (Lepidoptera) (Gange and West 1994) and transient on eucalyptus trees attacked by unidentified geometrid larvae (Lepidoptera) (Gange et al. 2005).

There are many possible explanations for these contrasting results, one being the repeatedly cited species specificity of different AM-plant associations effects. However, two more hypotheses have been formulated to justify either a positive or a negative influence of the mycorrhizal symbiosis on the performances of chewers, both based on the chemical alterations occurring in colonised plants. A detrimental effect on chewer development can be related to the production of toxic compounds induced by AM symbiosis, such as phenolics (Morandi 1996), terpenoids (Peipp et al. 1997) and isoflavonoids (Vierheilig et al. 1998). All these compounds are known to belong to the plant battery of defensive substances whose production is often associated to herbivore attack (Mullin et al. 1991; Dakora 1995). Conversely, the nutritional theory predicts that mycorrhizal plants are qualitatively (and often quantitatively) better that non-mycorrhizal ones, thus providing a better food for herbivores that results in better performances.

A contrasting scenario has been reported for the resistance response towards sap feeders induced in plant by AM symbiosis. This is probably due to the scarcity of studies, namely five, that have investigated this interaction (Pacovsky et al. 1985; Gange and West 1994; Gange et al. 1999; Guerrieri et al. 2004; Wurst et al. 2004).

Pacovsky et al. (1985) found no effect of colonization by *Glomus fasciculatum* on the reproduction of the aphid *Schizaphis graminum* developing on sorghum. In the field, with natural occurring populations of AM fungi, all the biological parameters considered (i.e. weight, embryo number and their development) of the generalist aphid species *Myzus persicae* on *Plantago lanceolata* were positively affected by the symbiosis (Gange and West 1994). These results were partially confirmed in a subsequent laboratory bioassay in which the same plant, two aphid species (*M. persicae* and *M. ascalonicus*) and a single symbiotic fungus (*Glomus intraradices*) were tested (Gange et al. 1999). In a more recent study on the same multitrophic system, the development time of *M. persicae* on *P. lanceolata* was delayed in the presence of the AM fungus *Glomus intraradices*, but was accelerated in the presence of both AM and earthworms (Wurst et al. 2004). Moreover, no effect of either AM or earthworms on aphid reproduction was recorded, and this was in clear contrast with the results of previous studies on the same system (Gange and West 1994; Gange et al. 1999). The sole difference between the two studies was the sterilization of the soil in the more recent one that could have mobilized more nutrients, thus attenuating the effects of either AM or earthworms on plant responses (Wurst et al. 2004).

In accordance with Wurst et al. 2004, Guerrieri et al. (2004) found that tomato plants colonised by the AM fungus *Glomus mosseae* became more resistant to the aphid *Macrosiphum euphorbiae* with respect to non-mycorrhizal plants, with only 16% of aphids reaching the adult stage and about 8% reproducing. One possible reason for this negative effect of mycorrhizae on aphid development is the synthesis of toxic compounds induced by AM in tomato plants that are known to be very rich in substances with defensive properties. However, another intriguing hypothesis can be formulated. It is known that in tomato, the aphid *M. euphorbiae* and the fungal pathogen *Phytophtora infestans* have similar effects on the expression of genes associated with plant defences, namely *P4* and *LOX* (Fidantsef et al. 1999) in accordance with modern theories that compare hemipteran herbivores (suborder Sternorrhyncha, including aphids, whiteflies, psyllids and scale insects) to plant pathogens (Kaloshian and Walling 2005). Also AM fungi do elicit a defensive response during the initial colonization, although this was noted to decline or to be subsequently downregulated as the symbiosis developed (Vierheilig 2004; Harrison 2005). However, this decline was not observed in the noncolonized regions of the root (Harrison and Dixon 1994).

5.4 The Effect of AM Symbiosis on Plant Indirect Defences Against Herbivore Insects

The possible interactions between AM symbiosis and the attraction of natural enemies of herbivore insects have been investigated by Guerrieri et al. (2004) (Fig. 5.2).

In the above-mentioned paper a multidisciplinary approach was followed by integrating the expertise of plant biologists, chemists and entomologists. In wind tunnel bioassays, the authors demonstrated that, in tomato plants, mycorrhizal symbiosis and aphid infestation produced similar results in terms of attractiveness towards an insect parasitoid, namely, the parasitic wasp *Aphidius ervi*, one of the most effective and studied natural enemies of the potato and tomato aphid (*Macrosiphum euphorbiae*)(Guerrieri et al. 1993, 1997, 1999, 2002). In detail, tomato plants colonised by the AM fungus *Glomus mosseae* became significantly more attractive towards *A. ervi* than control, non-mycorrhizal plants in complete absence of aphid infestation. Moreover, the percentage of female wasps landing on mycorrhizal plants was comparable to that recorded for tomato plants infested by *M. euphorbiae*. It was hypothesised that the possible basis of this similarity could be in the genes that are induced by both *M. euphorbiae* infestation and *Glomus mosseae* colonization (Guerrieri et al. 2004). A first step towards the characterization of tomato responses in terms of attractiveness towards parasitic wasps has recently led to the identification of the VOC released in response to aphid attack (Sasso et al. 2007). In accordance with other studies about tomato response to different herbivore species (Kant et al. 2004 and references therein), the differences

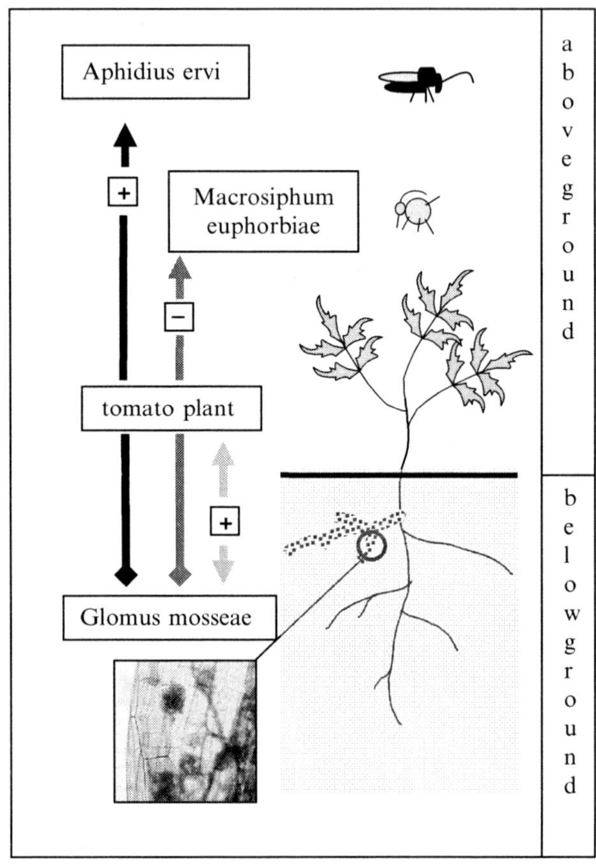

Fig. 5.2 Outcome of belowground-aboveground interactions in a multitrophic system. (+) indicates a positive interaction; (−) indicates a negative interaction (see text for explanation)

recorded between the emissions collected from uninfested and aphid-infested plants were only quantitative. Among the identified compounds whose release significantly increased following aphid attack, several terpenes, i.e. α-pinene, *E*-ß-ocimene and *E*-ß-caryophyllene, and methyl salicylate, were found to be involved in the long-range attractiveness of *A. ervi*. The identification of VOC released by tomato plants colonised by *G. mosseae* is currently in process and will help in understanding whether aphids and mycorrhizal fungi activate similar pathways of response (Guerrieri et al., in preparation).

The effect of mycorrhizal symbiosis on the final percentage of parasitism has been investigated in both field and laboratory tests by Gange et al. (2003). As reported for the attack rate by the leaf-miner *Chromatomyia syngensiae* (Diptera: Agromyzidae) on *Leucanthemum vulgare* (see above), in laboratory experiments the parasitism rate by *Digliphus isaea* (Hymenoptera: Eulophidae) was shown to be dependent on the AM fungal species involved (Gange et al. 2003). More precisely, mines on plants colonised with the single inoculum of *Glomus fasciculatum* suffered the highest rate of parasitism, whilst the inoculum of *G. mosseae* alone or together with *G. fasciculatum* caused a highly significant reduction in the parasitism rate (Gange et al. 2003). One possible explanation of these negative associations could have been the different plant response to diverse AMF species, in terms of growth and number of leaves that in turn could have affected the searching efficiency by an insect parasitoid (Cloyd and Sadof 2000). For example, the colonization by *G. mosseae* alone resulted in the highest number of leaves in *L. vulgare* plants, even though this was not associated to the lowest percentages of leaves mined, thus suggesting the existence of other mechanisms regulating these interactions (Gange et al. 2003).

5.5 Signal-Transduction Pathways Involved in Plant Response to AM and to Herbivore Insects

Induced responses to insects and pathogens rely on the circulation of signal molecules that alert the plant and eventually protect it from further attacks. Two sets of responses that regulate plant resistance have jasmonic acid (JA) and salicylic acid (SA) as key signal components (Fig. 5.3) (see Agrawal et al. 1999 and references

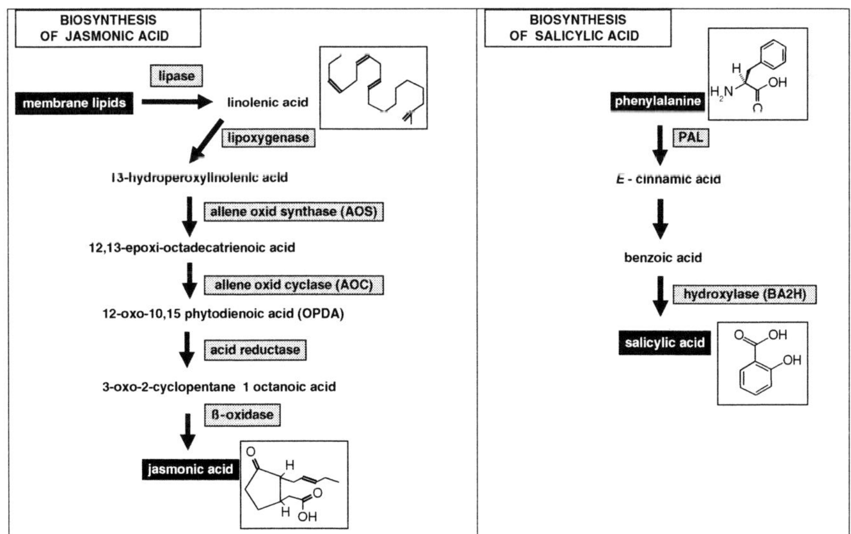

Fig. 5.3 Metabolic pathways of jasmonic acid and salicylic acid

therein). There is evidence that the JA pathway is mainly triggered by insects and heavy mechanical wounding of plant tissues whilst SA is switched on by pathogen infection.

Membrane disruption and liberation of lipids constitute the initial substrate of the octadecanoid pathway that starts with the production of linolenic and linoleic acids released from plastidial membranes by phospholypases and catabolized by enzymatic and nonenzymatic reactions to produce a set of oxygenated lipids (oxylipins), including JA and its methyl ester (MeJA) (Fig. 5.3) (see Schaller et al. 2005 and the entire special issue of J Plant Growth Regul 2005, volume 23, number 3). Circulation of JA induces the accumulation of proteinase inhibitors, polyphenol oxidase, and steroid glicoalkaloids, making the plant more resistant to further attacks by insects (Staswick and Lehman 1999). The activation of the octadecanoid pathway coordinates the contemporary release of VOC (terpenes, aldehydes and alcohols) that are known to be involved in the attractiveness of natural enemies of insect pests (Birkett et al. 2000; Thaler 2000).

The biosynthetic pathway of SA appears to begin with the conversion of the amino acid phenylalanine to *E*-cinnamic acid (*E*-CA) catalysed by phenylalanine ammonia lyase (PAL) (Fig. 5.3). The conversion of *E*-CA into SA proceeds via chain shortening to produce benzoic acid (BA), followed by hydroxylation to derive SA. The accumulation of SA is required for the induction of the systemic acquired resistance (SAR) which provides a protection against plant diseases (Ryals et al. 1996; Gozzo 2003 and references therein).

However, more recently it has been demonstrated that plant response towards invading organisms is far more complicated than this simplistic scenario. For example, insect suckers like aphids, whose intercellular stylets produce little or no mechanical damage, are perceived by the plant as if they were pathogens, which causes a concomitant activation of both SA and JA (Kaloshian and Walling 2005). In tomato it has been demonstrated that the attack of *Macrosiphum euphorbiae* or *Myzus persicae* induces the expression of both LOX that is correlated to JA pathway and pathogen related protein P4 that is linked to the activation of a SAR response following the accumulation of SA (Fidantsef et al. 1999).

The peculiarity of plant response to sap feeders can also be seen in the composition of volatiles released by plants attacked by aphids. In soybean and tomato, among the few compounds whose production is significantly higher in respect to that of uninfested plants, there are methyl salicylate, the volatile ester of SA, and several terpenes from the octadecanoid pathway (Zhu and Park 2005; Sasso et al. 2007). These findings confirm previous observations carried out on different systems, underlining the role of JA on direct and indirect defences against sap feeders (Birkett et al. 2000; Thaler 2000; Cooper and Goggin 2005) and have been recently reinforced by using tomato mutants (Corrado et al. 2007).

Recent evidence on the variegated role of jasmonates in plant responses have been reported (Peña-Cortés et al. 2005). In the model plant *Arabidopsis*, mutants defective in JA related processes showed susceptibility to normally non-pathogenic soil-borne oomycetes of the genus *Pythium*, increased susceptibility to *Fusarium oxysporum*, *Alternaria brassicola* and *Botrytis cynerea* and to the bacterial leaf pathogen *Erwinia carotovora*, as well as impairment of induced resistance against Cucumber mosaic virus (CMV) (Pozo et al. 2005). In accordance with these findings, several cases of increased resistance towards pathogens have been demonstrated in plants overexpressing JA related proteins (see Pozo et al. 2005 and references therein).

More interestingly, there is evidence that the recognition and the control of the intimate symbioses, such as AM and *Rhizobium*, involve the defence-related pathways. For example, it has been reported that treatments with JA significantly increased the percentage of AM infected root length and speeded up the process of colonization by *Glomus* inoculum in *Allium sativum* plants (Regvar et al. 1996). Moreover, the synergistic effect of AM+JA resulted in a greater shoot length and fresh weight in respect to either non-treated or non-inoculated plants (Regvar et al. 1996).

These findings were confirmed and reinforced later on by Hause et al. (2002). By following a molecular approach, these authors elegantly demonstrated that in barley roots the process of mycorrhization by *Glomus intraradices* is associated to a fivefold accumulation of endogenous jasmonates, as demonstrated by transcripts of AOS (see Fig. 5.3) and JIP23, a 23-kD protein that accumulates in barley leaves following JA treatment (Hause et al. 2002). The authors also demonstrated that transcripts and proteins of these two genes accumulate within arbuscule-containing cells, thus inferring a causal link to mycorrhization (Hause et al. 2002).

It must be noted that the bacterial phytotoxin *coratine* also elicits the expression of jasmonate-induced proteins in tomato (see Mithöfer et al. 2005 and references therein).

The increased attractiveness of mycorrhizal tomato towards the aphid parasitoid *Aphidius ervi* reported by Guerrieri et al. (2004) could represent a further demonstration of JA involvement in AM symbiosis. However, only the characterization of volatile compounds released by mycorrhizal tomato and their comparison with those released by aphid infested tomato could shed light on these intriguing interactions (Guerrieri et al., in preparation).

Far more complicated is the scenario in cases of multiple interactions such as those involving the contemporary presence of microbes (beneficial and/or pathogens) and insects. There have been a few studies investigating the possible cross-talk between JA and SA metabolic pathways in response to multiple "elicitation" on both model and agricultural plants (Kunkel and Brooks 2002; Thaler et al. 2002a,b; Glazebrook et al. 2003). Although it appears that the interaction between these two pathways is complex, there is evidence that in the majority of cases it results in a mutual antagonism (Gupta et al. 2000). For example, in tomato plants, SA and its related compound acetyl salicylic acid (ASA) have been shown not only to inhibit proteinase inhibitor synthesis induced by wounding and oligouronides (Doherty et al. 1988) and by linolenic acid (Peña-Cortés et al. 1993), but also to impair all the induced defensive mechanism based on JA and on the tomato hormone systemin (Doares et al. 1995). These SA effects have been reported to act at different sites of plant responses, either along the octadecanoid pathway (Fig. 5.3), thus stopping JA synthesis, or immediately after it, thus blocking the transcription of defensive genes (Doherty et al. 1988; Peña-Cortés et al. 1993; Doares et al. 1995). Similarly, it has been demonstrated that, in tobacco plants, JA inhibits the expression of SA-dependent genes encoding for pathogenesis-related (PR) proteins (Niki et al. 1998).

The same antagonism between SA and JA pathways has been further demonstrated in the model plant *Arabidopsis thaliana*. For example, *eds4* and *pad4* mutants, who are impaired in SA accumulation, exhibit enhanced responses to inducers of JA-dependent gene expression (Gupta et al. 2000). However, there are many parameters to be considered while assessing the final outcome of SA and JA interaction on plant defences against insect pests and pathogens. For example, concentration, timing of elicitation and life style of plant parasites all play a key role on tomato plant defensive performances (Thaler et al. 2002a,b). By using chemical elicitors, such as the SA functional analogue benzothiodiazole (BTH) and JA, these authors demonstrated that SA pathway had a stronger effect on JA pathway than did the reverse. Moreover, the negative interaction in the biochemical expression of the two pathways was most consistent in the case of simultaneous elicitation compared to when a two-day time lag passed between the applications of single elicitors. Interestingly, the application of BTH and JA at low concentration produced inconsistent antagonism (Thaler et al. 2002a).

Finally, it was demonstrated that the negative interaction between JA and SA pathways had biological consequences that varied among the herbivores and pathogens tested, thus making impossible to formulate a general theory that could be applicable in the control of plant parasites (Thaler et al. 2002b).

5.6   New Tools in the Study of Multitrophic Interactions

In this section we will indicate how the exploitation of techniques capable of identifying altered gene expression can point out genes possibly involved in plant defence. Actually, the exploitation of the emergent technologies made available for researchers of AM/plant interaction can easily accommodate one or more biotic levels, such as insect herbivores and their natural enemies.

An important instrument in the molecular approach to the study of plant/other organisms interactions is the crucifer *Arabidopsis thaliana*, whose genome has been elucidated. However, this classical model plant is not useful in the study of plant/AM fungi/insect interactions, because of its inability to host AM fungi, as the majority of Brassicaceae; therefore, the only model plant we can use with this purpose is *Medicago truncatula*, whose genome sequencing project is nearly completed and can be seen at the URL http://www.medicago.org. The roots of this legume can host, besides the nitrogen fixing rhizobial bacteria, additional symbiont microorganisms, including AM fungi. The presence of bacteria on the roots of legume species is certainly a further complication to be taken into account while studying belowground-aboveground interactions that could have beneficial effects on plant defence; indeed legume mutants resistant to AM fungus colonization (myc⁻) have been obtained, for example, through chemical induction (Duc et al. 1989). A number of cases of mutant myc⁻ plants species in legumes are reviewed by Peterson and Guinel (2000) and Rillig (2004).

Legumes, though, are not a typical mycorrhizae host, and non-legumes are a necessary instrument for the study of multilevel interactions. For this reason, plants resistant to AM fungus colonization (myc⁻) have been obtained in other plant families. For example, several mutagenized myc⁻ tomato plants are now available, preventing the establishment of AM symbiosis at different levels. In some cases, mutants are totally resistant to AMF infection and colonization and elicit a lower spore germination and appressoria formation in respect to wild type (myc⁺) (David-Schwarz et al. 2001) and these effects can be displayed with different intensity (David-Schwarz et al. 2003).

Other myc⁻ tomato plants stop the penetration of the root surface and symbiosis is associated with minimal accumulation of defence gene mRNAs, differently from myc⁺ plants (Gao et al. 2004).

In the symbiosis between maize and *Glomus mosseae*, a series of mutants has been identified, lacking the ability to form appressoria, or forming appressoria of normal morphology and activity but in reduced number and unpaired in colonization (Paszkowski et al. 2006).

The use of myc⁻ mutants could help in unravelling the intimate mechanisms regulating the plant resistance response towards herbivore insects that occurs at different stages of AM symbiosis. For example, it is still not clear whether it is the mechanical damage caused to plant tissues by the AM fungus penetration, or its growth into the roots or the formation of arbuscules and, when present, of vesicles, that triggers the cascade of events leading to plant resistance response, either direct and indirect.

In addition, experimental design can take advantage from the use of such mutants, allowing the cultivation in the same soil of mutant and control plant, as in the case of the same container (Neumann and George 2005) or also when upgrading from the pot to the field scale (Rillig 2004).

Tomato mutants have been collected over several decades and today more than a thousand monogenic stocks are described and deposited in the CM Rick Tomato Genetic Resource Center at University of California, Davis, and can be seen at the URL http://tgrc.ucdavis.edu/.These stocks are from several sources: spontaneous and induced mutants, natural variants from the edible tomato (*Lycopersicon esculentum*) and wild relatives (e.g. *L. hirsutum*). Tomato mutants have been particularly used to assess the metabolic pathways involved in plant response to biotic stresses (see above) and can equally be used to assess the genes involved in multilevel interactions (see Emmanuel and Levy 2002 and references therein).

All the advantages of the use of mutants apply equally to transgenic plants, in which either gene silencing or overexpression can help to dissect the outcomes of the plant response to the colonization of herbivore insects on the attacking organisms or on the attractiveness towards natural enemies.

Molecular biology techniques for the study of AM symbiosis are based on the establishment of cDNA libraries, deriving from mRNA extracted separately from fungal spores and from the roots of mychorrhizal plants and of non-mychorrhizal control plants. Several induced genes have been characterized by means of differential screening of cDNA libraries, differential RNA display or subtractive hybridization (Franken and Requena 2001). The use of DNA arrays is a powerful technique, allowing one to confront through hybridization the cDNAs from the three different organisms/symbionts and to analyse a large number of genes.

An interesting technique is based on the partial cloning of cDNAs, leading to the definition of expressed sequence tags (EST) that can be screened for similarity in DNA databases, in order to hypothesise a biological function. In the model plant *M. truncatula* EST analysis has been extensively applied, including root tissues colonized by AM fungi and other mutualistic or pathogenic microbes, and control roots. Similarly, Thompson and Goggin (2006) lately reviewed transcriptomics approaches to the study of the interactions between plant and phloem feeding insects (aphids, whiteflies, and planthoppers).

RNA probes constituted from 18 S rRNA deriving from plant and fungus can be used for Northern blot analysis; the use of 18 S rRNA probes in RNA protection assay (RPA) allows the discrimination and quantification of RNAs of fungal and plant origin (Maldonado-Mendoza et al. 2002).

The molecular approach in the study of multitrophic interactions even allows one to carry out *in silico* (i.e. computer-based) analyses and experiments (Strack et al. 2003). In the study of multilevel interactions, plant response to herbivore insects can be evaluated in presence and absence of AM fungi, in terms of overexpression or downregulation of genes associated with plant defence. Still another biotic level can be added to the system, as AM symbiosis positively affects the attraction of natural enemies of insect herbivores (often economic pests), through the altered profile of plant volatile emissions that constitute the necessary cues for parasitoid and predator insects to seek their hosts/prey (see the paragraph "The effect of AM symbiosis on plant indirect defences against herbivore insects"). Specific techniques are required for the study of the volatile compounds produced by the plant in presence or absence of AM fungi. In synthesis, plant odours are collected by air-tight systems (air entrainments) from the whole plant (head-space) or from a single leaf and directed towards a trap containing an adsorbent (e.g. Tenax). Volatile compounds are then released from the trap by either high temperature (thermal desorption) or solvents (chemical desorption) and analysed by a gas chromatograph usually coupled with a mass spectrometer (GC-MS). More recently, an innovative, solvent free technology has been made available, the solid phase microextraction (SPME). SPME utilises fiber coated with a liquid (polymer), a solid (sorbent), or a combination of both that removes the compounds from the sample by adsorption. The SPME fiber is then inserted directly into the gas chromatograph for desorption and analysis. Regardless of the type of collection and desorption, the peaks appearing in the chromatograms are identified by confrontation with available standards or databases (see, for example, Birkett et al. 2003). The ability to manipulate the searching behaviour of natural enemies is a most desirable prospect in modern plant protection (see below), and a better knowledge of the molecular basis for the changed plant odours would be very advantageous.

However, the power of the proteomic techniques depicted above might misleadingly suggest that, having once designed an experiment detecting AM and control plants, both infested and not, a search for the genes involved in plant defence is just a matter of technicalities. In fact, micro-array analysis, a very common gene profiling tool, requires the previous identification of a set of relevant transcripts (Thompson and Goggin 2006).

A possible start in the scrutiny of genes comes from the examination of the available literature. Balestrini and Lanfranco (2006) recently reviewed the genes transcriptionally induced or regulated in the different phases of the establishment of the symbiosis, from the fungal spores to the early stages of the interaction with root tissues, and to the symbiotic phase. AM symbiosis certainly has a huge impact on plant gene expression, and some of these might be appealing for plant protection applications; increased transcripts of chitinase have been reported in *M. truncatula* cells containing arbuscules, and a PR10 and a wound-induced protein are also reported (see Balestrini and Lanfranco 2006 and references therein). Those PR proteins, i.e. Pathogenesis Related proteins, are induced in the plant as a product of the metabolic Pathways activated in response to pest attack. The potential of chitinolytic enzymes in pest control is already established against pathogenic fungi, and more recently the possibility to damage the peritrophic matrix lining the insect midgut is being explored, as an adjuvant in conjunction with other toxins (Ding et al. 1998) or on its own (Gongora et al. 2001). Nonetheless, PR10 have been reported to be induced by the attack of herbivore mites and insects (Walling 2000).

5.7 Towards a Multilevel Approach of Pest Control in Agriculture

Sustainable control of insect pests in agriculture can be achieved by enhancing plant resistance and/or the activity of natural enemies, i.e. predators and parasitoids.

Plant resistance has been always considered the centre of the sustainable control of insect herbivores. For example, the use of a resistant rootstock (*Vitis labrusca* or other American native species), on which to graft a *Vitis vinifera* scion, as promoted by T.V. Munson at the beginning of the twentieth century, is still the most widespread technique to control the devastating infestations of phylloxera (*Daktulosphaira vitifoliae*, Aphidoidea). However, the concept of plant resistance has been recently widened to include those mechanisms which involve the activity of natural enemies, referred to as indirect defences (see above).

The study of direct resistance of plants towards insect is a very old discipline whilst the reverse is true for plant indirect resistance. These two mechanisms of plant resistance share some features. The first, and probably the most important, is that both are regulated by genes. As a consequence, we can select (or breed) to enhance both direct and indirect resistance in agricultural varieties although the characterization of genes related to plant attractiveness towards natural enemies is still in its infancy. To date, very few data are available about the "behaviour" of agricultural plant varieties towards either parasitoids and predators of insect pests (e.g. Lou et al. 2006 and references therein); hence it remains a crucial point to expand this kind of knowledge.

Moreover, as illustrated above, both these resistance mechanisms are frequently induced (see Fig. 5.1) and, in this optic, the study of multitrophic and multilevel interactions can be pivotal in unravelling the mechanisms of biological induction which can be used to increase the sustainable control of insect pests.

As for direct resistance, we have discussed above how there is often a species specificity of the effect of AM symbiont on a given herbivore species. For example, it has been reported that *Glomus mosseae* colonization dramatically reduced the fitness of the aphid *Macrosiphum euphorbiae* on tomato (Guerrieri et al. 2004) as did *Glomus intraradices* for insect chewers on soybean (Rabin and Pacovsky 1985). For all those associations whose final outcome is a reduction of herbivore fitness, a purified inoculum, of known composition, could be provided to plant roots in nurseries before transplanting them in the field, in order to induce a resistance response. In the case of perennial plants, the same "selection" of AM symbiont could be achieved by selective fungicide treatments and subsequent applications of purified inoculum.

Similarly, the higher attractiveness of mycorrhizal tomato plants towards the aphid parasitoid *A. ervi* opens new perspectives for the biological control of insect pests (Fig. 5.2). Following the same approach as reported for direct resistance, beneficial associations between AM fungal species and natural enemies could be artificially created to enhance the attractiveness of agricultural plants.

This point represents a bridge between the enhancement of plant resistance and natural enemies, given that among the classical biological control techniques there is the increase of natural enemies fitness (Van Driesche and Bellows 1996). However, a multidisciplinary approach to investigate plant-beneficial organisms interactions has only recently offered new tools to be used in the sustainable control of insect pests. For example, both jasmonic acid (JA) and salicylic acid (SA) are potent elicitors of plant responses, even though the former seems to be much involved not only in plant responses to insect herbivores (chewers and sap-feeders) but also in the identification and development of arbuscular endosymbiosis (Hause et al. 2002). There have already been field applications of JA (and derivates) that have determined a significant increase in the field presence of natural enemies of insect pests (James 2005). It could be possible to promote arbuscular endomycor-rhizal symbiosis through radical or foliar application of this elicitor which in turn results in the production/release of semiochemicals attractive for natural enemies of insect herbivores.

On an a wider perspective, we can use these multilevel interactions as a further tool in the application of *push and pull* strategies that have been proved to be highly sustainable and exploitable in the Integrated Pest Management (IPM) practice (Cook et al. 2007). These strategies involve the behavioural manipulation of insects (pests and natural enemies) by using an integration of stimuli. By using chemical cues we can make the plant unattractive or unsuitable to the pests (push) while driving them towards an attractive source from which they are subsequently removed (pull). A similar concentration (pull) of natural enemies can be attained in those fields where their activity is most required by using a combination of chemical and visual stimuli (James 2005; Cook et al. 2007).

The positive bottom-up interaction between belowground AM symbiosis and aboveground natural enemies is not only played on the field of host/prey location cues but also on an enhanced availability of shelter and/or food. For example, it has been widely reported that AM plants are usually larger than non-mycorrhizal control ones and this leads to a larger availability of possible shelters for the natural enemies of insect pests. Similarly, it has been demonstrated that mycorrhizal symbiosis positively affect the flower number and size, along with the amount of pollen produced by plants (Gange and Smith 2005). This larger availability of food leads to an increase of the visitation rate by pollinating insects (Gange and Smith 2005), but also constitutes a precious source of carbohydrates and proteins available for parasitoids and predators of insect pests. For example, a 15-fold increase in the longevity of *Diadegma semiclausum*, a parasitoid of the diamond back moth *Plutella xylostella*, is determined by the presence of flowers (Wratten et al. 2003), while pollen has been indicated as a fundamental component in the diet of egg parasitoids (Zhang et al. 2004).

Finally, there is a wide literature about the ability of insect parasitoids to "learn" chemical compounds involved in host location and thus improving their searching ability at further encounters with their victims (see Meiners et al. 2003 and references therein). We can "enhance" the parasitoid response towards VOC induced by AM symbiosis in the laboratory before its release in the field, thus improving its fitness and in turn pest control.

5.8 Synthesis and Future Directions

It is well established that there is an incredible variety of interactions between soil and aerial organisms, with the plant mediating their final outcome. AM symbiosis is just one, though extremely complicated, of these interactions that could be exploited to enhance a sustainable control of insect pests and pathogens. However, in the perspective of practical application, many questions remain unanswered, although more than two decades have passed from the first pioneering paper by Rabin and Pacovsky (1985).

The first, and possibly the most urgent one is the assessment of the species-specificity effect induced by AM fungal symbiont on plant response, given that there are preferential associations in agricultural soils (Mathimaran et al. 2006). For example, it has been demonstrated that the parasitism rate by *Diglyphus isaea* is mycorrhizal-species dependent (Gange et al. 2003) but it is not possible to draw any conclusion about the influence of AM symbiosis on plant defences against either insect chewers or suckers, neither to have a clear view of whether insect specificity could be an important parameter to be considered. We know that, through AM symbiosis, more nutrients are transferred to the plant, especially in soils with a low content of P, but whether this goes towards an enhancement (e.g. tolerance) or a reduction (better performance of herbivore) of plant defences needs to be evaluated case by case.

Similarly, it has been recently reported that the production of essential oils in an aromatic plant is mycorrhizal species dependent (Copetta et al. 2006), and this could turn into a multiple defence response, given that many of these substances have both direct (Digilio et al. 2008) and indirect effect (Corrado et al. 2007; Sasso et al. 2007) on herbivore performances. Nonetheless, it needs to be tested whether plant cultivars respond in different ways to AM symbiosis, as it has been demonstrated for maize, in respect to the emission of induced VOC that are attractive towards the parasitoids of caterpillars (Gouinguené et al. 2001).

Overall, a standardization of methods is required because different results have been recorded on similar (if not identical) multitrophic systems. In fact, these discrepancies can be due to a transient effect of defensive responses induced by AM symbiosis, and this is probably another key point that needs to be demonstrated experimentally.

Regardless of the total number of papers published on the interactions between AM symbiosis and aboveground insects (both herbivores and their natural enemies), it must be noted that none of them includes a model plant. For example, many aspects of plant/insect and plant/AM interactions have been separately elucidated by using *Medicago truncatula*, a plant whose genome is close to being completely characterized (Harrison 2005; Leitner et al. 2005) but the contemporary presence of mycorrhizae and insects (herbivores and/or natural enemies) has never been investigated on this plant.

The role of bacteria associated with mycorrhizae (Gamalero et al. 2004) on plant defences against herbivores is still a completely unexplored field as is that of AM and nitrogen fixing bacteria that coexist in legume species (see Scheublin and van der Heijden 2006 and references therein). For example, a reduction of plant defence against a plant root fungal pathogen induced by a mycorrhiza helper bacterium has recently been reported (Lehr et al. 2007).

Still largely unknown and unexploited remain the mechanisms regulating plant-to-plant communication that has been demonstrated both belowground and aboveground (Bais et al. 2006). For example, root exudates of *Vicia faba* plants infested by the aphid *Acyrthosiphon pisum* elicit the release of host-induced VOC in neighbouring uninfested plants (Guerrieri et al. 2002). Although the specific compound involved in this interaction was not characterized, it remains a fact that a change in root exudates composition was determined by aboveground herbivory.

A similar induction of plant defence was demonstrated to happen at aerial level. In detail, host-induced VOC elicited the secretion of extrafloral nectar in neighbouring plants of Lima bean (*Phaseolus lunatus*), and (3Z)-hex-3-enylacetate was indicated as the compound seemingly most involved in this response (Kost and Heil 2006). Can we expect the same beneficial outcome of the plant-to-plant interactions in the event that the roots of at least one plant are colonized by AM symbiont?

From an ecological point of view, it would be essential to assess the long-term effect of these interactions, because there is a continuous and mutual influence of insect herbivory on AM colonization and vice versa (Ghering and Whitham 2002).

As outlined above, we can manipulate the temporal interaction between AM and herbivores by transplanting colonised plants in the field but how long will the plant defensive performance last? There is still an incredible lack of long-term field tests to assess whether the outcome of these interactions is stable or not.

All these questions lead to the big one: which are the genes involved in AM symbiosis that play a role in plant defence against aerial herbivores?

On the plant side, cDNA libraries have been established from mycorrhizal RNA using suppressive subtractive hybridization, and a large number of clones are being sequenced to obtain expressed sequence tags (EST). This constituted the base to characterize plant gene expression in response to AM symbiosis that now can be investigated with increasing precision by using modern approaches including proteomic, forward and reverse genetics and transgenic plants (see the paragraph "New tools in the study of multitrophic interactions"). Linking these observations to in vivo bioassay on plant defences against herbivore insects will be the fundamental step towards the understanding of the potentiality of these interactions. For example, it has been demonstrated that gene modulation induced by AM symbiosis happens not only in root tissues but also in the leaves and that the same plant response is recorded following a pathogenic infection (García-Rodriguez et al. 2005).

The thorough understanding of these fascinating interactions can only be achieved by following a multidisciplinary approach that involves plant physiologists, geneticists, microbiologists, chemists, entomologists and ecologists, each one participating with his own expertise and interacting with each other to formulate plausible theories. A typical flow of experiments and results between these different components is outlined in Fig. 5.4.

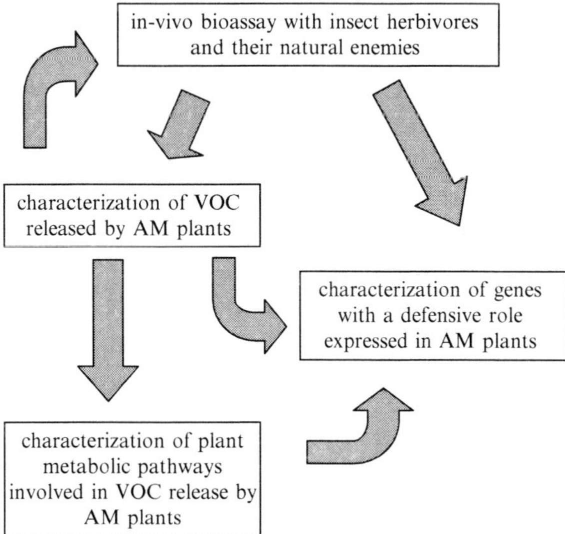

Fig. 5.4 Interdisciplinary approach to assess the role of AM symbiosis on plant defences against aboveground insect herbivores

The modern techniques available for plant breeding can be used to exploit the most updated findings in the field of beneficial interactions (see above). Apart from the controversial acceptance of GM plants by public and farmers of some countries (Herdt 2006), biotechnology can be used to mark the most promising genes for a sustainable control of insect pests to be transferred into commercial varieties even with conventional breeding programmes (Sharma et al. 2002). In this view, it is certainly "cheaper" to invest on single genes that are involved in multiple defence responses or on those whose expression, induced by beneficial microrganisms, results in a better resistance towards herbivores.

Only by unravelling the intimate mechanisms that regulate belowground symbiosis, plant resistance and biocontrol of insect pests, can we drive the plant responses and the behaviour of entomophagous insects towards a protection of agricultural crops that is environmentally friendly.

Acknowledgements We would like to thank Graziella Berta and Guido Lingua (Università del Piemonte Orientale) for useful discussions and for reviewing the manuscript. The editors are thanked for enthusiastic support and for precious comments on an earlier version of the manuscript. EG and MCD are financially supported by the Italian Consiglio Nazionale delle Ricerche (CNR) and Ministero dell'Università e della Ricerca.

References

Agrawal AA, Karban R (1997) Domatia mediate plant arthropod mutualism. Nature 387:562–563

Agrawal AA, Tuzun S, Bent E (1999) Induced plant defenses against pathogens and herbivores. APS Press, St Paul, Minnesota

Alborn HT, Turlings TCJ, Jones TH, Stenhagen G, Loughrin JH, Tumlinson JH (1997) An elicitor of plant volatiles from beet army worm oral secretion. Science 276:945–949

Augè RM (2001) Water relations, drought and VA mycorrhizal symbiosis. Mycorrhiza 11:3–42

Azcón-Aguilar C, Barea JM (1997) Arbuscular mycorrhizas and biological control of soil-borne plant pathogens – an overview of the mechanisms involved. Mycorrhiza 6:457–464

Bais PH, Weir TL, Perry LG, Gilroy S, Vivanco JM (2006) The role of root exudates in rhizosphere interactions with plants and other organisms. Annu Rev Plant Biol 57:233–266

Balestrini R, Lanfranco L (2006) Fungal and plant gene expression in arbuscular mycorrhizal symbiosis. Mycorrhiza 16:509–524

Bever JD, Morton JB, Antonovics J, Schultz PA (1996) Host dependent sporulation and species diversity of arbuscular mycorrhizal fungi in a mown grassland. J Ecol 84:71–82

Bezemer TM, De Deyn GB, Bossinga TM, van Dam NM, Harvey JA, van der Putten WH (2005) Soil community composition drives aboveground plant-herbivore-parasitoid interactions. Ecol Lett 8:652–661

Birkett MA, Campbell CAM, Chamberlain K, Guerrieri E, Hick AJ, Martin JL, Matthes M, Napier JA, Pettersson J, Pickett JA, Poppy GM, Pow EM, Pye BJ, Smart LE, Wadhams GH, Wadhams LJ, Woodcock C (2000) New roles for *cis*-jasmone as an insect semiochemical and in plant defence. Proc Natl Acad Sci USA 97:9329–9334

Birkett MA, Chamberlain K, Guerrieri E, Pickett JA, Wadhams LJ, Yasuda T (2003) Volatiles from whiteflies-infested plants elicit a host-locating response in the parasitoid *Encarsia formosa*. J Chem Ecol 29:1589–1600

Borowicz VA (1997) A fungal root symbiont modifies plant resistance to an insect herbivore. Oecologia 112:534–542

Borowicz VA (2001) Do arbuscular mycorrhizal fungi alter plant-pathogen relations? Ecology 82:3057–3068

Borowicz VA (2006) When enemies attack do plants get by with a little help from their friends? New Phytol 169:644–646

Cipollini D, Purrington CB, Bergelson J (2003) Costs of induced responses in plants. Basic Appl Ecol 4:79–85

Cloyd RA, Sadof CS (2000) Effects of plant architecture on the attack rate of *Leptomastix dactylopii* (Hymenoptera: Encyrtidae), a parasitoid of the citrus mealybug (Homoptera: Pseudococcidae). Environ Entomol 29:535–541

Colazza S, McElfresh JS, Millar JG (2004) Identification of volatile synomones, induced by *Nezara viridula* feeding and oviposition on bean spp., that attract the egg parasitoid *Trissolcus basalis*. J Chem Ecol 30:945–964

Cook SM, Khan ZR, Pickett JA (2007) The use of push-pull strategies in integrated pest management. Annu Rev Entomol 52:375–400

Cooper WR, Goggin FL (2005) Efects of jasmonate-induced defenses in tomato on potato aphid, *Macrosiphum euphorbiae*. Entomol Exp Appl 115:107–115

Copetta A, Lingua G, Berta G (2006) Effects of three AM fungi on growth, distribution of glandular hairs, and essential oil production in *Ocimum basilicum* L. var. *Genovese*. Mycorrhiza 16:485–494

Corrado G, Sasso R, Pasquariello M, Iodice L, Carretta A, Cascone P, Ariati L, Digilio MC, Guerrieri E, Rao R (2007) Systemin regulates both systemic and volatile signalling in tomato plants. J Chem Ecol 33:669–681

Dakora FD (1995) Plant flavonoids-biological molecules for useful exploitation. Aust J Plant Physiol 22:87–99

David-Schwartz R, Badani H, Smadar W, Levy AA, Galili G, Kapulnik Y (2001) Identification of a novel genetically controlled step in mycorrhizal colonization: plant resistance to infection by fungal spores but not extra-radical hyphae. Plant J 27:561–569

David-Schwartz R, Gadkar V, Wininger S, Bendov R, Galili G, Levy AA, Kapulnik Y (2003) Isolation of a premycorrhizal infection (pmi2) mutant of tomato, resistant to arbuscular mycorrhizal fungal colonization. Mol Plant Microbe Interact 16:382–388

Dicke M, van Poecke RP, de Boer JG (2003) Inducible indirect defence of plants: from mechanisms to ecological functions. Basic Appl Ecol 4:27–42

Digilio MC, Quaranta E, Voto E, De Feo V (2008) Insecticide activity of Mediterranean essential oils. J Plant Interact (in press)

Ding X, Gopalakrishnan B, Lowell BJ, White FF, Wang X, Morgan TD, Kramer KJ, Muthukrishnan S (1998) Insect resistance of transgenic tobacco expressing an insect chitinase gene. Transgenic Research 7:77 84

Doares SH, Narváez-Vásquez J, Conconi A, Ryan C (1995) Salicylic acid inhibits synthesis of proteinase inhibitors in tomato leaves induced by systemin and jasmonic acid. Plant Physiol 108:1741–1746

Doherty HM, Selvendran RR, Bowels DJ (1988) The wound response of tomato plants can be inhibited by aspirin and related hydroxy-benzoic acids. Physiol Mol Plant Pathol 33:377–384

van Driesche RG, Bellows TS (1996) Biological control. Chapman and Hall, International Thomson Publishing Company

Duc G, Trouvelot A, Gianinazzi-Pearson V, Gianinazzi S (1989) First report of non-mycorrhizal plant mutants (myc-) obtained in pea (*Pisum sativum* L.) and faba bean (*Vicia faba* L.). Plant Sci 60:215–222

Emmanuel E, Levy AA (2002) Tomato mutants as tools for functional genomics. Curr Opin Plant Biol 5:112–117

Fidantsef AL, Stout MJ, Thaler JS, Duffey SS, Bostock RM (1999) Signal interactions in pathogen and insect attack: expression of lipoxigenase, proteinase inhibitor II, and pathogenesis-related protein P4 in the tomato *Lycopersicon esculentum*. Physiol Mol Plant Path 54:97–114

Filion M, St Arnaud M, Fortin JA (1999) Direct interaction between the arbuscular mycorrhizal fungus *Glomus intraradices* and different rhizosphere micro-organisms. New Phytol 141:525–533

Franken P, Requena PA (2001) Analysis of gene expression in arbuscular mycorrhizas: new approaches and challenges. New Phytol 150:517–523

Fritz M, Jakobsen I, Lyngkjær MF, Thordal-Christensen H, Pons-Kühnemann J (2006) Arbuscular mycorrhiza reduces susceptibility of tomato to *Alternaria solani*. Mycorrhiza 16:413–419

Gamalero E, Trotta A, Massa N, Copetta A, Martinotti MG, Berta G (2004) Impact of two fluorescent pseudomonads and an arbuscular mycorrhizal fungus on tomato plant growth, root architecture and P acquisition. Mycorrhiza 14:85–192

Gange AC (2001) Species-specific responses of a root- and shoot-feeding insect to arbuscular mycorrhizal colonization of its host plant. New Phytol 150:611–618

Gange AC, Bower E (1997) Interactions between insects and mycorrhizal fungi. In: Gange AC, Brown VK (eds) Multitrophic interactions in terrestrial systems. Blackwell, Oxford, pp 115–132

Gange AC, Nice HE (1997) Performance of the thistle gall fly, *Urophora cardui*, in relation to host plant nitrogen and mycorrhizal colonization. New Phytol 137:335–343

Gange AC, Smith AK (2005) Arbuscular mycorrhizal fungi influence visitation rates of pollinating insects. Ecol Entomol 30:600–606

Gange AC, West HM (1994) Interactions between arbuscular mycorrhizal fungi and foliar-feeding insects in *Plantago lanceolata* L. New Phytol 128:79–87

Gange AC, Bower E, Brown VK (1999) Positive effects of an arbuscular mycorrhizal fungus on aphid life history traits. Oecologia 120:123–131

Gange AC, Bower E, Brown VK (2002) Differential effects of insect herbivory on arbuscular mycorrhizal colonization. Oecologia 131:103–112

Gange AC, Brown VK, Aplin DM (2003) Multitrophic links between arbuscular mycorrhizal fungi and insect parasitoids. Ecol Lett 6:1051–1055

Gange AC, Gane DR, Chen Y, Gong M (2005) Dual colonization of *Eucaliptus urophylla* S.T. Blake by arbuscular and ectomycorrhizal fungi affects levels of insect herbivore attack. Agric For Entomol 7:253–263

Gao LL, Knogge W, Delp G, Smith FA, Smith SE (2004) Expression patterns of defense-related genes in different types of arbuscular mycorrhizal development in wild-type and mycorrhiza-defective mutant tomato. Mol Plant Microbe Interact 17:1103–1113

García-Rodriguez S, Pozo MJ, Azcón-Aguilar C, Ferrol N (2005) Expression of tomato sugar transporter is increased in the leaves of mycorrhyzal or *Phytophtora parasitica*-infected plants. Mycorrhiza 15:489–496

Ghering CA, Whitham TG (2002) Mychorrizae-herbivore interactions: population and community consequences. In: van der Heijden MGA, Sanders I (eds) Mycorrhizal ecology. Springer, Berlin Heidelberg New York, pp 295–320

Glazebrook J, Chen W, Estes B, Chang H-S, Nawrath C, Métraux J-P, Zhu T, Katagiri F (2003) Topology of the network integrating salicylate and jasmonate signal transduction derived from global expression phenotyping. Plant J 34:217–228

Gongora CE, Wang S, Barbehenn RV, Broadway RM (2001) Chitinolytic enzymes from Streptomyces albidoflavus expressed in tomato plants: effects on *Trichoplusia ni*. Entomol Exp Appl 99:193–204

Gouinguené S, Degen T, Turlings TCJ (2001) Variability in herbivore-induced odour emissions among maize cultivars and their wild ancestors (teosinte). Chemoecology 11:9–16

Goverde M, van der Heijden MGA, Wiemken A, Sanders IR, Erhardt A (2000) Arbuscular mycorrhizal fungi influence life history traits of a lepidopteran herbivore. Oecologia 125:362–369

Gozzo F (2003) Systemic acquired resistance in crop protection: from nature to a chemical approach. J Agric Food Chem 51:4487–4503

Guerrieri E, Pennacchio F, Tremblay E (1993) Flight behaviour of the aphid parasitoid *Aphidius ervi* Haliday (Hymenoptera: Braconidae) in response to plant and host volatiles. Eur J Entomol 90:415–421

Guerrieri E, Pennacchio F, Tremblay E (1997) Effect of adult experience on in-flight orientation to plant and plant-host complex volatiles in *Aphidius ervi* Haliday (Hymenoptera, Braconidae). Biol Control 10:159–165

Guerrieri E, Poppy GM, Powell W, Tremblay E, Pennacchio F (1999) Induction and systemic release of herbivore-induced plant volatiles mediating in-flight orientation of *Aphidius ervi* (Hymenoptera: Braconidae). J Chem Ecol 25:1247–1261

Guerrieri E, Poppy GM, Powell W, Rao R, Pennacchio F (2002) Plant to plant communication mediating in-flight orientation of *Aphidius ervi*. J Chem Ecol 28:1703–1715

Guerrieri E, Lingua G, Digilio MC, Massa N, Berta G (2004) Do interactions between plant roots and the rhizosphere affect parasitoid behaviour? Ecol Entomol 29:753–756

Gupta V, Willits MG, Glazebrook J (2000) *Arabidopsis thaliana* EDS4 contributes to salicylic (SA)-dependent expression of defense responses: evidence for inhibition of jasmonic acid signaling by SA. Mol Plant Microbe Interact 13:503–511

Harrison MJ (2005) Signaling in the arbuscular mycorrhizal symbiosis. Annu Rev Microbiol 59:19–42

Harrison MJ, van Buuren ML (1995) A phosphate transporter from the mycorrhizal fungus *Glomus versiforme*. Nature 378:626–629

Harrison MJ, Dixon RA (1994) Spatial patterns of expression of flavonoid/isoflavonoid pathway genes during interactions between roots of *Medicago truncatula* and the mycorrhizal fungus *Glomus versiforme*. Plant J 6:9–20

Hart MM, Reader RJ (2002) Taxonomic basis for variation in the colonization strategy of arbuscular mycorrhizal fungi. New Phytol 153:335–344

Hause B, Maier W, Miersch O, Kramell R, Strack D (2002) Induction of jasmonate biosynthesis in arbuscular mycorrhizal barley roots. Plant Physiol 130:1213–1220

van der Heijden MGA, Boller T, Wiemkem A, Sanders IR (1998a) Different arbuscular mycorrhizal fungal species are potential determinants of plant community structure. Ecology 79:2082–2091

van der Heijden MGA, Klironomos JN, Ursic M, Moutoglis P, Streitwolf-Engel R, Boller T, Wiemkem A, Sanders IR (1998b) Mycorrhizal fungal diversity determines plant biodiversity. Nature 396:69–72

van der Heijden MGA, Streitwolf-Engel R, Riedl R, Siegrist S, Neudecker A, Ineichen K, Boller T, Wiemken A, Sanders IR (2006) The mycorrhizal contribution to plant productivity, plant nutrition and soil structure in experimental grassland. New Phytol 172:739–752

Herdt RW (2006) Biotechnology in agriculture. Annu Rev Environ Resourc 31:265–295

Hodge A, Campbell CD, Fitter AH (2001) An arbuscular mycorrhizal fungus accelerates decomposition and acquires nitrogen directly from organic material. Nature 413:297–299

James D (2005) Further field evaluation of synthetic herbivore-induced plant volatiles as attractants for beneficial insects. J Chem Ecol 31:481–495

Johanson JF, Paul LR, Finlay RD (2004) Microbial interactions in the mycorrhizosphere and their significance for sustainable agriculture. FEMS Microbiol Ecol 48:1–13

Joner EI, Ravnskov S, Jakobsen I (2000) Arbuscular mycorrhizal phosphate transport under monoxenic conditions using radio-labelled inorganic and organic phosphate. Biotechnol Lett 22:1705–1708

Kaloshian I, Walling LL (2005) Hemipterans as plant pathogens. Annu Rev Phytopatyol 43:491–521

Kant MR, Ament K, Sabelis MW, Haring MA, Schuurink RC (2004) Differential timing of spider mite-induced direct and indirect defenses in tomato plants. Plant Physiol 135:483–495

Kendrick B (1992) The fifth kingdom. Mycologue Publications, Waterloo

Klironomos JN (2000) Host specificity and functional diversity among arbuscular mycorrhizal fungi. In: Bell CR, Brylinsky M, Johnson-Green P (eds) Microbial biosystems: new frontiers. Proceedings of the 8th International Symposium of Microbial Ecology. Atlantic Canada Society for Microbial Ecology, Halifax, pp 845–851

Klironomos JN (2003) Variation in plant response to native and exotic arbuscular mycorrhizal fungi. Ecology 84:2292–2301

Klironomos JN, McCune J, Moutoglis P (2004) Species of arbuscular mycorrhizal fungi affect mycorrhizal reponses to simulated herbivory. Appl Soil Ecol 26:133–141

Kost C, Heil M (2006) Herbivore-induced plant volatiles induce an indirect defence in neighbouring plants. J Ecol 94:619–628

Kula AAR, Harnett DC, Wilson GWT (2005) Effects of mycorrhizal symbiosis on tallgrass prairie plant-herbivore interactions. Ecol Lett 8:61–69

Kunkel BN, Brooks DM (2002) Cross talk between signal pathways in pathogen defense. Curr Opin Plant Biol 5:325–331

Lehr NA, Schrey SD, Bower R, Hampp R, Tarkka MT (2007) Suppression of plant defence response by a mycorrhiza helper bacterium. New Phytol 174:892–903

Leitner M, Boland W, Mithöfer A (2005) Direct and indirect defences induced by piercing-sucking and chewing herbivores in *Medicago truncatula*. New Phytol 167:597–606

Liu J, Maldonado-Mendoza I, Lopez-Meyer M, Cheung F, Town CD, Harrison MJ (2007) Arbuscular mycorrhizal symbiosis is accompained by local and systemic alterations in gene expression and an increase in disease resistance in the shoots. Plant J 50:529–544

Lou Y, Hua X, Turlings TCJ, Cheng J, Chen X, Ye G (2006) Differences in induced volatile emissions among rice varieties result in differential attraction and parasitism of *Nilaparvata lugens* eggs by the parasitoid *Anagrus nilaparvatae* in the field. J Chem Ecol 32:2375–2387

Maldonado-Mendoza IE, Dewbre GR, van Buuren ML, Versaw WK, Harrison MJ (2002) Methods to estimate the proportion of plant and fungal RNA in an arbuscular mycorrhiza. Mycorrhiza 12:67–74

Marschner H, Dell B (1994) Nutrient uptake in mycorrhizal symbiosis. Plant Soil 159:89–102

Mathimaran N, Ruh R, Vullioud P, Frossard E, Jansa J (2006) *Glomus intraradices* dominates arbuscular mycorrhizal communities in a heavy textured agricultural soil. Mycorrhiza 16:61–66

Mattiacci L, Dicke M, Posthumus MA (1995) Beta-glucosidase – an elicitor of herbivore induced plant odor that attracts host-searching parasitic wasps. Proc Natl Acad Sci USA 92: 2036–2040

McNaughton SJ, Chapin FS III (1985) Effects of phosphorous nutrition and defoliation on C_4 graminoids from the Serengeti Plains. Ecology 66:1617–1629

Meiners T, Wäckers F, Lewis WJ (2003) Associative learning of complex odours in parasitoid host location. Chem Senses 28:231–236

Mithöfer A, Maitrejean M, Boland W (2005) Structural and biological diversity of cyclic octadecanoids, jasmonates and mimetics. J Plant Growth Regul 23:170–178

Morandi D (1996) Occurrence of phytoalexins and phenolic compounds in endomycorrhizal interactions, and their potential role in biological control. Plant Soil 185:241–251

Mullin CA, Al-Fatafta AA, Harman JL, Serino AA, Everett SL (1991) Corn rootworm feeding on sunflower and other Compositae – influence of floral terpenoid and phenolic factors. ACS Symp Ser 449:278–292

Neumann E, George E (2005) Does the presence of arbuscular mycorrhizal fungi influence growth and nutrient uptake of a wild-type tomato cultivar and a mycorrhiza-defective mutant, cultivated with roots sharing the same soil volume? New Phytol 166:601–609

Niki T, Mitsuhara I, Seo S, Ohtsubo N, Ohashi Y (1998) Antagonistic effect of salicylic acid and jasmonic acid on the expression of pathogenesis-related (PR) proteins. Plant Cell Physiol 39:500–507

Pacovsky RS, Rabin LB, Montllor CB, Waiss AC Jr (1985) Host plant resistance to insect pest altered by *Glomus fasciculatum* colonization. In: Molina R (ed) Proceedings of the 6th North American Conference on Mycorrhizae. Oregon State University, Corvallis, p 288

Paszkowski U, Jakovleva L, Boller T (2006) Maize mutants affected at distinct stages of the arbuscular mycorrhizal symbiosis. Plant J 47:165–173

Peipp H, Maier W, Schmidt J, Wray V, Strack D (1997) Arbuscular mycorrhizal fungus-induced changes in the accumulation of secondary compounds in barley roots. Phytochemistry 44:581–587

Peña-Cortés H, Albrecht T, Prat S, Weiler EW, Willmitzer L (1993) Aspirin prevents wound-induced gene expression in tomato leaves by blocking jasmonic acid biosynthesis. Planta 191:123–128

Peña-Cortés H, Barrios P, Dorta F, Polanco V, Sánchez C, Sánchez E, Ramírez I (2005) Involvement of jasmonic acid and derivates in plant responses to pathogens and insects and in fruit ripening. J Plant Growth Regul 23:246–260

Peterson RL, Guinel FC (2000) The use of plant mutants to study regulation of colonization by AM fungi. In: Kapulnick Y, Douds Jr DD (eds) Arbuscular mycorrhizas: physiology and function. Kluwer Academic Press, pp 147–172

Pinochet J, Calvet C, Camprubí C, Fernández C (1996) Interactions between migratory endoparasitic nematodes and arbuscular mycorrhizal fungi in perennial crops: a review. Plant Soil 185:183–190

Pozo MJ, van Loon LC, Pieterse CMJ (2005) Jasmonate-signals in plant-microbe interactions. J Plant Growth Regul 23:211–222

Price PW, Bouton CE, Gross P, McPheron BA, Thompson JN, Weis AE (1980) Interactions among three trophic levels: influence of plants on interactions between insect herbivores and natural enemies. Annu Rev Ecol Syst 11:41–65

Rabin LB, Pacovsky RS (1985) Reduced larva growth of two lepidoptera (Noctuidae) on excised leaves of soybean infected with a mycorrhizal fungus. J Econ Entomol 78:1358–1363

Rassman S, Köllner TG, Degenhardt J, Hiltpold I, Toepfer S, Kuhlmann U, Gershenzon J, Turlings TCJ (2005) Recruitment of entomopathogenic nematodes by insect-damaged maize roots. Nature 434:732–737

Regvar M, Gogala N, Zalar P (1996) Effects of jasmonic acid on mycorrhizal *Allium sativum*. New Phytol 134:144–152

Rillig MC (2004) Arbuscular mycorrhizae and terrestrial ecosystem processes. Ecol Lett 7:740–754

Rose USR, Manukian A, Heath RR, Tumlinson JH (1996) Volatile semiochemicals released from undamaged cotton leaves (a systemic response of living plants to caterpillar damage). Plant Physiol 111:487–495

Ryals JA, Neuenschwander UH, Willits MG, Molina A, Steiner H-Y, Hunt MD (1996) Systemic acquired resistence. Plant Cell 8:1809–1819

Sanders IR (2002) Specificity in the arbuscular mycorrhizal symbiosis. In: van der Heijden MGA, Sanders IR (eds) Mycorrhizal ecology. Springer, Berlin Heidelberg New York, pp 415–437

Sasso R, Iodice L, Carretta A, Digilio MC, Ariati L, Guerrieri E (2007) Identification of volatiles from tomato plants eliciting a host-locating response in the parasitoid *Aphidius ervi*. J Plant Interact 2:175–183

Schaller F, Schaller A, Stintzi A (2005) Biosynthesis and metabolism of jasmonates. J Plant Growth Regul 23:179–199

Scheublin TR, van der Heijden MGA (2006) Arbuscular mycorrhizal fungi colonize nonfixing root nodules of several legume species. New Phytol 172:732–738

Scheublin TR, van Logtestijn RSP, van der Heijden MGA (2007) Presence and identity of arbuscular mycorrhizal fungi influence competitive interactions between plant species. J Ecol 4:631–638

Schnee C, Köllner TG, Held M, Turlings TCJ, Gershenzon J, Degenhardt J (2006) The product of a single maize sesquiterpene synthase form a volatile defense signal that attracts natural enemies of maize herbivores. Proc Natl Acad Sci USA 103:1129–1134

Scutareanu PB, Drukker B, Bruin J, Posthumus MA, Sabelis MW (1997) Isolation and identification of volatile synomones involved in the interaction between *Psylla*-infested pear trees and two anthocorid predators. J Chem Ecol 23:2241–2260

Sharma HC, Crouch JH, Sharma KK, Seetharama N, Hash CT (2002) Applications of biotechnology for crop improvement: prospects and constraints. Plant Sci 163:381–395

Simms EL, Fritz RS (1990) The ecology and evolution of host plant resistance to insects. Trends Ecol Evol 5:356–360

Smith SE, Read DJ (1997) Mycorrhizal symbiosis. Academic Press, London

Soler R, Harvey JA, Kamp AFD, Vet LEM, van der Putten WH, van Dam NM, Stuefer JF, Gols R, Hodijk CA, Bezemer M (2007) Root herbivores influence the behaviour of an aboveground parasitoid through changes in plant-volatiles. Oikos 116:367–376

Staswick PE, Lehman CC (1999) Jasmonic acid-signaled responses in plants. In: Agrawal AA, Tuzun S, Bent E (eds) Induced plant defenses against pathogens and herbivores. APS Press, St Paul, Minnesota, pp 117–136

Strack D, Fester T, Hause B, Schliemann W, Walter MH (2003) Arbuscular mycorrhiza: biological, chemical and molecular aspects. J Chem Ecol 29:1955–1979

Thaler J (2000) Effect of jasmonate-induced plant responses on the natural enemies of herbivores. J Appl Ecol 71:141–150

Thaler J, Fidantsef AL, Bostock RM (2002a) Antagonism between jasmonate- and salicilate-mediated induced plant resistance:effects of concentration and timing of elicitation on defense-related proteins, herbivore, and pathogen performance in tomato. J Chem Ecol 28:1131–1159

Thaler J, Karban R, Ullman DE, Boege K, Bostock RM (2002b) Cross-talk between jasmonate and salicylate plant defense pathways:effects on several plant parasites. Oecologia 131:227–235

Thompson GA, Goggin FL (2006) Transcriptomics and functional genomics of plant defence induction by phloem-feeding insects. J. Exp Bot 57:755–766

Turlings TCJ, Tumlinson JH (1992) Systemic release of chemical signals by herbivore-injured corn. Proc Natl Acad Sci USA 89:8399–8402

Turlings TCJ, Tumlinson JH, Heat RR, Proveaux AT, Doolittle RE (1991) Isolation and identification of allelochemicals that attract the larval parasitoid *Cotesia marginiventris* (Cresson), to the microhabitat of one of its hosts. J Chem Ecol 17:1251–1253

Turlings TCJ, Lengwiller UB, Bernasconi M, Wechsler D (1998) Timing of induced volatile emissions in maize seedlings. Planta 207:146–152

Vandenkoornhuyse P, Ridgway KP, Watson IJ, Fitter AH, Young JP (2003) Co-existing grass species have distinctive arbuscular mycorrhizal communities. Mol Ecol 12:3085–3095

Vet LEM, Dicke M (1992) Ecology of infochemicals use by natural enemies in a tritrophic context. Annu Rev Entomol 37:141–172

Vierheilig H (2004) Regulatory mechanisms during the plant-arbuscular mycorrhizal fungus interaction. Can J Bot 82:1166–1176

Vierheilig H, Bago B, Albrecht C, Poulin MJ, Piche Y (1998) Flavonoids and arbuscular-mycorrhizal fungi. Adv Exp Med Biol 439:9–33

Wäckers FL, van Rijn PCJ, Bruin J (2005) Plant-provided food for carnivorous insects: a protective mutualism and its applications. Cambridge University Press

Walling L (2000) The myriad plant responses to herbivores. J Plant Growth Regul 19:195–216

Wamberg C, Christensen S, Jakobsen I (2003) Interaction between foliar-feeding insects, mycorrhizal fungi, and rhizosphere protozoa on pea plants. Pedobiologia 47:281–287

Wardle DA, Bardgett RD, Klironomos JN, Setälä H, van der Putten WH, Wall DH (2004) Ecological linkages between aboveground and belowground biota. Science 304:1629–1633

Wratten SD, Lavandero BI, Tylianakis J, Vattala D, Çilgi T, Sedcole R (2003) N Z Plant Prot 56:239–245

Wurst S, Dugassa-Gobena D, Lange R, Bonkowski M, Scheu S (2004) Combined effects of earthworms and vesicular–arbuscular mycorrhizas on plant and aphid performance. New Phytol 163:169–176

Zangerl AR (2003) Evolution of induced plant responses to herbivores. Basic Appl Ecol 4:91–103

Zhang G, Zimmermann O, Hassan SA (2004) Pollen as a source of food for egg parasitoids of the genus Trichogramma (Hymenoptera: Trichogrammatidae). Biocontrol Sci Techn 14:201–209

Zhu J, Park K-C (2005) Methyl salicylate, a soybean aphid-induced plant volatile attractive to the predator *Coccinella septempunctata*. J Chem Ecol 31:1733–1746

Part II
Coexistence Between Genomes

Chapter 6
Evolutionary Genomics: Linking Macromolecular Structure, Genomes and Biological Networks

Gustavo Caetano-Anollés

6.1 Introduction

What makes individuals, populations, species and organismal lineages unique? Are genetic complements enough to define phenotypic repertoires? Only 1.5% differences in nucleic acid sequence separate humans from chimpanzee, two species believed to have diverged from each other over six million years ago (Cheng et al. 2005). Yet humans differ notably from chimpanzees and other primates. Are nucleic acid sequence differences at the gene level important? A recent whole-genome analysis of concatenated gene sequences shows that higher organisms have been given more taxonomic resolution than microbes; organisms assigned to separate phyla in Eukarya would clearly belong to a same phylum in the prokaryotic classification (Ciccarelli et al. 2006). Yet they appear to be phenotypically more plastic expressing greater morphological diversity. We may be tempted to state that differences in phenotypes between species are due to limited sets of coding genes that make critical proteins, or to differential regulation of a larger number of protein coding genes. The discovery of a diverse modern RNA world with regulatory function could support the differential regulatory explanation (Bartel 2004). We could also argue that it is not the gene repertoire what counts but the encoded proteins. Protein sequence is extraordinarily diverse and so is the three-dimensional (3D) structure of proteins and their associated functions (Chothia et al. 2003). However, protein sequences encoded in the genomes of the millions of species that currently inhabit earth cover necessarily only a minute fraction (at most one in 10^{-300}) of the enormous permutational space defined by amino acid sequence. Yet the tools of structural genomics and protein structure determination reveal that this limited

G. Caetano-Anollés
Department of Crop Sciences, 332 NSRC, 1101 West Peabody Drive,
University of Illinois at Urbana-Champaign, Urbana, IL 61801, USA
e-mail: gca@uiuc.edu

C.S. Nautiyal, P. Dion (eds.) *Molecular Mechanisms of Plant and Microbe Coexistence*. Soil Biology 15, DOI: 10.1007/978-3-540-75575-3
© Springer-Verlag Berlin Heidelberg 2008

                                                                G. Caetano-Anollés

exploration of sequence space has uncovered considerable diversity in structure and biological function (e.g. enzymatic catalysis; Gutteridge and Thornton 2005). We could also argue that it is the unique modular structure of proteins that makes the difference. A substantial portion of proteins is made of multiple domains, units of compact structure that can combine in different ways to provide structural diversity (Vogel et al. 2004). Are differences at this level crucial?

In order to answer these and many other fundamental questions we need to draw from the vast information that has been accumulating since the first secrets of the genome were unveiled by the genomic revolution of this past decade. A first and fundamental recognition of modern biology is the need to survey component parts and their interaction. In fact, we have been very effective in this task (Fig. 6.1). Hundreds of genomes have been completely sequenced yielding tens of billions of base pairs, millions of protein sequences, and thousands of putative non-coding RNA molecules that serve a regulatory function and are likely to play important roles in species diversity. This effort outpaces structural genomics with its over 35,000 Protein Data Bank (PDB) entries of 3D molecular structure. A second important recognition following half a century of research into molecular evolution is that we can only understand the present if we can reconstruct our past effectively. Fundamental developments related to natural history reconstruction include the generation of a comprehensive tree of life, global phylogenetic analyses that help track evolutionary history at genome levels, and better understanding of evolution-

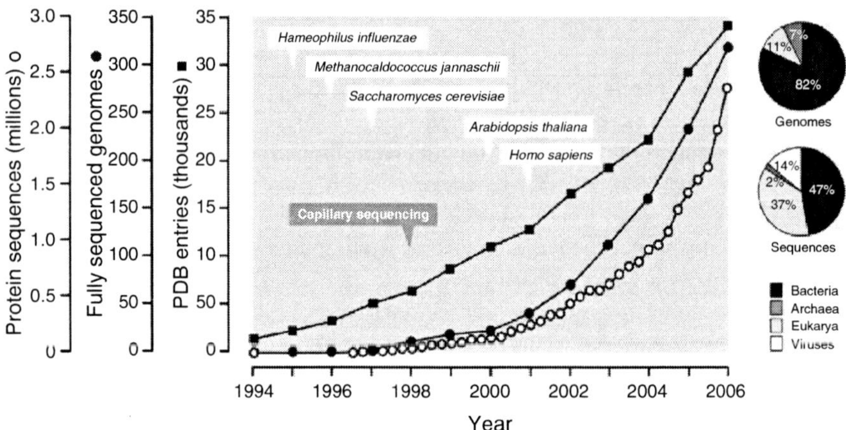

Fig. 6.1 The genomic revolution of this past decade provides hundreds of genomes, millions of protein sequences, and thousands of 3D models of molecular structure embedded in Protein Data Bank (PDB) entries. Fundamental milestones include the sequencing of the first bacterial, archaeal and eukaryotic genome, the genome of the first plant, and the human genome. All this was made possible by the technological development of capillary sequencing. Presently, the living world represented in genomes and sequences is highly biased towards microbial life. Data was retrieved from the PDB (http://www.rcsb.org/pdb), UniProt (http://www.ebi.ac.uk/uniprot), and GOLD (Kyrpides 1999) databases (April 25, 2006)

ary processes (Doolittle 2005; Kurland et al. 2006). A third fundamental recognition comes with the development of systems biology, with the tenet that cells and organisms are integrated systems and not collections of isolated parts. Currently, we use molecular survey components to define descriptive, graphical and mathematical models, confirm these models by perturbation (mutation, environment, etc.), and integrate information and models effectively (Kitano 2002; O'Malley and Dupré 2005). All this is made possible thanks to enhancements in computational power and development of efficient computational algorithms. Molecular survey, history reconstruction, and systems analysis are the fuel of evolutionary genomics and the three pillars of modern biology. Erected at the start of this new millennium, they promise deep understanding of life.

In this chapter I will discuss how evolutionary genomics is helping define new paradigms. Phylogenomic approaches will be described that take advantage of the opportunity to characterize unique sets of genes capable of defining lineages at different taxonomical levels, including species, populations, and organisms. I will also lay the principles of a general evolutionary framework capable of reconstructing evolutionary history directly from the structure of macromolecules.

6.2 Evolutionary Genomics, Networks and Systems

6.2.1 The Genomic Revolution

In the past few years, nucleic acid and protein sequences have been acquired in a massively parallel effort from a wide variety of organisms. Moreover, initiatives that seek to create a complete inventory of the structure of orthologous gene groupings across whole genomes, protein fold architectures from crystallographic data, and the tree of life itself offer unprecedented opportunities to understand genomic complexity (Zhang and Kim 2003; Doolittle 2005). A recent survey (April 25, 2006) showed there were 373 published genomes, with many being deposited in GenBank on a weekly basis (Fig. 6.1). Most genomes that were sequenced were prokaryotic (88%) and had small genome size. However, eukaryotic genomes represented a substantial portion (~38%) of the sequencing effort. There were also 46 finished and ongoing metagenomic projects that study genomic sequences present in a wide range of environments, including soil from Alaska and Minnesota, rice and poplar endophytic communities, root colonizing archaeal communities, and other complex environments. The number of ongoing genome sequencing projects (1605) was also an indicator of exponential increase in years to come. For the first time we have an opportunity to explore the evolution of entire sets of genomes representing a diverse range of organisms and environments, including plants and associated microbes, using the tools of computational biology, comparative genomics and molecular evolution Whole genome comparisons are now possible on a scale from which general principles of evolution can be derived. This has given rise to the new field of phylogenomics (Doolittle 2005).

6.2.2 Phylogenomics

Genomics has opened new avenues in evolutionary research. Evolutionary history
has been reconstructed using combined or concatenated genomic sequences, and
genomic features describing the survey (genomic demography) and arrangement
(genomic topography) of genomic component parts (reviewed in Wolf et al. 2002;
Delsuc et al. 2005; Doolittle 2005). In particular, phylogenomic (whole-genome)
trees were built effectively from features describing the occurrence and distribution
of protein folds in proteomes (Gerstein 1998, Gerstein and Hegyi 1998; Wolf et al.
1999, 2002; Lin and Gerstein 2000; House and Fitz-Gibbon 2002; Caetano-Anollés
and Caetano-Anollés 2003; Yang et al. 2005). In one implementation of this strategy, we measured the popularity (number of occurrences) of each protein fold in
sequenced genomes and used multi-state phylogenetic characters to reconstruct
intrinsically rooted proteome trees invoking the concept that being popular at the
molecular level is a favored evolutionary outcome (Caetano-Anollés and Caetano-
Anollés 2003, 2005). We have recently taken these approach further and reconstructed phylogenies from features describing the content and arrangement of
domains in proteins at a genomic level (Wang and Caetano-Anollés 2006).
Phylogenetic characters are here drawn from a molecular topography that describes
how evolutionary units of structure arrange in protein molecules and how popular
these arrangements are within each proteome. The reconstructed universal tree suggests dramatic diversification events in the history of life (Fig. 6.2). It also shows
that genomes in Eukarya were basal, suggesting a eukaryotic rooting of the tree of
life. However, phylogenetic trees also revealed early reductive tendencies in the
architectural repertoire of Archaea that suggest the very early split of this lineage
(Wang et al. 2007). Almost all pan-domain phylogenies generated from genomic
information support the tripartite (three-domain) nature of life already evident in
trees reconstructed from ribosomal RNA molecules, confirm accepted lineage relationships within major organismal groups, support disputed or preliminary classifi-
cations, and reveal novel evolutionary patterns (Doolittle 2005).

6.2.3 Network Biology: Understanding the Wiring
         Diagram of Life

Network biology characterizes and describes quantitatively the networks of molecular interactions that operate in biological systems (Barabási and Oltvai 2004).
These networks can be represented naturally as graphs and hypergraphs and their
study is supported by graph and percolation theory. There is considerable interest
in the processes underlying evolution of networks. Networks of different kinds
appear at different levels of molecular evolution (Schuster and Stadler 2003). We
can find networks embedded in biopolymer molecules through conformational
spaces that are highly complex and multidimensional and describe molecular and

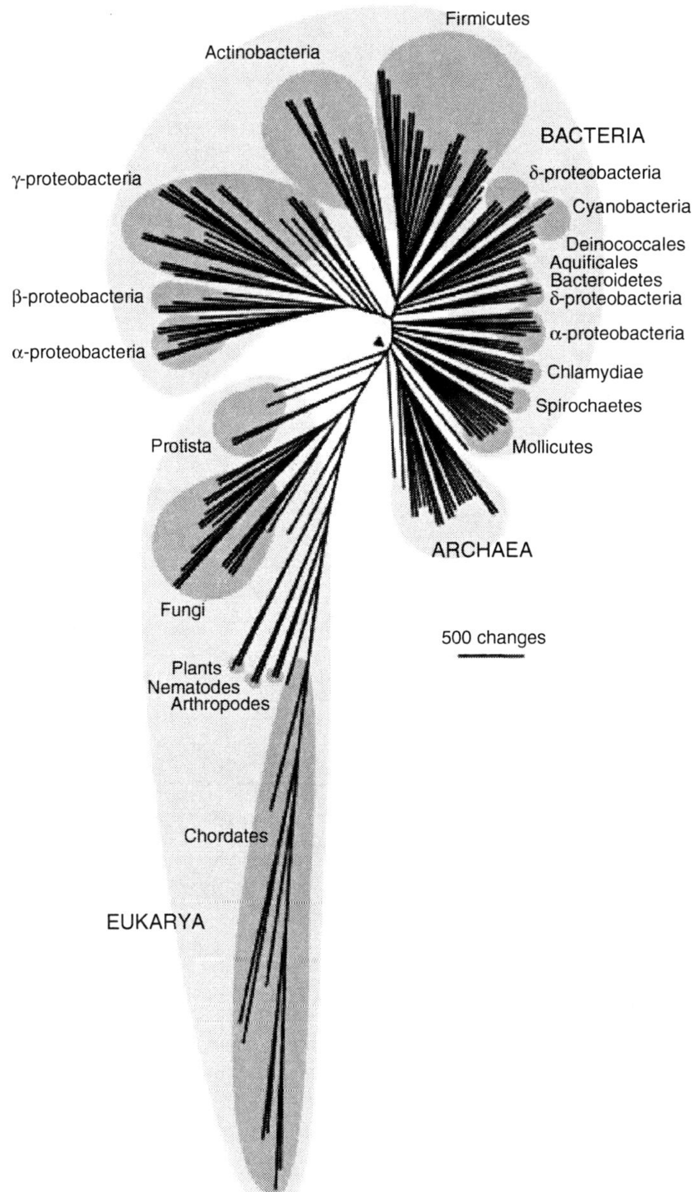

Fig. 6.2 Phylogenomic tree of life. The tree was reconstructed from an analysis of 35,559 domain combinations at fold superfamily level in proteins belonging to 185 organisms that have been completely sequenced. Only one optimal tree of 948,547 steps was obtained using maximum parsimony as the optimality criterion (CI=0.2714; RI=0.5375; RC=0.1459; g_1=−1.0334). Terminal leaves are not labeled as they would not be legible. The arrowhead shows the placement of the root. Note that character change is maximum in Eukarya, that Mollicutes, Spirochaetes and Chlamydiae are basal within the bacterial clade, and that plants and animals represent sister taxa

combinatorial diversity. Examples include RNA and multi-domain proteins. Thermodynamics and forces that stabilize molecular structure drive evolution of these networks through replication and mutation. We can also find inter-molecular networks expressed for example in metabolism, gene expression, protein-protein interaction, and signaling networks. These ubiquitous networks are generally scale-free (i.e. their degree distribution approximates a power law) and evolve by two fundamental processes, growth and preferential attachment (Barabási and Oltvai 2004). Growth arises when new network components (nodes) are added to the system, and preferential attachment results when nodes establish interactive connection (links or edges) preferentially with already well-connected nodes. Growth and preferential attachment are jointly responsible for the emergence of the scale-free ("rich get richer") property of complex networks, and probably, have an origin in duplication and mutational divergence of network components. Gene duplication has been postulated to drive evolution of networks in protein domain combinations (Rzhetsky and Gomez 2001), protein fold occurrence in genomes (Qian et al. 2001), gene expression (Bhan et al. 2002), and protein interactions (Pastor-Satorras et al. 2003). On the other hand, gene duplication may not be the only driver of evolution of networks, or the generator of power law behavior (Wagner 2003).

6.2.4 Molecular Mechanics and Evolution

Molecular machines made of protein and RNA can be considered the major operating components of the living world. The function of these molecules is largely determined by their structure. Consequently, structural conformations can be regarded as molecular phenotypes to which genotypes can be mapped. Because of their unique chemistries, the mapping of genotype (sequence) to phenotype (structure) in proteins and RNA biopolymers offers different challenges but share three properties: (i) the sequence-to-structure map is degenerate; i.e. there are orders of magnitude more sequences than structures; (ii) few common but many rare structures materialize in structure space; and (iii) extensive neutral networks that percolate sequence space define common structures and structural neighborhoods (Fontana 2002; Schuster and Stadler 2003). Because the distribution of sequences that fold into the same structure within neutral networks in RNA is approximately random, the mapping has "space covering" properties. This means that all structures can materialize within relatively few mutational changes in sequence space. This property has been confirmed experimentally using RNA functional switches (Schultes and Bartel 2000). Computational studies also predict the existence of neutral networks and space covering for polypeptides (Babajilde et al. 2001) and experiments support the model (Keefe and Szostak 2001). However, the sequence-to-structure mapping of proteins is much more complex and its landscape "holey", with protein conformations missing in vast segments of sequence space due to the effects of steric hindrance, hydrophobic and H-bonding interactions, and short-range dispersion forces.

6.3 Defining an Evolutionary Genomic Framework

Evolutionary genomics can be powerful when it interfaces with network biology thermodynamics, and molecular mechanics. The function of molecules is curved by evolution, generally resulting from natural selection operating at high levels of structural organization. We have therefore chosen to design a general evolutionary genomic framework that reconstructs evolutionary history directly from the structure of protein and nucleic acid molecules. In initial studies, structure, function, and genomic demography are embedded directly into phylogenetic analyses and molecules and genomes are compared at a wide range of evolutionary levels, from the subspecies analysis of laboratory strains of unicellular green algae to the universal tree of life (Caetano-Anollés 2001, 2002a,b, 2005; Caetano-Anollés and Caetano-Anollés 2003, 2005). This approach can be used to unravel evolutionary processes and uncover functional relationships in macromolecules, and the basis of molecular diversity and genome coexistence. The framework enables global bottom-up or top-bottom approaches of genomic analysis and is supported by three fundamental premises:

1. *Molecular structure is far more conserved than sequence and carries considerable phylogenetic signal.* Structure is directly linked to function and is therefore the subject of natural selection and strong evolutionary constraint (Bajaj and Blundel 1984; Vukmirovik and Tilghman 2000). Consequently, 3D structure is less prone to be affected by mutation than sequence and the information in structure is expected to persist longer than in primary sequence. Similarly, rare genomic processes such as intron indels, retrotransposon integrations and genome rearrangements can preserve deep phylogenetic information (Rokas and Holland 2000). Theoretical considerations suggest that sequence data may be inherently limited in its ability to uncover deep phylogenetic signatures and ancient relationships when the repeated accumulation of substitutions in nucleotide sites (site saturation) erases evolutionary history (Sober and Steel 2002; Penny et al. 2003; Mossell 2003). Convergent evolution of nucleotide sites, differing substitution rates among sites and lineages, and non-independent substitutions among sites, are just few of many other contributing factors (Philippe and Laurent 1998; Delsuc et al. 2005).
2. *Successfully implemented biological designs tend to be reused over and over again in nature.* Structural designs that had been successfully deployed will have more chances to be reused in other biological contexts, and consequently, are expected to become popular (Hartwell et al. 1999). Moreover, robust and well-evolved molecular designs have more chances of withstanding the effects of time. Evidence of this can be found in the redundant and modular nature of protein structure, where certain supersecondary structures and protein domains are highly ubiquitous (Söding and Lupas 2003).
3. *There is a universal tendency towards molecular order.* This very simple hypothesis of polarization depicts generalized trends applied to the structure of molecules, which have been supported by a considerable body of evidence. In

the case of RNA molecules, a tendency towards order was supported by: (i) the study of extant and randomized RNA sequences, showing that evolution enhances conformational order and diminishes frustration over that intrinsically acquired by self-organization (Stegger et al. 1984; Higgs 1993, 1995; Schultes et al. 1999; Seffens and Digby 1999; Gultyaev et al. 2002; Caetano-Anollés 2005); (ii) experimental verification of a molecular tendency towards order and stability using thermodynamic principles generalized to account for non-equilibrium conditions (Gladyshev and Ershov 1982); (iii) a large body of theoretical evidence that maps the structural repertoire of evolving RNA sequences from energetic and kinetic perspectives (Ancel and Fontana 2000; Higgs 2000; Fontana 2002); (iv) phylogenetic congruence in the reconstruction of trees generated from sequence, structure, and genomic rearrangements at different taxonomical levels (Billoud et al. 2000; Collins et al. 2000; Caetano-Anollés 2001, 2002a,b, 2005; Swain and Taylor 2003).

Bottom-up strategies unify phylogenetic analysis with structural biology using a cladistic approach based on shared and derived features descriptive of common descent that use features of molecular structure to generate phylogenetic trees. Cladistic methods offer explicit and general definitions of biological relationships proven to be powerful tools in phylogenetic systematics and molecular evolution. We applied this approach to the study of RNA molecules, generating histories of architectural and organismal diversification directly from their structure.

Top-bottom strategies study global diversification patterns in molecules using information embedded in entire genomic and proteomic complements. Since parsimony analysis has been one of the most widely used methods of phylogenetic inference and has mathematical attributes compatible with the complexity of these genomic datasets, we use this method to chart the protein world. We also explore the unique genomic regions that differentiate genomes from each other and shape, for example, the diversity of closely related species. The phylogenomic framework we have developed can be used to characterize the protein repertoire at the gene family, protein family, superfamily, and protein fold levels (Fig. 6.3), 'structuring' the evolutionary relationships between sequences and architectures, and revealing evolutionary patterns unique to individuals, species and organismal lineages.

6.4　Exploring the Evolution of Modern RNA

6.4.1　Diversity of Non-protein Coding RNA

RNA molecules are ubiquitous and highly sociable and exhibit defined structural, enzymatic and regulatory activities. They have been considered predecessors of DNA and protein in an ancestral RNA world (Gilbert 1986). In recent years, however, we came to realize that our 'modern' RNA world is not a 'relic' but a truly functional entity that is quite diverse (Eddy 2001; Storz 2002; Bartel 2004). Besides

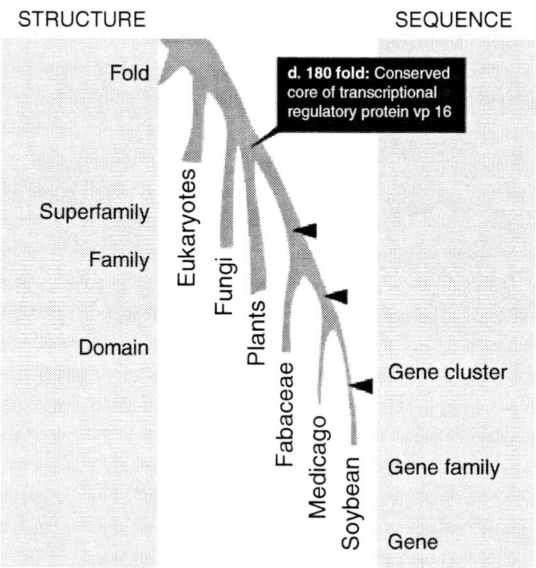

Fig. 6.3 Phylogenomic analysis of protein sequence and structure. Genes can be grouped into gene families and gene clusters using the tools of phylogenetic analysis. At the structural level, these families and clusters can be further defined by domains, and these can be unified into families, fold superfamilies and folds. This hierarchical scheme of molecular organization contains entities (from folds to genes) that can uniquely define organismal lineages. For example, there is currently only one protein fold that is unique to plants (d.180). We expect to find many superfamilies and families unique to individual plant lineages (arrowheads). In fact, the number of unique entities will increase at lower taxonomical levels

the classical three groups of molecules, tRNA, rRNA and mRNA, a repertoire of other RNA have been described. Collectively, these molecules have been termed non-protein coding RNA (ncRNA). ncRNAs are generally small. However, they range in size from ~21–25 nt (for regulatory RNAs) to ~10^3 10^4 nt (for ncRNAs involved in the maintenance of chromatin structure). ncRNAs play important roles in a number of cellular processes, such as those related to transcription, replication, RNA processing and modification, mRNA translation, and protein stability and translocation. Gene expression is modulated by micro RNA (miRNA) and small interfering RNA (siRNA). These molecules, discovered by their role in the control of developmental timing in *Caenorhabditis elegans* (Lee et al. 1993), are tiny and ubiquitous in animals and plants, and are present in all organismal domains (Bartel 2004). ncRNAs play roles in other cellular processes such as the translational tagging of proteins by tmRNA and the targeted mRNA degradation in RNA interference (RNAi) (e.g. Hutvágner and Zamore 2002). Other small ncRNA molecules are important for RNA processing, modification, and stability, such as the catalytic core of the universally conserved RNase P enzyme (~300–500 nt) that cleaves leader sequences from tRNA precursors (Frank and Pace 1998) or the small nucleolar RNA (snoRNA) (~70–250 nt) that are required for cleavage and processing of rRNA precursors (Eliceiri 1999). ncRNA molecules are also involved in protein translocation across membranes. One example is the signal recognition particle (SRP) that targets nascent secretory and membrane proteins (Keenan et al. 2001). Finally, ncRNA molecules have also been implicated in post-transcriptional gene silencing (siRNA) (Baulcombe 2004). Many other ncRNA molecules have been discovered that play structural roles, mimic the structure of other nucleic acids, or have very specific catalytic activities (Storz 2002).

Holistic views of the universe of RNA structure are missing. This is in part due to difficulties related to the study of RNA (Eddy 2001). For example, genes are identified by the proteins they encode. So genes that encode other molecules remain 'computationally' intractable when using standard tools that scan genome sequences. Novel systematic gene-discovery approaches are therefore needed to uncover effectively the RNA-encoding component of genomes (Washietl et al. 2005). There are no RNA taxonomies and the study of the evolution of RNA structure is still incipient.

6.4.2 *Phylogenetic Analysis of RNA Structure*

In our laboratory we search for evolutionary patterns embedded in the structure of functional RNA (Caetano-Anollés 2002a,b, 2005). Structures are first characterized using attributes that describe the overall geometry ('shape') of molecules and 'statistical' parameters that describe stability and statistical mechanic features quantitatively. Shape attributes measure for example the nucleotide length of each and every spatial component of secondary structure, such as double helical stems and unpaired sequences, and the number of loops in coaxial stem tracts. Note that unpaired nucleotides can form unusual base-pairings or establish non-covalent interactions (Hermann and Patel 1999). These base pairs and interactions are involved in high-order three-dimensional motifs that are not considered in the structural models of our analysis. Statistical parameters include the Shannon entropy of the base-pairing probability matrix (Q), base-pairing propensity (P), and mean length of helical stems (S) (Fontana et al. 1993; Schultes et al. 1999; Ancel and Fontana 2000). Q, P and S define a complete molecular morphospace, in which Q measures the number of conflicting inter- and intra-molecular interactions (frustration) during RNA folding, and P and S describe how extensively folded and ramified (multifurcated) are molecules (Schultes et al. 1999). In phylogenetic analysis, attributes are considered 'characters', and the numerical values they display 'character states' (Page and Holmes 1998). Characters that are homologous (i.e. share common ancestry) and have been appropriately coded (i.e. provide maximum phylogenetic signal) are compared. Structural characters used in this study transform from one state to another in linearly ordered and reversible pathways 'polarized' by superimposing an evolutionary tendency towards structural order (described above). This tendency should be interpreted as an evolutionary lock-in triggered by the branching of lineages in the trees (cladogenesis), resulting in molecules that are less plastic but more modular. Finally, hypotheses about character states and models of character evolution were transformed into hypothesis about evolutionary relationship of molecules using maximum parsimony (Page and Holmes 1998). Figure 6.4 describes the overall rationale.

We reconstructed structural phylogenies from several kinds of RNA, including tRNA, rRNA, spacer rRNA, SRP RNA, small mRNA molecules, and retroelements. We also generated a universal tree of life from the structure of rRNA that was rooted in the Eukarya (Caetano-Anollés 2002a,b). However, we decided to focus on tRNA, a molecule that bridges fundamental components of the translation machinery (Sun and Caetano-Anollés 2008). We analyzed the entire set of 571 tRNA molecules deposited as RNA sequences in the Bayreuth database. tRNA structural phylogenies placed tRNA molecules that coded for a group of four amino acids and harbored a variable loop (tRNA^{Sec}, tRNA^{Ser}, tRNA^{Leu}, and tRNA^{Tyr}) at the base of the tree of tRNA structure. These four amino acids were probably the first charged or coupled by tRNA in processes related to translation and/or RNA-world based replication that occurred before organismal diversification. Because our phylogenies did not reveal clearly the tripartite nature of life, or clear anticodon or amino acid-linked patterns, we used phylogenetic constraint to falsify alternative hypotheses about the origin of organismal diversification, amino acid specificities, and structural diversification in tRNA molecules. The results of these analyses suggest a sister-clade relationship between Bacteria and Archaea that is consistent with trees of life reconstructed from rRNA structure (Caetano-Anollés 2002a,b), protein fold architecture (Caetano-Anollés and Caetano-Anollés 2003), and domain combinations (Fig. 6.2). Results also show patterns of diversification of tRNA that developed once the cloverleaf structure was fully formed. Apparently, structural diversification preceded the establishment of amino acid and anticodon specificities, and these probably preceded organismal diversification.

We also designed a novel phylogenetic approach that reconstructs the evolution of substructural components of a molecule and generates *"contour maps"* capable of superimposing ancestral-derived relationships directly onto 3D RNA representations (Fig. 6.4). This involves defining new kinds of taxa (substructures) and characters (molecules), and a criterion of primary homology pertaining substructural repertoires based on molecular lineages. Phylogenetic trees of substructures describe here the evolutionary relationships of molecular substructural components that make up RNA molecules. These trees reveal evolutionary patterns of structural diversification, showing how RNA structure changes in the course of evolution. Patterns suggest by definition a structural origin and a relative timeline (a series of steps) describing how individual substructures are incorporated into the evolving RNA molecules. Analysis of tRNA molecules using this novel approach provided strong support to the 'two halves' hypothesis put forth by Maizels and Weiner (1994) that proposes that the anticodon/ dihydrouridine domain constitutes a refinement that was incorporated later in evolution (Fig. 6.5). However, our structural trees also support a more detailed structural transformation sequence. In this model, the tRNA molecule evolves by gradual addition of nucleotide pairs to a primordial hairpin stem loop

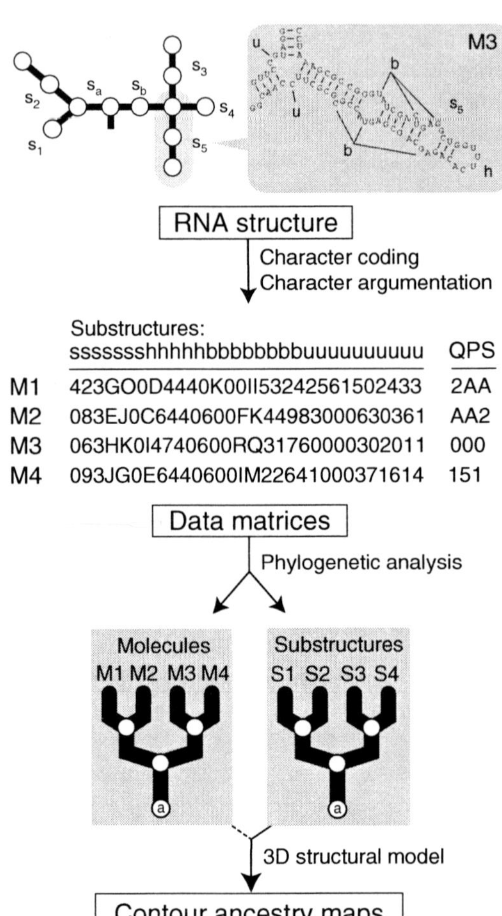

Fig. 6.4 Molecular structures (M1, M2, ...) and substructural repertoires (S1, S2, ...) of RNA molecules can be organized hierarchically in nested sets (*phylogenetic trees*) using cladistic principles. Trees describe structural diversification and allow identification of ancestors (e.g. nodes labeled *a*). The structure of an RNA molecule, such as signal recognition particle (SRP) RNA from rice (M3), can be decomposed for example into segments (S_1–S_5, S_a and S_b) and substructures (e.g., coaxial stem tracts and unpaired loop regions), and these substructural components studied using molecular features (*characters*) that describe their geometry [e.g., length of stems (s) and unpaired regions (h, b, and u)] or their stability and uniqueness (e.g., using morphospace parameters Q, P and S). These *shape* and *statistical* characters are coded and assigned 'character states' (in alphanumeric format) according to an evolutionary model that polarizes character transformation towards an increase in molecular order (character argumentation). Coded characters are arranged in data matrices and subjected to cladistic analysis, generating phylogenies of molecules and substructures. Rooted trees can be used to color 2D or 3D structural models of RNA (*contour ancestry maps*) that help infer models of structural evolution

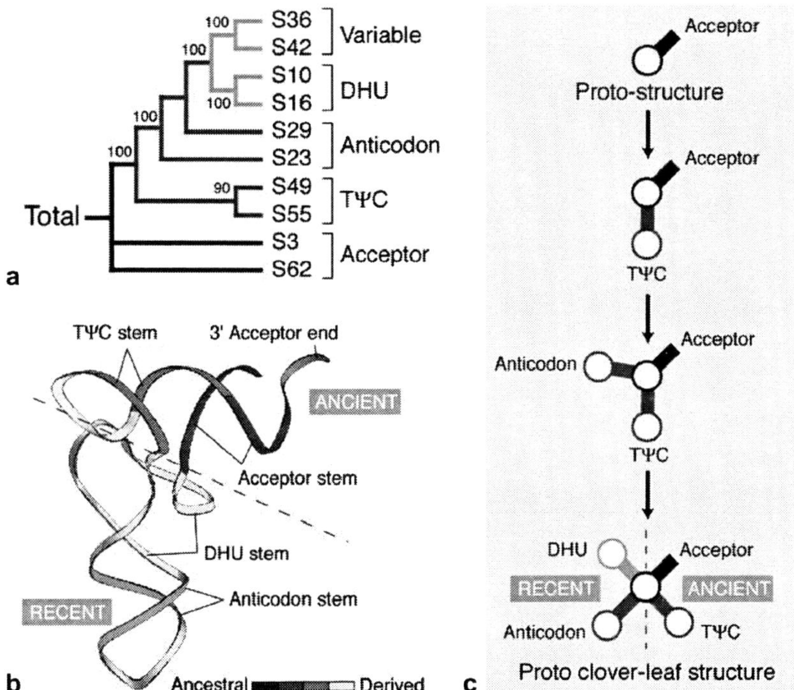

Fig. 6.5 Evolution of tRNA structure. **a** Trees of tRNA substructures show patterns of structural evolution inferred from the total tRNA dataset using maximum parsimony and branch-and-bound searches. Analysis of stabilizing stem characters produced two optimal trees of 4468 steps each (CI = 0.961; RC = 0.937; g_1=−1.25). The tree that is shown represents a strict consensus of these two trees and is labeled with bootstrap support values >50%. **b** Contour ancestry map showing the geometrical evolution of stem components that stabilize tRNA molecules. Trees were painted directly on the structural model using a color scale bar describing relative ancestry values. **c** A model of the early evolution of proto-tRNA molecules. The model is derived directly from trees of substructures and shows formation of substructures homologous to present-day acceptor, TΨC, anticodon and dihydrouridine arms. Substructures may have had different functions than those of extant tRNA molecules. Unpublished data from Sun and Caetano-Anollés (2008)

and then to its growing stems, ultimately resulting in a molecular arrangement that favors multiloop conformations and molecular multifurcation, an expected outcome when seeking to maximize molecular order.

We have extended our phylogenetic approach to the analysis of other interesting questions. For example, we used the structure of tRNA-derived transposable elements (SINEs) to study their evolution in plants (Sun et al. 2007). The exercise established a model of structural evolution of these transposable elements that explains the popularity of sequence families in the plant genome. We also found interesting patterns in the small (SSU) and large (LSU) subunits of rRNA (Harish and Caetano-Anollés, unpublished), including the ancestral placement of stem S49, the dominant SSU rRNA component of the subunit interface and the proposed ribosomal functional relay (Yusupov et al. 2001).

It is particularly noteworthy that ancient substructures were located in the middle of the rRNA ensemble and at the subunit interface. The origins of these ribosomal ancient substructures appear not associated with translation.

6.5   Exploring the Evolution of the Protein World

6.5.1   The Hierarchical Nature of Protein Structure

The protein world is extraordinarily diverse in sequence, structure and function (Ponting and Russell 2002). Most proteins (60%) fold compactly into more than one domain, and these domains can be repeated or combined in defined order. The number of available domains is considerable but appears finite (Chothia et al. 2003) and so does the repertoire of domain combinations in proteins (Vogel et al. 2004). When creating new functions, redundancy appears to be a favored outcome, with domains reused more often than discovered.

Domains are not only units of protein structure and function but also units of evolution (Riley and Labedan 1997). Taxonomies that attempt to provide a comprehensive description of structural and evolutionary relationship of proteins of known structure, such as the Structural Classification of Proteins (SCOP) (Murzin et al. 1995) and the CATH protein structure classification (Orengo et al. 1997) use these building blocks as units of classification. In SCOP, proteins that are evolutionarily closely related at the sequence level are clustered together into protein families. Proteins belonging to different families that exhibit low sequence identities but share structural and functional features suggesting a common evolutionary origin are further unified into fold superfamilies. Finally, fold superfamilies sharing secondary structures that are similarly arranged and topologically connected are unified into protein folds (Murzin et al. 1995). These folds sometimes have peripheral regions of secondary structure that differ in size and conformation and 'decorate' distinctly the central fold architecture.

While our knowledge of sequence space is far from complete (Kunin et al. 2003), it is apparent that protein diversity originated from a limited set of architectural designs (Koonin et al. 2000). Most proteins have been formed by gene duplication, recombination, and divergence and proteome evolution can be tracked by matching proteins of known folding structure to genome sequences (Chothia et al. 2003). While protein folds can be mapped onto about half of amino acid residues encoded in genome sequences, using hidden Markov models (HMMs) of structural recognition, it has become increasingly more difficult to find new folds in nature (Grant et al. 2004). Consequently, the world of protein molecules appears finite and its study feasible at global levels. However, fold categories should be regarded as "neighborhoods" defined by how much structural overlap exists between them (Harrison et al. 2002). In fact, some regions of the protein fold space represent a continuum for some architectural arrangements (sometimes linked by super-secondary motifs) while in other regions clearly distinct non-overlapping topologies are observed.

6.5.2 An Evolutionarily Structured Universe of Protein Architecture

A number of approaches have been used to characterize protein space, including fold family trees (Efimov 1997; Zhang and Kim 2000), a periodic table of structures (Taylor 2002), or taxonomies based on secondary structure (Przytycka et al. 1999). Recently, the metric comparison of structure similarity of proteins representing different protein fold categories provided measurements of distance between the different structures and a global representation of protein space (Hou et al. 2003). Four clear groups representing the α/β, $\alpha+\beta$ all-α, all-β protein classes were evident in this representation. These studies show that it is possible to generate global views of the protein universe. However, comparative genomic efforts have been largely confined to describing wide-encompassing features as similarities and differences. To be useful, however, strategies require methods capable of organizing the comparative data within an evolutionary perspective.

We recently reconstructed universal phylogenies of protein architecture (Caetano-Anollés and Caetano-Anollés 2003, 2005; Wang et al. 2006). These phylogenies depict the evolution of the protein world – they also bring a unique power to the identification of structurally orthologous gene families defining unique gene complements. The general strategy is depicted in Fig. 6.6. We counted the number of genes corresponding to particular protein architectures in genomes and used these measures of 'genomic demography' to map the world of proteins and track architectural and organismal history directly at the proteome level. Intrinsically rooted phylogenomic trees of proteomes and fold architectures were generated that described phylogenomic relationships, patterns of evolution, and information on the underlying evolutionary processes. Studies involved small and large subsets of protein folds, and complete datasets matching three releases of SCOP (1.39, 1.59 and 1.67). Figure 6.6 shows a tree of fold architectures generated using information embedded in 185 genomes.

6.5.3 Evolutionary Patterns and Transformation Pathways

The universal tree of protein architecture revealed interesting patterns. Folds that were widely distributed in nature were found at the base of the tree and were only missing in parasitic organisms with highly reduced genomes (Fig. 6.6). These organisms (e.g., *Mycoplasma, Nanoarchaeum, Encephalitozoon*) have discarded enzymatic and cellular machinery in exchange from resources provided by their hosts. In fact, the first nine folds to emerge in evolution are common to every genome analyzed and include folds widespread in metabolism. It is noteworthy that only 16 folds are universally shared and all of them originated deep in the tree. Similarly, all classes of protein architecture appeared very early in the tree of architectures. Folds in the α/β protein class arose first and were followed by those in the $\alpha+\beta$, all-α, all-β, small, and multi-domain classes, in that order. These folds accumulated at different levels. The α/β folds occurred at relatively constant rates and were prevalent in the bottom half of the tree. In contrast, the $\alpha+\beta$ folds started to

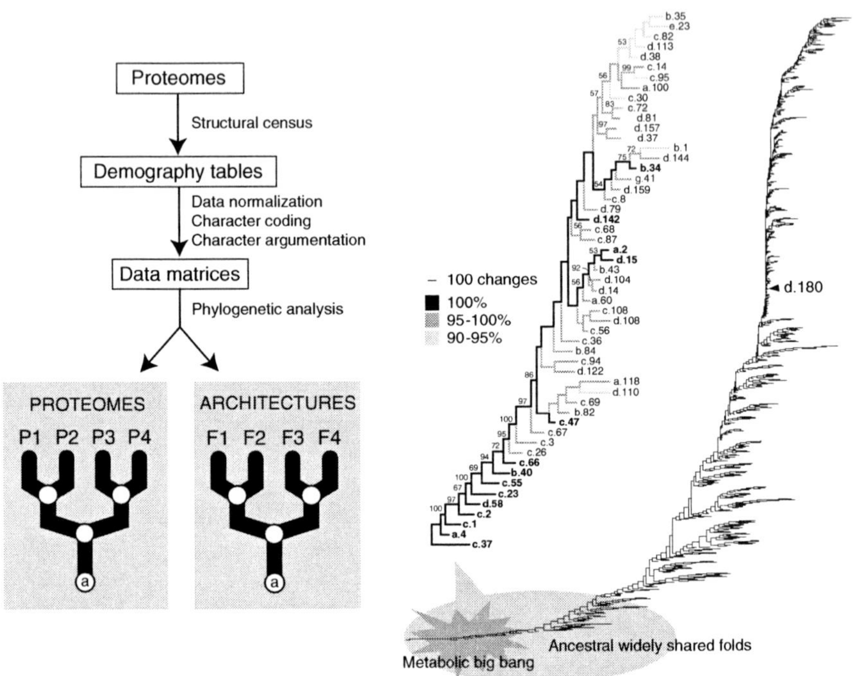

Fig. 6.6 Exploring the diversity of the protein world. The diagram (*left*) shows how fold architectures of extant proteins (F1, F2, …) and entire protein complements (P1, P2, …) can be used to generate hypotheses (phylogenies) about groups of folds and proteomes. This involves a structural census defined by advanced HMMs that assign domain structure to genomic sequences, normalization of data, and phylogenetic analysis. A phylogenomic tree of protein architecture generated from a protein domain census in 185 completely sequenced genomes (*right*) was recovered from an heuristic maximum parsimony search with branch swapping and 100 replicates of random addition sequence. The tree had 85,644 steps (CI=0.043, RI=0.770) and is well supported by measures of skewness in tree distribution ($g_1 = -0.138$; $P < 0.01$). Terminal leaves are not labeled except for SCOP fold d.180 (see Fig. 8.3), which is unique to plants. The subtree drawn above describes the evolution of the 53 most ancestral folds and has branches labeled with different shades indicating percentage of genomes sharing folds. The subtree shows folds labeled according to SCOP nomenclature and bootstrap support values >50% above nodes. It is noteworthy that folds shared by >90% genomes are missing almost exclusively in parasitic organisms with reduced genomes. Data from Wang et al. (2006)

accumulate significantly later but with increasing rates until these folds became the most prevalent class. Folds in all other classes followed this same pattern of accumulation but with lower rates. Maximum rates diminished in the order of fold appearance, i.e. all-α, all-β, small and multi-domain proteins. These patterns suggest that the most primitive proteins contained interspersed α-helical and β-sheet elements (as in the α/β class). In the course of evolution, these elements were first segregated within their structure (α+β class) and then confined to separate molecules (all-α and all-β classes). This hypothesis is consistent with the suggestion that diversity in protein architecture originated by stochastic processes expressed both in protein sequence and structure (the random origin hypothesis; White 1994).

Remarkably, the most ancestral folds harbored interleaved β-sheets and α-helices and barrel structures (Fig. 6.7). Many important structural designs were derived in the tree, including polyhedral folds in the all-α class and β-sandwiches, β-propellers and β-prisms in the all-β class. Protein transformation pathways that describe

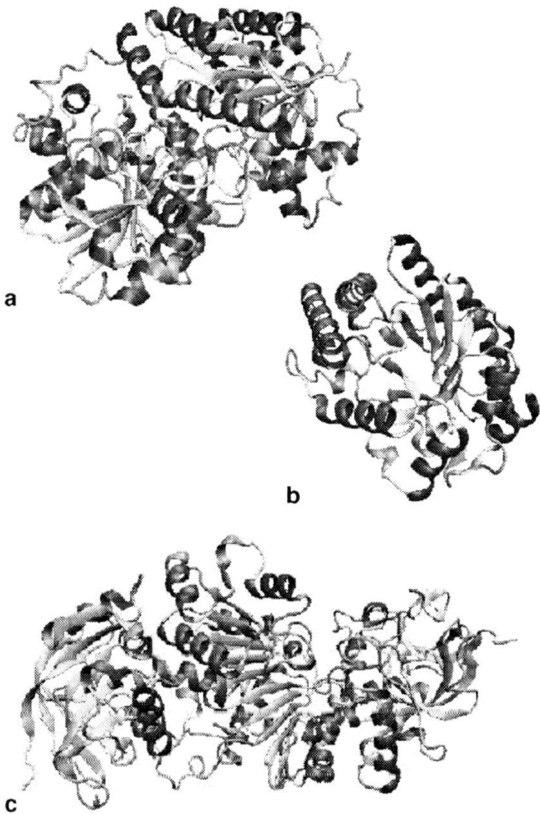

Fig. 6.7 Ancient protein folds share a common architecture of sheets and helices that form either barrels or are interleaved and are highly symmetrical. The structural models of selected structures show the arrangement of α-helices (described by dark helical ribbons) and β-strands (described by arrows that point towards the C-terminus of the protein). Structures were visualized in 3D using the new cartoon format of the VMD (Visual Molecular Dynamics; http://www.uiuc.edu/research/vmd) visualization package. **a** The nitrogenase iron protein from *Azotobacter vinelandii* (PDB entry 1fp6), an enzyme important for nitrogen fixation, harboring the P-loop hydrolase fold (c.37), the most ancient protein architecture with three layers in which a parallel or mixed β-sheet is sandwiched by α-helices. **b** The xylanase from *Penicillium simplicissimum* (1bg4), a protein exhibiting the TIM β/α-barrel fold (c.1), a α/β protein architecture with a parallel β-sheet closed barrel. **c** The gluthatione-dependent formaldehyde dehydrogenase enzyme from humans (1m6h) with the NAD(P)-binding Rossmann fold (c.2) that harbors two layers of α-helices sandwiching a parallel β-sheet of 6 β-strands. These three ancient architectures are very common in modern metabolism

likely scenarios of structural evolution (Murzin 1998; Grishin 2001) could be traced in our tree of architectures. For example, the conversion of an α-helix into a three-stranded β-meander causes Rossmann fold proteins to change to the FAD/NAD(P)-binding domain architecture. Both folds are ancient and are closely related, so this putative transformation must have already occurred very early during evolution. In contrast, circular permutations in protein phosphatases resulted in changes that were quite derived in the history of protein diversification. Figure 6.8 shows how an all-α protein containing a three-helical bundle transforms by indels and substitutions into a β-sheet structure that is part of an all-β barrel-like architecture, probably through an $\alpha+\beta$ protein intermediary. Transformations from all-α to all-β proteins may be quite common and follow general tendencies of architectural transformation (Caetano-Anollés and Caetano-Anollés 2003). Clear transformation pathways were also evident in structural families of fold architectures. For example, the popular β-barrels increased the tilt of the β-strands, the frequency of open barrel structures, and the complexity of strand topology. These tendencies suggest barrel architectures with increased curl and stagger of β-sheets (sensu Taylor 2002) are favored evolutionary outcomes.

6.5.4   Sharing Patterns of Fold Architecture in Life

We found that tracing features depicting organismal diversity along the branches of the evolutionary tree of protein architecture provided interesting information (Caetano-Anollés and Caetano-Anollés 2005). We were able to infer a relative timing for the emergence of prokaryotes, congruent episodes of architectural loss and diversification in Archaea and Bacteria, and a late and quite massive rise of architectural novelties in Eukarya probably linked to the rise of multicellularity. Folds associated with processes related to multicellularity (e.g. apoptosis, cell death, adhesion and recognition, and extracellular matrix remodelling) contained multiple domains and appeared both immediately after prokaryotic diversification (mostly folds common to all domains of life) and during eukaryotic diversification (mostly eukaryotic-specific).

Our observations indicate that protein novelties unique to organismal lineages appeared late and in defined order during evolution. The proteomes of these diversified organisms originated apparently from ancestors that shared already an arrangement of quite complicated molecular architectures and biological functions. This view is consistent with a proto-eukaryote (Poole et al. 1998; Kurland et al. 2006) responsible for 'crystallizing' diversified life (Woese 2000).

6.6   Exploring the Evolution of Networks

Our phylogenomic analysis is quite novel (Doolittle 2005) and offers the opportunity to identify and trace architectures unique to organisms or organismal groups, unique to functions and ontologies, and unique to biological networks. Since proteins

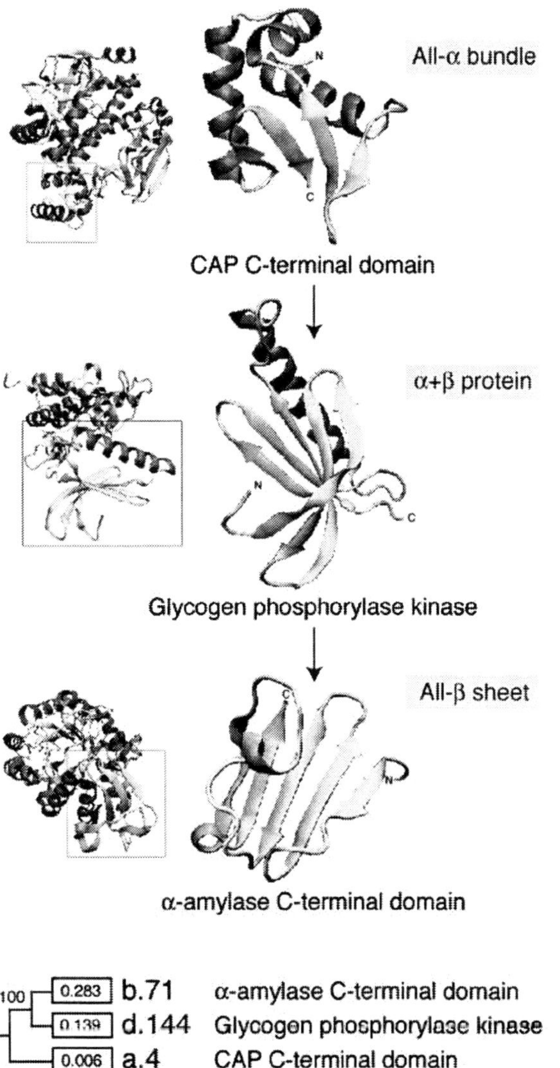

Fig. 6.8 Evolutionary transformation pathway from an all-α to an all-β protein architecture induced by indels and substitutions. The figure shows how the winged helix-turn-helix (HTH) domain characteristic of nucleic-acid-binding domains such as the C-terminal domain of the catabolite gene activator protein (CAP) (1cgp) transforms into the γ-subunit of the glycogen phosphorylase kinase (1phk), and this structure then transforms into the C-terminal domain of the G4-α-amylase (1,4-α-D-glucan maltotetrahydrolase) (2amg). The CAP C-terminal domain has a DNA/RNA-binding three-helical bundle fold (a.4), in which three α-helices form a partly opened right-handed bundle. The glycogen phosphorylase kinase has a protein kinase-like fold (d.144) with two α+β domains, one of which (the C-terminal) is almost α-helical. The α-amylase C-terminal domain has a glycosyl hydrolase domain (b.71) with a β-sheet that follows the catalytic β/α barrel domain. The entire multidomain proteins are shown in the left with the relevant domain enclosed by rectangles. The transformation from a three-helical bundle to a β-sheet seen in the structural models of the domains is confirmed by the phylogenomic tree shown below with terminal nodes indicating ancestry values of individual folds derived from the tree of fold architecture in Fig. 6.6

are generally components of biological networks, protein structure can be used to study network evolution.

Cellular metabolism is the best-studied biological network. It represents one of the greatest achievements of science, resulting from almost two centuries of biochemical research. However, we do not know its origin or how it has evolved. In an initial study, we explored the relationship between protein architecture and function by tracing the total number of enzymatic functions associated with folds in the tree of architectures (Caetano-Anollés and Caetano-Anollés 2003). As expected, the most ancestral folds had the most enzymatic functions associated with them. This supports the proposal that during metabolic evolution enzymatic multifunctionality was replaced by specialized function (Kacser and Beeby 1984). We also explored the origins and evolution of modern metabolism using phylogenomic information embedded in protein structure. We first painted the ancestries of enzymes derived from rooted phylogenomic trees directly onto over one hundred metabolic subnetworks in mesonetworks defined by the Kyoto Encyclopedia of Genes and Genomes (KEGG) (Kim et al. 2006). This evolutionary tracing exercise involved linking metabolic enzymes to fold architectures and an analysis of 860,000 genomic sequences with HMMs (Fig. 6.9). To our knowledge, this represents the first global attempt to map evolutionary relationships directly onto biological networks. Careful analysis of evolutionarily painted subnetworks revealed patchy distribution patterns indicative of widespread enzymatic recruitment, consistent with previous evidence (Schmidt et al. 2003). It is noteworthy that the distribution of abundance of folds with various ancestries showed that mesonetworks differ in mean ancestry, with amino acids oldest and lipids and glycans youngest. We also revealed patterns of origin of modern metabolism (Caetano-Anollés et al. 2007). Apparently, a "big bang" of enzymatic diversification occurred at the base of the tree of protein architectures (Fig. 6.6). In fact, most enzymatic reactions at all levels of Enzyme Commission (EC) classification were associated with the nine most ancestral and widespread folds. Furthermore, phylogenetic trees reconstructed from enzymatic sharing of fold architectures and other information indicated that metabolism originated in the purine and pyrimidine subnetworks. Consequently, the first enzymatic take-over of a prebiotic chemistry involved the synthesis of nucleotides for the RNA world.

6.7 Evolutionary Genomics and Organismal Coexistence

An important focus of genomic research has been the identification of differences between genomes (Koonin et al. 2000) and the systematic grouping of hundreds of thousands of protein sequences into protein clusters based on sequence and structural similarities (Grant et al. 2004). Initial studies uncovered a diverse genetic repertoire and a large proportion of genes that were uniquely species-characteristic

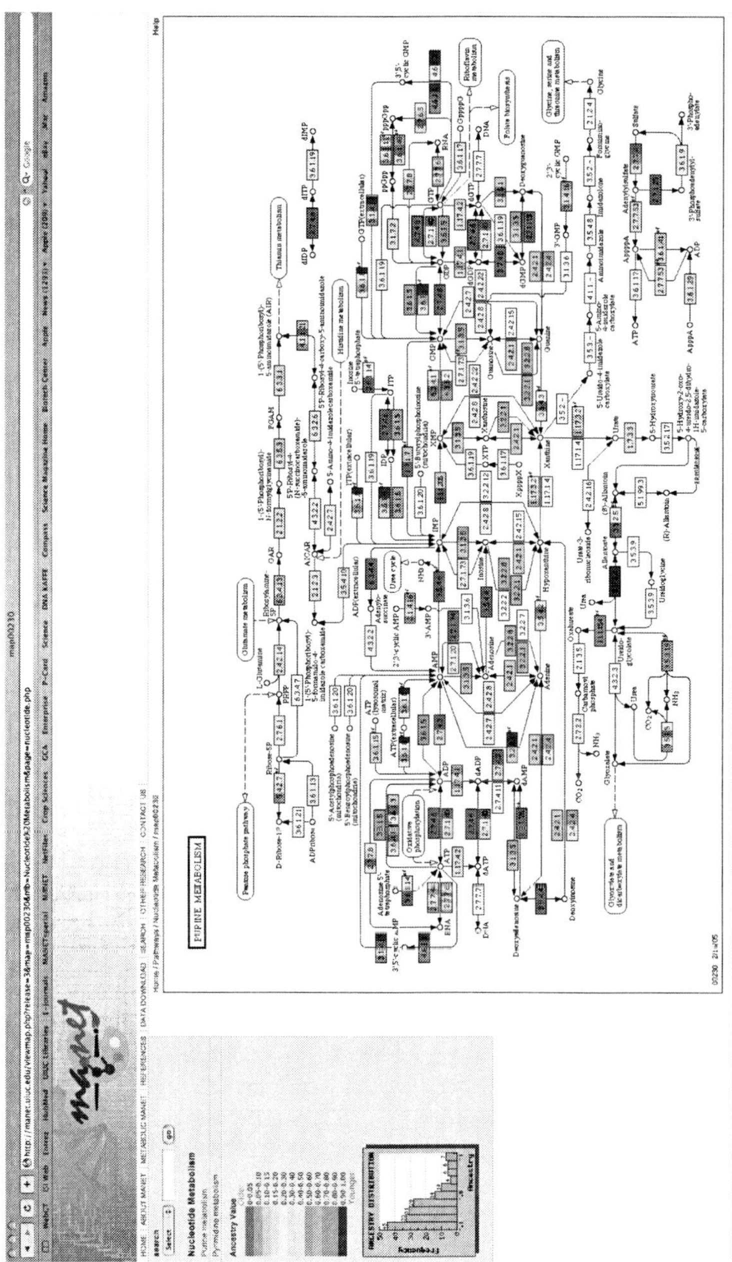

Fig. 6.9 Web page showing a representative subnetwork diagram in the metabolic MANET (Molecular Ancestry Networks) database (http://manet.uiuc.edu). A colored scale is used to assign binned ancestry values of protein folds to enzyme nodes named with Enzyme Commission (EC) numbers. Note that some enzymes have more than one structural assignment. They are multidomain proteins or exhibit different structures in different organisms. Colored enzymes with ancestry assignments from HMM-based predictions are labeled with an 'sf' marking at the top right of KEGG-hyperlinked rectangles depicting enzymatic nodes. Each subnetwork diagram also shows a frequency distribution plot of ancestries (Kim et al. 2006)

(Doolittle 2005). With the advent of evolutionary genomics, the focus of research now shifts heavily towards molecular evolution and the mechanisms that fuel genomic sequence and structural divergence.

Evolutionary genomics places the comparative relationship of organisms within an evolutionary perspective, and does so at the genomic level. The interaction between organisms and the interaction of organisms with the environment are curbed by ecology and evolution and are therefore expected to affect complements defined by the survey of genomic component parts. A substantial body of evidence suggests complex interactions of gene products are responsible for the establishment of pathogenic or symbiotic interactions. For example, plants and pathogenic microbes interact in an endless race to cause disease, and this interplay dominates many important issues in plant pathology (Schumann and D'Arcy 2006). However, our knowledge of how plant and microbial coexistence shapes genomic composition is limited (Ochman and Moran 2001). We know changes in microbes can be large and involve instances of lateral transfer events that exchange considerable genetic material and occur pervasively but not indiscriminately. The existence of fully sequenced genomes from pathogenic and non-pathogenic organisms as well as organisms that have different lifestyles now offer the opportunity to explore the specific effects of organismal coexistence on genomic repertoires. For example, in a recent study the proteomes of several parasites and symbionts exhibiting highly reduced genomes were compared (Chandonia and Kim 2006). The study showed that proteins performing essential functions closely related to transcription and translation exhibited a higher degree of fold usage than proteins in other functional categories. In a systematic and global study of 185 fully sequenced genomes exhibiting free-living, parasitic and obligate parasitic lifestyles, we revealed very specific effects of lifestyle on proteome composition at protein fold level (Wang et al. 2007). For example, Fig. 6.10 shows how protein folds are used and how fold abundance distributes along the tree of fold architectures in genomes from free-living (FL), parasitic (P) and obligate parasitic (OP) bacteria that establish interactions with plants.

The representative organisms analyzed illustrate the general tendency observed in genomes from organisms with P and OP lifestyles to diminish the number of folds used as well as their abundance, regardless of whether the lifestyle causes genomes to be reduced in size. These tendencies are general and are also observed in Archaea and Eukarya. Interestingly, even folds that are ancient and common to all fully sequenced genomes (ABE_o) were considerably under-represented in P and OP bacterial genomes. This and other evidence suggests strongly that establishing parasitic (or symbiotic) interactions results in either protein architectural specialization or the forfeit of protein architectures in exchange of resources from their hosts.

These and many other studies suggest biotic and abiotic interactions impact the makeup and evolution of genomes. I anticipate that patterns and processes uncovered by evolutionary genomics will explain these and other phenomena, benefiting the study of molecular diversity embodied in genome coexistence.

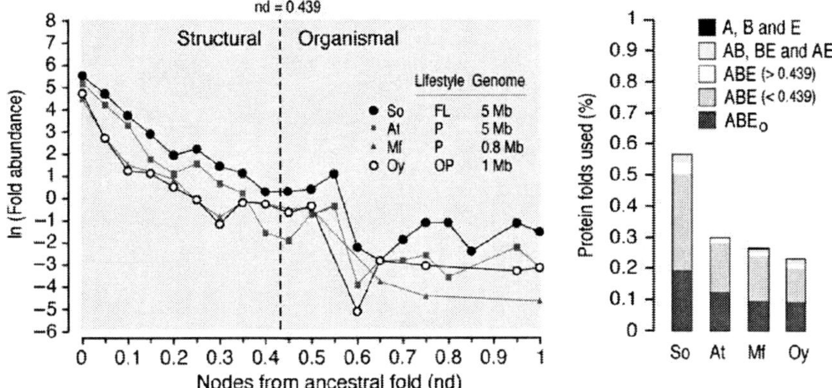

Fig. 6.10 Impact of organismal lifestyle on protein fold architectures in bacterial proteomes. The genomes of four representative bacterial species known to interact with plants and harbor either free-living (FL), parasitic (P) or obligate parasitic (OP) lifestyles were studied and both fold genomic abundance and fold use determined (Wang et al., in preparation). *Shewanella oneidensis* (So) is a FL bacterial species that is present in freshwater sediments and is known to inhabit a wide range of environments and utilize a wide variety of electron acceptors during anaerobic respiration. *Agrobacterium tumefaciens* (At) is a pathogenic bacteria that produces tumors (crown galls) on dicotyledoneous plants. *Mesoplasma florum* (Mf) is a mollicute that establishes P interactions with plants, insects and mammals and has a highly reduced genome. Finally, onion yellows phytoplasma (Oy) is an OP organism that inhabits phloem sieve elements causing a variety of plant diseases. The bacterium has a highly reduced genome and interestingly, lacks the phosphotransferase system, the pentose phosphate pathway and ATP synthases. Fold abundance was studied as a function of the ancestry of individual folds measured by the number of nodes from the most ancestral fold (nd) in the tree of fold architectures. Folds specific to Archaea (A), Bacteria (B) and Eukarya (E) start appearing at an nd value of 0.439 that signals the transition between architectural and organismal diversification in the evolution of the protein world (Caetano-Anollés and Caetano-Anollés 2005). Fold usage in individual genomes was depicted in bar diagrams as the percentage of protein folds used that are either common to all fully sequenced genomes (ABE_o), common to all organismal domains (ABE) or specific to individual or sets of domains

Acknowledgements I would like to thank Jay E. Mittenthal for invigorating discussions and team members Fengjie Sun, Minglei Wang, Hee Shin Kim, Ajith Harish, Liudmila Yafremava, and Vegeir Knudsen for their effort and support. Research was funded by the National Science Foundation (NSF), the Office of Naval Research, Illinois Council for Food and Agricultural Research (C-FAR), the International Atomic Energy Agency (IAEA) in Vienna, and the Critical Research Initiative of the University of Illinois. Any opinions, findings, and conclusions or recommendations expressed in this material are those of the author and do not necessarily reflect the views of the funding agencies.

References

Ancel LW, Fontana W (2000) Plasticity, evolvability, and modularity in RNA. J Exp Zool (Mol Dev Evol) 288:242–283

Babajilde A, Farber R, Hofacker IL, Inman J, Lapedes AS, Stadler PF (2001) Exploring protein sequence space using knowledge based potentials. J Theor Biol 212:35–46

Bajaj M, Blundell T (1984) Evolution and the tertiary structure of proteins. Annu Rev Biophys Bioeng 13:453–492

Barabási AL, Oltvai ZN (2004) Network biology: understanding the cell's functional organization. Nature Rev 5:101–113

Bartel DP (2004). MicroRNAs: genomics, biogenesis, mechanism, and function. Cell 116:281–297

Baulcombe D (2004) RNA silencing in plants. Nature 431:356–363

Bhan A, Galas DJ, Dewey TG (2002) A duplication growth model of gene expression networks. Bioinformatics 18:1486–1493

Billoud B, Guerrucci MA, Masselot M, Deutsch JS (2000) Cirripede phylogeny using a novel approach: molecular morphometrics. Mol Biol Evol 17:1435–1445

Caetano-Anollés G (2001) Novel strategies to study the role of mutation and nucleic acid structure in evolution. Plant Cell Tissue Org Culture 67:115–132

Caetano-Anollés G (2002a) Evolved RNA secondary structure and the rooting of the universal tree of life. J Mol Evol 54:333–345

Caetano-Anollés G (2002b) Tracing the evolution of RNA structure in ribosomes. Nucleic Acids Res 30:2527–2587

Caetano-Anollés G (2005) Grass evolution inferred from chromosomal rearrangements and geometrical and statistical features in RNA structure. J Mol Evol 60:635–652

Caetano-Anollés G, Caetano-Anollés D (2003) An evolutionarily structured universe of protein architecture. Genome Res 13:1563–1571

Caetano-Anollés G, Caetano-Anollés D (2005) Universal sharing patterns in proteomes and evolution of protein fold architecture and life. J Mol Evol 60:484–498

Caetano-Anollés G, Kim H-S, Mittenthal JE (2007) The origins of modern metabolism inferred from phylogenomic analysis of protein architecture. Proc Natl Acad Sci USA 104:9358–9363

Chandonia J-M, Kim S-H (2006) Structural proteomics of minimal organisms: conservation of protein fold usage and evolutionary implications. BMC Struct Biol 6:7

Cheng Z, Ventura M, She X, Khaitovich P, Graves T, Osoegawa K, Church D, DeJong P, Wilson RK, Paabo S, Rocchi M, Eichler EE (2005) A genome-wide comparison of recent chimpanzee and human segmental duplications. Nature 437:88–93

Chothia C, Gough J, Vogel C, Teichmann SA (2003) Evolution of the protein repertoire. Science 300:1701–1703

Ciccarelli FD, Doerks T, von Mering C, Creevey CJ, Snel B, Bork P (2006) Towards automatic reconstruction of a highly resolved tree of life. Science 311:1283–1287

Collins LJ, Moulton V, Penny D (2000) Use of RNA secondary structure for studying the evolution of RNase P and RNase MRP. J Mol Evol 51:194–204

Delsuc F. Brinkmann H, Philippe H (2005) Phylogenomics and the reconstruction of the tree of life. Nature Rev Genet 6:361–375

Doolittle RF (2005) Evolutionary aspects of whole-genome biology. Curr Opin Struct Biol 15:248–253

Eddy SR (2001) Non-coding RNA genes and the modern RNA world. Nat Rev Genet 2:919–929

Efimov AV (1997) Structural trees for protein superfamilies. Proteins 28:241–260

Eliceiri GL (1999) Small nucleolar RNAs. Cell Mol Life Sci 56:22–31

Fontana W (2002) Modelling 'evo-devo' with RNA. BioEssays 24:1164–1177

Fontana W, Konings DA, Stadler PF, Schuster P (1993) Statistics of RNA secondary structures. Biopolymers 33:1389–1404

Frank DN, Pace NR (1998) Ribonuclease P: unity and diversity in a tRNA processing ribozyme. Annu Rev Biochem 67:153–180

Gerstein M (1998) Patterns of protein-fold usage in eight microbial genomes: a comprehensive structural census. Proteins Struct Funct Genet 33:518–534

Gerstein M, Hegyi H (1998) Comparing genomes in terms of protein structure: Surveys of a finite parts list. FEMS Microbiol Rev 22:277–304

Gilbert W (1986) The RNA world. Nature 319:618

Gladyshev GP, Ershov YA (1982) Principles of the thermodynamics of biological systems. J Theor Biol 94:301–343

Grant A, Lee D, Orengo C (2004) Progress towards mapping the universe of protein folds. Genome Biol 5:107

Grishin NV (2001) Fold change in evolution of protein structures. J Struct Biol 134:167–185

Gultyaev PA, van Batenburg FHD, Pleij CWA (2002) Selective pressures on RNA hairpins in vivo and in vitro. J Mol Evol 54:1–8

Gutteridge A, Thornton JM (2005) Understanding nature's catalytic toolkit. Trends Biochem Sci 30:622–629

Harrison A, Pearl F, Mott R, Thornton J, Orengo C (2002) Quantifying the similarities within fold space. J Mol Biol 323:909–926

Hartwell LH, Hopfield JJ, Leibler S, Murray AW (1999) From molecular to modular cell biology. Nature 401:C47–C52

Hermann T, Patel DJ (1999) Stitching together RNA tertiary architectures. J Mol Biol 294:829–849

Higgs PG (1993) RNA secondary structure: a comparison of real and random sequences. J Phys I France 3:43–59

Higgs PG (1995) Thermodynamic properties of transfer RNA: a computational study. J Chem Soc Faraday Trans 91:2531–2540

Higgs PG (2000) RNA secondary structure: physical and computational aspects. Q Rev Biophys 33:199–253

Hou J, Sims GE, Zhang C, Kim S-H (2003) A global representation of the protein fold space. Proc Natl Acad Sci USA 100:2386–2390

House CH, Fitz-Gibbon ST (2002) Using homolog groups to create a whole-genomic tree of free-living organisms: an update. J Mol Evol 54:539–547

Hutvágner G, Zamore PD (2002) RNAi: nature abhors a double-strand. Curr Opin Genet Develop 12:225–232

Kacser H, Beeby R (1984) On the origin of enzyme species by means of natural selection. J Mol Evol 20:38–51

Keefe AD, Szostak JW (2001) Functional proteins from a random-sequence library. Nature 410:715–718

Keenan RJ, Freymann DM, Stroud RM, Walter P (2001) The signal recognition particle. Annu Rev Biochem 70:755–775

Kim H-S, Mittenthal J, Caetano-Anollés G (2006) MANET: tracing evolution of protein architecture in metabolic networks. BMC Bioinformatics 7:351

Kitano H (2002) Computational systems biology. Nature 420:206–210

Koonin EV, Aravind L, Kondrashov AS (2000) The impact of comparative genomics on our understanding of evolution. Cell 101:573–576

Kunin V, Cases I, Enright AJ, de Lorenzo V, Ouzounis CA (2003) Myriads of protein families, and still counting. Genome Biol 4:401

Kurland CG, Collins LJ, Penny D (2006) Genomics and the irreducible nature of eukaryote cells. Science 312:1011–1014

Kyrpides N (1999) Genomes Online Database (GOLD): a monitor of complete and ongoing genome projects worldwide. Bioinformatics 15:773–774

Lee RC, Feinbaum RL, Ambros V (1993) The *C. elegans* heterochronic gene *lin-4* encodes small RNAs with antisense complementarity to lin-14. Cell 75:843–854

Lin J, Gerstein M (2000). Whole-genome trees based on the occurrence of folds and orthologs: implications for comparing genomes on different levels. Genome Res 10:808–818

Maizels N, Weiner AM (1994) Phylogeny from function: evidence from the molecular fossil record that tRNA originated in replication, not translation. Proc Natl Acad Sci 91:6729–6734

Mossell E (2003) On the impossibility of reconstructing ancestral data and phylogenies. J Comp Biol 10:669–678

Murzin A (1998) How far divergent evolution goes in proteins. Curr Op Struct Biol 8:380–387

Murzin A, Brenner SE, Hubbard T, Clothia C (1995) SCOP: a structural classification of proteins for the investigation of sequences and structures. J Mol Biol 247:536–540

O'Malley MA, Dupré J (2005) Fundamental issues in systems biology. BioEssays 27:1270–1276

Ochman H, Moran NA (2001) Genes lost and genes found: evolution of bacterial pathogenesis and symbiosis. Science 292:1096–1098

Orengo CA, Michie AD, Jones S, Jones DJ, Swindells MB, Thornton JM (1997) CATH: a hierarchic classification of protein domain structures. Structure 5:1093–1108

Page RDM, Holmes EC (1998) Molecular evolution: a phylogenetic approach. Blackwell Science, Oxford

Pastor-Satorras R, Smith E, Sole R (2003) Evolving protein interaction networks through gene duplication. J Theor Biol 222:199–210

Penny D, Hendy MD, Poole AM (2003) Testing fundamental evolutionary hypotheses. J Theor Biol 223:377–385

Philippe H, Laurent J (1998) How good are deep phylogenetic trees? Curr Opin Genet Dev 8:616–623

Ponting CP, Russell RR (2002) The natural history of protein domains. Annu Rev Biophys Biomol Struct 31:45–71

Poole A, Jeffares DC, Penny D (1998) The path from the RNA world. J Mol Evol 46:1–17

Przytycka T, Aurora R, Rose GD (1999) A protein taxonomy based on secondary structure. Nat Struct Biol 6:672–682

Qian J, Luscombe NM, Gerstein M (2001) Protein family and fold occurrence in genomes: power-law behavior and evolutionary model. J Mol Biol 313:673–681

Riley M, Labedan B (1997) Protein evolution viewed through *Escherichia coli* protein sequences: introducing the notion of a structural segment of homology, the module. J Mol Biol 268:857–868

Rokas A, Holland PWK (2000) Rare genomic changes as a tool for phylogenetics. Trends Ecol Evol 15:454–459

Rzetsky A, Gomez SM (2001) Birth of scale-free molecular networks and the number of distinct DNA and protein domains per genome. Bioinformatics 17:988–996

Schmidt S, Sunyaev S, Bork P, Dandekar T (2003) Metabolites: a helping hand for pathway evolution? Trends Biochem Sci 28:336–341

Schultes EA, Bartel DP (2000) One sequence, two ribozymes: implications for the emergence of new ribozyme folds. Science 289:448–452

Schultes EA, Hraber PT, LaBean TH (1999) Estimating the contributions of selection and self-organization in RNA secondary structure. J Mol Evol 49:76–83

Schumann GL, D'Arcy CJ (2006) Essential plant pathology. APS Press, St Paul, Minnesota

Schuster P, Stadler PF (2003) Networks in molecular evolution. Complexity 8:34–42

Seffens W, Digby D (1999) mRNA have greater negative folding free energies than shuffled or codon choice randomized sequences. Nucleic Acids Res 27:1578–1584

Sober E, Steel M (2002) Testing the hypothesis of common ancestry. J Theor Biol 218:395–408

Söding J, Lupas AN (2003) More than the sum of their parts: on the evolution of proteins from peptides. BioEssays 25:837–846

Stegger G, Hofman H, Fortsch J, Gross HJ, Randles JW, Sanger HL, Riesner D (1984) Conformational transitions in viroids and virusoids: comparison of results from energy minimization algorithm and from experimental data. J Biomol Struct Dynam 2:543–571

Storz G (2002) An expanding universe of noncoding RNAs. Science 296:1260–1263

Sun F-J, Caetano-Anollés G (2008) The origin and evolution of tRNA inferred from phylogenetic analysis of structure. J Mol Evol 66:21–35

Sun F-J, Fleudépine S, Bousquet-Antonelli C, Caetano-Anollés G, Deragon J-M (2007) Common evolutionary trends for tRNA-derived SINE RNA structures. Trends Genet 23:26–33

Swain TD, Taylor DJ (2003) Structural rRNA characters support monophyly of raptorial limbs and paraphyly of limb specialization in water fleas. Proc R Soc London B 270:887–896

Taylor WR (2002) A 'periodic table' for protein structures. Nature 416:657–660

Vogel C, Bashton M, Kerrison ND, Chothia C, Teichmann SA (2004) Structure, function and evolution of multidomain proteins. Curr Opin Struct Biol 14:208–216

Vukmirovic OG, Tilghman SM (2000) Exploring genome space. Nature 405:820–822

Wagner A (2003) How the global structure of protein interaction networks evolves. Proc R Soc Lond B 270:457–466

Wang M, Caetano-Anollés G (2006) Evolution inferred from domain combination in proteins. Mol Biol Evol 23:2444–2454

Wang M, Boca SM, Kalelkar R, Mittenthal JE, Caetano-Anollés G (2006) A phylogenomic reconstruction of the protein world based on a genomic census of protein fold architecture. Complexity 12:27–40

Wang M, Yafremava LS, Caetano-Anollés D, Mittenthal JE, Caetano-Anollés G (2007) Reductive evolution of architectural repertoires in proteomes and the birth of the tripartite world. Genome Res 17:1572–1585

Washietl S, Hofacker IL, Lukasser M, Hüttenhofer A, Stadler PF (2005) Mapping of conserved RNA secondary structures predicts thousands of functional noncoding RNAs in the human genome. Nat Biotechnol 23:1383–1389

White SH (1994) Global statistics of protein sequences: implications for the origin, evolution, and prediction of structure. Annu Rev Biophys Biomol Struct 23:407–439

Woese CR (2000) The universal ancestor. Proc Natl Acad Sci USA 95:6854–6859

Wolf YI, Brenner SE, Bash PA, Koonin EV (1999) Distribution of protein folds in the three superkingdoms of life. Genome Res 9:17–26

Wolf YI, Rogozin IB, Grishin NV, Koonin EV (2002) Genome trees and the tree of life. Trends Genet 18:472–479

Yang S, Doolittle RF, Bourne PE (2005) Phylogeny determined by protein domain content. Proc Natl Acad Sci USA 102:373–378

Yusupov MM, Yusupova GZ, Baucom A, Lieberman K, Earnest TN, Cate JHD, Noller HF (2001) Crystal structure of the ribosome at 5.5 Å resolution. Science 292:883–896

Zhang C, Kim SH (2000) A comprehensive analysis of the Greek key motifs in protein β-barrels and β-sandwiches. Proteins 40:409–419

Zhang C, Kim SH (2003) Overview of structural genomics: from structure to function. Curr Op Chem Biol 7:28–32

Chapter 7
Evolutionary Genomics of the Nitrogen-Fixing Symbiotic Bacteria

Víctor González(✉), Luis Lozano, Santiago Castillo-Ramírez,
Ismael Hernández González, Patricia Bustos, Rosa I. Santamaría,
José L. Fernández, José L. Acosta, and Guillermo Dávila

7.1   Introduction

Soil contains the most complex communities of microorganisms (Tringe et al. 2005). The earth's global ecology depends largely on the metabolic activities of different soil bacteria. One of these processes is biological nitrogen fixation. Nitrogen from the atmosphere is made available to plants either by free-living bacteria or symbionts associated with leguminous plants. There are ample bacterial species that are able to establish nitrogen fixing symbiosis. Ordinarily known as rhizobia, they are grouped in different taxonomic families of the α-proteobacteria classified as Rhizobiaceae, Phylobacteriaceae and Bradyrhizobiaceae (Table 7.1) (Garrity et al. 2002). Recently, it was reported that some strains of *Burkholderia*, a member of the β-proteobacteria is also able to nodulate tropical legumes like *Aspalathus carnosa* (Moulin et al. 2001). The fact that the common nodulation genes *nodA* and *nodB* of *Burkholderia* spp STM678 are phylogenetically close to *nod* genes of rhizobia suggests a horizontal gene transfer mechanism for acquisition of the nodulating ability (Moulin et al. 2001).

Many years of research have been dedicated to different aspects of nitrogen fixation in symbiosis. This has given a good general view of the process, but it is far from complete (Palacios and Newton 2005). Nowadays, the ability to sequence the complete genome of almost any organism has produced an integral view of the physiology and evolution of the bacterial cell (Dávila and Palacios 2005). Several genomes of rhizobia and related bacteria have been

V. González
Centro de Ciencias Genómicas, Universidad Nacional Autónoma de México Av.
Universidad N/C, Col. Chamilpa CP 62210. Apdo. Postal 565-A, Cuernavaca,
Morelos, México
e-mail: vgonzal@ccg.unam.mx

C.S. Nautiyal, P. Dion (eds.) *Molecular Mechanisms of Plant
and Microbe Coexistence.* Soil Biology 15, DOI: 10.1007/978-3-540-75575-3
© Springer-Verlag Berlin Heidelberg 2008

Table 7.1 General features of the complete genomes of nitrogen fixing symbiotic bacteria

Family	Rhizobiaceae			Phylo-bacteriaceae	Brady-rhizobiaceae
Features	*R. etli*	*R. leguminosarum*	*S. meliloti*	*M. loti*	*B. japonicum*
Size, bp	6,535,229	7,751,309	6,691,694	7,036,071	9,105,828
Number of replicons	1cc*, 6p**	1cc, 6p	1cc, 2p	1cc, 2p	1cc
GC average %	60.54%	60.86%	62.10%	60%	64.10%
Ribosomal RNA operons	3	3	3	2	1
tRNAs	50	52	54	50	50
Total CDS	6,034	7,263	6,204	6,752	8,317
CDS in functional classes	(71%)	–	3,703 (60%)	3,675 (54%)	4,348 (52%)
Hypothetical CDS	1,389 (23%)	–	1,993 (32%)	1,423 (21%)	2,506 (30%)
Orphan CDS	358 (6%)	–	508 (8%)	1,654 (25%)	1,463 (18%)
Transcriptional regulators	536 (8%)	–	539 (7.4%)	539 (7.4%)	567 (6.8%)
Transporters	837 (13.7%)	816	744 (12%)	764 (12%)	752 (9%)
External elements	157 (2.6%)		136 (2.2%)	171 (2.5%)	167 (2%)
Sigma subunits	23	16	13	23	23

*Circular chromosome

**Plasmids

sequenced, and others are in process. To date, complete sequences are available for *Mesorhizobium loti* MAFF303099 (Kaneko et al. 2000), *Sinorhizobium meliloti* 1021 (Galibert et al. 2001), *Bradyrhizobium japonicum* USDA 110 (Kaneko et al. 2002), *Rhizobium etli* CFN42 (González et al. 2003, 2006), and *Rhizobium leguminosarum* 3841 (Young et al. 2006). There is also information about the symbiotic plasmid of *Rhizobium* spp NGR234 (Freiberg et al. 1997) and the chromosomal symbiotic island of *M. loti* R7A (Sullivan et al. 2002). All of these rhizobia are of great economic importance because their hosts are domestic legumes like alfalfa, pea and beans. In addition, very recently it was reported the complete genome sequence of two strains of *Bradyrhizobium* spp that make stem nodules in the aquatic leguminous plant *Aeschynomene*. Remarkably, no canonical *nod*ABC genes neither typical lipochitoligosaccarides Nod factors are present in these strains (Giraud et al. 2007). This chapter is aimed at highlighting the main features of the complete genomes of nitrogen fixing symbiotic rhizobia available so far, concurrently with genome comparisons and discussion on the current views on the evolution of these complex genomes.

7.2 Origin, Taxonomy and Phylogeny

Several important reviews concerning the diversity and phylogenetic relationships of rhizobia have been published during the recent years (Lloret and Martínez-Romero 2005; Martínez-Romero 2003; Sessitsh et al. 2002). Analyses and simulations of the earth's early atmosphere indicate that nitrogen fixation is an ancient function estimated to have originated in the archean period about 3000 million years ago (MYA) (Kasting and Siefert 2001; Navarro-González et al. 2001). The scattered distribution of *nif* genes among different species of bacteria and archea suggests two possible scenarios about the origin. As an ancient function, nitrogen fixation may have originated from the last common ancestor, and was then inherited by all species but not all of these retained the trait. Alternatively, lateral transfer of *nif* genes among lineages may have resulted in the same phylogenetic pattern. Nevertheless, this is still a controversial issue.

Estimates of the divergence dates of rhizobia based on paralogous glutamine synthetase genes (GSI and GSII) suggest that fast-growing rhizobia lineages were established 203–324 MYA (Turner and Young 2000). Bradyrhizobia, the slow-growing rhizobia, is the more ancient branch calculated to have diverged about 507–553 MYA (Turner and Young 2000). In contrast, nodulation may have emerged when terrestrial plants appeared, 400 MYA and flourished as symbiosis with the expansion of the Leguminosae family 100 MYA (Lloret and Martínez-Romero 2005).

Rhizobia diversity has been assessed by a variety of methods. All indicate that rhizobia are highly heterogeneous since they differ in growth rates, biosynthetic pathways, catabolic activities, habitats, plasmid content, and the lipo and exopolysacharides structures. Phylogenetic reconstructions fail to place rhizobia as a coherent clade. Instead, phylogenies locate rhizobia species intermingled with non-symbiotic species. Actually, seven different symbiotic genera are recognized among the α-proteobacteria, *Rhizobium, Sinorhizobium, Bradyrhizobium, Mesorhizobium, Azorhizobium, Devosia,* and *Methylobacterium.* The whole genome sequence of model species of four genera has been achieved (Table 7.1). These are included in the order Rhizobiales for which a phylogeny based on the complete genomes available is shown here (Fig. 7.1). The tree constructed by using 636 orthologous proteins located in the chromosome matches very well phylogenetic reconstructions based on 16S ribosomal RNA genes. As seen in Fig. 7.1, symbiotic species are in the same evolutionary branches of species without symbiotic traits. Moreover, these non-symbiotic species have very different lifestyles that in some cases contrast with symbiosis. More specifically, *B. japonicum* is related to *Rhodopseudomonas palustris,* a photoautothrophic and chemoautothrophic organism; *S. meliloti, R. etli* and *R. leguminosarum* are near relatives of the plant pathogen *A. tumefaciens*; and *M. loti* seems to be in the same branch that leads to mammal pathogens *Bartonella* and *Brucella.* Such distribution indicates that

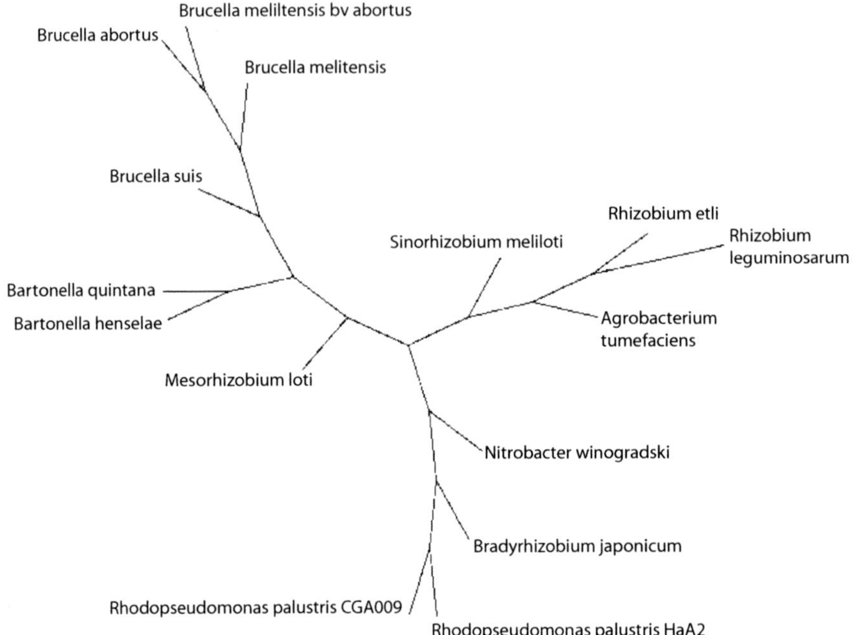

Fig. 7.1 Phylogenetic relationships among 15 species of the order Rhizobiales for which complete genomes are available. The unrooted tree was constructed by using 636 concatenated orthologous proteins common to all the species and parsimony methods (Protpars program of the Phylip package). Orthologous proteins were defined as reciprocal best-hits between pairs of species. Branch length is not equivalent to evolutionary distance

symbiosis, once acquired, has evolved in different and divergent genomic backgrounds.

7.3 Genome Structure

Rhizobia genomes are partitioned into several replicons, variable in size and number (Jumas-Bilak et al. 1998). A circular chromosome contains the majority of the essential genes, whereas plasmids encode diverse functions that are considered auxiliary. The exception to this arrangement is *B. japonicum* which has only one replicon. *A. tumefaciens*, closely related to *Rhizobium* genus, has an unusual genome organization including a linear chromosome (Allardet-Servent et al. 1993). Differences in genome architectures could be explained by rearrangements, horizontal transfer and cointegration of replicons. Some recent experiments have shown that *S. meliloti* and *Rhizobium* spp NGR234 could fuse their replicons into a single molecule without altering the symbiotic or growth properties of the strains (Guo

                187

et al. 2003; Mavingui et al. 2002). However, this arrangement is unstable and reverts to its original state after several replications (Guo et al. 2003). It has been proposed that a subdivided genome might confer adaptive advantages to rhizobia allowing the redistribution of dispensable genes to cope with challenging environments (González et al. 2006).

Plasmids have long been considered unessential molecules, or, as proposed by Campbell, needed only occasionally under certain conditions during the life of the bacterium (Campbell 1981). Genome sequences have revealed that some genes that are essential in different bacteria, in rhizobia are present in plasmids. For instance, the *minCDE* genes required for cell division in *E. coli* have been found on the pSymB of *S. meliloti*, the p42e of *R. etli*, the pRL11 of *R. leguminosarum,* and on the linear chromosome of *A. tumefaciens* and chromosome II of *Brucella*. Recent genetic analysis of the *min*CDE genes of *S. meliloti* has shown that they are dispensable for cell viability. Mutants in *min*CDE do not alter the growth properties and only mutations in *min*E produce branched cells and diminished symbiotic performance (Cheng et al. 2007). Undoubtedly, plasmids represent an important source of variation in natural rhizobia population, and we know only a fraction of the functions encoded by them.

Almost all plasmids found in rhizobia replicate using RepABC proteins, a system only found in α-proteobacteria (Ramírez-Romero et al. 1997). The linear chromosome of *A. tumefaciens* and chromosome II of *Brucella* species also have this type of replicator. The proteins RepA and RepB are involved in partition and RepC is the initiator protein for replication (Ramírez-Romero et al. 2000). Because of the high number of plasmids present in every rhizobia isolated from the field, efficient incompatibility systems are expected to be operating (Soberón et al. 2004). Necessarily, plasmids that coexist in the same genomic background belong to distinct incompatibility groups. They could have originated by duplication and divergence of the RepABC system or, as proposed by Cevallos and colleagues, they have moved among different species by horizontal transfer evolving independently (Cevallos et al. 2002).

The complexity of the rhizobial genomes raises important questions about advantages related to such large genomes. How have the replicons present in the same strain coevolved and coadapted? What is the origin of multiple replicons? Some insights on these matters are emerging from the analysis of *R. etli* and *R. leguminosarum,* which possess the most complex system of replicons of any sequenced genome so far (González et al. 2006; Young et al. 2006). *R. etli* contains six plasmids, four of them showing similar features such as GC content, codon usage, and comparable distribution of groups of orthologous genes (COGs) that resemble the chromosome. The other two plasmids, called the p42a and p42d (pSym) differ completely in these features and also harbor most of the ISs found in the genome (González et al. 2006). Furthermore, bioinformatic analyses of the predicted protein associations among replicons show that p42a and p42d are the replicons that are connected least with the rest of the genome (González et al. 2006). These data are consistent with the notion that plasmids p42a and p42d were acquired recently, while the rest of the plasmids share long-term coevolution with the chromosome. Like that of *R. etli*, the genome of *R. leguminosarum* is subdivided into seven replicons. By means of analysis of the nucleotide composition and quartop phylogenies, Young et al. (2006) have shown that the *R. leguminosarum* genome can be formed by two components: a "core", which is higher in GC and mainly placed in the chromosome, and an "accessory" component, lower in GC and located on the plasmids and chromosomal islands. This observation leads to the suggestion that horizontal gene transfer by phages and plasmids may have been occurred frequently in the history of *R. leguminosarum* and related bacteria.

7.4   Symbiotic Genome Compartments

The majority of the genes needed for establishing symbiosis are in specific genome compartments (Symbiotic Genome Compartments, SGCs), either in plasmids or in islands integrated in the chromosome (González et al. 2003). They are very heterogeneous in size and gene content and represent a mosaic of genes possibly assembled from different sources. Comparative analysis of the symbiotic plasmids pNGR234a of *Rhizobium* NGR234 spp, pSymA of *S. meliloti*, p42d of *Rhizobium etli* CFN42, and the symbiotic islands of *Mesorhizobium loti* and *Bradyrhizobium japonicum* has evidenced the lack of synteny. As few as 20 homologous genes are common among the SGCs already compared. These include the *nodABC*, *nodIJ*, *nodD*, *nifHDKENXAB*, *fixABCX*, *fdxN*, and *fdxB*. A few other genes like *nodN*, *fixK*, *fixNOQP*, *fixGHIS* and *fixL* are also common symbiotic genes, but they are not always in SGCs. In contrast, there are more than 30 genes involved in symbiosis that are particular to certain rhizobial species (*nodP*, *nodM*, *nodU*, *nod0*, *noeI*, *noeK*, *noeL*, *nolR*, *nolG*, *nfeD*).

Current ideas about the origin of nodulation point to a probable recruitment of genes of the biosynthetic pathways of lipids and exopolysacharides. In fact, some nodulation genes like *nodM* and *nodG* are recent paralogs of the housekeeping *glsM* and *fabG* genes which participate in the biosynthesis of glucosamine and fatty acids respectively (López-Lara and Geiger 2001; Marie et al. 1992). Other nodulation proteins encoded by *nodL* and *nodJ* have been shown to be homologous to proteins for the transportation of capsular polysaccharides in gram-negative bacteria (Vázquez et al. 1993). In *R. etli*, nodulation genes are included into large clusters of genes coding for components of the external surface of the bacteria (González et al. 2006). The high number of paralogous genes found in rhizobia (see below) may have contributed to the origin and diversification of symbiosis.

Rhizobium symbiotic plasmids are related to *Agrobacterium* plasmids pTi and pRi that induce crown-galls or hairy-roots in dicotyledonous plants. Several authors have reported on the close evolutionary relationship between these classes of plasmids, suggesting that they may have chimeric origin (Moriguchi et al. 2001). Supporting this idea, it has been found, in different SGCs, complete or partial genetic systems for conjugation (*tra/trb*) or the type IV transport system (*vir*) that in *Agrobacterium* plasmids are responsible for transporting the T-DNA to plant cells. For instance, the pSymA of S. *meliloti* has retained partially the *vir*, the pNGR234a has complete *tra* and *trb* systems but lacks the *vir* genes (Barnett et al. 2001). Moreover, the p42d (the pSym of *R. etli*) has an entire set of *virB* genes and partially the *tra* region, but lacks *trb* and other *vir* genes (González et al. 2003).

The genetic heterogeneity of the SGCs suggests that they have been shaped during evolution by rearrangements, recombination, horizontal transfer, and transposition (González et al. 2003). Genetic rearrangements have been shown to occur in *R. etli* involving both plasmids and chromosome (Brom et al. 1991; Flores et al. 1988). In particular, the pSym of *R. etli* is prone to undergo deletions and amplifications by recombination of the reiterated *nifH* genes (Romero et al. 1991). The high number of repeated sequences present in the SGCs suggests that the occurrence of rearrangements is very common, as was shown for the pSym of *Rhizobium* spp NGR234 (Flores et al. 2000). Furthermore, rearrangements involving the symbiotic region of *R. tropici* lead to increased symbiotic capabilities, such as competition for nodulation (Mavingui et al. 1997).

With the SGCs available so far, only general comparisons can be made. Nevertheless, to understand the evolution of SGCs, comparisons of SGCs from strains of the same specie are needed. Such studies have been done for the symbiotic islands of two strains of *M. loti* and for sequences of the pSym of various strains of *R. etli* (Flores et al. 2005; Sullivan et al. 2002). The symbiotic islands of the *M. loti* strains R7A and MAFF303099 have probably derived from an ancestral island. A conserved backbone of 248 kb shows clear colinearity, and about 98% nucleotide identity. In both islands the backbone is interrupted by insertions and deletions of diverse DNA segments. These specific regions harbor ISs (that are strain specific) and hypothetical genes. Interestingly, some segments of the islands are homologs of regions of the plasmids pMLa and pMLb of *M. loti*. Like in other SGCs, in the symbiotic island R7A there is a complete set of *vir* genes for the type IV secretion system, whereas the symbiotic island MAFF303099 lacks *vir* genes but has the genes for the type III secretion system.

Further insights on the mechanisms that contribute to the diversification of SGCs were published recently (Flores et al. 2005). A comparison of the variations in nucleotide sequences among pSym of different *R. etli* strains shows an asymmetric distribution of single nucleotide polymorphisms (Flores et al. 2005). There are regions with few changes and regions with a high number of nucleotide substitutions. Moreover, some highly polymorphic sites share exactly the same changes in several strains. The authors propose that the majority of the nucleotide substitutions are produced in the population by recombination, and that the contribution of mutations to polymorphism is relatively low (Flores et al. 2005).

7.5 Horizontal Gene Transfer and Mobile Elements

Horizontal gene transfer is one of the major evolutionary forces in most bacteria (Gogarten et al. 2002; Lawrence and Hendrickson 2003). The rhizobial species are no exception. Different methods have been employed to infer cases of horizontal transfer, like GC composition, codon usage, GC composition in the third base of the codon, dinucleotide frequency, and phylogenetic testing for congruency (Ragan et al. 2006). In the rhizobial chromosome there are regions of lower GC content or dinucleotide composition that differ from the average. These islands contain genes related to integrases, transposons, DNA transfer and plasmid stabilization genes (Capela et al. 2001; González et al. 2006; Young et al. 2006). Insertion sequences belonging to distinct families like IS66, IS630, IS110, IS3, IS4, and others are particularly abundant in the SGCs. The prevalence of an IS family in a particular *Rhizobium* contrasts with its poor representation in others. For instance, the IS66 is the most abundant IS in *R. etli* but it is poorly represented in *B. japonicum* (González et al. 2006). Even though the role of horizontal transfer in rhizobial evolution is widely recognized, precise estimates of the degree and rates of gene acquisitions are still lacking.

There is good evidence suggesting that symbiotic plasmids and islands might have been acquired during the evolution of rhizobia. In general, SGCs have a lower GC content and different codon usage compared with the rest of the genome. As mentioned before, they also have numerous insertion sequences and, in some cases, phage sequences on their borders, which may indicate the acquisition of these regions by lateral transfer. Because of these characteristics, it has been proposed that the symbiotic regions correspond to entire mobile genetic elements (Kaneko et al. 2002).

It is well known that rhizobia have conjugative plasmids and phages that could be the means of gene dissemination in the bacterial population. The role of phages in gene transfer has been poorly studied, whereas self-transmissible symbiotic plasmids have been described in *Rhizobium leguminosarum*, *Rhizobium* spp NGR234, *R. etli*, and *S. meliloti* (Jhonston 1978; Pérez-Mendoza et al. 2004; Tun-Garrido et al. 2003; Herrera-Cervera et al. 1998). The plasmids pRL1J1 of *R. leguminosarum* and pNGR234a of *Rhizobium* spp are able to conjugate at relatively high frequencies in laboratory conditions by a mechanism that involves quorum sensing (Danino et al. 2003; He et al. 2003). Indeed, genomic sequences of the symbiotic plasmids and islands have revealed the presence of conjugative systems encoded by *tra* and *trb* genes (Freiberg et al. 1997; González et al. 2003; Sullivan et al. 2002). In *R. etli* there are two well-characterized conjugative plasmids. One is the 194-kb p42a plasmid that is transmissible at high frequencies depending on cell-density (Tun-Garrido et al. 2003). Moreover, p42a can mobilize the symbiotic plasmid p42d of *R. etli* by a mechanism that involves cointegration of both replicons (Brom et al. 2004). Recently, it has been shown that the symbiotic plasmid p42d of *R. etli* transfers itself, this process requiring a protein encoded by the gene *yp028* (Pérez-Mendoza et al. 2004). Furthermore, it was also shown that the conjugal transfer of the pSym of *R. etli* and *S. meliloti* are repressed in laboratory conditions by the product of *rctA*, a DNA winged-helix DNA binding transcriptional regulator (Pérez-Mendoza et al. 2005). These data indicate that conjugative transfer of the symbiotic plasmids could be activated in natural conditions once an environmental signal, as yet unknown, is present.

The most persuasive evidence of horizontal transfer of symbiotic plasmids comes from experiments carried out in New Zealand (Sullivan et al. 1995). After the introduction of an inoculant strain of *M. loti* into a field devoid of native *Mesorhizobium*, a set of genetically diverse strains that contain symbiotic regions identical to that of the introduced strain was recovered (Sullivan et al. 1995). Furthermore, it was shown that the symbiotic region that was transferred is about 500 kb-long and inserted into the chromosomal loci for tRNA-phe (Sullivan and Ronson 1998).

7.6 Genetic and Metabolic Redundancy

Rhizobia face many challenging situations in the soil and in the nodule. They compete for energy sources with other microorganisms and cope with variable environmental conditions like humidity, drought, salinity, pH, temperature and others. We may think that some strategies exist for successful survival in such an environment. Rhizobia are heterotrophic obligate microaerophiles that can assimilate a wide range of rhizosphere carbon and nitrogen sources (Galibert et al. 2001; González et al. 2003; Goodner et al. 2001; Kaneko et al. 2000, 2002; Wood et al. 2001). They can exploit many sugars (mannose and rhamnose) present in plant root exudates as well as other rhizosphere compounds such as rhizopines (Prell and Poole 2006). They can also uptake many nutrients that exist in low concentrations in the soil by means of a large number of ABC transporters (Prell and Poole 2006). Such metabolic plasticity can be accounted for by the extensive genetic redundancy already evidenced by the high number of gene duplications in rhizobia (González et al. 2006). For instance, global genomic analysis shows that there are more isozymes in the genomes of *R. etli* and *S. meliloti* than in those of *E. coli* (González et al. 2006). In fact, in *R. etli* there are many isozymes for aminoacid biosynthesis distributed in both plasmids and chromosome. There are also an important number of pathways for fermentation, degradation and assimilation of aminoacids, aromatic compounds, carboxylates, sugars and polysaccharides. Such metabolic plasticity might be related to different degrees of physiological responses and the alternative regulation needed to be successful in the soil. In effect, large families of transcriptional regulators exist in rhizobia accounting for 9% of the total gene content, including members of the families LysR, TetR, AraC, LacI, and GntR. In addition, substantial amounts of two-component regulators indicate that signal transduction is a frequent mean for coordinating environmental responses. Moreover, Rhizobia harbor a high number of sigma factors of unknown function (Table 7.1). There are three basic sigma factors, RpoD, RpoN, and RpoH and several sigma factors belonging to the class of extracytoplasmic factors (ECFs). Various ECFs have been studied in species other than rhizobia, having diverse roles like iron uptake, toxin secretion, alginate biosynthesis, and others. With the exception of ECF RpoI, that in *R. leguminosarum* controls the siderophore synthesis, the physiological role of the other ECFs is unknown. RpoS, the sigma factor that controls the transition to the stationary phase of growth in enterobacteria is notably absent from rhizobia.

7.7 An Ancestral Chromosome?

It has been suggested that an ancestral chromosome was present in the origin of the α-proteobacteria group and consequently in species of the order Rhizobiales (Boussau et al. 2004; Galibert et al. 2001). Such an ancestral chromosome might have had 3000–5000 genes and could have been well adapted to saprophytic life. Thus, the ability to interact with animal or vegetal cells is a derived character specific for some species. Interestingly, two opposite phenomena have taken place in the α-proteobacteria: massive genome expansions in Rhizobiales associated to plants, and genome contractions in several branches of Rhizobiales and Ricketssiales, pathogens of mammals (*Ricketssia, Brucella, and Bartonella*) or endosymbionts of insects (*Wolbacchia*) (Boussau et al. 2004). In fact, rhizobia have a high amount of paralogous genes. Most of them are ancient duplications. The number of paralogous genes in the five complete genomes of rhizobia represents 30–40% of the total gene content (Galibert et al. 2001; González et al. 2006). As was shown for *R. etli*, paralogous families with the largest number of members belong to diverse ABC transporters, transcription regulators, and a variety of genes related to the metabolism of amino acids, nucleotides, carbohydrates, coenzymes, lipids, inorganic ions, and secondary metabolites (González et al. 2006). It has been proposed that bacteria with large genomes are more ecologically successful in environments where resources are scarce but diverse, and there is no disadvantage for slow growth (Konstantinidis and Tiedje 2004). This is the expected situation that rhizobia confront in the soil.

It has been shown that gene order is well correlated with the phylogenetic distance that separates the different clades (Tamames 2001). Synteny maps of Rhizobiales chromosomes reveal a very high degree of colinearity between pairs of related species (Fig. 7.2) (Boussau et al. 2004; González et al. 2006; Goodner et al. 2001). Indeed, they reveal the close relationship of symbiotic species with plant or mammal pathogens. For instance, *S. meliloti* and *R. etli* chromosomes are very conserved among themselves and with *A. tumefaciens* (plant pathogen), but they also are very syntenic with the chromosome I of *Brucella* species. In contrast, plasmids or secondary chromosomes do not display any synteny pattern (González et al. 2006).

As has been discussed, rhizobia are not a homogeneous group, and the proportion of orthologous genes shared between closer species like *R. etli* and *S. meliloti*, is nearly 60% of the total gene content, whereas *B. japonicum*, the most divergent lineage, shares 42% of the orthologs with other rhizobia (González et al. 2006).

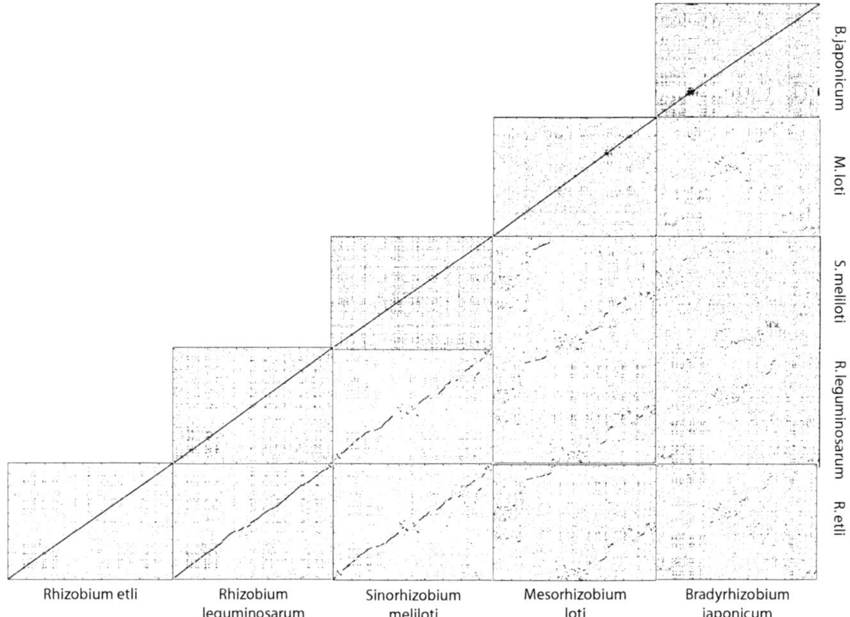

Fig. 7.2 Syntenic relationships between the chromosomes of pairs of rhizobia species. Complete chromosomes were aligned using the PROMER application of the MUMMER package (Delcher et al. 1999). All matching and alignments were performed on the six frame aminoacid translation of the DNA input sequences. These are used to perform pairwise alignments. Diagonals represent maximal matches between pairs of sequences. Points or lines outside the diagonals are matches due to repeated aminoacid stretches of paralogous genes or domains. Recently, two additional complete genomes of *Bradyrhizobium* spp have been reported and comparisons with *B. japonicum* USDA110 show several syntenic regions but also extensive rearrangements of other segments. Authors suggest that *Bradyrhizobium* genomes are highly plastic by the presence of numerous mobile elements (Giraud et al. 2007)

These observations suggest that not only gene duplications or gene acquisitions by horizontal transfer have had a role in rhizobia evolution but also severe gene losses have occurred. Conversely, it may be expected that a set of common genes would be present in rhizobia but absent from close relatives like *A. tumefaciens* or *Brucella*. In fact, by discounting the common orthologs between the plant pathogen *A. tumefaciens* and rhizobia, a set of about 500 genes constitute a core shared exclusively by *R. etli* and *S. meliloti* or by *R. leguminosarum*, *B. japonicum* and *S. meliloti*, and are not found in close relatives. These genes include the well-known *nif*, *fix* and *nod* genes, large families of adenylate cyclases and glutathione reductases, and several sigma factors that are not found in *A. tumefaciens* (González et al. 2006). Moreover, the core includes genes of all functional classes but not essential for cell survival, as well as hypothetical genes, perhaps related to symbiosis (González et al. 2006; Young et al. 2006).

7.8 Conclusions

The advent of genomics has changed our point of view about biology. Integrative approaches based on the knowledge of all genetic elements of the cell are currently used to answer fundamental questions. For many years, researchers in the field of nitrogen fixation have had questions about the origin and evolution of symbiosis. How do *nod* and *nif* genes unite to perform a complex function? Why is the distribution of the symbiotic trait so scattered? What is the functional connection between the symbiotic metabolism and the rest of the cellular metabolism? What are the commonalities and differences within rhizobia? Surely, field researchers have many other questions, but we can conclude on some suggestive ideas derived from genomic analysis. The ancestor of the present day rhizobia species likely had a large genome with broad biosynthetic capabilities to survive in variable environments where energy sources were scarce (Boussau et al. 2004). The high number of paralogous families presently found in the rhizobia genomes is the reflection of ancient mechanisms (Galibert et al. 2001; González et al. 2006). Thus, the ancestor of rhizobia was well-adapted to the saprophytic life-style, and the ability to perform symbiosis might have been a function acquired by rhizobia along the course of the evolution. Gene recruitment, through gene duplication and horizontal transfer, not only enriched the evolution of symbiosis but contributed to expanding the metabolic repertoire of rhizobia.

Acknowledgements The authors wish to thank to José Espíritu and Víctor del Moral for technical assistance during the writing of this chapter. This work is supported by grants from PAPIIT IN223005-3 and CONACyT U46333-Q.

References

Allardet-Servent A, Michaux-Charachon S, Jumas-Bilak E, Karayan L, Ramuz M (1993) Presence of one linear and one circular chromosome in the *Agrobacterium tumefaciens* C58 genome. J Bacteriol 175:7869–7874

Barnett MJ, Fisher RF, Jones T, Komp C, Abola AP, Barloy-Hubler F, Bowser L, Capela D, Galibert F, Gouzy J, Gurjal M, Hong A, Huizar L, Hyman RW, Kahn D, Kahn ML, Kalman S, Keating DH, Palm C, Peck MC, Surzycki R, Wells DH, Yeh K-C, Davis RW, Ederspiel NA, Long SR (2001) Nucleotide sequence and predicted functions of the entire *Sinorhizobium meliloti* pSymA megaplasmid. Proc Natl Acad Sci USA 98:9883–9888

Boussau B, Karlberg EO, Frank AC, Legault BA, Andersson SG (2004) Computational inference of scenarios for alpha-proteobacterial genome evolution. Proc Natl Acad Sci USA 101:9722–9727

Brom S, García de los Santos A, de Lourdes Girard M, Dávila G, Palacios R, Romero D (1991) High-frequency rearrangements in *Rhizobium leguminosarum* bv. *phaseoli* plasmids. J Bacteriol 173:1344–1346

Brom S, Girard L, Tun-Garrido C, Santos AG, Bustos P, González V, Romero D (2004) Transfer of the symbiotic plasmid of *Rhizobium etli* CFN42 requires cointegration with p42a, which may be mediated by site-specific recombination. J Bacteriol 186:7538–7548

Campbell A (1981) Evolutionary significance of accessory DNA elements in bacteria. Annu Rev Microbiol 35:55–83

Capela D, Barloy-Hubler F, Gouzy J, Bothe G, Ampe F, Batut J, Boistard P, Becker A, Boutry M, Cadieu E, Dréano S, Gloux S, Godrie T, Goffeau A, Kahn D, Kiss E, Lelaure V, Masuy D, Pohl T, Portetelle D, Pühler A, Purnelle B, Ramsperger U, Renard C, Thébault P, Vandenbol M, Weidner S, Galibert F (2001) Analysis of the chromosome sequence of the legume symbiont *Sinorhizobium meliloti* strain 1021. Proc Natl Acad Sci USA 98:9877–9882

Cevallos MA, Porta H, Izquierdo J, Tun-Garrido C, García-de-los-Santos A, Dávila G, Brom S (2002) *Rhizobium etli* CFN42 contains at least three plasmids of the repABC family: a structural and evolutionary analysis. Plasmid 48:104–116

Cheng J, Sibley CD, Zaheer R, Finan TM (2007) A *Sinorhizobium meliloti minE* mutant has an altered morphology and exibits defects in legume symbiosis. Microbiology 153:375–387

Danino VE, Wilkinson A, Edwards A, Downie JA (2003) Recipient-induced transfer of the symbiotic plasmid pRL1JI in *Rhizobium leguminosarum bv. viciae* is regulated by a quorum-sensing relay. Mol Microbiol 50:511–525

Dávila G, Palacios R (2005) Origins of genomics in nitrogen fixation research. In: Palacios R, Newton WE (eds) Genomes and genomics of nitrogen-fixing organisms. Springer, Berlin Heidelberg New York, pp 1–6

Delcher AL, Kasif S, Fleishman RD, Peterson J, White O, Salzberg SL (1999) Alignment of whole genomes. Nucleic Acids Res 27:2369–2376

Flores M, González V, Pardo MA, Leija A, Martínez E, Romero D, Piñero D, Dávila G, Palacios R (1988) Genomic instability in *Rhizobium phaseoli*. J Bacteriol 170:1191–1196

Flores M, Mavingui P, Perret X, Broughton WJ, Romero D, Hernández G, Dávila G, Palacios R (2000) Prediction, identification, and artificial selection of DNA rearrangements in *Rhizobium*: toward a natural genomic design. Proc Natl Acad Sci USA 97:9138–9143

Flores M, Morales L, Avila A, González V, Bustos P, García D, Mora Y, Guo X, Collado-Vides J, Piñero D, Dávila G, Mora J, Palacios R (2005) Diversification of DNA sequences in the symbiotic genome of *Rhizobium etli*. J Bacteriol 187:7185–7192

Freiberg C, Fellay R, Bairoch A, Broughton WJ, Rosenthal A, Perret X (1997) Molecular basis of symbiosis between *Rhizobium* and legumes. Nature 387:394–401

Galibert F, Finan TM, Long SR, Puhler A, Abola P, Ampe F, Barloy-Hubler F, Barnett MJ, Becker A, Boistard P, Bothe G, Boutry M, Bowser L, Buhrmester J, Cadieu E, Capela D, Chain P, Cowie A, Davis RW, Dreano S, Federspiel NA, Fisher RF, Gloux S, Godrie T, Goffeau A, Golding B, Gouzy J, Gurjal M, Hernandez-Lucas I, Hong A, Huizar L, Hyman RW, Jones T, Kahn D, Kahn ML, Kalman S, Keating DH, Kiss E, Komp C, Lelaure V, Masuy D, Palm C, Peck MC, Pohl TM, Portetelle D, Purnelle B, Ramsperger U, Surzycki R, Thebault P, Vandenbol M, Vorholter FJ, Weidner S, Wells DH, Wong K, Yeh KC, Batut J (2001) The composite genome of the legume symbiont *Sinorhizobium meliloti*. Science 293:668–672

Garrity GM, Jhonson KL, Bell JA, Searles DB (2002) Taxonomic outline of the Procaryotes. Release 3.0. In: Bergey's manual of systematic bacteriology. Springer, Berlin Heidelberg New York, p 365

Giraud E, Moulin L, Vallenet D, Barbe V, Cytryn E, Avarre JC, Jaubert M, Simon D, Cartieaux F, Prin Y, Bena G, Hannibal L, Fardoux J, Kojadinovic M, Vuillet L, Lajus A, Cruveiller S, Rouy Z, Mangenot S, Segurens B, Dossat C, Franck WL, Chang WS, Saunders E, Bruce D, Richardson P, Normand P, Dreyfus B, Pignol D, Stacey G, Emerich D, Vermeglio A, Medigue C, Sadowsky M (2007) Legumes symbiosis: abscence of *nod* genes in photosynthetic Bradyrhizobia. Science 316:1307–1311

Gogarten JP, Doolittle WF, Lawrence JG (2002) Prokaryotic evolution in light of gene transfer. Mol Biol Evol 19:2226–2238

Gonzalez V, Bustos P, Ramirez-Romero MA, Medrano-Soto A, Salgado H, Hernandez-Gonzalez I, Hernandez-Celis JC, Quintero V, Moreno-Hagelsieb G, Girard L, Rodriguez O, Flores M, Cevallos MA, Collado-Vides J, Romero D, Davila G (2003) The mosaic structure of the symbiotic plasmid of *Rhizobium etli* CFN42 and its relation to other symbiotic genome compartments. Genome Biol 4:R36

Gonzalez V, Santamaria RI, Bustos P, Hernandez-Gonzalez I, Medrano-Soto A, Moreno-Hagelsieb G, Janga SC, Ramirez MA, Jimenez-Jacinto V, Collado-Vides J, Davila G (2006) The partitioned

Rhizobium etli genome: genetic and metabolic redundancy in seven interacting replicons. Proc Natl Acad Sci USA 103:3834–3839

Goodner B, Hinkle G, Gattung S, Miller N, Blanchard M, Qurollo B, Goldman BS, Cao Y, Askenazi M, Halling C, Mullin L, Houmiel K, Gordon J, Vaudin M, Iartchouk O, Epp A, Liu F, Wollam C, Allinger M, Doughty D, Scott C, Lappas C, Markelz B, Flanagan C, Crowell C, Gurson J, Lomo C, Sear C, Strub G, Cielo C, Slater S (2001) Genome sequence of the plant pathogen and biotechnology agent *Agrobacterium tumefaciens* C58. Science 294:2323–2328

Guo X, Flores M, Mavingui P, Fuentes SI, Hernández G, Dávila G, Palacios R (2003) Natural genomic design in *Sinorhizobium meliloti*: novel genomic architectures. Genome Res 13:1810–1817

He X, Chang W, Pierce DL, Seib LO, Wagner J, Fuqua C (2003) Quorum sensing in *Rhizobium* sp. strain NGR234 regulates conjugal transfer (*tra*) gene expression and influences growth rate. J Bacteriol 185:809–822

Herrera-Cervera JA, Sanjuan-Pinilla JM, Olivares J, Sanjuan J (1998) Cloning and identification of conjugative transfer origins in the *Rhizobium meliloti* genome. J Bacteriol 180:4583–4590

Jhonston AWB, Beynon JL, Buchanan-Wollaston AV, Setchell SM, Hirsh PR, Beringer JE (1978) High frequency transfer of nodulating ability between strains and species of *Rhizobium*. Nature 276:635–636

Jumas-Bilak E, Michaux-Charachon S, Bourg G, Ramuz M, Allardet-Servent A (1998) Unconventional genomic organization in the alpha subgroup of the Proteobacteria. J Bacteriol 180:2749–2755

Kaneko T, Nakamura Y, Sato S, Asamizu E, Kato T, Sasamoto S, Watanabe A, Idesawa K, Ishikawa A, Kawashima K, Kimura T, Kishida Y, Kiyokawa C, Kohara M, Matsumoto M, Matsuno A, Mochizuki Y, Nakayama S, Nakazaki N, Shimpo S, Sugimoto M, Takeuchi C, Yamada M, Tabata S (2000) Complete genome structure of the nitrogen-fixing symbiotic bacterium *Mesorhizobium loti*. DNA Res 7:331–338

Kaneko T, Nakamura Y, Sato S, Minamisawa K, Uchiumi T, Sasamoto S, Watanabe A, Idesawa K, Iriguchi M, Kawashima K, Kohara M, Matsumoto M, Shimpo S, Tsuruoka H, Wada T, Yamada M, Tabata S (2002) Complete genomic sequence of nitrogen-fixing symbiotic bacterium *Bradyrhizobium japonicum* USDA110 (supplement). DNA Res 9:225–256

Kasting JF, Siefert JL (2001) Biogeochemistry: the nitrogen fix. Nature 412:26–27

Konstantinidis KT, Tiedje JM (2004) Trends between gene content and genome size in prokaryotic species with larger genomes. Proc Natl Acad Sci USA 101:3160–3165

Lawrence JG, Hendrickson H (2003) Lateral gene transfer: when will adolescence end? Mol Microbiol 50:739–749

Lloret L, Martínez-Romero E (2005) Evolución y Filogenia de *Rhizobium*. Rev Latinoam Microbiol 47:43–60

López-Lara IM, Geiger O (2001) The nodulation protein NodG shows the enzymatic activity of an 3-oxoacyl-acyl carrier protein reductase. Mol Plant Microbe Interact 14:349–357

Marie C, Barny MA, Downie JA (1992) *Rhizobium leguminosarum* has two glucosamine synthases, GlmS and NodM, required for nodulation and development of nitrogen-fixing nodules. Mol Microbiol 6:843–851

Martínez-Romero E (2003) Diversity of *Rhizobium-Phaseolus vulgaris* symbiosis:overview and perspectives. Plant Soil 252:11–23

Mavingui P, Flores M, Romero D, Martínez-Romero E, Palacios R (1997) Generation of *Rhizobium* strains with improved symbiotic properties by random DNA amplification (RDA). Nat Biotechnol 15:564–569

Mavingui P, Flores M, Guo X, Dávila G, Perret X, Broughton WJ, Palacios R (2002) Dynamics of genome architecture in *Rhizobium* sp. strain NGR234. J Bacteriol 184:171–176

Moriguchi K, Maeda Y, Satou M, Hardayani NS, Kataoka M, Tanaka N, Yoshida K (2001) The complete nucleotide sequence of a plant root-inducing (Ri) plasmid indicates its chimeric structure and evolutionary relationship between tumor-inducing (Ti) and symbiotic (Sym) plasmids in Rhizobiaceae. J Mol Biol 307:771–784

Moulin L, Munive A, Dreyfus B, Boivin-Masson C (2001) Nodulation of legumes by members of the beta-subclass of Proteobacteria. Nature 411:948–950

Navarro-González R, McKay CP, Mvondo DN (2001) A possible nitrogen crisis for Archaean life due to reduced nitrogen fixation by lightning. Nature 412:61–64

Palacios R, Newton WE (2005) Nitrogen fixation: origins, applications, and research progress. In: Palacios R, Newton WE (eds) Genomes and genomics of nitrogen-fixing organisms Springer, Berlin Heidelberg New York, pp ix–xiv

Pérez-Mendoza D, Domínguez-Ferreras A, Muñoz S, Soto MJ, Olivares J, Brom S, Girard L, Herrera-Cervera JA, Sanjuán J (2004) Identification of functional *mob* regions in *Rhizobium etli*: evidence for self-transmissibility of the symbiotic plasmid pRetCFN42d. J Bacteriol 186:5753–5761

Pérez-Mendoza D, Sepúlveda E, Pando V, Muñoz S, Nogales J, Olivares J, Soto MJ, Herrera-Cervera JA, Romero D, Brom S, Sanjuán J (2005) Identification of the *rctA* gene, which is required for repression of conjugative transfer of rhizobial symbiotic megaplasmids. J Bacteriol 187:7341–7350

Prell J, Poole P (2006) Metabolic changes of rhizobia in legume nodules. Trends Microbiol 14:161–168

Ragan MA, Harlow TJ, Beiko RG (2006) Do different surrogate methods detect lateral genetic transfer events of different relative ages? Trends Microbiol 14:4–8

Ramírez-Romero MA, Bustos P, Girard L, Rodríguez O, Cevallos MA, Dávila G (1997) Sequence, localization and characteristics of the replicator region of the symbiotic plasmid of *Rhizobium etli*. Microbiology 143(8):2825–2831

Ramírez-Romero MA, Soberón N, Pérez-Oseguera A, Téllez-Sosa J, Cevallos MA (2000) Structural elements required for replication and incompatibility of the *Rhizobium etli* symbiotic plasmid. J Bacteriol 182:3117–3124

Romero D, Brom S, Martínez-Salazar J, Girard ML, Palacios R, Dávila G (1991) Amplification and deletion of a *nod-nif* region in the symbiotic plasmid of *Rhizobium phaseoli*. J Bacteriol 173:2435–2441

Sessitsh A, Howieson JG, Perret X, Antoun H, Martínez E (2002) Advances in *Rhizobium* research. Crit Rev Plant Sci 21:323–378

Soberón N, Venkova-Canova T, Ramírez-Romero MA, Téllez-Sosa J, Cevallos MA (2004) Incompatibility and the partitioning site of the repABC basic replicon of the symbiotic plasmid from *Rhizobium etli*. Plasmid 51:203–216

Sullivan JT, Ronson CW (1998) Evolution of rhizobia by acquisition of a 500-kb symbiosis island that integrates into a phe-tRNA gene. Proc Natl Acad Sci USA 95:5145–5149

Sullivan JT, Patrick HN, Lowther WL, Scott DB, Ronson CW (1995) Nodulating strains of *Rhizobium loti* arise through chromosomal symbiotic gene transfer in the environment. Proc Natl Acad Sci USA 92:8985–8989

Sullivan JT, Trzebiatowski JR, Cruickshank RW, Gouzy J, Brown SD, Elliot RM, Fleetwood DJ, McCallum NG, Rossbach U, Stuart GS, Weaver JE, Webby RJ, De Bruijn FJ, Ronson CW (2002) Comparative sequence analysis of the symbiosis island of *Mesorhizobium loti* strain R7A. J Bacteriol 184:3086–3095

Tamames J (2001) Evolution of gene order conservation in prokaryotes. Genome Biol 2(6): research0020.1–0020.11

Tringe SG, von Mering C, Kobayashi A, Salamov AA, Chen K, Chang HW, Podar M, Short JM, Mathur EJ, Detter JC, Bork P, Hugenholtz P, Rubin EM (2005) Comparative metagenomics of microbial communities. Science 308:554–557

Tun-Garrido C, Bustos P, González V, Brom S (2003) Conjugative transfer of p42a from *Rhizobium etli* CFN42, which is required for mobilization of the symbiotic plasmid, is regulated by quorum sensing. J Bacteriol 185:1681–1692

Turner SL, Young JP (2000) The glutamine synthetases of rhizobia: phylogenetics and evolutionary implications. Mol Biol Evol 17:309–319

Vázquez M, Santana O, Quinto C (1993) The NodL and NodJ proteins from *Rhizobium* and *Bradyrhizobium* strains are similar to capsular polysaccharide secretion proteins from Gram-negative bacteria. Mol Microbiol 8:369–377

Wood DW, Setubal JC, Kaul R, Monks DE, Kitajima JP, Okura VK, Zhou Y, Chen L, Wood GE, Almeida NF Jr, Woo L, Chen Y, Paulsen IT, Eisen JA, Karp PD, Bovee D Sr, Chapman P,

Clendenning J, Deatherage G, Gillet W, Grant C, Kutyavin T, Levy R, Li MJ, McClelland E, Palmieri A, Raymond C, Rouse G, Saenphimmachak C, Wu Z, Romero P, Gordon D, Zhang S, Yoo H, Tao Y, Biddle P, Jung M, Krespan W, Perry M, Gordon-Kamm B, Liao L, Kim S, Hendrick C, Zhao ZY, Dolan M, Chumley F, Tingey SV, Tomb JF, Gordon MP, Olson MV, Nester EW (2001) The genome of the natural genetic engineer *Agrobacterium tumefaciens* C58. Science 294:2317–2323

Young JP, Crossman LC, Johnston AW, Thomson NR, Ghazoui ZF, Hull KH, Wexler M, Curson AR, Todd JD, Poole PS, Mauchline TH, East AK, Quail MA, Churcher C, Arrowsmith C, Cherevach I, Chillingworth T, Clarke K, Cronin A, Davis P, Fraser A, Hance Z, Hauser H, Jagels K, Moule S, Mungall K, Norbertczak H, Rabbinowitsch E, Sanders M, Simmonds M, Whitehead S, Parkhill J (2006) The genome of *Rhizobium leguminosarum* has recognizable core and accessory components. Genome Biol 7:R34

Chapter 8
Genetic and Epigenetic Nature
of Transgenerational Changes
in Pathogen Exposed Plants

Alex Boyko and Igor Kovalchuk(✉)

8.1   Introduction

Adverse environmental conditions named stresses are constantly shaping genomes of living organisms. Most of these external stimuli have a negative influence on growth, development, and reproduction (Arnholdt-Schmitt 2004; Madlung and Comai 2004). Avoidance represents the most common response to severe or long lasting environmental conditions. The organisms with a sedentary life style are unable to escape stress, and thus utilize mechanisms of tolerance and resistance. Plants integrate into the environment through the efficient use of adaptation mechanisms that depends on the constant exchange of signaling molecules (reviewed in Cronk 2001).

Plants are the organisms that continue their development throughout the entire life cycle. The germ line in plants is not predetermined but is established during the development. This allows plants to percept stress and integrate the memory of it through multiple feedback mechanisms. The only possible way of transmitting the memory of stress is via the epigenetic regulations involving DNA methylation, histone modifications and chromatin restructuring. These changes lead to the differential gene expression and allow to establish new epialleles, thus resulting in the destabilization of defined loci. The majority of these changes are neutral or deleterious, but in some rare cases they could be beneficial.

Constant exposure to certain stresses should lead to the selection of adaptive traits beneficial to these conditions. The retention and fixation of a necessary trait requires the selection from a number of neutral random changes in the genome.

I. Kovalchuk
Department of Biological Sciences, University of Lethbridge, Lethbridge, Alberta,
T1K 3M4, Canada
e-mail: igor.kovalchuk@uleth.ca

C.S. Nautiyal, P. Dion (eds.) *Molecular Mechanisms of Plant
and Microbe Coexistence.* Soil Biology 15, DOI: 10.1007/978-3-540-75575-3
© Springer-Verlag Berlin Heidelberg 2008

Typically, plants don't have time for all these changes to occur. Plants are also capable of acclimation, altering their homeostasis on a reduced time scale (Shinozaki et al. 2003; Sung et al. 2003). The fact that plants similarly respond to unrelated physical, chemical, or temporal environmental factors suggests the existence of complex intercrossing perception and response mechanisms (Shinozaki et al. 2003; Chinnusamy et al. 2004; Ludwig et al. 2004).

In the current chapter we discuss what is known about the stress-induced epigenetic changes resulting in differential genome rearrangements and present several experiments indicating the role of methylation changes and homologous recombination in the plant response to pathogen stress.

8.2 Genome Stability is Regulated via Multiple Pathways

Genome instability is defined by the susceptibility of a genome to mutations and rearrangements, whereby a stable genome impedes these mechanisms. Major mechanisms of the genome protection involve the different arrangement of methylated and unmethylated histones and attachment of chromatin loops to the nuclear matrix. Changes in the chromatin structure represent the natural ability of plants to respond to stress (Takeda et al. 2004; Buchanan et al. 2005). DNA repair mechanisms including two double strand-break repair pathways, non-homologous end-joining (NHEJ) and homologous recombination (HR) represent a more direct way of the genome protection (Jeggo 1998; Hays 2002).

Plants regulate the genome rearrangements by applying different DNA repair mechanisms at different developmental stages. *Arabidopsis* plants have higher HR frequency (HRF) at early developmental stages (Boyko et al. 2006c). In contrast, the frequency of NHEJ employment does not change. It is possible that the older cells suppress the activity of HR in order to decrease the chances of deleterious rearrangements in the polyploidy cells (Boyko et al. 2006c). The younger cells can have an elevated HRF in order to allow the inheritance of genome rearrangements. In this case, the exposure to mutagen early at the developmental stage should result in a higher increase of HR frequency. Recent data indeed showed that the exposure to UVB early in the development resulted in a higher increase in HRF when compared to the exposure at a later stage (Boyko et al. 2006a).

8.3 The Homologous Recombination as a Mechanism Supporting Rearrangements

The homologous recombination is the primary mechanism responsible for crossing over events during meiosis (Gerton and Hawley 2005), and as such could serve as the main mechanism for creating diversity. The HR can prove dangerous to cells, as it can quickly generate the recessive genotypes from heterozygous loci. The recessive traits, however, are not necessarily deleterious, as they may appear to be useful under certain environmental conditions. The organisms with a large proportion of traits in the heterozygous stage could have the advantage from the evolutionary point of view because of having "evolutionary flexible" genomes. It can be suggested that the genome stability is closely monitored to balance risks of negative events with the need for genome diversity.

The addition of methyl groups to DNA as well as histone deacetylation and specific changes in histone methylation stabilize the genome and prevent the recombination events, whereas the loss of methyl groups, termed hypomethylation and histone acetylation allow such events to occur (Engler et al. 1993; Madlung and Comai 2004).

8.4 Regulation of Gene Expression via Chromatin Modifications

Gene expression is regulated by various mechanisms including kinase-activated transcription factors, RNA turnover, posttranscriptional gene silencing as well as changes in protein half-life (Jover-Gil et al. 2005). One more mechanism involved in the establishment of new chromatin structures is small regulatory RNAs, named short interfering (si)RNAs, involved in targeting specific genome areas and establishing a new chromatin structure (Mathieu and Bender 2004; Steimer et al. 2004; Kapoor et al. 2005). This process starts by the generation of sequence specific siRNAs via the developmental regulation or change in environmental conditions that are capable of spreading to specific tissue or plant organs and promoting changes in a methylation status followed by changes in methylation/acetylation of specific histones (Steimer et al. 2004; Sunkar and Zhu 2004; Borsani et al. 2005). Such changes in chromatin structure reinforced by the small regulatory RNAs allow one to establish and maintain new patterns of chromatin modifications required for the proper leaf development and transition to flowering (Fransz and de Jong 2002; Liu et al. 2004; Peragine et al. 2004; Grigg et al. 2005; Kandasamy et al. 2005).

Gametes are produced in plants from meristemic cells. The ability of these cells to accumulate the information from the developing plant makes it possible to incorporate epigenetic signals into a unique methylation pattern. The newly developed progeny will likely have changes in the pattern of gene expression at differently methylated loci. This variation in the gene expression can be easily found in the population of plants, and it represents a heritable epimutation event. The advantage of having epimutations is that they are often reversible, making the resulting phenotypes more variable and less severe as compared to those resulting from sequence changes (Cronk 2001).

8.5 Exposure to Stress Influences Methylation Status and HRF

Plants – as any other organisms – have the ability to perceive stress and respond to it via differential changes in methylation pattern. Indeed, it has been shown that cold treatment promotes tissue-specific hypomethylation of the particular genome regions, including those specific to retrotransposon sequences (Steward et al. 2002). Similarly, the exposure of tobacco plants to a tobacco mosaic virus (TMV) induces demethylation of the *NtAlix1* stress-responsive gene, resulting in the continuous accumulation of the gene transcript (Wada et al. 2004).

Many experiments have shown that such stresses as changes in growth conditions, exposure to salt, heavy metals, ultraviolet and ionizing radiation, herbicides and even pathogens influence the HR frequency in plants (Kovalchuk et al. 1998, 2000, 2003a,b, 2004; Filkowski et al. 2003; Besplug et al. 2004; Molinier et al. 2005; Boyko et al. 2006a; Boyko et al., unpublished data). Some of these stresses are associated with the changes in a DNA methylation pattern in the progeny (Kovalchuk et al. 2003b, 2004). The association of other stresses with the changes in methylation and genome instability remains to be established.

The most straightforward reliable assay that can be used for the analysis of the HRF in plants is the one that uses visible markers. We have used two reporter genes, either β-glucuronidase (*uid*A, or GUS) or luciferase (LUC), that are integrated into plants as two non-functional overlapping copies (Kovalchuk et al. 1998; Filkowski et al. 2004; Boyko et al. 2006b). The repair of a double strand break in one of the homologous regions (Fig. 8.1) using the second intact homologous copy results in the restoration of the marker gene structure. The activity of these genes is readily visualized either via histochemical staining or observing with a CCD camera for GUS and LUC, respectively (Fig. 8.1).

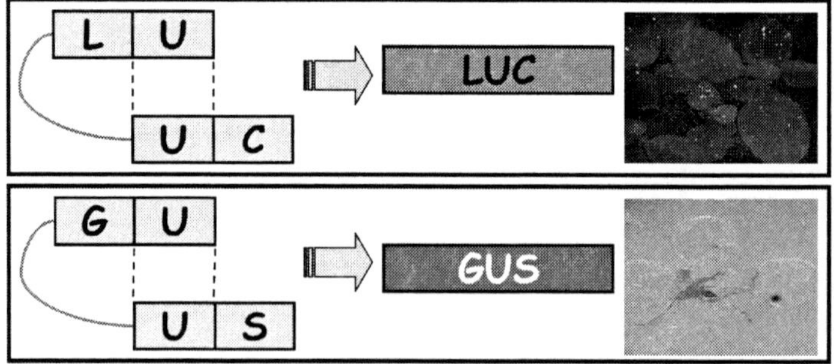

Fig. 8.1 Recombination reporter assays. Homologous recombination reporter lines carry in the genome two non-functional overlapping copies ("GU" and "US" or "LU" and "UC") of the transgene in the genome. The activity of the gene is restored via homologous recombination between the homology regions ("U"). The transgene activation is observed either as shiny sectors (LUC) or grey sectors (GUS) on the transparent background (chlorophyll washed with ethanol)

8.6   Systemic Signaling in Plants

In plants every single cell has the ability to communicate with the adjacent cells through the plasmodesmata openings, and with the distant cells through the phloem. These communication channels allow plants to exchange information about any changes in environmental conditions. Often, local stimuli are integrated by plants into a broad responsive network resulting in a global response like a systemic acquired resistance (SAR) (Dong 2001), systemic wound signaling (Pearce et al. 1991), systemic acquired acclimation to light (Karpinski et al. 1999), systemic post-transcriptional RNA silencing (Mlotshwa et al. 2002; Waterhouse et al. 2001), and the photoperiodic induction of flowering (Colasanti and Sundaresan 2000). It is likely to be a small portion of the great variety of responses that plants are capable of. All these processes depend on the ability of plants to respond to stress and to produce the mobile signals that can activate specific reactions in distant tissues.

8.7   Systemic Recombination Signal Is One of the Mechanisms of Stress Response

Recent experiments in our laboratory suggest that plants respond to local stresses in a systemic way. We found that exposure of a single tobacco leaves to UVC or rose Bengal (singlet oxygen producer) resulted in the HRF increase throughout the whole plant (Filkowski et al. 2004). It should be noted that the pre-treatment of a tissue to be stressed with radical scavenging enzymes decreased but not totally abolished the global HRF increase. This suggests that stress results in the production of a systemic signal that at least is partially dependent on the radical production. The nature of this signal, named the systemic recombination signal (SRS), is still enigmatic. The other questions that still remain to be answered are: whether other stresses are capable to generate the SRS, whether all the plants respond to biotic stress in a similar manner, and what the role of SRS in plants is?

8.8   Can Pathogen Induce Genome Rearrangements?

The result of plant pathogen interaction depends on the presence of avirulence (*Avr*) gene in a pathogen and resistance (*R*) gene in a plant (Dong 2001). The absence of one of the components results in a "compatible" interaction and systemic spread of a pathogen. If both are present, this results in an "incompatible" interaction and leads to a local hypersensitive response, immediately followed by a systemic acquired resistance (SAR) to any future encounter with a similar or different pathogen (Dong 2001). To be a systemic process, SAR has to be dependent on the spread of a signal throughout the plant. The nature of this signal remains enigmatic.

Plants that are not able to mount a hypersensitive response and SAR are not necessarily completely vulnerable. The typical response of such plants is the increase in the level of unspecific resistance/innate immunity. It is believed, however, that the evolution of response to pathogen is a constant arm race, where plants are "trying" to create a new *R* gene, and pathogens are trying to modify the *Avr* gene to escape recognition. This process is constant and thus requires a complex signaling mechanism.

8.9 Compatible Viral Infection Leads to the Production of SRS and Global Increase in HRF

We previously reported that a compatible interaction between the pathogen Tobacco mosaic virus (TMV) and the plant *Nicotiana tabacum* (tobacco) results in the production of a signal that leads to local and systemic changes in the frequency of somatic recombination (Kovalchuk et al. 2003b). We assume that this signal has the same nature as the one produced upon abiotic stress. Since the signal was produced upon the viral infection, it was important to check whether the virus presence was required for the HRF increase. First, we checked how long it takes for a virus to move from the infection site systemically, and we found that cutting leaves at 24 h after infection does not allow the virus to spread. The same experiment showed that cutting the infected leaves as early as 8 h after infection still results in the HRF increase. Thus we concluded that the SRS moves to non-infected leaves faster than the virus (Kovalchuk et al. 2003b).

Next we compared the generation of a signal in tobacco cultivars that do or do not have the gene of resistance to TMV, the *N*-gene. This experiment showed that the infection of SR1 plants that do not have the *N*-gene results in the SRS production, whereas the infection of Big Havana plants that do have the *N*-gene does not. We hypothesized that the absence of *Avr:R* gene interaction results in a compatible interaction, and this allows the production of the SRS. We also found that the conditions inactivating the function of the N-gene, like the increase in temperature to over 30 °C, result in a compatible interaction and allows the pathogen to spread. We confirmed that the SRS is indeed generated in such conditions. This suggests that the signal is generated as soon as the resistance gene is either absent (SR1) or inactive ("Havana" at >30 °C).

8.10 SRS Results in Heritable Changes in HRF and Methylation Pattern

We hypothesized that the generation of SRS is a part of a plant adaptive response. Hence, it should lead to the changes that can be observed in the following generation. We call such changes "transgenerational". To check whether these types of events are possible, we collected the seeds from the infected plants and plants

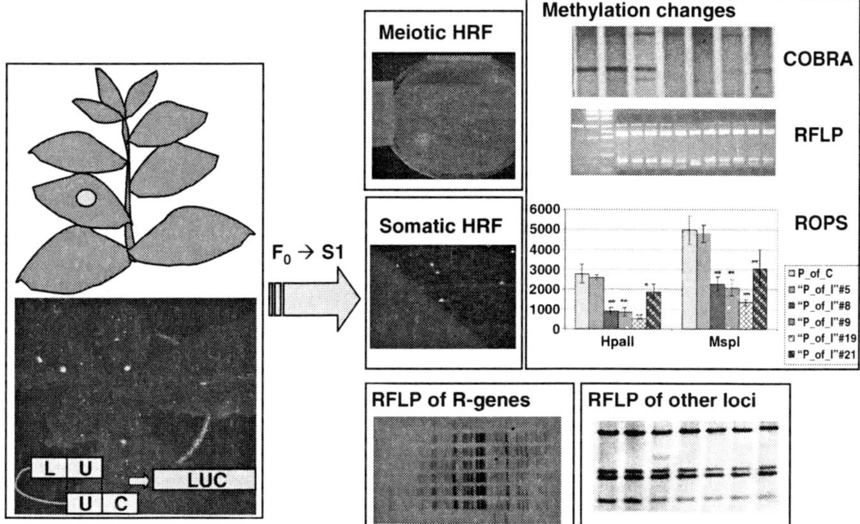

Fig. 8.2 Major changes found in the progeny of infected plants. Schematic representation of the experimental set up. Briefly, in a previous experiment, single leaves of 10-week old SR1 tobacco plants were inoculated with 300 ng of TMV RNA and 24 h after inoculation the infected leaves were removed. Seeds called from these plants were virus free. In the next generation (shown as "S1", or "stressed#1") we analyzed the meiotic and somatic recombination frequency, RFLP of the *N*-gene like and control (actin, 5.8 S and RENT loci) as well as global genome and locus-specific methylation patterns (analyzed with combined bisulfite restriction analysis or COBRA, methylation sensitive RFLP and random oligonucleotide-primed synthesis, or ROPS assays)

treated with buffer (control for wounding stress). First, we checked whether infection leads to the increase in the number of plants with a completely recombined transgene (Fig. 8.2). These plants express the luciferase in all cells. The experiments showed a close to threefold increase in these types of plants (Kovalchuk et al. 2003b). There are two possible reasons for the occurrence of the plants with the completely recombined transgenes. It could be either the somatic recombination event inherited early, or most likely the meiotic recombination event. It is plausible to think that SRS triggers the increase in meiotic recombination frequency, contributing to the diversity in the progeny. The similar increase in the meiotic HRF has been observed previously upon the constant exposure to elevated UVB levels (Ries et al. 2000).

The increase in the meiotic HRF observed in the transgene suggests that there could be similar changes in the other loci. This, however, can be harmful, and plants should employ various strategies to protect the important housekeeping genes essential to the proper plant function. In contrast, rearrangements in the loci carrying the homology to *R* genes might prove to be beneficial. The instability of the transgene could suggest that plants "recognize" the "neutrality" or the "importance" of a gene. It is also possible that the newly formed loci (such as a transgene) do not establish similar levels of chromatin structure, and thus do not follow the same type of the control over the rearrangements. If our hypothesis is correct, an increase in *R* gene rearrangements triggered by a compatible infection could be seen as an attempt to formulate novel *R* genes for the next generation (Richter et al. 1995; Tornero et al. 2002).

Such a form of adaptive response to stress through the genomic alterations was previously proposed by Barbara McClintock (McClintock 1984). Extensive research has supported this model, showing that salicylic acid, methyl jasmonate, oxidative stress, wounding, pathogen attack, $CuCl_2$, cell subculture, and protoplast isolation activate transposons (reviewed in Arnholdt-Schmitt 2004). Since the activation of transposons is associated with a substantial decrease of the genome stability (Dennis and Brettell 1990; Miura et al. 2001), to regulate their activity is a very important task. The activation of transposons is in part associated with the decrease in DNA methylation (Wang et al. 1996; Neidhart et al. 2000; Cui and Fedoroff 2002). It remains to be established, however, whether there is a link between biotic stress, loci-specific hypomethylation, and changes in the genome stability of pathogen-infected plants.

8.11 Viral Infection Leads to Global Genome Hypermethylation in the Progeny of Infected Plants

The role of epigenetic control in the adaptation and acclimation process is hard to underestimate, as transgenerational changes in DNA methylation, histone modifications patterns and in the regulation of chromatin binding proteins are powerful tools of reversible changes in the gene expression. We have revealed that the progeny of plants that are constantly exposed to ionizing radiation have hypermethylated genomes (Kovalchuk et al. 2003a). Thus changes in DNA methylation patterns could be a part of the plant stress protection mechanism (Rizwana and Hahn 1999).

We have analyzed the global genome methylation status of progenies of infected and buffer-treated plants by digesting their genomic DNA using methylation sensitive enzymes *Hpa*II and *Msp*I (Fig. 8.2; ROPS assay). Four out of five progeny of infected lines showed the significantly increased global genome methylation levels (Boyko et al. 2007). Furthermore, we have found that the changes in DNA methylation were mostly due to the methylation of cytosines at the symmetrical CG and CNG sites rather than at the asymmetrical cytosing methylation (Boyko et al. 2007).

8.12 Viral Infection Results in Loci-Specific Methylation Changes

The high level of genome rearrangements is generally associated with the low level of methylation (Bassing et al. 2002; Bender 1998). The hypermethylation observed in the progeny of infected plants should then be associated with low recombination frequency. In contrast, we observed a higher somatic and meiotic HRF. Previous reports showed a higher HRF in the progeny of stressed plants (Ries et al. 2000; Kovalchuk et al. 2003b; Boyko et al., unpublished data). The only explanation we could come up with was that there may be a differential pattern of methylation throughout the genome of progeny of infected plants. In this case, the majority of loci have higher methylation levels, while some loci, including the loci containing the transgene, could be hypomethylated.

If the aforementioned hypothesis is right, the R gene loci that carry homology to the N gene should have a lower methylation level to allow more frequent rearrangements to occur. Hypomethylation of these loci would allow them to have a higher degree of freedom for rearrangements (Bassing et al. 2002; Engler et al. 1993). As a control, we tested the methylation of actin, repetitive elements in *Nicotiana tabacum* (RENT) and 5.8 S rRNA loci in both progenies. The actin and 5.8 S rRNA loci are neutral to stress and thus should not undergo hypomethylation. The RENT loci contain regulatory elements, thus alterations of these loci could potentially change the expression of the neighboring genes (Foster et al. 2003). As such, a decrease in the methylation status at any of these loci could potentially be detrimental to plants.

The methylation sensitive RFLP analysis has indeed shown a differential pattern in the aforementioned loci (Fig. 8.2). Out of 22 loci that carried homology to the N-gene visible on the gel, 4 were drastically hypomethylated in the progeny of infected plants; the rest were similarly methylated among the 2 progenies. In contrast, 3 out of 9 actin loci were heavily hypermethylated, and the rest were equally methylated. At the same time, the methylation level of RENT and 5.8 S rRNA was similar in both progenies.

These experiments showed that the methylation status of the progeny of plants infected with a virus was significantly changed, whereby several N-gene-like loci were severely hypomethylated, and several actin loci were strongly hypermethylated. This differential methylation could possibly allow more flexibility in rearrangements of N-gene-like loci and for less flexibility of actin loci to occur.

8.13 Viral Infection Results in the Destabilization of R-Gene Loci in the Progeny of Infected Plants

The experiments describing a methylation pattern in the resistance gene, actin, RENT and 5.8 S loci represent only a portion of all the changes that occur in various areas of the genome. It is known that the increase in methylation in certain genome loci is correlated with a lower frequency of recombination, while the contrary is true for the loci that have a decrease in methylation (Bassing et al. 2002; Bender 1998; Engler et al. 1993). We hypothesized that the change in the methylation status of the N-gene-like R-gene loci would change its stability, whereby hypomethylation would lead to more frequent rearrangement events.

A rich variety of polymorphic *R*-gene families apparently have evolved by extensive rearrangement mechanisms such as gene and chromosomal duplications, unequal crossing over, and deletions/insertions incited in plants challenged by pathogens. In fact, all of the above have been shown to exist in various clusters of *R* gene loci (Mauricio et al. 2003; Stahl et al. 1999; Tian et al. 2002; Van der Hoorn et al. 2002). Therefore, in order to survive the constant battle with highly mutable pathogens, plants must continuously modify their *R* genes in order to recognize the pathogen *Avr* genes via gene-for-gene interactions (Madsen et al. 2003).

The RFLP analysis of the *N*-gene like loci in both progenies has indeed revealed a greater than fivefold higher frequency of rearrangements in the progeny of infected plants (Boyko et al. 2007). Most of the rearrangements analyzed by RFLP included the appearance of extra fragment(s) (Fig. 8.2). This is not surprising, as most of the events would be occurring in only one allele, and thus the original fragments would also be retained. In two cases, we observed the disappearance of fragments coupled to the appearance of several others. This suggests that the rearrangements either occur very early in the embryo development, or that the rearrangements occur in both alleles. Taking into consideration the fact that the loci containing the *R*-genes are complex in structure and are often unstable, it is not surprising that changes could be observed in both loci.

8.14 Stability of the Actin, RENT and 5.8S Loci is not Changed

It is known that *R*-gene loci are generally more unstable than other loci. Spontaneous rearrangements in these loci rather than those induced by pathogen or any other stress, resulting in changes in pathogen resistance, have previously been shown (Sudupak et al. 1993). In contrast, we present the first case when such reshuffling is induced by pathogens. The fact that we observed a significantly high frequency of rearrangements in both transgene (Kovalchuk et al. 2003b) and *R* gene loci (Boyko et al. 2007) does not imply that every genome locus has such a high frequency of reshuffling. In fact, it would be harmful if these events were so frequent throughout the genome. We hypothesized that the majority of loci should have a "normal" frequency of rearrangements in the genome.

The most important task was to show that the genes "neutral" to a pathogen attack but essential for a correct plant function were equally stable in the progeny of both infected and control plants. Our analysis has revealed no difference in the RFLP of actin, 5.8S rRNA or RENT. The comparable stability of the RENT loci is an important finding. RENT contains cryptic gene regulatory elements that are inactive at their native locations in the genome, but have the capacity to regulate the gene expression when positioned adjacent to genes (Foster et al. 2003). Thus the increased frequency of rearrangements of RENT loci in the genome could lead to undesired changes in the expression of neighboring genes. Profiling of 5.8 S loci has revealed more rearrangements than profiling of either actin or RENT loci. Higher than normal recombination rates in the clusters of rRNA coding genes is a well-known phenomenon (Kobayashi et al. 2004). Despite the significant variations in the rDNA clusters observed in bacteria and yeasts, these loci have the reasonably uniform expression levels and are not significantly influenced by stress (Brunner et al. 2004). This is the main reason why these loci are frequently used as internal controls (Brunner et al. 2004).

DNA methylation is a well-explored process of genome maintenance, whereby methyl groups tend to make chromatin less accessible to various remodeling processes. It can thus be suggested that hypomethylation is the mechanism that facilitates the rearrangement of *R*-gene loci.

Plant genomes (especially those that are relatively large) are known to be highly repetitive and thus highly methylated. In this case, the methylation could be a mechanism that stabilizes the genome and prevents rearrangements (Puchta and Hohn 1996). The hypermethylation of the rest of the genome, as could be assumed from the global genome methylation data, prevents the deleterious effect of this genome reshuffling at unfavorable loci. Additionally, this can explain why some loci, such as the actin loci, seem to be less duplicated or found in clusters, since these configurations would result in higher conservation (Kroymann et al. 2003; Mauricio et al. 2003).

The chromatin structure of an organism is established after fertilization, and is specific to each individual organism. The changes in DNA methylation observed upon pathogen infection suggest that exposure to a stress apparently "rewrites" the methylation pattern according to the needs of an organism dictated by a particular stress (Wada et al. 2004; Weaver et al. 2004).

The information about pathogen-induced changes in the plant genome is scarce. Similarly, very little is known on how stresses in general influence the stability of the genomes in the progeny. There are only few reports on the changes observed in flax exposed to certain environmental pressures. Heritable changes in inbred flax in response to specific and defined environmental changes, such as nutrients balance and temperature regimes (Schneeberger and Cullis 1991; Cullis et al. 1999) have been reported. These conditions activate a transposon-like sequence, LIS1, that assembles and inserts itself into the genome of stressed plants (Chen et al. 2005). The newly established "genotrophs" appear to be stabilized, as no further changes in LIS1 activity occur in plants upon the exposure to additional stresses (Cullis et al. 1999). No further studies on this phenomenon have been carried out, and thus it is not known whether the transposon activation was methylation dependent.

Recent publication by Molinier et al. (2006) reports the induction of the genome destabilization in the progeny of *Arabidopsis thaliana* plants exposed to UVC or flagellin. The latter is a bacterial elicitor that can mimic a pathogen attack. The fact that flagellin induces both somatic and meiotic recombination frequency is exciting as it supports our finding with TMV-infected tobacco plants. It is not clear, however, whether flagellin is capable of inducing the SRS as the authors applied the chemical systemically, whereas we performed local TMV infection. The authors were also able to show the epigenetic nature of the phenomenon by crossing the stressed plants to non-stressed plants and showing the increase in HRF in all the progeny. If this is a transgenerational phenomenon (presumably SRS-induced) was a "mendelian event", then the authors would observe either no increase in HRF if this event was recessive (Ns × NN = 0% of ss, where "s" is "stressed" and "N" is non-stressed "allele") or an increase in 50% of the population if the event was dominant (SN × NN = 50% of SN, where "S" is "stressed" and "N" is non-stressed "allele"). It is important to note that both maternal and paternal alleles were able to contribute to the change, since reciprocal crosses resulted in the increase in HRF in the progeny. This means that these transgenerational events were not the result of cytoplasmic (mitochondrial or chloroplast) inheritance but rather of nuclear changes in the chromatin structure or perhaps in the pool of small regulatory RNAs (miRNA, siRNA or others).

There is still a lot to be learned about pathogen-induced changes in the progeny of infected plants. It remains to be determined whether changes observed in the progeny were virus-specific or pathogen-wide. Future studies are needed to show whether the infection with bacterial pathogens leads to rearrangements in *N*-gene-like loci or perhaps any other type of *R*-gene loci, and whether the methylation status and genome stability return to normal in the progeny of infected plants that are grown under normal conditions. It also remains to be established whether plants with higher instabilities and methylation changes acquire any changes in their tolerance to similar (TMV) or different pathogens, or even other stresses.

8.15 Conclusion

We discussed various genetic and epigenetic mechanisms of plant response to stress. We focused mainly on pathogen stress and introduced the phenomenon of SRS-induced somatic and transgenerational changes in genome stability and DNA methylation. We described the existence of a specific, epigenetically controlled mechanism that promotes rearrangements in *R* gene loci in the progeny of plants infected with a compatible pathogen. This phenomenon suggests a second, more flexible level of inheritance regulated by stress that leads to the changes in the progeny of treated plants. Future studies are needed to understand the specificity of such regulation including signal production, maintenance and "inheritance".

Acknowledgements We would like to thank Franz Zemp and for critical comments on this manuscript and Valentina Titova for proofreading the manuscript. The Alberta Ingenuity, Human Frontier Science Program and NSERC Discovery to IK are acknowledged for financial support.

References

Arnholdt-Schmitt B (2004) Stress-induced cell reprogramming. A role for global genome regulation? Plant Physiol 136:2579–2586

Bassing CH, Swat W, Alt FW (2002) The mechanism and regulation of chromosomal V(D)J recombination. Cell 109:S45–S55

Bender J (1998) Cytosine methylation of repeated sequences in eukaryotes: the role of DNA pairing. Trends Biochem Sci 23:252–256

Besplug J, Filkowski J, Burke P, Kovalchuk I, Kovalchuk O (2004) Atrazine induces homologous recombination but not point mutation in the transgenic plant-based biomonitoring assay. Arch Environ Contam Toxicol 46(3):296–300

Borsani O, Zhu J, Verslues PE, Sunkar R, Zhu JK (2005) Endogenous siRNAs derived from a pair of natural cis-antisense transcripts regulate salt tolerance in Arabidopsis. Cell 123(7):1279–1291

Boyko A, Greer M, Kovalchuk I (2006a) Acute exposure to UVB has a more profound effect on plant genome stability than chronic exposure. Mutat Res 602:100–109

Boyko A, Hudson D, Bhomkar P, Kathiria P, Kovalchuk I (2006b) Increase of homologous recombination frequency in vascular tissue of Arabidopsis plants exposed to salt stress. Plant Cell Physiol 47:736–742

Boyko A, Zemp F, Filkowski J, Kovalchuk I (2006c) Double strand break repair in plants is developmentally regulated. Plant Physiol 141(2):488–497

Boyko A, Kathiria P, Zemp FJ, Yao Y, Pogribny I, Kovalchuk I (2007) Transgenerational changes in the genome stability and methylation in pathogen-infected plants: (virus-induced plant genome instability). Nucleic Acids Res 35:1714–1725

Brunner A, Yakovlev I, Strauss S (2004) Validating internal controls for quantitative plant gene expression studies. BMC Plant Biol 4:14

Buchanan CD, Lim S, Salzman RA, Kagiampakis I, Morishige DT, Weers BD, Klein RR, Pratt LH, Cordonnier-Pratt MM, Klein PE, Mullet JE (2005) Sorghum bicolor's transcriptome response to dehydration, high salinity and ABA. Plant Mol Biol 58(5):699–720

Chen Y, Schneeberger RG, Cullis CA (2005) A site-specific insertion sequence in flax genotrophs induced by environment. New Phytol 167:171–180

Chinnusamy V, Schumaker K, Zhu JK (2004) Molecular genetic perspectives on cross-talk and specificity in abiotic stress signalling in plants. J Exp Bot 55:225–236

Colasanti J, Sundaresan V (2000) 'Florigen' enters the molecular age: long-distance signals that cause plants to flower. Trends Biochem Sci 25:236–240

Cronk QC (2001) Plant evolution and development in a post-genomic context. Nat Rev 2:607 619

Cui H, Fedoroff NV (2002) Inducible DNA demethylation mediated by the maize suppressor-mutator transposon-encoded TnpA protein. Plant Cell 14:2883–2899

Cullis CA, Swami S, Song Y (1999) RAPD polymorphisms detected among the flax genotrophs. Plant Mol Biol 41:795–800

Dennis E, Brettell R (1990) DNA methylation of maize transposable elements is correlated with activity. Philos Trans R Soc Lond B Biol Sci 326:217–229

Dong X (2001) Genetic dissection of systemic acquired resistance. Curr Opin Plant Biol 4:309–314

Engler P, Weng A, Storb U (1993) Influence of CpG methylation and target spacing on V(D)J recombination in a transgenic substrate. Mol Cell Biol 13:571–577

Filkowski J, Besplug J, Burke P, Kovalchuk I, Kovalchuk O (2003) Genotoxicity of 2,4-D and dicamba revealed by transgenic Arabidopsis thaliana plants harboring recombination and point mutation markers. Mutat Res 542(1/2):23–32

Filkowski J, Kovalchuk O, Kovalchuk I (2004) Genome stability of vtc1, tt4, and tt5 Arabidopsis thaliana mutants impaired in protection against oxidative stress. Plant J 38(1):60–69

Foster E, Hattori J, Zhang P, Labbé H, Martin-Heller T, Li-Pook-Than J, Ouellet T, Malik K, Miki B (2003) The new RENT family of repetitive elements in Nicotiana species harbors gene regulatory elements related to the tCUP cryptic promoter. Genome 46:146–155

Fransz PF, de Jong JH (2002) Chromatin dynamics in plants. Curr Opin Plant Biol 5(6):560–567

Gerton JL, Hawley RS (2005) Homologous chromosome interactions in meiosis: diversity amidst conservation. Nat Rev Genet 6:477–487

Grigg SP, Canales C, Hay A, Tsiantis M (2005) SERRATE coordinates shoot meristem function and leaf axial patterning in Arabidopsis. Nature 437(7061):1022–1026

Hays JB (2002) Arabidopsis thaliana, a versatile model system for study of eukaryotic genome-maintenance functions. DNA Repair (Amst) 1(8):579–600

Jeggo PA (1998) DNA breakage and repair. Adv Genet 38:185–218

Jover-Gil S, Candela H, Ponce MR (2005) Plant microRNAs and development. Int J Dev Biol 49:733–744

Kandasamy MK, Deal RB, McKinney EC, Meagher RB (2005) Silencing the nuclear actin-related protein AtARP4 in Arabidopsis has multiple effects on plant development, including early flowering and delayed floral senescence. Plant J 41(6):845–858

Kapoor A, Agius F, Zhu JK (2005) Preventing transcriptional gene silencing by active DNA demethylation. FEBS Lett 579(26):5889–5898

Karpinski S, Reynolds H, Karpinska B, Wingsle G, Greissen G, Mullineaux P (1999) Systemic signaling and acclimation in response to excess excitation energy in Arabidopsis. Science 284:654

Kobayashi T, Horiuchi T, Tongaonkar P, Vu L, Nomura M (2004) SIR2 regulates recombination between different rDNA repeats, but not recombination within individual rRNA genes in yeast. Cell 117:441–453

Kovalchuk I, Kovalchuk O, Arkhipov A, Hohn B (1998) Transgenic plants are sensitive bioindicators of nuclear pollution caused by the Chernobyl accident. Nat Biotechnol 16(11):1054–1059

Kovalchuk O, Arkhipov A, Barylyak I, Karachov I, Titov V, Hohn B, Kovalchuk I (2000) Plants experiencing chronic internal exposure to ionizing radiation exhibit higher frequency of homologous recombination than acutely irradiated plants. Mutat Res 449:47–56

Kovalchuk O, Burke P, Arkhipov A, Kuchma N, James SJ, Kovalchuk I, Pogribny I (2003a) Genome hypermethylation in Pinus silvestris of Chernobyl – a mechanism for radiation adaptation? Mutat Res 529(1/2):13–20

Kovalchuk I, Kovalchuk O, Kalck V, Boyko V, Filkowski J, Heinlein M, Hohn B (2003b) Pathogen-induced systemic plant signal triggers DNA rearrangements. Nature 423:760–762

Kovalchuk I, Abramov V, Pogribny I, Kovalchuk O (2004) Molecular aspects of plant adaptation to life in the Chernobyl zone. Plant Physiol 135(1):357–363

Kroymann J, Donnerhacke S, Schnabelrauch D, Mitchell-Olds T (2003) Evolutionary dynamics of an Arabidopsis insect resistance quantitative trait locus. Proc Natl Acad Sci USA 100:14587–14592

Liu J, He Y, Amasino R, Chen X (2004) siRNAs targeting an intronic transposon in the regulation of natural flowering behavior in Arabidopsis. Genes Dev 18(23):2873–2878

Ludwig A, Romeis T, Jones JD (2004) CDPK-mediated signalling pathways: specificity and cross-talk. J Exp Bot 55:181–188

Madlung A, Comai L (2004) The effect of stress on genome regulation and structure. Ann Bot 94:481–495

Madsen LH, Collins NC, Rakwalska M, Backes G, Sandal N, Krusell L, Jensen J, Waterman EH, Jahoor A, Ayliffe M et al. (2003) Barley disease resistance gene analogs of the NBS-LRR class: identification and mapping. Mol Genet Genomics 269:150–161

Mathieu O, Bender J (2004) RNA-directed DNA methylation. J Cell Sci 117(21):4881–4888

Mauricio R, Stahl EA, Korves T, Kreitman M, Bergelson J (2003) Natural selection for polymorphism in the disease resistance gene Rps2 of Arabidopsis thaliana. Genetics 163:735–746

McClintock B (1984) The significance of responses of the genome to challenge. Science 226:792–801

Miura A, Yonebayashi S, Watanabe K, Toyama T, Shimada H, Kakutani T (2001) Mobilization of transposons by a mutation abolishing full DNA methylation in Arabidopsis. Nature 411:212–214

Mlotshwa S, Voinnet O, Mette M, Matzke M, Vaucheret H, Ding S, Pruss G, Vance V (2002) RNA silencing and the mobile silencing signal. Plant Cell 14:289–301

Molinier J, Oakeley EJ, Niederhauser O, Kovalchuk I, Hohn B (2005) Dynamic response of plant genome to ultraviolet radiation and other genotoxic stresses. Mutat Res 571(1/2):235–247

Molinier J, Ries G, Zipfel C, Hohn B (2006) Transgeneration memory of stress in plants. Nature 442(7106):1046–1049

Neidhart M, Rethage J, Kuchen S, Kunzler P, Crowl RM, Billingham ME, Gay RE, Gay S (2000) Retrotransposable L1 elements expressed in rheumatoid arthritis synovial tissue: association with genomic DNA hypomethylation and influence on gene expression. Arthritis Rheum 43:2634–2647

Pearce G, Strydom D, Johnson S, Ryan C (1991) Systemic wound signaling in plants: a new perception. Science 253:895

Peragine A, Yoshikawa M, Wu G, Albrecht HL, Poethig RS (2004) SGS3 and SGS2/SDE1/RDR6 are required for juvenile development and the production of trans-acting siRNAs in Arabidopsis. Genes Dev 18(19):2368–2379

Puchta H, Hohn B (1996) From centiMorgans to basepairs: homologous recombination in plants. Trends Plant Sci 1:340–348

Richter T, Pryor T, Bennetzen J, Hulbert S (1995) New rust resistance specificities associated with recombination in the Rp1 complex in maize. Genetics 141:375–381

Ries G, Heller W, Puchta H, Sandermann H, Seidlitz HK, Hohn B (2000) Elevated UV-B radiation reduces genome stability in plants. Nature 406:98–101

Rizwana R, Hahn PJ (1999) CpG methylation reduces genomic instability. J Cell Sci 112:4513–4519

Schneeberger RG, Cullis CA (1991) Specific DNA alterations associated with the environmental induction of heritable changes in flax. Genetics 128:619–630

Shinozaki K, Yamaguchi-Shinozaki K, Seki M (2003) Regulatory network of gene expression in the drought and cold stress responses. Curr Opin Plant Biol 6:410–417

Stahl EA, Dwyer G, Mauricio R, Kreitman M, Bergelson J (1999) Dynamics of disease resistance polymorphism at the Rpm1 locus of Arabidopsis. Nature 400:667–671

Steimer A, Schob H, Grossniklaus U (2004) Epigenetic control of plant development: new layers of complexity. Curr Opin Plant Biol 7(1):11–19

Steward N, Ito M, Yamaguchi Y, Koizumi N, Sano H (2002) Periodic DNA methylation in maize nucleosomes and demethylation by environmental stress. J Biol Chem 277: 37741–37746

Sudupak M, Bennetzen J, Hulbert S (1993) Unequal exchange and meiotic instability of disease resistance genes in the Rp1 region of maize. Genetics 133:119–125

Sung DY, Kaplan F, Lee KJ, Guy CL (2003) Acquired tolerance to temperature extremes. Trends Plant Sci 8:179–187

Sunkar R, Zhu JK (2004) Novel and stress-regulated microRNAs and other small RNAs from Arabidopsis. Plant Cell 16(8):2001–2019

Takeda S, Tadele Z, Hofmann I, Probst AV, Angelis KJ, Kaya H, Araki T, Mengiste T, Mittelsten Scheid O, Shibahara K, Scheel D, Paszkowski J (2004) BRU1, a novel link between responses to DNA damage and epigenetic gene silencing in Arabidopsis. Genes Dev 18:782–793

Tian D, Araki H, Stahl EA, Bergeslson J, Kreitman M (2002) Signature of balancing selection in Arabidopsis. Proc Natl Acad Sci USA 99:11525–11530

Tornero P, Chao R, Luthin W, Goff S, Dangl J (2002) Large-scale structure-function analysis of the Arabidopsis RPM1 disease resistance protein. Plant Cell 14:435–450

Van der Hoorn R, De Wit P, Joosten M (2002) Balancing selection favors guarding resistance proteins. Trends Plant Sci 7:67–71

Wada Y, Miyamoto K, Kusano T, Sano H (2004) Association between up-regulation of stress-responsive genes and hypomethylation of genomic DNA in tobacco plants. Mol Gen Genomics 271:658–666

Wang L, Heinlein M, Kunze R (1996) Methylation pattern of activator transposase binding sites in maize endosperm. Plant Cell 8:747–758

Waterhouse P, Wang M-B, Finnegan J (2001) Role of short RNAs in gene silencing. Trends Plant Sci 6:297

Weaver IC, Cervoni N, Champagne FA, D'Alessio AC, Sharma S, Seckl JR, Dymov S, Szyf M, Meaney MJ (2004) Epigenetic programming by maternal behavior. Nat Neurosci 7:847–854

Chapter 9
Recent Advances in Functional Genomics and Proteomics of Plant Associated Microbes

P. Nannipieri(✉), J. Ascher, M.T. Ceccherini, G. Guerri,
G. Renella, and G. Pietramellara

9.1 Introduction

Research in soil microbiology has concerned the determination of the presence of gene sequences so as to assess microbial diversity rather than the determination of gene expression. Generally these molecular techniques are based on the specific amplification of the target nucleic acid by polymerase chain reaction (PCR) with either restriction analysis or separation by denaturing or conformational properties of the resulting amplicons (Lynch et al. 2004). On the other hand microbial activities in soil have been measured by classical techniques such as those for determining soil respiration, enzyme activities, N mineralization, adenylate energy charge, leucine and thymidine incorporation, etc., with no idea of gene expression. The rhizosphere effects on microbial diversity and activity are discussed in Chap. 14 of this book.

Rapid progress in genomics has led to the availability of full genome sequences of hundreds of microorganisms, mostly bacteria (DeLong 2002). Combinations of new molecular methodology and genomics have been used successfully to link microbial phylogeny with function in several ecological studies and the same approach could provide significant insights into plant–microbe interactions in the rhizosphere. The functional genomics is based on a holistic or systemic approach with studying information flow within a cell and this requires the application of high throughput methods using automated technologies, which allow functional analysis of genome, proteome and metabolome of an organism (Wren 2000). In this way it is possible to get an insight on interplay of a large number of gene products and the relative consequences of this communication to the physiology of a cell. In contrast, molecular biologists have followed the reductionistic approach by studying single genes, and the individual actions of genes with a step-by-step characterization of metabolic pathways. Thus, an efficient DNA extraction with

P. Nannipieri
Dipartimento della Scienza del Suolo e Nutrizione della Pianta, Universita' degli Studi
di Firenze, 50144 Firenze, Italy
e-mail: paolo.nannipieri@unifi.it

C.S. Nautiyal, P. Dion (eds.) *Molecular Mechanisms of Plant
and Microbe Coexistence.* Soil Biology 15, DOI: 10.1007/978-3-540-75575-3   215

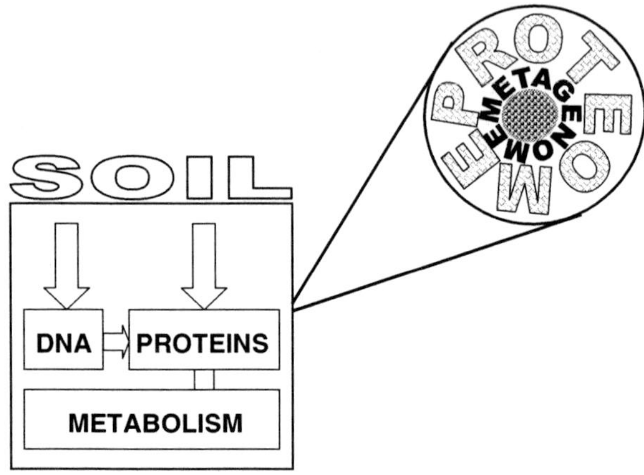

Fig. 9.1 Holistic view of genomics and proteomics of rhizosphere

a high quality DNA sequence is followed by the analysis of the synthesised proteins.

It is possible to apply the functional genomics to the rhizosphere if we assume that rhizosphere soil can be seen as an organism and the genome of this organism includes genomes of all organisms inhabiting the rhizosphere soil (metagenome), the proteome includes all proteins associated with these organisms and the metabolism includes all metabolic reactions of these organisms (Fig. 9.1). However, this holistic view must consider that:

1. Both abiotic reactions and reactions catalysed by extracellular enzymes adsorbed by soil colloids can occur in soil and they are not under the control of any gene (Nannipieri et al. 2003).
2. Several methodological problems exist due to the complexity of soil. For example, as will be discussed later, only recently techniques for determining mRNA has been set up for soil due to problems in extracting and manipulating mRNA molecules, which are less stable than DNA molecules (Nannipieri et al. 2003). Methodological problems also exist in analysing the final products of gene expression, the proteins (proteomic approach), in soil (Nannipieri 2006; Ogunseitan 2006).

The aim of this chapter is to discuss the different approaches and relative techniques for studying the functional genomics of the rhizosphere soil. Gene expression at both transcription and translational levels will be discussed. Since the matter is relatively new and the relative bibliography is scarce, the discussion will mainly concern the potential applications by underlying the potential advantages and drawbacks of the various methods. Studies of gene expression in bulk soil will be also reviewed by considering both advantages and disadvantages of the employed assays and discussing the meaning of the relative results with the aim to evaluate the potential insights that can be got by carrying out these studies in the rhizosphere soil.

9.2 Plant and Genetic Diversity in the Rhizosphere Soil

Plants can influence the genetic diversity of microorganisms inhabiting the rhizosphere; for example, plants select for a higher diversity of ammonia-oxidizing and nitrogen-fixing bacteria in the rhizosphere, when compared with bulk soil (Briones et al. 2003; Cocking 2003); on the other hand soil microorganisms may affect plant diversity through providing competing advantages to specific plants (Bever 2003). It is not known if the higher diversity of functional genes depends on a higher diversity and number of microorganisms in the root regions as a consequence of the accessibility of C source or the results of a natural selection, whereby plant supplies C and in turn the microbes provide N, P, other nutrients and protection against pathogens and herbivores. Obviously these hypotheses cannot be verified by research only based on fingerprinting techniques but it is important to study the utilization of plant C by microbial species with a particular functional role by using advanced techniques, which allow correlating a specific metabolic activity with phylogenetically identifiable units in natural environments (Gray and Head 2001; Wagner 2004).

Bacteria frequently exchange genetic material and the acquired DNA can be integrated in the recipient cell genome by recombination catalyzed by bacterial integrases, and the selective advantages acquired by the new genetic material may be maintained or lost depending on the characteristics of the environment. Particular genes required for effective root colonization by bacteria can be subjected to phase variation, a DNA rearrangement process, regulated by site-specific recombinase (Dekkers et al. 1998a). A rhizosphere-incompetent *Pseudomonas* strain was capable of colonising root tips after transfer of the site-specific recombinase from a rhizosphere-competent *Pseudomonas fluorescens* (Dekkers et al. 1998b). The presence of plasmids with novel gene clusters containing open reading frames (ORFs) encoding small proteins are advantageous for *Pseudomonas* spp. to colonise roots during particular periods of plant growth (Liley and Bailey 1997). The encoded proteins are assumed to play a role in the cell-to-cell contact mechanism leading to horizontal gene transfer in the natural habitat.

Studies on genetic traits of *P. fluorescens* and *P. putida* responsible for successful colonization of roots include those encoding molecules involved in iron uptake, NADH dehydrogenase, extracellular proteins, surface proteins, lipopolysaccharides, and constituents of flagella (Espinosa-Urgel 2004). Some of these factors and the cell-cell communication through quorum sensing are responsible for the formation of biofilms, which seems essential for the root colonization.

Horizontal gene transfer in nature can occur between same or different species which are similar taxonomically (Veal et al. 1992; Nielsen et al. 1998; Jain et al. 1999) and it is not limited to bacteria but may concern fungi (group I introns located in mitochondrial genome and nuclear ribosomal RNA genes), which can be found in different plants (Cho et al. 1998). Plant surfaces, including rhizoplane, are considered hot spot for conjugation probably because the close proximity of bacterial cells in biofilms facilitates the exchange of genetic material (van Elsas et al. 2003; Espinosa-Urgel 2004). Transfer by conjugation does not seem related to the microbial activity, determined by incorporation of ^{3}H-leucine into cells, since, in spite of the fact that this activity was similar in the rhizosphere of barley, wheat and pea, the conjugal transfer of plasmid between *P. fluorescens* and *S. plymuthica* was the highest in the pea rhizosphere (Schwaner and Kroer 2001).

Of the three horizontal gene processes (conjugation, transduction and transformation) occurring in bacteria, transformation is the only one involving extracellular DNA. Under natural soil conditions, transformation is likely to be due to illegitimate recombination (Mercier et al. 2006). It has been calculated that natural transformation of bacteria in soil is an extremely rare event and no specific reports on the frequency of transformation in the rhizosphere have been reported; thus there is no conclusive evidence of enhanced competence or more availability of extracellular DNA in the rhizosphere than in the bulk soil (Mercier et al. 2006). It is not clear whether under peculiar rhizosphere conditions, such as reduced pH and Eh, enhanced microbial activity, the extracellular DNA may be more or less available to soil microorganisms. The extracellular DNA is thought to be rapidly degraded in soil environment (Blum et al. 1997) unless it is stabilized through adsorption by soil colloids (Paget et al. 1992); DNA adsorbed by soil colloids can persist for a long time (Nielsen et al. 1997; Gebhard and Smalla 1999), maintaining its transforming capacity (Pietramellara et al. 1997; Nielsen et al. 2000; Ceccherini et al. 2003).

9.3   Gene Expression in the Rhizosphere Soil

9.3.1   Reporter Gene

Both rDNA and rRNA techniques do not allow determining the phenotypic expression in the rhizosphere soil of the specific genes and this shortcoming can be overcome by a reporter gene, expressed from a constitutive or inducible promoter resulting in a detectable phenotype. It is important to select a representative soil microorganism or a specific microorganism for the specific purpose. There are several available reporter genes such as those encoding substances responsible for the production of bioluminescence (*lux*), green fluorescent protein (*gfp*), β-galactosidase (*lacZ*) and ice nucleation promoting proteins (*inaZ*) (Sørensen and Nybroe 2006; Burmølle et al. 2006). The proper selection depends on: i) the type of detection used to quantify the reporter gene expression; ii) the sensitivity of the used reporter; iii) the type of sensing promoter (Saleh-Lakha et al. 2005).

Whole-cell bacterial biosensors have been used for monitoring the availability of C, N and P in the rhizosphere soil. This information can be obtained using *lux*-marked *Pseudomonas fluorescens* strains in which bioluminescence is regulated by the cellular levels of $FMNH_2$, in the presence of sufficient aldehyde and oxygen (Amin-Hanjani et al. 1993; Kragelund et al. 1995, 1997; Yeomans et al. 1999). Such strains are bioluminescent under starving conditions whereas assimilation of nutrients reduces the bioluminescence (Yeomans et al. 1999;

Roca and Olson 2001). Using the whole cell biosensor *Pseudomonas fluorescens* 10586 pUCD607 responding to a range of compounds present in the root exudate, Darwent et al. (2003) reported that a reduced supply of NO_3^- to winter barley induced a greater root exudation. DeAngelis et al. (2005) found a significantly lower NO_3^- availability in the rhizosphere of wild oat (*Avena fatua*) than in bulk soil; in addition, competition for NO_3^- between roots and the whole-cell bioreporters could be removed by soil amendment with NO_3^-; they used two bacterial biosensors set up by fusion of promoterless ice nucleation *ina*Z from *P. syringae* and *gfp* with the nitrate-regulated promoter of *nar*G in *E. cloacae* EcCT501R.

A simultaneous measurement of C, N and P in the rhizosphere soil solution has been conducted by Standing et al. (2003) by using a tripartite reporter gene system. However, responses of bacterial biosensors to N and P limitations in rhizosphere may depend on both the concentrations and forms of nutrients; for example, the *Pseudomonas* N-reporter strains reacted towards limitation by both NH_4^+ and common amino acids (e.g. glutamate). However, it is important to know the reaction of the reporter to other N-containing compounds of root exudates. A partial answer has been obtained by using reporter bacteria responding specifically to individual amino acids. The induction of a lysine-responsive *P. putida* reporter was demonstrated in rhizosphere of corn, but not in the bulk soil (Espinosa-Urgel and Ramos 2001) and a tryptophan-reporter strain showed significant induction in older root segments of Avena grass, but not at the root tip (Jaeger et al. 1999). Kuiper et al. (2001) demonstrated that uptake regulation of putrescine, commonly released in root exudates of tomato, was important for growth and root colonization of rhizobacteria using a *P. fluorescens* reporter strain. By using a whole cell biosensor constructed in *Pseudomonas* spp. by fusing *lux*-reporter genes to the promoters of operons active in the synthesis of ribosomal RNA, Marschner and Crowley (1996) demonstrated that rDNA genes were highly expressed and thus growth of *P. fluorescens* was higher in the rhizosphere of pepper than in the bulk soil. Ramos et al. (2000), using a ribosomal promoter inserted with a *gfp*-reporter encoding for an unstable GFP variant, demonstrated that the growth of a *P. putida* strain was greater at the root tip than in mature or senescent root areas.

The iron stress response of pseudomonads has been monitored by using an iron-regulated, ice nucleation gene reporter (*ina*Z) (Loper and Henkels 1997; Marschner and Crowley 1997, 1998). It has been shown that iron stress was greater in the zone behind root tips than in older lupine (*Lupinus albus*) and barley (*Hordeum vulgare* L.) root zones and bulk soil (Loper and Henkels 1997). In addition, the iron stress response of *P. putida* Pf-5 could be shut down by the production of phytosiderophores after induction of the iron stress response in plant roots of rice and barley (Marschner and Crowley 1998). Ice nucleation activity expressed by rhizosphere populations of *Pseudomonas fluorescens* Pf-5 decreased from one to two days after the bacterium was inoculated onto root surfaces, suggesting that iron became more available to rhizosphere populations of Pf-5 once they were established in the rhizosphere.

Whole cell (usually enteric bacteria such as *E. coli* and *S. typhimurium*) biosensors, whose response is controlled by the expression of global regulatory genes (regulons), have been also used to monitor microbial activity as affected by fluctuations of temperature, osmolarity, moisture, radicals concentrations, etc. in the rhizosphere soil due to plant activity (van Dyk et al. 1995; Ptitsyn et al. 1997; Vollmer et al. 1997; Bechor et al. 2002). Other whole cell biosensors have been used to detect antibiotics (Bahl et al. 2004), endocrine disruptors (Desbrow et al. 1998) and quorum sensing molecules (Andersen et al. 2001). Using a *Rhizobium* strain carrying a *nod*C-*lac*Z fusion, Bolanos Vasquez and Warner (1997) studied the activation of the *nod* gene by different flavonoids from bean plants. Since this pioneer observation, a large number of bioreporters for the study of chemical colloquia involved in bacterial growth and bacterial activity in rhizosphere and bulk soil have been constructed (Rainey 1999; Timms-Wilson et al. 2000; Allaway et al. 2001; Marco et al. 2003). For example, whole-cell bacterial biosensors, based on constructs with *gfp* as a reporter system, have been set up for detection of *N*-acyl homoserine lactones (AHLs) signals involved in quorum sensing phenomena (Steidle et al. 2001). It is believed that these signals are common among rhizosphere bacteria colonising roots (Zhang and Pierson 2001). Identification of promoters activated upon the exposure of bacteria such as *Rhizobium* or *Pseudomonas* species to the rhizosphere environment can be carried out by inserting a promoterless reporter system into a host strain, with the following recovery of the activated cells to identify the activated promoter by gene sequencing at the site of reporter insertion. Thus, the use of a *Rhizobium* strain with a promoterless *gfp* allowed one to identify rhizosphere-activated promoters controlling the synthesis of thiamine and cyclic glucan, or surface growth-activated promoters controlling methionine synthesis or putrescine uptake (Allaway et al. 2001).

The use of whole cell biosensors has increased our knowledge on the interactions between plants and plant pathogens and on the relative biological control strategies. For example, Smith et al. (1999), using the biosensor *P. fluorescens* F113, found that the fungus *Pythium ultimum* releases molecules which down-regulate the *rrn* promoter of ribosomal RNA synthesis in the bioreporter and Lee and Cooksey (2000) reported that *P. putida* activates promoters controlling the synthesis of ATP binding cassette (ABC) transporter proteins in the presence of the fungus *Phytophthora parasitica*. These findings suggest that in some cases the pathogenic attack of a plant root is also connected to a reduced activity of beneficial rhizobacteria.

9.3.2 *Extraction and Characterization of mRNA*

The relationship between the amount of rRNA and cell activity is not valid for soil microorganisms because most of them under dormancy show high ribosome level so as to be ready to respond to suitable environmental conditions, such as the appearance of substrate (Kowalchuk et al. 2006). In order to determine the expression of a gene it is necessary to monitor its transcript; this can be done by determining the relative mRNA or indirectly through the detection of the signal of a reporter gene fused to the target gene, as discussed above (Krsek et al. 2006). Given the rapid degradation of non-functional mRNA in the environment, the analysis of the target mRNA represents one of the best opportunities to analyse the activity of the target gene. However, as it is discussed below, a complete functional assessment of microbial communities in soil should also involve the relative encoded proteins.

Successful extraction and characterization of mRNA from soil has been delayed with respect to that of DNA, in spite of the fact that extraction procedures of RNA and DNA from soil are similar in principle. Unfortunately, RNA extraction requires several precautions such as the inhibition of RNase activity, using, for example, diethylpyrocarbonate in the RNA extraction solution (Bakken and Frostegård 2006). Since both skin and many microorganisms can be sources of RNases, care should be taken in keeping solutions and equipments free from these enzymes. The fast turnover rate of mRNA in prokaryotic cells is another problem; indeed, the half-life of specific *E. coli* mRNA ranges from 0.5 to 20 min (Nierlich and Murakawa 1996). Fortunately, slower growing microorganisms, the majority of microorganisms inhabiting soil, have a longer mRNA half-life (Krsek et al. 2006). The situation is completely different for mRNA of eukaryotic cells where transcription and translation are physically separated by nucleus membrane with temporal separation of the two processes. In addition, eukaryotic mRNA often undergoes post-transcriptional modifications, such as splicing of intervening sequences or methylation before being translated, and for this reason the half-life of this mRNA can be prolonged to several hours (Sarkar 1997).

Several methods exist to characterize mRNA so as to assess differences in gene expression (Krsek et al. 2006). Subtractive hybridization involves synthesis of cDNA from total RNA extracted from a wild type of mutant cultivated under certain conditions and then this cDNA is hybridised with RNA from the respective strain or mutant cultivated under different conditions. After elimination of double-stranded cDNA-mRNA hybrids, the composition of single-stranded cDNA is analysed (Utt et al. 1995). Subtractive hybridisation has been used to compare the activity of genes responsible for the nodulation of two strains of *Bradyrhizobium japonicum* in culture with or without water extract of a sandy loam soil or root exudates from aseptically grown soybean seedlings (Bhagwat and Keister 1992).

Another assay involves the formation of an hybrid between RNA and a labelled RNA probe, complementary to the gene of interest and prepared by a transcription anti-sense. The double-stranded RNA hybrid is analysed after removal of the non-hybridised RNA by degradation by ribonuclease (Fleming et al. 1993). This technique was used to quantify transcript levels of NAH7 naphthalene dioxygenase (*nah*A) gene with mRNA extracted by a modified hot phenol method from a polycyclic aromatic hydrocarbons (PAHs) contaminated soil. The level of transcripts correlated significantly with [^{13}C] naphthalene mineralization rate and soil naphthalene concentration; the use of a naphthalene-lux reporter system showed that the hydrocarbon was available to microorganisms (Sanseverino et al. 1993–1994).

A differential display method is based on the fact that 100% of eukaryotic mRNA and 2–50% of prokaryotic mRNA present polyadenylate tails (Krsek et al.

2006). In contrast to other methods, this assay does not require one to know gene sequence to design the primers. Indeed, the 3′ primer used in the reverse transcription is made of a poly(T) region with two additional bases recognising a minor part of the target mRNA and the 5′ primer is short and should anneal with 500 bp of the 3′ primer. Then the PCR products are characterised by techniques generating a fingerprint, and single cDNA bands can be excised, sequenced or cloned (Krsek et al. 2006).

Genes involved in production and resistance of streptothricin (ST), an antibiotic from *Streptomyces rochei* F20, have been studied in liquid culture, in soil and in the rhizosphere of spring wheat (Anukool et al. 2004). RNA was extracted and amplified by RT-PCR with primers specific for either ST resistance gene (*sst*R) or ST biosynthesis gene (*sst*A) coding for peptide synthetase. The *sst*R mRNA was detected in both sterile and non-sterile rhizosphere whereas both transcripts were not detected in the rhizoplane.

Transcripts of ligninolytic enzymes (*lip*) of *Phanerochaete chrysosporium* involved in the degradation of wood have been monitored by extracting poly (A) RNA by magnetic capture (Janse et al. 1998). Transcription patterns of all ten *lip* genes and three manganese peroxidase (*mnp*) genes, both quantified by competitive RT-PCR with full-length genomic subclones, were markedly different in wood when compared with those got with defined media or from soil cultures. It was suggested that this depended on the specialisation of different enzymes to different environmental conditions. Indeed transcription of *lip* gene in *Phanerochaete chrysosporium* is regulated by the availability of C and N sources as determined by competitive RT-PCR (Stewart and Cullen 1999). In addition, iron has been found to regulate the transcription of 97 differentially expressed cDNA fragments in *Phanerochaete chrysosporium*, most of them encoding proteins involved in iron uptake (Assmann et al. 2003). Obviously the regulation of genes in *Phanerochaete chrysosporium* in the rhizosphere depends on the competition between plants and microbes for nutrients and in the case of iron availability, it can play a role in the biological control (Crowley 2001).

9.3.3   *Linking Enzyme Activity to Gene Expression*

Measurements of enzyme activity represent one of the classical determinations carried out in bulk and rhizosphere soil (Nannipieri et al. 2003). As discussed in Chap. 14, the interpretation of these measurements is not simple due to the many locations of enzymes contributing to the overall measured enzyme activity. Only a few studies have been carried out to link enzyme activity with expression of genes encoding the protein molecules responsible for the target enzyme activity (Fig. 9.2). The reverse transcription has been used to monitor the expression of three manganese peroxidase (*mn*P) genes during removal of PAHs by *Phanerochaete chrysosporium* grown in presterilized soil (Bogan et al. 1996a). The maximum level of transcripts preceded for one to two days the highest manganese peroxidase extracted from soil

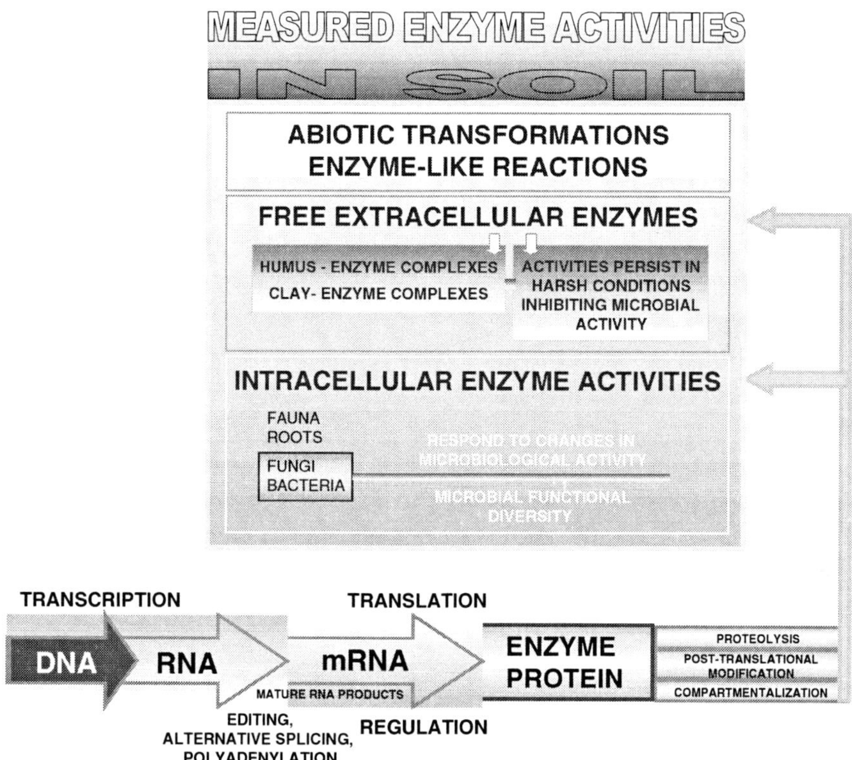

Fig. 9.2 Linking enzyme activity with expression of genes

by the method of Bollag et al. (1987) and both peaks occurred during the maximum rate of two PAHs, fluorene and chrysene, degradation. Bogan et al. (1996b) quantified transcripts of all ten known lignin peroxidase genes (*lip*A-*lip*J) of *Phanerochaete chrysosporium* by competitive RT-PCR, after extraction of mRNA from soil microcosms amended with anthracene. Lip proteins were also extracted from soil microcosms, purified and applied to a nitrocellulose membrane before their quantification by Western blotting with specific monoclonal antibodies. Levels of transcripts were significantly correlated with levels of proteins, even if there was a two-day delay in the peak of the latter respect to the peak of the former. Transcripts of nine *lip* were detected in a soil contaminated by a waste. Studies on both manganese and lignin peroxidase are important not only for the implications on bioremediation of polluted soils but also for a better understanding of the degradation of litter since the importance of these enzymes in the degradation of the lignocellulose complex, which is the main component of the plant residues.

However, the best study covering all events starting from gene presence, through gene expression and up to the detection of target enzyme in soil, has been carried so far by Metcalfe et al. (2002). Chitinase activity was measured by loss of chitin in buried litter-bags and by a luminescence assay by using as a substrate 4-methylambelliferyl-(GlcNAc)$_2$, whereas the composition of community was evaluated by DGGE. In addition, samples from community DNA was amplified with primers targeted to a gene fragment from family 18 group A chitinases. The addition of sludge to the pasture soil increased the chitinase activity and the number of actinobacteria with prevalence of actinobacterium-like chitinase sequences. Unfortunately extraction of transcripts was unsuccessful probably due to the adsorption of mRNA by soil colloids and the target enzyme proteins were not monitored.

9.3.4 Proteomic Approach

The proteomic approach involves the study of all proteins of a cell so as to get an integrated view of the cell itself. It is well established that more protein isoforms can be synthesised by a single gene because mRNA molecules can be subjected to post-transcriptional control such as alternative splicing, polyadenylation and mRNA editing (Graves and Haystead 2002). The recent availability of extensive metagenomic sequences from various environmental microbial communities has extended the postgenomic era to the field of environmental microbiology. Although still restricted to a small number of studies, metaproteomic investigations have revealed interesting aspects of functional gene expression within microbial habitats that contain limited microbial diversity. These studies highlight the potential of proteomics for the study of microbial consortia. However, the application of proteomic investigations to complex microbial assemblages such as seawater and soil still presents considerable challenges. Nonetheless, metaproteomics will enhance the understanding of the microbial world and link microbial community composition to function (Wilmes and Bond 2006). Although the number of proteins that have been identified and separated using available proteomic analysis methods is encouraging, the diversity of other microbial ecosystems still poses enormous challenges.

Recently it has been said that we need to study microbial proteomics in soil and thus to go beyond DNA and mRNA characterization in order to have a better understanding of soil functionality (Nannipieri 2006; Ogunseitan 2006). Unfortunately, soil proteomics is still in its infancy. According to Nannipieri (2006), one of the main methodological problems is to extract specifically intracellular proteins in the presence of a large background of extracellular proteins; these extracellular proteins are stabilised, that is protected against proteolysis, by their association with soil colloids. On average only 4% of the total organic N is present as microbial N whereas amino acid N released after acid hydrolysis can account for 30–45% of the total organic N in soil mostly as extracellular protein N or peptides N (Stevenson 1986). By considering the protein N distribution in soil it has been hypothesised that soil proteomics might be subdivided in *functional proteomics* and *structural proteomics* (Nannipieri 2006). The latter concerns the characterization of the extracellular protein N stabilised by soil colloids so as to improve our understanding of the mechanisms responsible for such a stabilization and gives insight on the dominant extracellular proteins before soil sampling. The combination of this proteomic approach with carbon dating might also reveal the period in which these extracellular proteins were dominant. On the contrary *functional proteomics* concerns proteins synthesised by microbial cells involved in biochemical processes occurring in soil in lab experiments or in the field at the moment of soil sampling. The study of functional proteomic can improve our understanding of degradation of organic pollutants and organic debris, nutrient cycling, blockage of inorganic pollutants, molecular colloquia between microorganisms, between plant roots and microorganisms and between plant roots. As is done for the extraction of nucleic acids from soils, two different approaches have been proposed to extract proteins from soil: i) separation of microbial cells from soil particles and successive cell lysis with release of proteins; ii) cell lysis in situ with extraction of proteins (Nannipieri 2006).

A pioneeristic proteomic approach in the rhizosphere was followed by Lambert et al. (1987) with characterization of rhizobacteria, isolated by cultivation from the rhizosphere soil of maize, by dodecylsulfate-polyacrylamide gel electrophoresis of total cell proteins These proteins were extracted by treating cultivated bacterial cells with lysozyme, followed by boiling with a buffer (a mixture of sodium dodecyl sulfate, β-mercaptoethanol, Tris, EDTA, sucrose, bromophenol blue). A similar approach was also carried out by Lambert et al. (1990) to characterize fast-growing, aerobic and heterotrophic bacteria isolated from the root surface of young sugar beet; in this case proteins were extracted from bacterial cells as mentioned above after sonication. The main bacterial species were *Pseudomonas fluorescens, Xanthomonas maltophilia, Pseudomonas paucimobilis,* and *Phyllobacterium sp.*

Recently the proteomic approach has been applied to characterize cell surface proteins of *Pseudomonas putida,* a plant-growth promoting rhizobacteria active against pathogens, capable of degrading pollutants and able to form biofilms (Arevalo-Ferro et al. 2005). The latter property and the cell-to-cell communication system (quorum sensing, QS), that enables the bacteria to co-ordinate the expression of special phenotypes in a cell density manner, seem to be involved in bacterial rhizosphere competence. Mature biofilms, complex three-dimensional structures where cells are embedded in a thick matrix of extracellular polymeric substances (EPS), show channels allowing the passage of nutrients to the interior parts and the transport of wastes to the external side, both moved by fluids (O'Toole et al. 2000; Hall-Stoodley et al. 2004). Initial steps in biofilm formation involve attachment to a surface, followed by formation of microcolonies and maturation of microcolonies in EPS-encased three-dimensional biofilms. Arevalo-Ferro et al. (2005) compared two-dimensional gel electrophoresis (2-DE) protein profiles of a strain and the QS-deficient mutant grown either in planktonic cultures or in 60 h old mature biofilms. The spots of differentially expressed proteins were excised from the 2-DE gel and the relative proteins were identified by peptide mass fingerprinting and database search. Almost 40% of the surface proteins were affected by the sessile life suggesting that an important fraction of the bacterial genome was involved in biofilm physiology; in addition, almost 10% of the protein spots were controlled by the QS system and differed if the cells were grown in surface or in suspension.

The state-of-the-art of soil proteomic has been recently reviewed by Nannipieri (2006) and Ogunseitan (2006). Schulze et al. (2005) have used the mass spectrometry based proteomics to characterize proteins of leachate from some soils with the aim to determine the phylogenetic origin and the potential catalytic functions of proteins. It was observed that the number of bacterial proteins in the forest soil was greater in winter than in summer; in addition, enzymes, such as peroxidase, involved in the degradation of complex molecules were present in water from the forest soil whereas enzymes such as transferases, which might be involved in methane production, were present in waters from peat bog.

9.4 Linking Gene Expression to Functions: The Use of Stable Isotope Probes (SIP)

A major advance in linking functional activity to community structure came with the development of stable isotope probing (SIP) (Radajewski et al. 2000), which involves tracking of a stable isotope atom from a particular substrate into components of microbial cells that provide phylogenetic and functional information, such as lipid, DNA or RNA. Indeed the major advantage of the SIP technique is that ^{13}C-enriched DNA will contain the entire genome of each functionally active microbe of the community. Detailed methodology, potential and future improvements needed for the SIP technique have been already reviewed (Radajewski et al. 2003; Wellington et al. 2003; Manefield et al. 2006). Successful applications of this technique are restricted to reactions of microbial anabolism, because SIP is based on assimilatory processes (Manefield et al. 2006). Thus non-assimilatory chemical transformations, which also occur in soil, fall outside the applicability of SIP. Even if the SIP technique can theoretically be applied to trace the assimilation of any element of biological importance that has a stable isotope, it has almost exclusively been restricted to the use of ^{13}C. The labelling and detection of DNA using $H_2^{18}O$ can allow characterising active community members by using a substrate-independent technique.

The SIP was first applied to monitor microorganisms responsible for the assimilation of C from the greenhouse gas methane in a freshwater sediment environment (Boschker et al. 1998). Stable isotope labelled methane ($^{13}CH_4$) was pulsed into the microbial community of the sediment and labelled polar lipid derived fatty acids (PLFAs) from methane assimilating organisms were separated from extracted PLFAs and analysed for the ^{13}C enrichment by isotope ratio mass spectrometry (IRMS). Species belonging to the *Methylobacter* and *Methylomicrobium* genera, were the organisms responsible of methane oxidation.

Radajewski et al. (2000, 2002) were the first to use the ^{13}C labelled substrates (methane and methanol) to label nucleic acids of soil. Equilibrium density centrifugation in CsCl gradients was used to separate 'heavy' (labelled) from 'natural' (unlabelled) DNA and 16S rRNA gene clone libraries constructed from 'heavy'

DNA were sequenced to obtain the identity of organisms assimilating the used substrates. By amplifying functional genes involved in the oxidation of one carbon compound it was found that not only were methanol dehydrogenase and methane monooxygenase genes involved in methanol and methane assimilation, but also species encoding ammonia monooxygenase had assimilated ^{13}C as $^{13}CO_2$ generated by the methylotrophs (Radajewski et al. 2002). Indeed a serious drawback of the SIP approach is the possibility of secondary feeding on breakdown products of the primary substrate. Another drawback of this technique is the presence of unlabelled substrates native to the system that will compete for assimilation. This has led researchers to apply artificially high concentrations of labelled substrates into soil microcosms for extended periods of time (Manefield et al. 2006) but the relevance of this approach at the real situation in situ has been questioned and the use of pulse ^{13}C labelled compounds has been suggested (Jeon et al. 2003; Padmanabhan et al. 2003). Another important issue of any SIP methodology concerns the degree of labelling, which depends on number of organisms using the established substrate (Manefield et al. 2006). When the substrate is consumed by a broad diversity of organisms, the degree of labelling in any substrate-using taxa will be low, making separation by density problematic, whereas if the substrate is consumed by a small number of taxa then the degree of labelling in the specific taxa will be high, facilitating isolation by density (Manefield et al. 2006).

There are differences in using fatty acids or nucleic acids as biomarkers in the SIP technique; PLFAs are more rapidly labelled and give more quantitative information when analysed with IRMS than nucleic acids (Manefield et al. 2006). However, extraction of PLFAs from soil is more laborious that nucleic acid extraction. In addition, the PLFA based SIP gives an inferior phylogenetic resolution than that offered by nucleic acid based biomarkers and signature PLFAs have to be identified from close culturable relatives (Manefield et al. 2006).

Accurate results can be obtained by applying SIP to soil if the delivery of a pulse is carefully planned by considering the ability of soil colloids to adsorb biological molecules, and the solubility and volatility of the used substrates. In addition, since cell replication in soil is slow there are limitations to the isotopic enrichment of DNA with pulse labelling unless the duration of a pulse is extended. To solve this problem, 16S rRNA based SIP methodologies have been developed for studying phenol degradation in the activated sludge community of an industrial waste water treatment plant since rates of RNA synthesis are always higher than those of DNA due to the fact that RNA is turned over in bacteria independently of replication (Manefield et al. 2002a,b). The RNA-SIP technique has also been applied to soil to identify methylotrophs responsible for methanol assimilation in a rice field soil (Lueders et al. 2004a) and bacteria and archaea responsible for syntrophic propionate degradation in flooded soil (Lueders et al. 2004b). In spite of the fact that the use of RNA as a biomarker and the precise quantitative examination of gradient profiles has enhanced the sensitivity of NA-SIP, the application of the technique to rhizosphere and bulk soil still presents some drawbacks such as the ability to extract clean and intact DNA or RNA from the soil or to sufficiently label nucleic acids of microorganisms involved in the metabolism of plant root exudates (Manefield et al. 2006). Grassaland monoliths (400 mm diameter × 200 mm deep) were pulsed with $^{13}CO_2$ to promote the release of ^{13}C labelled root exudates into soil (Ostle et al. 2003) but the analysis of 16S rRNA from root associated soil by IRMS and equilibrium density centrifugation showed that the degree of labelling was too low to get meaningful results (Griffiths 2003; Manefield et al. 2006). On the other hand the PLFA-SIP analysis of soil samples derived from a $^{13}CO_2$ plant pulse showed the assimilation of root exudates by Gram-negative bacteria and fungi, but the phylogenetic resolution was low (Treonis et al. 2004). Therefore, it is problematic to use $^{13}CO_2$ to label soil microbes via root exudates because a broad range of labelled organic compounds are released as root exudates and several microbial species can use the labelled root exudates. According to Manefield et al. (2006) it can be more rewarding to pulse directly soil with labelled root exudate compounds and monitoring microorganisms of rhizosphere soil involved in the assimilation of the target compound by the use of any SIP technique. This can be done in experimental systems simulating the delivery of root exudates into soil (Badalucco and Kuikman 2001; Falchini et al. 2003).

9.5 Other Techniques Used to Link Activity to Phylogenetic Information

Radioactive isotope based methods can be used to study the link between the function and the presence of any taxa. The fluorescence in situ hybridisation (FISH) allows the phylogenetic identification of uncultured bacteria in natural environments using fluorescent group specific phylogenetic probes (targeting rRNA) and fluorescence microscopy. FISH targeting intracellular rRNA has been widely used over the past decade as a type of 16S rRNA targeting method to identify and quantify certain bacteria in soils, sediments, and activated sludge. Although FISH is useful, difficulties are often encountered when it is used on complex community samples. Problems encountered include variable probe binding to different rRNA target sites and highly autofluorescent samples or cell backgrounds. Another shortcoming of this approach is that cells must be counted visually under a microscope, which is laborious. Quantitative membrane hybridization of labeled DNA probes with bacterial community RNA can avoid these problems. By combining FISH with microautoradiography (MAR), it is possible to identify individual cells, to determine the three-dimensional position of a cell and quantify the active population utilizing a specific substrate. The combination of FISH–MAR has been used successfully to monitor the substrate utilization by bacterial groups of an activated sludge system (Lee et al. 1999; Gray and Head 2001) and in various natural environments (Gray et al. 2000). This technique might be used to study plant-microbial interactions, even if it may be difficult to apply this technique to soil. Indeed, it is necessary to extract cells from soil in order to eliminate interferences by soil particles and increase the sensitivity of the technique (Krsek et al. 2006).

Both chemical and physical methods have been used and they present drawbacks such as incomplete recovery, rupture of cells and changes in cell physiology during the separation of cells from soil particles. In addition, the FISH-MAH is a time-consuming and labor intensive technique, and because of the vast diversity of microorganisms in soil, it is almost impossible to design probes and analyse data to resolve all individual species. An advancement in the simultaneous monitoring of the diversity and substrate incorporation by complex microbial communities has been obtained by combining FISH-microautoradiography with an isotope microarray (Adamczyk et al. 2003). The RNA, extracted from ammonia-oxidizing bacteria in pure culture or in the activated-sludge samples both grown in the presence of ^{14}C bicarbonate, was fluorescence labeled and microarray hybridized. It was shown that incorporation of ^{14}C into rRNA could be detected by scanning all probe spots for fluorescence and radioactivity. Therefore, the isotope array enables the application of many probes in parallel, whereas this cannot be done with FISH-MAH, and to measure directly the substrate incorporation into target nucleic acids in contrast to DNA or RNA-SIP.

Another technique allowing detection of active microbial species in environmental samples involves incubating these samples with bromodeoxy uridine (BrdU); the BrdU in the DNA of active microorganisms can be separated from unlabelled DNA by immunocapturing. Then the labelled DNA can be characterised by profiling or cloning and sequenced (Borneman 1999; Yin et al. 2000). Unfortunately not all microbial cells are capable of taking up BrdU, and with FISH-MAH and SIP techniques the conditions of the assay can be different respect to those occurring in situ.

9.6　The Metagenome

The total number of prokaryotic cells on earth has been estimated at $4–6\times10^{30}$, thought to comprise between 10^6 and 10^8 separate genospecies (distinct taxonomic groups based on gene sequence analysis). It is widely accepted that this diversity presents an enormous (and largely untapped) genetic and biological pool that can be exploited for the recovery of novel genes, entire metabolic pathways and their products. Observations showing that culturing yields a fraction of the microbial diversity evident from microscopic analysis have been consistently supported by the results of phylotypic analyses on community DNA preparations, leading to the concept of 'unculturables'. The apparent underestimation of true microbial diversity derives largely from a reliance on culture based enumeration methods. There is a growing belief that the term 'unculturable' is inappropriate and that in reality we rather have yet to discover the correct culture conditions (Cowan et al. 2005). The development of metagenomic technologies over the past 10 years tries to overcome this bottleneck by developing and using culture-independent approaches.

One of the metagenomic approaches which generates a massive amount of data on microbial ecology is the reconstruction of the metagenome of an ecosystem using random shotgun sequencing. The basic steps of DNA library construction (generation of suitably sized DNA fragments, cloning of fragments into an appropriate vector and screening for the gene of interest) have been extensively and successfully used for over three decades. As there are no obvious limitations in translating the technologies of genomic library construction and screening to metagenomic libraries, it is perhaps surprising that metagenomics only developed in the mid-1990s with the successful application of library construction to marine metagenomes. The metagenomic approach is based on cloning the total microbial genome (the metagenome) extracted from natural environments in culturable bacteria such as *Eschericha coli* (Handelsman et al. 1998; Rondon et al. 2000). The recent development of technologies designed to access this wealth of genetic information through environmental nucleic acid extraction has provided a means of avoiding the limitations of culture-dependent genetic exploitation. Since most of bacteria living in natural environments such as soil are unculturable, the metagenomic approach can allow characterizing the unknown genome of these unculturable bacteria with the probability of finding novel microbial products such as antibiotics and enzymes and assessing specific metabolic and ecological functions (Schloss and Handelsman 2003). By virtue of their low copy number, bacterial artificial chromosome (BAC) vectors can be used to propagate large DNA fragments (around 100 kb), which would otherwise be unstable (Wellington et al. 2003). The major criticism of this technique is that it might miss members of the microbial community which are low in number but perform essential processes.

Microbial DNA extracted directly from soil has been used successfully to construct a BAC library with an average insert size of between 37 kb and 150 kb (Béjà et al. 2000). It is probably too early to state that metagenomic gene discovery is a technology that has 'come of age'. New approaches and technological innovations are reported on a regular basis and many of the technical difficulties have yet to be fully resolved. However, there can be little doubt that the field of metagenomic gene discovery offers enormous scope and potential for both fundamental microbiology and biotechnological development (Cowan et al. 2005). Approaches that enrich for a portion of the microbial community or for a collection of metagenomic clones will enhance the power of metagenomic analysis to address targeted questions in microbial ecology and to discover new biotechnological applications. Metagenomics in plant–microbial interactions has not yet been attempted, but this approach holds great promise to link phylogeny of rhizosphere microbes to function, especially when a functional gene is known.

9.7 Microarray

DNA microarrays have recently been developed for quantifying and monitoring bacteria members of communities in various environments. Sequence analysis of large insert libraries with environmental DNA combined with genetic and functional analysis has the potential to provide significant insight into the genomic potential and ecological roles of cultured and uncultured microbes. Microarrays (or microchips) technique is a powerful tool that is becoming increasingly used to monitor gene expression under different growth conditions, to detect specific mutations and to characterize microorganisms in environmental samples (Zhou 2003). Microarrays thus represent a powerful high-throughput system for analysis of genes. They are typically used to monitor differential gene expression, to quantify the environmental bacterial diversity and to catalogue genes involved in key processes. This approach identifies only those genes transcribed in the analysed community and is therefore very valuable for the identification of functional genes with relevance to the ecosystem. Compared with traditional nucleic acid hybridisations on porous membranes, glass slide based microarrays offer the additional advantage of high density, high sensitivity, rapid (real-time) detection, lower costs, automation and low background levels. The methodology, advantages and challenges in applying microarrays to environmental studies have been reviewed by Zhou and Thompson (2002). Microarrays can be classified, by considering the type of probe arrayed and the potential application, in: i) Phylogenetic Oligonucleotide Arrays (POAs), used for assessing microbial diversity and based on probes from rRNA genes; ii) Functional Gene Arrays (FGAs), used to evaluate microbial activities and based on the use of probes for gene coding proteins involved in specific functions; iii) Community Genome Arrays (CGA) containing probes of large fragments of DNA and used for detecting target genes and organisms (Schadt and Zhou 2006). The probes designed for POAs are shorter than those designed for FGAs because the latter should target protein-coding genes. In order to detect the presence of functional genes, total RNA is extracted from environmental samples, reverse transcribed and probed against microarrays (Saleh-Lakha et al. 2005; Schadt and Zhou 2006). However, the RNA extraction efficiency from environmental samples is much lower than that from pure culture of bacteria and thus signals cannot often be detected on microarrays. Microarrays have been used successfully to differentiate microbial community composition in environmental samples (Peplies et al. 2003) and to detect and quantify the functional diversity of genes involved in nitrogen cycling, such as *amoA*, *nifH*, *nirK*, *nirS* (Taroncher-Oldenburg et al. 2003) and *nirS* (Cho and Tiedje 2002) and methane oxidation (*pmoA*) (Stralis-Pavese et al. 2004). However, the application of this technique to soil microbiology is still potential.

In addition to low mRNA extraction efficiencies from environmental samples, the application of microarrays to environmental samples presents the following shortcomings: i) presence of background noise in environmental samples and false positive hybridisation signals, with problems in quantifying microbial populations from environmental samples; ii) even though this is a high throughput technology with one chip being used for a thousand probes, the current capacity of microarrays is limited with respect to the number of genes needed for monitoring the enormous diversity of the soil microbial community; iii) the hybridization of probes with target nucleic acids can be difficult due to stable secondary structure of DNA or RNA; iv) in spite of the fact that it is assumed that target sequences containing single base mismatches can be differentiated by microarray hybridization, generally discrimination can occur in the case of more than one single base mismatch; v) both humic and clay substances can interfere with hybridization; vi) the functional analysis is based on genes and pathways that have been revealed through isolates studied in the laboratory (Saleh-Lakha et al. 2005; Schadt and Zhou 2006).

Though several challenges still remain to be overcome to apply microarray technology most effectively to monitoring gene expression in complex microbial ecosystems (e.g. increasing the sensitivity), the potential and capability of this promising in situ technology appears to be highly attractive. In spite of these shortcomings, microarrays have been used to monitor functions of sulfate reducing and nitrate-fixing bacteria from soil (Loy et al. 2004; Stewart et al. 2004), for detecting genes involved in the biodegradation processes (Rhee et al. 2004) and methanotrophs in the environment (Bodrossy et al. 2006).

Avarre et al. (2007) have recently reviewed the use of DNA microarrays for detection and identification of bacteria and genes of interest from various environments. So far, most of the genomic methods that have been described rely on the use of taxonomic markers (such as 16 S rRNA) that can easily be amplified by PCR prior to hybridization on microarrays. However, taxonomical markers are not always informative on the functions present in these bacteria. Moreover, genes for which sequence database is limited or that lack any conserved regions will be difficult to amplify and thus to detect in unknown samples. Furthermore, PCR amplification often introduces biases that lead to inaccurate analysis of microbial communities. An alternative solution to overcome these strong limitations is to use genomic DNA (gDNA) as target for hybridisation, without prior PCR amplification (Avarre et al. 2007). Though hybridization of gDNA is already used for comparative genome hybridization or sequencing by hybridization, in addition to the high cost of tiling strategies and important data filtering, its adaptation for use in environmental research poses great challenges in terms of specificity, sensitivity and reproducibility of hybridization. This specificity problem is inevitable as long as many probes are hybridized to targets at the same time.

In view of these problems, Hoshino et al. (2007) have recently developed an RNA microarray in which total RNA from a microbial community is attached to a glass slide, and specific rRNAs are detected by fluorescently labeled oligonucleotide probes. In this RNA array method, hundreds of RNA samples can be analysed at the same time, and direct attachment of RNA onto the slide glass can omit PCR bias, while it cannot detect a low amount of target gene as functional gene. Moreover, any specificity problems of hybridization cannot arise because only a few probes are used for hybridization. The RNA microarray requires only 4 h for hybridization and enables double staining and estimating relative abundance of rRNA.

9.8 Conclusions

The study of gene expression in the rhizosphere soil is only just starting because of the limitations in the use of most of the relative techniques, including those for characterising synthesised proteins in the rhizosphere. For this reason the link

between the composition of the microbial community and the biochemical processes occurring in the rhizosphere is still vague. New cultivation methods continue
to be developed to improve our ability to capture a greater diversity of microorganisms within the environment. However, once these techniques are fully developed
for studying soil samples, it will be possible to assess simultaneously in a single
assay microbial diversity and most of the processes occurring in the rhizosphere by
using, for example, functional microarray techniques. This will permit a better
understanding of processes such as biological control by rhizobacteria, stimulation
of microbial activity by root exudates, competition between microorganisms and
plant roots for nutrients, molecular colloquia between microorganisms and between
plant root and microorganisms. Development of soil functional genomics will also
clarify whether the higher diversity of functional genes (and microbial species) of
the rhizosphere than bulk soil is solely linked to the greater C availability, or is the
result of a selection mechanism where plant actively select species capable of supplying nutrients.

Although microarrays can provide rapid generation of large functional datasets,
this technique has several shortcomings including the fact that the functional analysis described is based on genes revealed in isolates studied in the laboratory with
exclusion of genes of unculturable microorganisms.

The use of techniques such as SIP have permitted a better understanding of the
link between the function and microbial cells involved in the specific function.
However, these techniques not only need to be methodologically improved but also
applied in a suitable way to the rhizosphere soil. According to Manefield et al.
(2006) it can be more rewarding to use labelled root exudate compounds and monitoring microorganisms of rhizosphere soil involved in the assimilation of the target
compound by the use of any SIP technique than to pulse the whole seedlings and
then monitoring labelled nucleic acids or PLFA of microorganisms of rhizosphere
metabolising the labelled root exudates.

Reporter technology has been used to assess several functions in the rhizosphere
soil including gene expression even at the single cell level The ever increasing
knowledge of the promoter and regulator gene along with the refinement of reporter
gene insertion techniques will allow use of the reporter gene technique for monitoring induction, expression and regulation of virtually any gene in the rhizosphere. In
addition, in this case the methodological improvement of the technique will also
allow the design of new reporter bacteria to respond to specific root exudates, as it
already occurs for specific signals involved in molecular colloquia (Sørensen and
Nybroe 2006).

Nucleic acid based methods present some limitations for getting information on
functions expressed by microbial communities in situ. In this context, the large
scale study of proteins expressed by indigenous microbial communities (metaproteome) should provide information to gain insights into the functioning of the
microbial component in ecosystems. Characterization of the metaproteome is
expected to provide data linking genetic and functional diversity of microbial communities. Studies on the metaproteome together with those on the metagenome and
the metatranscriptome will contribute to progress in our knowledge of microbial

communities and their contribution in ecosystem functioning (Maron et al. 2007). Looking to increase our understanding of the role that members of a microbial community play in ecological processes, several techniques have been developed that are enabling greater in depth analysis of environmental metagenomes.

Microarrays for investigating the content of microbial communities in environmental samples are still in their infancy. One of the main reasons is the technical difficulty inherent in the hybridization of gDNA. Avarre et al. (2007) have proposed few strategies that may contribute to improve detection of gDNA with microarrays based on a simple and low-cost design, like isothermal probes, sensitivity enhancement, competitive hybridization or the use of the promising gold nanoparticles technology. Once this limitation is overcome, we can predict an explosion of microarray based investigations of microbial communities, which will eventually lead to a dramatic increase in our knowledge of the microbial genetic structure and composition of samples of high environmental and ecological value.

RNA microarray protocol developed by Hoshino et al. (2007) for determining the abundances of individual bacterial members in complex microbial communities in terms of requirement of much less RNA than conventional membrane hybridization methods is quite useful. Because such a small amount of RNA is required, sufficient signal could be obtained beginning just 1 h after hybridization, whereas overnight hybridization is required to obtain sufficient signal in membrane based methods. This RNA microarray enables hundreds of samples to be easily and rapidly processed to determine the relative abundances of individual members in complex bacteria communities. This protocol will be useful in determining the abundances of bacteria in, for example, wastewater treatment processes and bioremediation of soils.

Proteomics thus still has a great potential for the functional analysis of microbial communities of rhizosphere soil because protein expression is linked to specific microbial activities in a given ecosystem. Unfortunately this technique also has several shortcomings that prevent its successful application to soil, as unlike other cellular macromolecules, proteins exist in many different biological and physical conformations. Consequently, there is no universal extraction protocol that can be followed up to now. Despite the limited number of environmental proteomic investigations that has been carried out, it is clear that metaproteomics has huge potential in the field of environmental microbiology. It is necessary to set up experimental protocols allowing extraction of intracellular proteins without extracting extracellular proteins, the latter largely prevailing over the former in soil. Although it remains a daunting task to elucidate all of the functional proteins that are contained within an environmental sample, metaproteomics will find immediate use in studies focusing only on parts of expressed environmental proteomes. Investigations that focus on limited numbers of highly expressed proteins can have immediate impacts on developments in the field (Wilmes and Bond 2006). However, because protein expression is a reflection of specific microbial activities in a given ecosystem, proteomics has great potential for the functional analysis of microbial communities. We believe that the elucidation of metaproteomic expression will be central to functional studies of microbial consortia. Metaproteomics will gain momentum with the advent of further environmental sequencing projects and it will provide a useful tool with which to focus research direction.

References

Adamczyk J, Hesselsoe M, Iversen N, Horn M, Lehner A, Nielsen PH, Scloter M, Roslev P, Wagner M (2003) The isotope array, a new tool that employs substrate-mediated labeling of rRNA for determination of microbial community structure and function. Appl Environ Microbiol 69:6875–6887

Allaway D, Schofield NA, Leonard ME, Gilardoni L, Finan TM, Poole PS (2001) Use of differential fluorescence induction and optical trapping to isolate environmentally induced genes. Environ Microbiol 3:397–406

Amin-Hanjani S, Meikle A, Glover LA, Prosser JI, Killham K (1993) Plasmid and chromosomally encoded luminescence marker systems for detection of *Pseudomonas fluorescens* in soil. Mol Ecol 2:47–54

Andersen JB, Heydorn A, Hentzer M, Eberl L, Geisenberger O, Christensen BB, Molin S, Givskov M (2001) *gfp*-Based *N*-acyl homoserine-lactone sensor systems for detection of bacterial communication. Appl Environ Microbiol 67:575–585

Anukool U, Gaze WH, Wellington EMH (2004) In situ monitoring of streptothricin production by *Streptomyces rochei* F20 in soil and rhizosphere. Appl Environ Microbiol 70:5222–5228

Arevalo-Ferro C, Reil G, Görg A, Eberl L, Riedel K (2005) Biofilm formation of *Pseudomonas putida* IsoF: the role of quorum sensing as assessed by proteomics. Syst Appl Microbiol 28:87–114

Assmann EM, Ottoboni LM, Ferraz A, Rodriguez J, De Mello MP (2003) Iron-responsive genes of *Phanerochaete chrysosporium* isolated by differential display reverse transcription polymerase chain reaction. Environ Microbiol 9:777–786

Avarre JC, Lajudie P, Bén G (2007) Hybridization of genomic DNA to microarrays: a challenge for the analysis of environmental samples. J Microbiol Method 69:242–248

Badalucco L, Kuikman P (2001) Mineralization and immobilization in the rhizosphere. In: Pinton R, Varanini Z, Nannipieri P (eds) The rhizosphere. Biochemistry and organic substances at the soil-plant interface. Marcel Dekker, New York, pp 159–196

Bahl MI, Hestbjerg Hansen L, Rask Licht T, Sorensen SJ (2004) In vivo detection and quantification of tetracycline by use of a whole-cell biosensor in the rat intestine. Antimicrob Agents Chemother 48:1112–1117

Bakken LR, Frostegård A (2006) Nucleic acid extraction from soil. In: Nannipieri P, Smalla K (eds) Nucleic acids and proteins in soil. Springer, Berlin Heidelberg New York, pp 49–73

Bechor O, Smulski DR, Van Dyk TK, LaRossa RA, Belkin S (2002) Recombinant microorganisms as environmental biosensors: pollutants detection by *Escherichia coli* bearing *fab*A:*lux* fusions. J Biotechnol 94:125–132

Béjà O, Suzuki MT, Koonin EV, Aravind L, Hadd A, Nguyen LP, Villacorta R, Amjadi M, Garrigues C, Jovanovich SB, Feldman RA, DeLong EF (2000) Construction and analysis of bacterial artificial chromosome libraries from a marine microbial assemblage. Environ Microbiol 2:516–529

Bever JD (2003) Soil community feedback and the coexistence of competitors: conceptual framework and empirical tests. New Phytol 157:465–473

Bhagwat AA, Keister DL (1992) Identification and cloning of *Bradyrhizobium japonicum* genes expressed strain selectively in soil and rhizosphere. Appl Environ Microbiol 58:1490–1495

Blum SAE, Lorenz MG, Wackernagel W (1997) Mechanisms of retarded DNA degradation and prokaryotic origin of DNases in non-sterile soil. Syst Appl Microbiol 20:513–521

Bodrossy L, Stralis-Pavese N, Konrad-Köszler M, Weilharter A, Reichenauer TG, Schöfer D, Sessitsch A (2006) mRNA-based parallel detection of active methanotroph populations by use of a diagnostic microarray. Appl Environ Microbiol 72:1672–1676

Bogan BW, Schoenike B, Lamar RT, Cullen D (1996a) Manganese peroxidase mRNA and enzyme activity levels during bioremediation of polycyclic hydrocarbon-contaminated soil with *Phanerochaete chrysosporium*. Appl Environ Microbiol 62:2381–2386

Bogan BW, Schoenike B, Lamar RT, Cullen D (1996b) Expression of lip genes during growth in soil and oxidation of anthracene by *Phanerochaete chrysosporium*. Appl Environ Microbiol 62:3697–3703

Bolanos Vasquez MC, Warner D (1997) Effects of *Rhizobium tropici*, *R. etli*, and *R. leguminosarum* bv *phaseoli* on *nod* gene-inducing flavonoids in root exudates of *Phaseolus vulgaris*. Mol Plant-Microbe Interact 10:339–346

Bollag J-M, Chen C-M, Sarkar JM, Loll MJ (1987) Extraction and purification of a peroxidase from soil. Soil Biol Biochem 19:61–67

Borneman J (1999) Culture-independent identification of microorganisms that respond to specified stimuli. Appl Environ Microbiol 65:3398–3400

Boschker HTS, Nold SC, Wellsbury P, Bos D, de Graaf W, Pel R, Parkes RJ, Cappenberg TE (1998) Direct linking of microbial populations to specific biogeochemical cycles by ^{13}C-labeling of biomarkers. Nature 392:801–805

Briones AM, Okabe S, Umemiya Y, Ramsing NB, Reichardt W, Okuyama H (2003) Ammonia-oxidising bacteria on root biofilms and their possible contribution to N use efficiency of different rice cultivars. Plant Soil 250:335–348

Burmölle M, Hestbjerg Hansen L, Sørensen JS (2006) Reporter gene technology in soil ecology; detection of bioavailability and microbial interactions. In: Nannipieri P, Smalla K (eds) Nucleic acids and proteins in soil. Springer, Berlin Heidelberg New York, pp 397–419

Ceccherini MT, Potè J, Kay E, Tran Van V, Maréchal J, Pietramellara G, Nannipieri P, Vogel TM, Simonet P (2003) Degradation and transformability of DNA from transgenic leaves. Appl Environ Microbiol 69:673–678

Cho JC, Tiedje JM (2002) Quantitative detection of microbial genes by using DNA microrrays. Appl Environ Microbiol 68:1425–1430

Cho Y, Qiu YL, Kuhlman P, Palmer JD (1998). Explosive invasion of plant mitochondria by a group I intron. Proc Nat Acad Sci 95:14244–14249

Cocking EC (2003) Endophytic colonisation of plant roots by nitrogen-fixing bacteria. Plant Soil 252:169–175

Cowan D, Meyer Q, Stafford W, Muyanga S, Cameron R, Wittwer P (2005) Metagenomic gene discovery: past, present and future. TRENDS Biotechnol 23:321–329

Crowley D (2001) Function of siderophores in the plant rhizosphere. In: Pinton R, Varanini Z, Nannipieri P (eds) The rhizosphere. Biochemistry and organic substances at the soil-plant interface. Marcel Dekker, New York, pp 223–261

Darwent MJ, Paterson E, McDonald AJS, Tomos AD (2003) Biosensor reporting of root exudation from Hordeum vulgare in relation to shoot nitrate concentration. J Exp Bot 54: 325–334

DeAngelis KM, Ji P, Firestone MK, Lindow SE (2005) Two novel bacterial biosensors for detection of nitrate availability in the rhizosphere. Appl Environ Microbiol 71:8537–8547

Dekkers LC, Muelders IH, Phoelich CC, Chin-A-Woeng TFC, Wijfjes AH, Lugtenberg BJJ (1998a) The *sss* colonization gene of the tomato-*Fusarium oxysporum*f. Sp. *Radicis-lycopersici* biocontrol strain *Pseudomonas fluorescen* WCS365 can improve root colonization of other wild-type *Pseudomonas* spp bacteria. Mol Plant Microbe Interact 13:1177–1183

Dekkers LC, Phoelich CC, Van der Fits L, Lugtenberg BJJ (1998b) A site-specific recombinase is required for competitive root colonization by *Pseudomonas fluorescens* WCS365 Proc Natl Acad Sci 95:7051–7056

DeLong EF (2002) Microbial population genomics and ecology. Curr Opin Microbiol 5:520–524

Desbrow C, Routledge EJ, Brighty GC, Sumpter JP, Waldock M (1998) Identification of estrogenic chemicals in STW effluent. 1. Chemical fractionation and *in vitro* biological screening. Environ Sci Technol 34:1548–1558

Espinosa-Urgel M (2004) Plant-associated *Pseudomonas* populations: molecular biology, DNA dynamcs, and gene transfer. Plasmid 52:139–150

Espinosa-Urgel M, Ramos JL (2001) Expression of a *Pseudomonas putida* involved in lysine metabolism is induced in the rhizosphere. Appl Environ Microbiol 67:5219–5224

Falchini L, Naumova N, Kuikman PJ, Bloem J, Nannipieri P (2003) CO_2 evolution and denaturing gradient gel electrophoresis profiles of bacterial communities in soil following addition of low molecular weight substrates to simulate root exudation. Soil Biol Biochem 36:775–782

Fleming JT, Sanseverino J, Sayler GS (1993) Quantitative relationship between naphthalene catabolic gene frequency and expression in predicting PAH degradation in soil at town gas manufacturing sites. Environ Sci Technol 27:1068–1074

Gebhard F, Smalla K (1999). Monitoring field releases of genetically modified sugar beets for persistence of transgenic plant DNA and horizontal gene transfer. FEMS Microbiol Ecol 28:261–272

Graves PR, Haystead TAJ (2002) Molecular biologist's guide to proteomics. Microbiol Mol Biol Rev 66:39–63

Gray ND, Head IM (2001) Linking genetic identity and function in communities of uncultured bacteria. Environ Microbiol 3:481–492

Gray ND, Howarth R, Pickup RW, Gwyn Jones J, Head IM (2000) Use of combined microoautoradiography and fluorescence in situ hybridization to determine carbon metabolism in mixed natural communities of uncultured bacteria from the genus *Achromatium*. Appl Environ Microbiol 66:4518–4522

Griffiths RI (2003) Carbon flux and soil bacteria: the correlation with diversity and perturbation. PhD Thesis, The University of Newcastle

Hall-Stoodley L, Costerton JW, Stoodley P (2004) Bacterial biofilms: from the natural environment to infectious diseases. Nat Rev 2:95–108

Handelsman J, Rondon MR, Brady SP, Clady J, Goodman RM (1998) Molecular biological access to the chemistry of unknown soil microbe: a new frontier for natural products. Chem Biol 5R:245–249

Hoshino T, Furukawa K, Tsuneda S, Inamori Y (2007) RNA microarray for estimating relative abundance of 16S rRNA in microbial communities. J Microbiol Meth 69:406–410

Jaeger CH III, Lindow SE, Miller W, Clark E, Firestone MK (1999) Mapping of sugar and amino acid availability in soil around roots with bacterial sensors of sucrose and tryptophan. Appl Environ Microbiol 65:2685–2690

Jain, R, Rivera MC, Lake JA (1999) Horizontal gene transfer among genomes: the complexity hypothesis. Proc Nat Acad Sci USA 96:3801–3806

Janse BJH, GaskellJ, Aktar M, Cullen D (1998) Expression of *Phanerochaete chrysosporium* genes encoding lignin peroxidase, manganese peroxidase, and glyoxal oxidase in wood. Appl Environ Microbiol 64:3536–3538

Jeon CO, Park W, Padmanabhan P, DeRito C, Snape JR, Madsen EL (2003) Discovery of a bacterium, with distinctive dioxygenase, that is responsible for in situ biodegradation in contaminated sediment. Proc Natl Acad Sci USA 100:13591–13596

Kowalchuk GA, Drigo B, Yergeau E, Van Veen JA (2006) Assessing bacterial and fungal community structure in soil using ribosomal RNA and other structural gene markers. In: Nannipieri P, Smalla K (eds) Nucleic acids and proteins in soil. Springer, Berlin Heidelberg New York, pp 159–188

Kragelund L, Christofersen B, Nybroe O, de Bruijn FJ (1995) Isolation of lux reporter gene fusions in *Pseudomonas fluorescens* DF57 inducible by nitrogen or phosphorus starvation. FEMS Microbiol Ecol 17:95–106

Kragelund L, Hosbond C, Nybroe O (1997) Distribution of metabolic activity and phosphate starvation response of *lux*-tagged *Pseudomonas fluorescens* reporter bacteria in the barley rhizosphere. Appl Environ Microbiol 63:4920–4928

Krsek M, Gaze WH, Morris NZ, Wellington EMH (2006) Gene detection, expression and related enzyme activity in soil. In: Nannipieri P, Smalla K (eds) Nucleic acids and proetins in soil. Springer, Berlin Heidelberg New York, pp 217–255

Kuiper I, Bloemberg GV, Noreen S, Thomas-Oates JE, Lugtenberg BJJ (2001) Increased uptake of putrescine in the rhizosphere inhibits competitive root colonization by *Pseudomonas fluorescens* strain WCS365. Mol Plant Microbe Interact 14:1096–1104

Lambert B, Leyns F, Van Rooyen L, Gosele F, Papon Y, Swings J (1987) Rhizobacteria of maize and their antifungal activities. Appl Environ Microbiol 53 1866–1871

Lambert B, Meire P, Joos H, Lens P, Swings J (1990) Fast-growing, aerobic, heterotrophic bacteria from the rhizosphere of young sugar beet plants Appl Environ Microbiol 56:3375–3381

Lee N, Nielsen PH, Andreasen KH, Juretschko S, Nielsen JL, Schleifer KH, Wagner M (1999) Combination of fluorescent in situ hybridization and microautoradiography – a new tool for structure-function analyses in microbial ecology. Appl Environ Microbiol 65:1289–1297

Lee SW, Cooksey DA (2000) Genes expressed in *Pseudomonas putida* during colonization of a plant-pathogenic fungus. Appl Environ Microbiol 66:2764–2772

Liley AK, Bailey MJ (1997) Impact of pQBR103 acquisition and carriage on the phytosphere fitness of *Pseudomonas fluorescens* SBW25: burden and benefit. Appl Environ Microbiol 63:1584–1587

Loper JE, Henkels MD (1997) Availability of iron to *Pseudomonas fluorescens* in rhizosphere and bulk soil evaluated with an ice nucleation reporter gene. Appl Environ Microbiol 63:99–105

Loy A, Kusel K, Lehner A, Drake HL, Wgner M (2004) Microrray and functional gene analysis of sulfate reducing prokaryote in low sulfate acidic fens reveal coocurrence of recognized-general and novel linages. Appl Environ Microbiol 70:6998–7009

Lueders T, Wagner B, Claus P, Friedrich MW (2004a) Stable isotope probing of rRNA and DNA reveals a dynamic methylotroph community and trophic interactions with fungi and protozoa in oxic rice field soil. Environ Microbiol 6:60–72

Lueders T, Pommerenke B, Friedrich MW (2004b) Stable isotope probing of microorganisms thriving at thermodynamic limits: syntrophic propionate oxidation in flooded soil. Appl Environ Microbiol 70:5778–5786

Lynch JM, Benedetti A, Insam H, Nuti PM, Smalla K, Torsvik V, Nannipieri P (2004) Microbial diversity in soil: ecological theories, the contribution of molecular techniques and the impact of transgenic plants and transgenic microorganisms. Biol Fertil Soils 40:363–385

Manefield M, Whiteley AS, Ostle N, Ineson P, Bailey MJ (2002a) Technical considerations for RNA-based stable isotope probing: an approach to associating microbial diversity with microbial community function. Rapid Commun Mass Spectrom 16:2179–2183

Manefield M, Whiteley AS, Griffiths RJ, Bailey MJ (2002b) RNA stable isotope probing, a novel means of linking microbial community function to phylogeny. Appl Environ Microbiol 68:5367–5373

Manefield M, Griffiths RI, Whiteley A, Bailey M (2006) Stable isotope probing: a critique of its role in linking phylogeny and function. In: Nannipieri P, Smalla K (eds) Nucleic acids and proteins in soil. Springer, Berlin Heidelberg New York, pp 205–255

Marco ML, Legac J, Lindow SE (2003) Conditional survival as a selection strategy to identify plant-inducible genes of *Pseudomonas syringae*. Appl Environ Microbiol 69:5793–5801

Maron PA, Ranjard L, Mougel C, Lemanceau P (2007) Metaproteomics: a new approach for studying functional microbial ecology. Microb Ecol 53:486–493

Marschner P, Crowley DE (1996) Physiological activity of a bioluminescent *Pseudomonas fluorescens* (strain 2–79) in the rhizosphere of mycorrhizal and non-mycorrhizal pepper (*Capsicum annuum* L.). Soil Biol Biochem 28:869–876

Marschner P, Crowley DE (1997) Iron stress and pyoverdin production by a fluorescent pseudomonad in the rhizosphere of white lupine (*Lupinus albus* L.) and barley (*Hordeum vulgare* L. Appl Environ Microbiol 63:277

Marschner P, Crowley DE (1998) Phytosiderophores decrease iron stress and pyoverdine production of *Pseudomonas fluorescens* PF-5 (PVD-INAZ), Soil Biol Biochem 30:1275–1280

Mercier A, Kay E, Simonet P (2006) Horizontal gene transfer by natural transformation in soil environment. In: Nannipieri P, Smalla K (eds) Nucleic acids and proteins in soil. Springer, Berlin Heidelberg New York, pp 355–373

Metcalfe AC, Krsek M, Gooday GW, Prosser JI, Wellington EM (2002) Molecular analysis of a bacterial chitinolytic community in an upland pasture. Appl Environ Microbiol 68:5042–5050

Nannipieri P (2006) Role of stabilised enzymes in microbial ecology and enzyme extraction from soil with potential applications in soil proteomics. In: Nannipieri P, Smalla K (eds) Nucleic acids and proteins in soil. Springer, Berlin Heidelberg New York, pp 75–94

Nannipieri P, Ascher J, Ceccherini MT, Landi L, Pietramellara G, Renella G (2003) Microbial diversity and soil functions. Eur J Soil Sci 54:655–670

Nielsen KM, Bones AM, Smalla K, van Elsas JD (1998) Horizontal gene transfer from transgenic plants to terrestrial bacteria – a rare event? FEMS Microbiol Rev 22:79–103

Nielsen KM, Smalla K, van Elsas JD (2000) Natural transformation of *Acinetobacter* sp. Strain BD413 with cell lysates of *Acinetobacter* sp., *Pseudomonas fluorescens*, and *Burkholderia cepacia* in soil microcosms. Appl Environ Microbiol 66:206–212

Nielsen KM, van Weerelt DM, Berg TN, Bones AM, Hageler AN, van Elsas JD (1997) Natural transformation and availability of transforming chromosomal DNA to *Acinetobacter calcoaceticus* in soil microcosms. Appl Environ Microbiol 63:1945–1952

Nierlich DP, Murakawa GJ (1996) The decay of bacterial messanger RNA. Prog Nucl Acids Res Mol Biol 52:153–216

O'Toole GA, Kaplan HB, Kolter R (2000) Biofilm formation as microbial development. Ann Rev Microbiol 54:49–79

Ogunseitan OA (2006) Soil proteomics: extraction and analysis of proteins from soil. In: Nannipieri P, Smalla K (eds) Nucleic acids and proteins in soil. Springer, Berlin Heidelberg New York, pp 95–115

Ostle N, Whiteley AS, Bailey MJ, Sleep D, Ineson P, Manefield M (2003) Active microbial RNA turnover in a grassland soil estimated using a $^{13}CO_2$ spike. Soil Biol Biochem 35:877–885

Padmanabhan P, Padmanabhan S, DeRito C, Gray A, Gannon D, Snape JR, Tsai CS, Park W, Jeon C, Madsen EL (2003) Respiration of ^{13}C-labeled substrates added to soil in the field and subsequent 16S rRNA gene analysis of ^{13}C-labeled soil DNA. Appl Environ Microbiol 69:1614–1622

Paget E, Jocteur-Monrozier L, Simonet P (1992) Adsorption of DNA on clay minerals: protection against DNase I and influence on gene transfer. FEMS Microbiol Lett 97:31–40

Peplies J, Glöckner FO, Amann R (2003) Optimization strategies for DNA microarray based detection of bacteria with 16S rRNA-targeting oligonucleotide probes. Appl Environ Microbiol 69:1397–1407

Pietramellara G, Dal Canto L, Vettori C, Gallori E, Nannipieri P (1997) Effects of air-drying and wetting cycles on the transforming ability of DNA bound on clay minerals. Soil Biol Biochem 29:55–61

Ptitsyn LR, Horneck G, Komova O, Kozubek S, Krasavin EA, Bonev M, Rettberg P (1997) A biosensor for environmental genotoxin screening based on an SOS lux assay in recombinnt *Escherichia coli* cells. Appl Environ Microbiol 63:4377–4384

Radajewski S, Ineson P, Parekh NR, Murrell JC (2000) Stable-isotope probing as a tool in microbial ecology. Nature 403:646–649

Radajewski S, Webster G, Reay DS, Morris SA, Ineson P, Nedwell DB, Prosser JI, Murrell JC (2002) Identification of active methylotroph populations in an acidic forest soil by stable-isotope probing. Microbiology 148:2331–2342

Radajewski S, McDonald IR, Murrell JC (2003) Stable-isotope probing of nucleic acids: a window to the function of uncultured microorganisms. Curr Opin Biotechnol 14:296–302

Rainey PB (1999) Adaptation of *Pseudomonas fluorescens* to the plant rhizosphere. Environ Microbiol 1:243–257

Ramos C, Mølbak L, Molin S (2000) Bacterial activity in the rhizosphere analysed at the single-cell level by monitoring ribosome contents and synthesis rates. Appl Environ Microbiol 66:801–809

Rhee S-K, Liu X, Wu L, Song CC, Wan X, Zhou J (2004) Detection of genes involved in biodegradation and biotransformation in microbial communities by using 50-mer oligonucleotide microarrays. Appl Environ Microbiol 70:4303–4317

Roca C, Olson L (2001) Dynamic responses of *Pseudomonas fluorescens* DF57 to nitrogen or carbon source addition. J Biotechnol 86:39–50

Rondon MR, August PR, Bettermann AD, Brady SF, Grossman TH, Liles MR, Loiacono KA, Lynch BA, Mac Neil IA, Minor C, Tiong CI, Gilman M, Osburne MS, Clardy J, Handelsman J,

Goodman RM (2000) Cloning the soil metagenome: a strategy for accessing the genetic and functional diversity of uncultured microorganisms. Appl Environ Microbiol 66:2541–2547

Saleh-Lakha S, Miller M, Campbell RG, Schneider K, Elahimanesh P, Hart MM, Trevors JT (2005) Microbial gene expression in soil: methods, applications and challenges. J Microbiol Methods 63:1–9

Sanseverino J, Werner C, Fleming J, Applegate B, King JM, Sayler GS (1993–1994) Molecular diagnostics of polycyclic aromatic hydrocarbons biodegradation in manufactured gas plant soils. Biodegradation 4:303–321

Sarkar N (1997) Polyadenylation of mRNA in prokaryotes. Annu Rev Biochem 66:173–197

Schadt CW, Zhou J (2006) Advances in microarray-based technologies for soil microbial community analyses. In: Nannipieri P, Smalla K (eds) Nucleic acids and proteins in soil. Springer, Berlin Heidelberg New York, pp 189–203

Schloss D, Handelsman J (2003) Biotechnological prospects from metagenomics. Curr Opin Biotechnol 14:303–310

Schulze WX, Gleixner G, Kaiser K, Guggenberger G, Mann M, Schulze E-D (2005) A proteomic fingerprinting of dissolved organic carbon and of soil particles. Oecologia 142:335–343

Schwaner NE, Kroer N (2001) Effect of plant species in the kinetics of conjugal transfer in the rhizosphere and relation to bacterial metabolic activity. Microb Ecol 42:458–465

Smith LM, Tola E, deBoer P, O'Gara F (1999) Signalling by the fungus *Pythium ultimum* represses expression of two ribosomal RNA operons with key roles in the rhizosphere ecology of *Pseudomonas fluorescens* F113. Environ Microbiol 1: 495–502

Sørensen S, Nybroe O (2006) Reporter genes in bacterial inoculants can monitor life conditions and functions in soil. In: Nannipieri P, Smalla K (eds) Nucleic acids and proteins in soil. Springer, Berlin Heidelberg New York, pp 375–395

Standing D, Meharg AA, Killham K (2003) A tripartite microbial reporter gene system for real-time assays of soil nutrient status. FEMS Microbiol Lett 220:35–39

Steidle A, Sigl K, Schuhegger R, Ihring A, Schmid M, Gantner S, Stoffels M, Riedel K, Givskov M, Hartmann A, Langebartels C, Eberl L (2001) Visualization of *N*-acylhomoserine lactone-mediated cell-cell communication between bacteria colonizing the tomato rhizosphere, Appl Environ Microbiol 67:5761–5770

Stevenson FJ (1986) Cycles of soil. Carbon, nitrogen, phosphorus, sulfur and micronutrients. Wiley, New York

Stewart GF, Bettany D, Bees J, Ward B, Zehr JP (2004) Development and testing of a DNA microarray to assess nitrogenase (*nif*H) gene diversity. Appl Environ Microbiol 70:455–465

Stewart P, Cullen D (1999) Organization and differential regulation of a cluster of lignin peroxidase genes of *Phanerochaete chrysosporium*. J Bacteriol 181:3427–3432

Stralis-Pavese N, Sessitsch A, Weilharter A, Reichenauer T, Riesing J, Csontos J, Murrell JC, Bodrossy L (2004) Optimisation of diagnostic microarray for application in analysing landfill methanotroph communities under different plant covers. Environ Microbiol 6:347–363

Taroncher-Oldenburg G, Griner EM, Francis CA, Ward BB (2003) Oligonucleotide microarray for the study of functional gene diversity in the nitrogen cycle in the environment. Appl Environ Microbiol 69:1159–1171

Timms-Wilson TM, Ellis RJ, Bailey MJ (2000) Immuno-capture differential display method (IDDM) for the detection of environmentally induced promoters in rhizobacteria. J Microbiol Methods 41:77–84

Treonis AM, Ostle NJ, Stott AW, Primrose R, Grayston SJ, Ineson P (2004) Identification of groups of metabolically-active rhizosphere microorganisms by stable isotope probing of PLFAs. Soil Biol Biochem 36:533–537

Utt EA, Brousal JP, Kikut-Oshima LC, Quinn FD (1995) The identification of bacterial gene expression differences using mRNA-based subtractive hybridisation. Can J Microbiol 41:152–156

van Dyk TK, Smulski DR, Reed TR, Belkin S, Vollmer AC, LaRossa R (1995) Responses to toxicants of an *Escherichia coli* strain carrying a *usp*A:*lux* genetic fusion and an *E. coli* strain carrying a *grp*E:*lux* fusion are similar. Appl Environ Microbiol 61:4124–4127

Van Elsas JD, Turner S, Bailey MJ (2003) Horizontal gene transfer in the phytosphere. New Phytol 157:525–537

Veal DA, Stokes HW, Daggard G (1992) Genetic exchange in natural microbial communities. Adv Microbiol Ecol 12:383–430

Vollmer AC, Belkin S, Smulski DR, Van Dyk TK, LaRossa RA (1997) Detection of DNA damage by use of *Escherichia coli* carrying *rec*A:*lux*, *uvr*A:*lux*, or *alk*A:*lux* reporter plasmids. Appl Environ Microbiol 63:2566–2571

Wagner M (2004) Deciphering functions of uncultured microorganisms. ASM News 70:63–70

Wellington EMH, Berry A, Krsek M (2003) Resolving functional diversity in relation to microbial community structure in soil: exploiting genomics and stable isotope probing. Curr Opin Microbiol 6:295–301

Wilmes P, Bond PL (2006) Metaproteomics: studying functional gene expression in microbial ecosystems. Trends Microbiol 14:92–97

Wren BW (2000) Microbial genome analysis: insights into virulence, host adaptation and evolution. Nat Genet 1:30–39

Yeomans C, Porteous F, Paterson E, Meharg AA, Killham K (1999) Assessment of *lux*-marked *Pseudomonas fluorescens* for reporting on organic carbon compounds. FEMS Microbiol Lett 176:79–83

Yin B, Crowley D, Sparovek G, De Melo WJ, Borneman J (2000) Bacterial functional redundancy along a soil reclamation gradient. Appl Environ Microbiol 66:4361–4365

Zhang ZG, Pierson LS (2001) A second quorum sensing system regulates cell surface properties but not phenazine antibiotic production in *Pseudomonas aureofaciens*. Appl Environ Microbiol 67:4305–4315

Zhou J (2003) Microarrays for bacterial detection and microbial community analysis. Curr Opin Microbiol 6:1–7

Zhou J-Z, Thompson D (2002) Microarrays: applications in environmental microbiology. In: Bitton G (ed) Encyclopedia of environmental microbiology, vol 4. Wiley, New York, pp 1968–1979

Chapter 10
Molecular Mechanisms of Biocontrol by *Trichoderma* spp.

P.K. Mukherjee(✉), C.S. Nautiyal, and A.N. Mukhopadhyay(✉)

10.1   Introduction

Trichoderma spp. are ubiquitous soil fungi. By virtue of their ability to decompose organic matter, they are free-living in soil as saprophytes. However, these species also have the capability to live on other fungi, and the ability to colonize plant roots and rhizosphere. *Trichoderma* spp. produce a range of hydrolytic enzymes that make them useful in industry (Mach and Zeilinger 2003). These fungi are capable of parasitizing some plant pathogenic fungi that makes them useful as biofungicides (Mukhopadhyay et al. 1992; Chet et al. 1998; Mukhopadhyay and Mukherjee 1996; Harman and Bjorkmann 1998; Hjeljord and Tronsmo 1988) (Fig. 10.1).

Trichoderma spp. produce various kinds of secondary metabolites in abundance, including antibacterial and antifungal antibiotics (Sivasithamparam and Ghisalberti 1998). Some of the species/strains are reported to be plant growth promoters and inducers of systemic resistance in plants (Harman et al. 2004). Faster metabolic rates, anti-microbial metabolites, and physiological conformation are key factors which chiefly contribute to antagonism of these fungi. Mycoparasitism, spatial and nutrient competition, antibiosis by enzymes and secondary metabolites, and induction of plant defence system are typical biocontrol actions of these fungi. On the other hand, *Trichoderma* spp. have also been used in a wide range of commercial enzyme productions, namely, cellulases, hemicellulases, proteases, and ß-1,3-glucanase. Information on the classification of the genus, *Trichoderma*, mechanisms of antagonism and role in plant growth promotion has been well documented. All these qualities have made *Trichoderma* spp. popular in industry as sources of enzymes, and in agriculture as biofungicides/growth promoters. Even though several commercial formulations based on *Trichoderma* are available in the world

P.K. Mukherjee
Nuclear Agriculture and Biotechnology Division, Bhabha Atomic Research Centre, Mumbai, India

A.N. Mukhopadhyay
Sangini, 151 Akansha, Udhay-II, Raibareilly Road, Lucknow, India
e-mail: *mukhopadhyayamar@indiatimes.com*

C.S. Nautiyal, P. Dion (eds.) *Molecular Mechanisms of Plant and Microbe Coexistence.* Soil Biology 15, DOI: 10.1007/978-3-540-75575-3

243

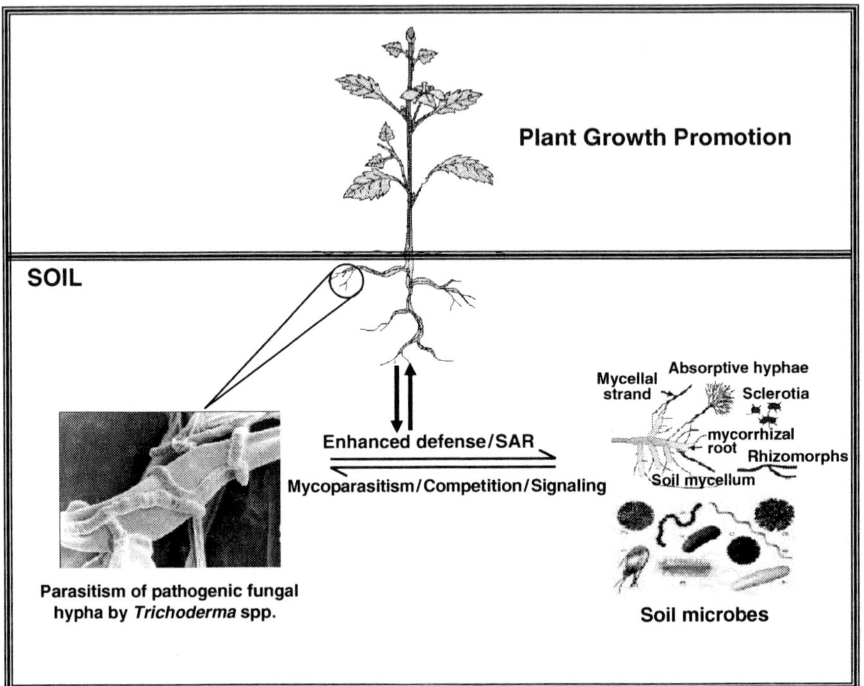

Fig. 10.1 Parasitization of *Rhizoctonia solani* hypha by *Trichoderma virens*. **A** chemotropical attraction of T, virens towards *R. solani* and formation of appressorium. **B** Parasitization of *R. solani* hypha by *T. virens*. Note intensive coiling of the *R. solani* by *T. virens*. **C** Rupture of R. solani hypha after intensive coiling. **D** Lysis of *R. solani* after parasitization by *T. virens*

market for use as biofungicides, their efficacies, in most cases, are not comparable with those of chemical pesticides. This is expected, because *Trichoderma* spp. are living entities, the activity and survival of which are dependent on biotic and abiotic environmental factors. This limitation has been overcome, to some extent, by combining *Trichoderma* spp. with chemical fungicides in the form of an integrated plant disease management (Mukhopadhyay et al. 1992). However, the growing demand for a ban on many chemical fungicides is likely to make the issue complicated, unless we have strains of *Trichoderma* with improved biocontrol potential, as well as survival under ability adverse environmental conditions. The genus *Trichoderma* is able to colonize every different niches because of its metabolic versatility and to its tolerance to stress conditions. These properties make *Trichoderma* a widespread biocontrol agent for management of plant diseases. Studies concerning this phenomenon have mainly focused on characterization of actively attacking processes (i.e. lytic enzyme production), but defense mechanisms by which *Trichoderma* tolerate biotic and abiotic stresses have been poorly addressed. One of the most interesting aspects of the science of biocontrol is the study of the mechanisms employed by the biocontrol agents to effect disease control (Howell 2003). Understanding the mechanisms of biocontrol at the molecular level would be useful to improve the

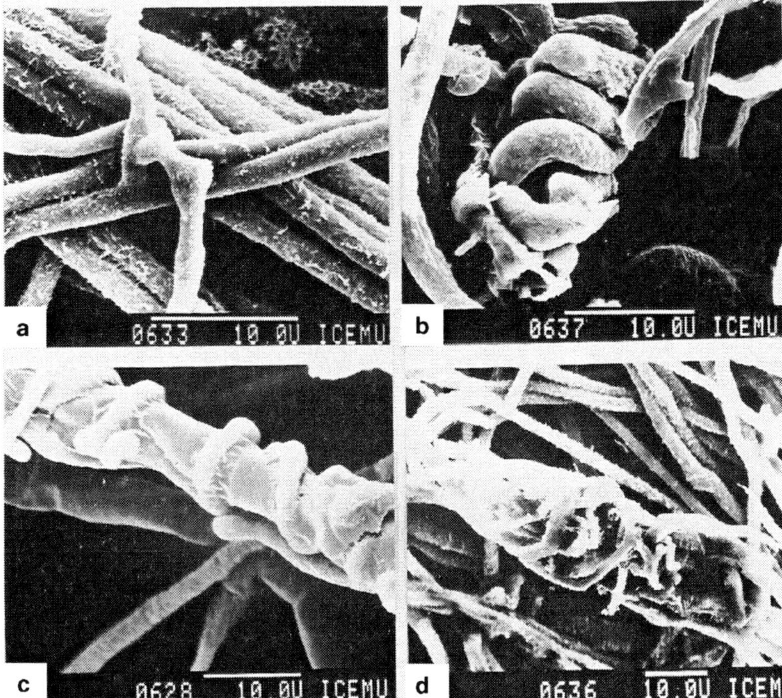

Fig. 10.2 Mechanisms of plant growth promotion by *Trichoderma* spp

potential of *Trichoderma* spp. as biocontrol agents. This is an emerging field; however, some outstanding work has been done in the recent past on understanding how *Trichoderma* spp. work against the pathogens, both directly and indirectly, at the molecular level (Fig. 10.2). This review is intended to consider recent developments, which we consider are just the beginning, in understanding the molecular mechanisms of biocontrol by *Trichoderma* spp.

10.2   Mechanisms of Biocontrol: An Overview

Classically, three principal mechanisms of action of *Trichoderma* spp. have been recognized – mycoparasitism (parasitism of one fungus by another fungus), antibiosis (production of antimicrobial metabolites, and thus inhibiting other fungi) and the universal phenomenon of competition for food, space or oxygen. In a typical mycoparasitic interaction, the parasite (e.g., *Trichoderma*) receives the chemical stimulus released by the host (e.g., *Rhizoctonia solani*) and gets chemotropically attracted towards the host. This is followed by coiling of the host hyphae, running adpressed to the host, production of appressoria-like structures, penetration of the host and derivation of nutrients, and finally, lysis of the host.

The work from the Ilan Chet group at the Hebrew University of Jerusalem, Israel, established the role of recognition by a biomimetic experiment where nylon fibres coated with lectins were coiled by *Trichoderma harzianum*, but not the fibres that have not been coated (Inbar and Chet 1992). This experiment underlined the possible role of signal interplay in this novel host-parasite interaction. It has been rather ironic that the role of parasitism of the surviving structures in biocontrol has been largely neglected. Since mycoparasitism and antibiosis are easy to assay, plenty of work, mostly in vitro, has been done describing how *Trichoderma* spp. kill other fungi through mycoparasitism and what the enzymes are which are secreted during the act of mycoparasitism. However, direct evidence for the exact role of mycoparasitism in in vivo biocontrol is rare. Howell (1987) generated a mutant of *T. virens* deficient in mycoparasitism (hyphal coiling); the mutant was as effective as the parental strain in control of *R. solani* in cotton. He thus questioned the role of mycoparasitism in biocontrol in this system. Mukherjee et al. (1995b), by comparison of an isolate of both *T. harzianum* and *T. virens*, postulated that parasitism of the sclerotia, rather than hyphal coiling or antibiosis, is the principal mechanism of biocontrol of *S. rolfsii* and *R. solani*, when *Trichoderma* is applied to soil. Similarly, the role of antibiosis has been established through the analysis of mutants deficient for the production of a particular antibiotic substance. Based on the biosynthesis of either gliotoxin (Q) or gliovirin (P), Howell et al. (1993) classified *T.* virens strains into two groups; "Q" groups were effective against *R. solani* and "P" groups against *Pythium ultimum*. Later, using gliotoxin or gliovirin-deficient mutants, gliotoxin was implicated to be involved in biocontrol of *Pythium* damping-off (Wilhite et al. 1994), but not for *R. solani* in cotton (Howell and Stipanovic 1995). In other studies (Howell et al. 2000; Howell 2002), a mutant of *T. virens* deficient for both mycoparasitism and gliotoxin biosynthesis still retained the biocontrol potential against *P. ultimum* and *R. solani*. Taken together, these experiments failed to establish whether mycoparasitism or antibiosis is the principal mechanism that effects biocontrol. The phenomenon of competition is universal and it is difficult to assay for its role in biocontrol. Nevertheless, pre-colonization of the spermosphere/rhizosphere, and thus pre-emptying the possibility of a subsequent colonization by the pathogen could be a strong factor responsible for bringing down the infection level. The rapid colonization of the dead/necrotic tissues by *Trichoderma*, thus preventing further spread, has been demonstrated by Mukherjee et al. (1995a) in the case of foliar application of *T. viride* against *B. cinerea* in chickpea. It is very likely that more than one mechanism is involved, and the biocontrol could be the outcome of host-pathogen-antagonist interactions under a given set of biotic and abiotic environmental conditions. In addition to direct effects on the plant pathogen, *Trichoderma* spp. can colonize plant roots and induce systemic resistance against root and foliar pathogens; they have been described as opportunistic, avirulent plant symbionts (Harman et al. 2004). This is a relatively recent discovery and the mechanisms will be discussed in later sections. Some other mechanisms that deserve special mention, but will not be discussed in detail, are the protection of rice plants against sheath blight by degradation of a host-specific phytotoxin (RS toxin) through the production of an extracellular α-glucosidase

(Shanmugam et al. 2001), inhibition of 5′-hydroxyaverantin dehydrogenase, an enzyme involved in aflatoxin biosynthesis by *T. harzianum* (Sakuno et al. 2000), suppression of fumonisin B1 production by *Fusarium moniliforme* by *T. viride* (Yates et al. 1999), inhibition of cell wall synthesis of *B. cinerea* by *T. harzianum* through the production of the peptaibols trichorzianin TA and TB (Lorito et al. 1996), and enhancing the production of nematicidal compounds by *Pseudomonas fluorescens* against *Meloidogyne incognita*, by *T. harzianum* (Siddiqui and Shaukat 2004).

10.3 Role of Hydrolytic Enzymes

The key factor to the ecological success of this genus is the combination of very active mycoparasitic mechanisms plus effective defense strategies induced in plants. The production and regulation of hydrolytic enzymes, particularly the chitinases, glucanases and proteases, have been studied very widely (reviewed by Viterbo et al. 2002b). However, direct evidence on their role in mycoparasitism/biocontrol by selective inactivation or overexpression of a particular enzyme is relatively scarce. *Trichoderma* spp. produce several types of chitinases, both endo and exo, that exhibit mycolytic activities. Many new chitinases are being discovered, and a genome-wide search would reveal more. In this section, we will focus on some aspects of regulation of these enzymes, and discuss direct evidence on the role of these proteins in mycoparasitism/biocontrol. In general, the hydrolytic enzymes are induced by the specific substrates (like chitin, glucan, fungal cell wall), and repressed by glucose. The first report on the systematic study of regulation of an endochitinase-encoding gene ech42 during mycoparasitism came in 1994 (Carsolio et al. 1994); it was shown that this gene is expressed during mycoparasitic interactions, on chitin and by exposure to light. In a significant finding, Lorito et al. (1996) observed that mycoparasitic interaction relieves binding of the Cre1 catabolite repressor protein to promoter sequences of the ech42 gene in *T. harzianum*. The proteinase encoding gene prb1 and the endochitinase-encoding gene ech42 in dual culture with *R. solani* get induced before physical contact, indicating that the induction is contact-independent, and is triggered by a diffusible factor, which was subsequently identified to be soluble chitooligosaccharides (Cortes et al. 1998, Zeilinger et al. 1999). In contrast, chit33 is induced only during the stage of overgrowth on *R. solani* (de las Mercedes Dana et al. 2001). Like chit33, chit36 in *T. asperellum* is also expressed before contact with the host fungus, and predicted to be triggered by a diffusible factor (Viterbo et al. 2002a). Expression of ech42 gene of *T. atroviride* under carbon starvation is antagonized via a BrlA-like *cis*-acting element (Brunner et al. 2003). The regulation of expression of two major chitinase genes (ech42 and nag1, encoding CHIT73) of *T. atroviride* is triggered by different regulatory signals (Mach et al. 1999). The expression of a gene encoding an antifungal glucan 1,3-β-glucosidase is repressed by glucose and induced by laminarin and other glucans (Donzelli et al. 2001). The basic, antifungal exo-α-1,3-glucanase from the biocontrol fungus *T. harzianum* is induced by fungal cell walls and autoclaved mycelium (Ait-Lahsen et al. 2001). A gene encoding an α-1,3-glucanase is induced during mycoparasitic interactions with B. cinerea. (Sanz et al. 2005). BGN16.3, a novel acidic β-1,6-glucanase from *T. harzianum*, is induced by fungal cell walls, indicating a possible role in biocontrol (Montero et al. 2005). Olmedo-Monfil et al. (2002) showed that the expression of prb1 gene is subject to nitrogen catabolite repression, and is induced by *Rhizoctonia solani* cell walls and osmotic stress. Overexpression of the proteinase-encoding gene *prb1* in *T. harzianum* improved the biocontrol activity against *Rhizoctonia solani* (Flores et al. 1997). The direct evidences on the role of hydrolytic enzymes in biocontrol came from the gene knockout studies, where a gene is selectively deleted through homologous recombination or inactivated by antisense/RNAi. Disruption of ech42 in *T. harzianum* resulted in almost no endochitinase42 activity, whereas strains carrying multicopies of this gene exhibited up to a 42-fold increase in enzyme activity (Carsolio et al. 1999). However, no significant difference in disease control ability of the wild type, or strains with no ech42 gene or harboring multiple copies was observed against S. rolfsii and R. solani. These results indicated that the 42-kDa endochitinase may not play a significant role in biocontrol in situ in this system. Woo et al. (1999) disrupted ech42 in *T. harzianum* P1 (*T. atroviride*) and showed reduced biocontrol activity against B. cinerea on bean leaves. However, interestingly, the biocontrol activities of the disruptants was enhanced against R. solani, and remained unaltered against P. ultimum. In *T. virens*, biocontrol of knockout and over-expression strains against R. solani in cotton were significantly decreased and enhanced, compared with the wild type strain, indicating that the CHIT42 is involved in biocontrol in this host-parasite interaction (Baek et al. 1999). This is interesting because the same enzyme is involved in biocontrol of same pathogen in *T. virens*, but not in *T. harzianum*. Disruption of nag-1, encoding a 73-kDa *N*-acetyl-β-D-glucosaminidase resulted in 30% reduced ability of *T. atroviride* to protect bean seedlings against infection by R. solani. An interesting observation in this experiment was that nag1 is essential for induction of ech42, and hence, the reduced biocontrol in the disruptant could have been due to reduced expression of ech42 or nag1 or both (Brunner et al. 2003). Through gene deletion and overexpression, Pozo et al. (2004) proved that a serine protease TVSP1 plays role in biocontrol of R. solani by *T. virens*. Recently, using similar genetic approach, Djonovic et al. (2006b) showed that a β-1,6-glucanase is involved in mycoparasitism and biocontrol of P. ultimum by *T. virens*, while the cellulose formation of *T. reesei* was found to be dispensable for the biocontrol of P. ultimum on zucchini plants (Seidl et al. 2006b).

10.4 Antibiosis

Compared to the bacterial antagonists, molecular evidences on the role of antibiosis in biocontrol in *Trichoderma* is scanty, and hardly any molecular biology data are available. The secondary metabolism in fungi is a very actively researched area, as many of the useful and toxic metabolites are produced by fungi, like *Gibberella* and *Aspergillus*. Extensive researches in this field led to the identification of gene clusters, and the biosynthetic pathways elucidated through gene deletion analysis (Keller et al. 2005). In contrast, there are only a few reports on the identification of genes responsible for the antifungal metabolite production in *Trichoderma* spp. *T. virens* produces many antifungal peptide metabolites. Wilhite et al. (2001) cloned a 5-kb partial cDNA encoding a putative peptide synthase (Psy1). The disruption of psy1 indicated a role in siderophore production in *T. virens*. However, the disrupted strains exhibited normal biocontrol properties against *R. solani* and *P. ultimum*, indicating that the iron competition may not play an important role in biocontrol in this system. Wiest et al. (2002) identified a 62.8-kb continuous open reading frame encoding a peptaibol synthetase from *T. virens*; the mutation of the gene eliminated the production of all the peptaibol isoforms, thus confirming that this gene is responsible for the synthesis of the peptaibols. A putative peptide synthetase gene has recently been identified in *T. harzianum* (Vizcaino et al. 2005). *T. virens* produces four major metabolites – gliotoxin, gliovirin, viridin and viridiol (Howell et al. 1993). A strain of *T. virens* (IMI 304061) has been found to produce plenty of viridin and its derivative viridiol in culture, while a mutant M7 did not produce these metabolites. Using this mutant, and suppression subtractive hybridization (SSH), Mukherjee M et al. (2006) identified several genes known to be involved in secondary metabolism in fungi. By sequencing a cosmid clone, a gene cluster was identified that consisted of cytochrome P450s and a cyclase. Based on the expression pattern and by comparison of the gene organization vis-à-vis other fungi, it was predicted to be involved in viridin synthesis. However, a gene knockout study would be required to confirm the role of this cluster in secondary metabolite production, as well as antagonistic properties.

One potential problem, which may affect the acceptance of *Trichoderma* spp. as useful biocontrol agents, is the possibility of activity against non-target species. Because Pr1 from insect pathogens has similar properties to prb1 from *Trichoderma* spp. it is possible that the proteinases may play a key role in both entomopathogenicity and antifungal action. Shakeri and Foster (2007) have recently reported on two strains of *Trichoderma harzianum*, 101645, an insect pathogen and 206040, used for biological control of fungal plant pathogens, which were investigated for the production of serine protease, chitinase and antibiotic activity in relation to entomopathogenicity. Both strains produced serine protease with a Mr of 31 kDa and chitinase with a Mr of 44 kDa. Enzymes from both strains had similar characteristics and were produced during the growth phase. Both strains also produced peptaibols active against fungi in late growth and stationary phases which differed in their amino-alcohol content. The peptaibols were insecticidal when fed to larvae of *Tenebrio molitor* or when applied to the cuticle together with the serine protease. The results suggest that the virulence factors involved in biocontrol are the same as those for insect pathogenicity. This may affect the use of *Trichoderma* spp. for biocontrol as there may be effects on non-target insect species.

10.5 Induced Resistance

The observation that *Trichoderma* spp. colonize plant roots and induce systemic resistance against a wide range of fungal, bacterial and viral pathogens can be considered a breakthrough in biocontrol research (reviewed by Harman et al. 2004). Inoculation of roots of cucumber seedlings with conidia of *T. harzianum* T-203 (*T. asperellum*) in an aseptic hydroponic system resulted in induction of defense responses (Yedidia et al. 1999). Electron microscopy of ultra-thin sections from *Trichoderma* treated roots revealed penetration of the mycoparasite into the roots, restricted mainly to the epidermis and outer cortex. *Trichoderma* colonization resulted in strengthening of the epidermal and cortical cell walls and deposition of newly formed barriers, these typical host reactions being found even beyond the sites of potential fungal penetration. The inoculation of *Trichoderma* initiated increased peroxidase and chitinase activities, both in roots and leaves. Later on, the authors showed that inoculation of cucumber roots with *Trichoderma* induced an array of PR proteins (Yedidia et al. 2000). Inoculation of cucumber roots with *T. asperellum* reduced the inoculum load of *Pseudomonas syringe* pv *lachrymans* to the extent of 80%, when challenge inoculated on leaves (Yedidia et al. 2003), thus providing direct evidence on induced defense-mediated protection of crop plants in response to *Trichoderma* inoculation. The protection afforded by the biocontrol agent was associated with the accumulation of mRNA of two defense-related genes: the phenylpropanoid pathway gene encoding phenylalanine ammonia lyase (PAL) and the lipooxygenase pathway gene encoding hydroxyperoxidase lyase (HPL). Recently, using the gene knockout approach, a hydrophobin TasHyd1 has been demonstrated to be involved in root colonization by *T. asperellum* (Viterbo and Chet 2006). In a significant finding, Shoresh et al. (2006) identified a MAPK (TIPK – *Trichoderma* induced MAPK) in cucumber, antisense – mediated silencing of this gene made plants susceptible even after inoculation of roots with *T. asperellum*. It was thus proved that *Trichoderma* exerts its positive effects on plants through the activation of a MAPK gene involved in signaling the pathway of defense response. In studies with *T. virens*, Howell et al. (2000) demonstrated that seed treatment of cotton with the antagonist or application of the culture filtrate to seedling radicles induced synthesis of much higher concentrations of the terpenoids deoxyhemigossypol, hemigossypol and gossypol in developing roots than those found in untreated controls. All these compounds were toxic to *R. solani*. Biocontrol activity was highly correlated with induction of terpenoid synthesis in cotton roots by *Trichoderma*. *T. virens* also induced significantly higher levels of peroxidase activity. Subsequently, Hanson and Howell (2004) identified an 18-kDa protein (a serine proteinase) from *T. virens* that stimulated terpenoid and peroxidase activity in cotton radicles. A definite role of phytoalexin induction in biocontrol has recently been demonstrated by Howell and Puckhaber (2005), who showed that the "P" strains of *T. virens* failed to stimulate phytoalexin synthesis in cotton and were ineffective as biocontrol, while the "Q" strains that stimulated phytoalexin biosynthesis were effective. This difference was attributed to the ability of "Q" strains to produce the

18-kDa elicitor protein. Recently, three groups independently identified a homologue of SnodProt proteins, variously named as SnodProt1 (GV Sible and PK Mukherjee, unpublished; GenBank Acc. no. DQ494198), Sm1 (Djonovic et al. 2006a) from *T. virens*, and Epl1 (Seidl et al. 2006a) from *T. atroviride*. Purified Sm1 protein triggered the production of reactive oxygen species in rice and cotton and induced expression of defense-related genes both locally and systemically in cotton (Djonovic et al. 2006a). Pre-treatment of cotton cotyledons with this protein also produced high levels of protection to the foliar pathogen *Colletotrichum* sp. These results indicated that Sm1, is involved in the induction of resistance by *Trichoderma* spp. through the activation of plant defense mechanisms.

Recently Olson and Benson (2007) have studied three root-colonizing fungi, binucleate *Rhizoctonia* (BNR) isolates BNR621 and P9023 and *T. hamatum* isolate 382 (T382), for suppression of Botrytis blight in geraniums by induction of host systemic resistance. Resistance to Botrytis blight was observed in geraniums transplanted into potting mix amended with formulations of P9023 and T382 two weeks prior to inoculation with *Botrytis cinerea* when grown under environments either highly or less conducive to disease development. Restriction of lesion development may play a role in the suppression of Botrytis blight in geraniums. This may be the first to demonstrate induced systemic resistance by BNR fungi to a foliar pathogen and support additional research into use of T382 in an integrated management program for *B. cinerea* (Olson and Benson 2007). Research on the specific effects of induced systemic resistance should be continued with additional pathogens since there is some indication of pathogen specificity in the method of suppression.

10.6 Signal Transduction and Biocontrol

Signal transduction through the G-protein/cAMP and MAP kinase pathways have long been known to be involved in the parasitism of plants by pathogenic fungi (Xu 2000; Lengeler et al. 2000). Since eukaryotic signaling mechanisms are well conserved, it is interesting to examine the role of these signaling elements in mycoparasitism, and hence biocontrol. The first direct evidence on the role of a G-protein came from Rocha-Ramirez et al. (2002). Antisense-mediated gene silencing of Tga1 attenuated mycoparasitism of *T. atroviride* against *R. solani*. On the other hand, transgenic strains carrying multicopies of the gene overgrew *R. solani* colonies at a faster rate. Using gene knockout, Reithner et al. (2005) demonstrated that Tga1 modulates chitinase formation and secondary metabolism in *T. atroviride*. The deletion of another G protein, Tga3 resulted in loss of mycoparasitism in *T. atroviride* (Zeilinger et al. 2005). Mukherjee et al. (2004) studied the role of the G-proteins TgaA and TgaB in *T. virens*. Deletion of these genes individually had no effect on hyphal coiling of R. solani, but TgaA was involved in the parasitism of sclerotia of *S. rolfsii*. Deletion of the MAPK TmkA in *T. virens* resulted in attenuation of sclerotial parasitism of *S. rolfsii* and *R. solani*, while the hyphal parasitism was unaltered (Mukherjee et al. 2003). The TmkA mutants also had reduced ability to induce resistance in cucumber seedlings, even though there was no effect on root colonization (Viterbo et al. 2005). The mutants also had reduced biocontrol of *S. rolfsii* in greenhouse tests. In contrast, however, Mendoza-Mendoza et al. (2003) reported improvement in biocontrol potential of *T. virens* through inactivation of the MAP kinase Tvk1. Whether this apparent contradiction is due to strain differences needs to be examined carefully. Recently, a *T. harzianum* stress-response MAPK ThHOG1 has been identified to be involved in osmotic and oxidative stress response (Delgado-Jarana et al. 2006). Recently, Mukherjee et al. (2007) has cloned the adenylate cyclase-encoding gene *tac1* of *T. virens* and obtained knockout mutants through homologous recombination. The mutants grew extremely slowly, failed to germinate in water, were impaired in mycoparasitism and produced lower amounts of secondary metabolites. This study proved that the cAMP signaling is involved in growth, germination and biocontrol properties in *T. virens*. Using suppression subtractive hybridization, the genes regulated by signaling genes like the TmkA/Tvk1 in *T. virens* have been identified (Mukherjee M et al. 2006; Mendoza-Mendoza et al. 2007). These target genes include some novel genes like the mrsp1 (MAPK Repressed Secreted Protein) with an expansin-like domain that could be involved in Trichoderma-plant root interactions (Mukherjee PK et al. 2006). Recently Reithner et al. (2007) have examined the function of the tmk1 gene encoding a MAPK during fungal growth, mycoparasitic interaction, and biocontrol was examined in *T. atroviride*. Dtmk1 mutants exhibited altered radial growth and conidiation, and displayed de-regulated infection structure formation in the absence of a host-derived signal. In confrontation assays, tmk1 deletion caused reduced mycoparasitic activity although attachment to *R. solani* and *B. cinerea* hyphae was comparable to the parental strain. Under chitinase-inducing conditions, nag1 and ech42 transcript levels and extracellular chitinase activities were elevated in a Dtmk1 mutant, whereas upon direct confrontation with *R. solani* or *B. cinerea* a host-specific regulation of ech42 transcription was found and nag1 gene transcription was no more inducible over an elevated basal level. Dtmk1 mutants exhibited higher antifungal activity caused by low molecular weight substances, which was reflected by an over-production of 6-pentyl-a-pyrone and peptaibol antibiotics. In biocontrol assays, a Dtmk1 mutant displayed a higher ability to protect bean plants against *R. solani* (Reithner et al. 2007). These findings strongly suggest the presence of further, still unknown, mycoparasitism related factors which are missing in our Dtmk1 mutants and which are therefore affected by a signaling pathway involving Tmk1.

10.7   The Genomics and Proteomics

With the advent of genomics, it is not long before the whole genome sequences of many *Trichoderma* spp. would be available, thus allowing the analysis of gene expression and gene functions on a genome-wide scale, rather than looking at the individual genes. The first step in characterizing genes potentially involved in biocontrol is to isolate and sequence them. Which strategy is most appropriate depends on previous results and on the objectives of the study. The targeted strategy should be privileged if the aim is to acquire further genetic knowledge on a well-studied mechanism of action. Differential gene expression techniques are very useful in original research aiming to isolate new genes potentially related to biocontrol properties when there is no a priori reason to suspect their involvement. Large-scale sequencing techniques, finally, support a broader endeavor: to improve genetic knowledge on a strain by sequencing numerous genes. Putative biocontrol genes may emerge from this endeavor. The genome sequencing of *T. virens* and *T. atroviride* is already under way, and is expected to be published soon (Charles Kenerley and Christian Kubicek, personal communication). In the meantime, quite a few ESTs database are available that identify hundreds of genes in various *Trichoderma* spp. Liu and Yang (2005) identified genes with biocontrol functions in *T. harzianum* mycelium using an ESTs approach – out of the 3298 clones sequenced, 673 represented novel genes. Suarez et al. (2005) identified a fungal cell wall induced aspartic protease from *T. virens* using the SSH approach. Carpenter et al. (2005) identified 19 novel genes in *T. hamatum* that are induced during mycoparasitism on *Sclerotinia sclerotiorum*. These included monooxygenases, metallopeptidases, gluconate dehydrogenase and endonuclease and a proton ATPase. Seidl et al. (2005) did a genome-wide search for chitinase genes in the *T. reesei* genome and identified three distinct subgroups of family 18 chitinases. Recently, an ESTs of 502 unique gene sequences of *T. virens* IMI 304061 have been deposited at the GenBank (PD Sherkhane, GV Sible and PK Mukherjee, unpublished) – this database includes many genes known to be involved in biocontrol and stress response. One of the earliest attempts to identify proteins from *T. harzianum* using the proteomic approach was by Grinyer et al. (2004a,b) who identified 25 proteins using 2D gel electrophoresis and LC MS/MS, either from mitochondria or whole cells. Using a proteomic approach, Grinyer et al. (2005) identified several novel proteins that were produced in response to mycoparasitism by *T. atroviride* on *B. cinerea* and *R. solani*. These included several hydrolytic enzymes also. Using a novel proteomic approach, Marra et al. (2006) studied the three-way interaction between *T. atroviride*, plant pathogens (*B. cinerea* and *R. solani*) and bean plant, in order to identify the proteins expressed in the various combination of host-pathogen-mycoparasite. This approach resulted in the identification of numerous differential proteins.

Trichoderma mycoparasitic activity depends on the secretion of complex mixtures of hydrolytic enzymes able to degrade the host cell wall. Suarez et al. (2005) have analysed the extracellular proteome secreted by *T. harzianum* CECT 2413 in the presence of different fungal cell walls. This allowed them to overcome the problems associated with the lack of genome sequence data for the identification of non-conserved *Trichoderma* spp. proteins. Optimized 2DE protein profiles were compared to that obtained on chitin and the most abundant protein induced by fungal cell walls was identified as the novel pepsin-like aspartic protease P6281 by a combination of matrix-assisted laser desorption/ionization time-of-flight mass spectrometry (MALDI-TOF), liquid chromatography mass spectrometry (LC–MS/MS) and in silico analysis of the available EST library. Significant differences were detected in 2DE maps, depending on the use of specific cell walls or chitin. A combination of MALDITOF and liquid chromatography mass spectrometry allowed the identification of a novel aspartic protease (P6281: MW 33 and pI 4.3) highly induced by fungal cell walls. A broad EST library from *T. harzianum* CECT 2413 was used to obtain the full-length sequence. The protein showed 44% identity with the polyporopepsin (EC 3.4.23.29) from the basidiomycete Irpex lacteus. Lower identity percentages were found with other pepsin-like proteases from filamentous fungi (<31%) and animals (<29%). Northern blot and promoter sequence analyses support the implication of the protease P6281 in mycoparasitism (Suarez et al. 2005).

In the future, numerous genes of various biocontrol agents will be identified by means of open strategies and with the help of advanced gene isolation and sequencing methods. The isolated genes will have to be characterized, and characterization will remain a bottleneck requiring much more time and effort than gene identification and sequencing.

10.8 The Transgenic Approach

Trans-kingdom transfer of genes for biocontrol from *Trichoderma* to plants to enhance disease resistance was, for the first time, demonstrated by Lorito et al. (1998). An endochitinase-encoding gene ech42, expressed in tobacco and potato provided near total protection against *Alternaria alternata*, *A. solani*, *B. cinerea* and *R. solani*. The high degree of broad-spectrum resistance was attributed to high fungitoxicity of *Trichoderma* chitinase, relative to the plant endogenous chitinases. This was followed by several reports on transfer of *Trichoderma* genes to plants. Bolar et al. (2001) produced transgenic apple resistant to *Venturia inaequalis* by expression of both an endochitinase and an exochitinase, and observed synergistic interaction. *T. harzianum* endochitinase, transferred to broccoli produced transgenic plants resistant to *Alternaria* leaf spot (Mora and Earle 2001). Liu and Yang (2005) produced transgenic rice resistant to blast and sheath blight by expressing ech42, nag70 and gluc78, in different combinations. A *T. virens* endochitinase gene ech42 transferred to cotton enhanced resistance against *A. alternata* and *R. solani* (Emani et al. 2003). Noël et al. (2005) introduced an endochitinase gene (ech42) from the biocontrol fungus *T. harzianum* into black spruce (*Picea mariana*) and hybrid poplar (*Populus nigraXP. maximowiczii*) by Agrobacterium-mediated transformation. Fifteen transgenic black spruce lines and six poplar lines were obtained. Northern hybridization analysis showed an increased accumulation of the transcript encoding the recombinant endochitinase gene in all the transgenic plants tested. Endochitinase activity 55–115 times the level of the control was detected in transformed poplar leaves. Embryogenic tissue of transgenic black spruce showed endochitinase activity two to eight times that of the non-transgenic line, despite stronger basal endogenous activity. In vitro assays using inoculated leaf disks demonstrated that the transgenic poplars had increased resistance to the leaf rust pathogen *Melampsora medusae*. Seedlings of transgenic spruce lines showed an increased resistance to the spruce root pathogen *Cylindrocladium floridanum* in vitro. These results suggest that constitutive expression of the ech42 gene from *T. harzianum* could be exploited to enhance resistance to fungal pathogens in important forest tree species. Thus tree genetic engineering with endochitinase genes could provide an alternative to the use of fungicides and help reduce tree growth losses caused by phytopathogenic fungi. In a very recent report, expression of *T. harzianum* endochitinases to tobacco was shown to enhance resistance not only against pathogens, but also against abiotic stress, presumably through the release of some elicitors (de las Mercedes Dana et al. 2006). Contrary to these reports, expression of *T. atroviride* ech42 in transgenic alfalfa did not yield resistance against *Phoma medicaginis* var *medicaginis*, even though there was a 50- to 2650-fold greater chitinase activity in transgenic plants (Samac et al. 2004). The protection provided by expression of mycoparasitism-related genes in plants thus may not be universal.

Promoter analysis can be used to confirm molecular models of gene regulation deriving from studies carried out under various in vitro conditions. Published studies of the promoter regions of genes involved in biocontrol have focused on either promoter sequences or regulatory proteins. Some investigators have studied the promoter sequence of a gene in order to confirm the involvement of previously identified motifs in the regulation of its transcription under biocontrol conditions. Electromobility Shift Assays (EMSAs), in vivo footprinting, and/or promoter deletion analysis (Peterbauer et al. 2002a) are the techniques used. Regulatory proteins can influence gene transcription either directly (by binding to the promoter sequence or indirectly via signal transmission). The molecular tools are also used to inactivate genes coding for regulatory proteins. Peterbauer et al. (2002b) found that inactivation of the seb1 gene does not modify transcription of the nag1, chit33, and ech42 genes. They also showed that other proteins can bind to the 5′-AGGGG-3′ promoter motifs of nag1 and cch42 in the disrupted strain. In in vitro studies, Mukherjee et al. (2003) examined how inactivating two mitogen activated protein kinases (MAPKs) affect the mycoparasitic properties of *T. virens*. In many fungal species, MAPK proteins participate in cascade signals involved, e.g., in plant parasitism. The role of two G-protein α-subunits, TgaA and TgaB, in biocontrol by *T. virens* has been studied by Mukherjee et al. (2004). G-proteins play an important role in intracellular signaling. They amplify receptor responses and influence the amplitude and duration of cellular signals. Using null-TgaA and null-TgaB strains, these authors showed that TgaA is involved in the biocontrol activity against *S. rolfsii*, but that neither TgaA nor TgaB is required for its activity against *R. solani*. The authors conclude that the involvement of G-proteins in biocontrol by *T. virens* depends on the plant pathogen with which the biocontrol agent is in contact. Zeilinger et al. (2005) have shown that the tga3 gene of *T. atroviride*, also coding for a G-protein α-subunit, is involved in this biocontrol agent's vegetative growth and mycoparasitic activity.

Zhou et al. (2007) have recently used restriction enzyme mediated integration (REMI) technique to construct mutants with improved cyanide-degradation ability from biocontrol fungus *T. koningii* strain T30. This successful insertional mutagenesis of the cyanide-biodegrading agent, *Trichodema* spp., led to the creation of mutants with deficient and enhanced cyanide-degrading properties. Liu et al. (2007) transformed three genes encoding for fungal cell wall degrading enzymes (CWDE), ech42, nag70 and gluc78 from the biocontrol fungus *T. atroviride* into rice mediated by *Agrobacterium tumefaciens* singly and in all possible combinations. These results indicated that expression of several genes in one T-DNA region interfered with each other and expression of exogenous gene in recipient plant was a complex behavior. It has been suggested that target gene must avoid being lost in transgenic process so as to be sure of expressing in transgenic plants and on the other hand, gene breaking and segregation in transgenic process can be used to delete selective gene so as to enhance transgenic security. This approach and the biological materials thus obtained could find a variety of applications in the discovery and manipulation of genes and gene products from *Trichoderma*.

10.9 Conclusion

Trichoderma spp. are a group of very useful fungi with many commercial applications in agriculture and industry. The popularity of *Trichoderma*-based biofungicides is growing by the day and hence, *Trichoderma*-based formulations have become an integral part of the crop management practices. However, an often-inconsistent performance vis-à-vis chemical fungicides is the major limitation associated with biological control. In order to improve the performance of *Trichoderma*-based formulations, we need to improve the strains for increasing its disease control potential (i.e., should be more effective than the existing strains against a particular target pathogen – this could be possible by generating strains with higher degree of mycoparasitism, competition, antibiosis and induced resistance), spectrum of activity (i.e., a single strain should be effective against a wide range of plant pathogens), as well as its survival ability (i.e., the ability to survive and perform under adverse environmental conditions). All these could be achieved if the physiology and genetics of these species are fully understood. However, until recently, this field of research remained largely neglected, which is apparent from the small proportion of the literature published on the mechanisms of biocontrol compared to the huge amount of work done on the biocontrol studies in laboratory, greenhouse and fields. Molecular techniques have been used to study the genetic basis of biological mechanisms and to identify partial or complete molecular pathways regulating gene expression. This is also true in the field of biocontrol. Whatever the technique used, it is paramount to choose an appropriate experimental model, as this will determine the reliability and validity scope of any conclusions drawn from an experiment. Molecular techniques have shed light on the antagonistic properties of numerous biological control agents, but they have also underlined the complexity of genetic regulation. They are now essential to studying the mechanisms of action of biocontrol agents and must be included in comprehensive studies that should also include microbiological, biochemical, and microscopic approaches.

It is a matter of relief that this trend is changing with quite a few classical works published in recent years on this aspect that have contributed greatly to our understanding the system at the molecular level. For example, the work on the negative regulation of conidiation (conidia are important in survival of *Trichoderma* spp., and also form the major biomass of the formulation products) by a MAP kinase, positive regulation of secondary metabolism (antibiotics production) by the adenylate cyclase Tac1 of *T. virens,* and the identification of the elicitor protein Sm1 (when applied to seedlings, it offered protection against infection) could be cited as major recent contributions, at the basic level, that would directly help in genetically improving *Trichoderma* spp. In addition, several new genes have been identified in the *Trichoderma* ESTs database, like many heat shock proteins/pH response proteins, regulators of heat shock response and genes involved in signal transduction, that could directly be used to genetically transform *Trichoderma* spp. for improving the biocontrol/survival potential. Recently, Montero-Barrientos et al. (2007) demonstrated such a possibility – they have successfully imparted thermotolerance in *T. harzianum* by heterologous expression of a small heat shock protein gene *hsp23* from *T. virens.* Massart and Jijakli (2007) have reviewed the techniques used in such studies, with their potential and limitations. It should provide a guide for researchers wanting to study the molecular basis of the biocontrol in diverse biocontrol agents.

The ability of a biocontrol agent to respond quickly and adequately to an environmental signal such as the presence of a potential host is a key factor in the development of mycoparasitism or in the metabolization of plant nutrients. The study of regulatory proteins is thus essential to an in-depth understanding of the genetic basis of biocontrol properties. It notably highlights relationships between the environment and biocontrol gene expression. The induction or repression of gene expression in response to environmental signals may occur through various pathways. So far research has focused on G-proteins and MAPK pathways. Other candidate genes, like the Abc transporters or the OPT protein family, should be studied for their possible involvement in biocontrol. Once the whole genome sequences are available, we will know about the genetic blue-print of these organisms, and, by comparison with many other fungal genome sequences that are already available, it would be possible to have an idea of what makes *Trichoderma* spp. effective as biocontrol agents. By functional genomics approaches, it would also be possible to identify the functions of the genes that are unique to *Trichoderma* in structure and function. In addition to direct genetic manipulation, once we identify the genes responsible for biocontrol, it would also be possible to discover/design drugs that could act as stimulants for genes/gene products, thus improving their bio-efficacy. Finally, we should bear in mind that biological control is an outcome of very complex interactions between plant, pathogen, antagonist and the environment, and there is unlikely to be a quick-fix, single-step solution to the major problems associated with biocontrol, especially the low efficiency and inconsistency. A consorted approach taking into account all the relevant parameters, including improvement of the antagonist for better inhibition of the pathogen, strengthening of the host plant, and improved survival potential of antagonists, would be helpful in bringing the biocontrol at par or even more effective than the chemical pesticides in terms of applicability under the field conditions.

References

Ait-Lahsen H, Soler A, Rev M, de la Cruz J, Monte E, Llobell A (2001) An antifungal exo-alpha-1,3-glucanase from the biocontrol fungus *Trichoderma harzianum*. Appl Env Microbiol 67:5833–5839

Baek JM, Howell CR, Kenerley CM (1999) The role of an extracellular chitinase from *Trichoderma* virens Gv29-8 in the biocontrol of *Rhizoctonia solani*. Curr Genet 35:41–50

Bolar JP, Norelli JL, Harman GE, Brown SK, Aldwinckle HS (2001) Synergistic activity of endo-chitinase and exochitinase from *Trichoderma atroviride* (*T harzianum*) against the pathogenic fungus (*Venturia inaequalis*) in transgenic apple plants Transgenic Res 10:533–543

Brunner K, Montero M, Mach RL, Peterbauer CK, Kubicek CP (2003) Expression of the ech42 (endochitinase) gene of *Trichoderma atroviride* under carbon starvation is antagonized via a BrlA-like cis-acting element. FEMS Microbiol Lett 218:259–264

Carpenter MA, Stewart A, Ridgway HJ (2005) Identification of novel *Trichoderma hamatum* genes expressed during mycoparasitism using subtractive hybridization. FEMS Microbiol Lett 251:105–112

Carsolio C, Benhamou N, Haran S, Cortes C, Gutierrez A, Chet I, Herrera-Estrella A (1994) Characterization of ech42, a *Trichoderma harzianum* endochitinase gene expressed during mycoparasitism. Proc Natl Acad Sci 10903–10907

Carsolio C, Benhamou N, Haran S, Cortes C, Gutierrez A, Chet I, Herrera-Estrella A (1999) Role of the *Trichoderma harzianum* endochitinase gene, ech42, in mycoparasitism. Appl Environ Microbiol 65:929–935

Chet I, Benhamou N, Haran S (1998) Mycoparasitism and lytic enzymes. In: Harman GE, Kubicek CP (eds) *Trichoderma* and *Gliocladium*, vol 2. Enzymes, biological control and commercial applications. Taylor and Francis London, United Kingdom, pp 153–171

Cortes C, Gutierrez A, Olmedo V, Inbar J, Chet I, Herrera-Estrella A (1998) The expression of genes involved in parasitism by *Trichoderma harzianum* is triggered by a diffusible factor. Mol Gen Genet 260:218–225

de las Mercedes Dana M, Limon MC, Mejias R, Mach RL, Benitez T, Pintor-Toro JA, Kubicek CP (2001) Regulation of chitinase 33 (chit33) gene expression in *Trichoderma harzianum*. Curr Genet 38:335–342

de las Mercedes Dana M, Pintor-Toro JA, Cubero B (2006) Transgenic tobacco plants overexpressing chitinases of fungal origin show enhanced resistance to biotic and abiotic stress agents. Plant Physiol 142:722–730

Delgado-Jarana J, Sousa S, Gonzalez F, Rey M, Llobell A (2006) ThHog1 controls the hyperosmotic stress response in *Trichoderma harzianu*. Microbiol 152:1687–1700

Djonovic S, Pozo MJ, Dangott LJ, Howell CR, Kenerley CM (2006a) Sm1, a proteinaceous elicitor secreted by the biocontrol fungus *Trichoderma* virens induces plant defense responses and systemic resistance Mol Plant Microbe Interact. 19:838–853

Djonovic S, Pozo MJ, Kenerley CM (2006b) Tvbgn3, a beta-1,6-glucanase from the biocontrol fungus *Trichoderma virens*, is involved in mycoparasitism and control of *Pythium ultimum*. Appl Environ Microbiol 72:7661–7670

Donzelli BG, Lorito M, Scala F, Harman GE (2001) Cloning, sequence and structure of a gene encoding an antifungal glucan 1,3-beta-glucosidase from *Trichoderma atroviride* (*T harzianum*). Gene 277:199–208

Emani C, Garcia JM, Lopata-Finch E, Pozo MJ, Uribe P, Kim DJ, Sunilkumar G, Cook DR, Kenerley CM, Rathore KS (2003) Enhanced fungal resistance in transgenic cotton expressing an endochitinase gene from *Trichoderma virens*. Plant Biotechnol J 1:321–336

Flores A, Chet I, Herrera-Estrella A (1997) Improved biocontrol activity of *Trichoderma harzianum* by over-expression of the proteinase-encoding gene prb1 Curr Genet 31:30–37

Grinyer J, McKay M, Nevalainen H, Helbert BR (2004a) Fungal proteomics: initial mapping of biological control strain *Trichoderma harzianum*. Curr Genet 45:163–169

Grinyer J, McKay M, Helbert BR, Nevalainen H (2004b) Fungal proteomics: mapping of the mitochondrial proteins of a *Trichoderma harzianum* strain applied for biological control. Curr Genet 45:170–175

Grinyer J, Hunt S, McKay M, Herbert BR, Nevalainen H (2005) Proteomic response of the biological control fungus *Trichoderma atroviride* to growth on the cell walls of *Rhizoctonia solani*. Curr Genet 47:381–388

Hanson LE, Howell CR (2004) Elicitors of plant defense responses from biocontrol strains of *Trichoderma virens*. Phytopathol 94:171–176

Harman GE, Bjorkmann T (1998) Potential and existing uses of *Trichoderma* and *Gliocladium* for plant disease control and plant growth enhancement. In: Harman GE, Kubicek CP (eds) *Trichoderma* and *Gliocladium*, vol 2. Enzymes, biological control and commercial applications Taylor and Francis London, UK, pp 229–265

Harman GE, Howell CR, Viterbo A, Chet I, Lorito M (2004) *Trichoderma* species-opportunistic, avirulent plant symbionts. Nat Rev Microbiol 2:43–56

Hjeljord L, Tronsmo A (1998) *Trichoderma* and *Gliocladium* in biological control: an overview. In: GE Harman, CP Kubicek (eds) *Trichoderma* and *Gliocladium*, vol 2. Enzymes, biological control and commercial applications Taylor and Francis London, UK, pp 129–155

Howell CR (1987) Relevance of mycoparasitism in the biological control of *Rhizoctonia solani* by *Gliocladium virens*. Phytopathology 77:992–994

Howell CR (2002) Cotton seedling preemergence damping-off incited by *Rhizopus oryzae* and *Pythium* spp and its biological control with *Trichoderma* spp. Phytopathology 92:177–180

Howell CR (2003) Mechanisms employed by *Trichoderma* species in the biological control of plant diseases: the history and evolution of current concepts. Plant Dis 87:4–10

Howell CR, Puckhaber LS (2005) A study of the characteristics of P and Q strains of *Trichoderma virens* to account for differences in biological control efficacy against cotton seedling diseases. Biol Control 33:217–222

Howell CR, Stipanovic RD (1995) Mechanisms in the biocontrol of Rhizoctonia solani- induced cotton seedling disease by Gliocladium virens: antibiosis. Phytopathology 85:469–472

Howell CR, Stipanovic RD, Lumsden RD (1993) Antibiotic production by strains of *Gliocladium virens* and its relation to the biocontrol of cotton seedling diseases. Biocontrol Sci Technol 3:435–441

Howell CR, Hanson LE, Stipanovic RD, Puckhaber LS (2000) Induction of terpenoid synthesis in cotton roots and control of *Rhizoctonia solani* by seed treatment with *Trichoderma virens*. Phytopathology 90:248–252

Inbar J, Chet I (1992) Biomimics of fungal cell-cell recognition by use of lectin-coated nylon fibres. J Bacteriol 174:1055–1059

Keller NP, Turner G, Bennett JW (2005) Fungal secondary metabolism- from biochemistry to genomics. Nat Rev Microbiol 3:937–947

Lengeler KB, Davidson RC, D'Souza C, Harashima T, Shen WC, Wang P, Pan X, Waugh M, Heitman J (2000) Signal transduction cascades regulating fungal development and virulence. Microbiol Mol Biol Rev 64:746–785

Liu M, Zhu J, Sun Z, Xu T (2007) Possible suppression of exogenous β-1,3-glucanase gene gluc78 on rice transformation and growth. Plant Sci 172:888–896

Liu PG, Yang Q (2005) Identification of genes with a biocontrol function in *Trichoderma harzianum* mycelium using the expressed sequence tag approach. Res Microbiol 156:416–423

Lorito M, Mach RL, Sposato P, Strauss J, Peterbauer CK, Kubicek CP (1996) Mycoparasitic interaction relieves binding of the Cre1 carbon catabolite repressor protein to promoter sequences of the ech42 (endochitinase-encoding) gene in *Trichoderma harzianum*. Proc Natl Acad Sci USA 93:14868–14872

Lorito M, Woo SL, Garcia I, Colucci G, Harman GE, Pintor-Toro JA, Filippone E, Muccifora S, Lawrence CB, Zoina A, Tuzun S, Scala F (1998) Genes from mycoparasitic fungi as a source for improving plant resistance to fungal pathogens. Proc Natl Acad Sci USA 95:7860–7865

Mach RL, Zeilinger S (2003) Regulation of gene expression in industrial fungi: *Trichoderma*. Appl Microbiol Biotechnol 60:515–522

Mach RL, Peterbauer CK, Payer K, Jaksits S, Woo SL, Zeilinger S, Kullnig CM, Lorito M, Kubicek CP (1999) Expression of two major chitinase genes of *Trichoderma atroviride* (*T harzianum* P1) is triggered by different regulatory signals. Appl Environ Microbiol 65:1858–1863

Marra R, Ambrosino P, Carbone V, Vinale F, Woo SL, Ruocco M, Ciliento R, Lanzuise S, Ferraioli S, Soriente I, Gigante S, Turra D, Fogliano V, Scala F, Lorito M (2006) Study of the three-way interaction between *Trichoderma atroviride*, plant and fungal pathogens by using a proteomic approach. Curr Genet 50:307–321

Massart S, Jijakli HM (2007) Use of molecular techniques to elucidate the mechanisms of action of fungal biocontrol agents: a review. J Microbiol Met 69:229–241

Mendoza-Mendoza A, Pozo MJ, Grzegorski D, Martinez P, Garcia JM, Olmedo-Monfil V, Cortes C, Kenerley C, Herrera-Estrella A (2003) Enhanced biocontrol activity of *Trichoderma* through inactivation of a mitogen-activated protein kinase. Proc Natl Acad Sci USA 100:15965–15970

Mendoza-Mendoza A, Rosales-Saavedra T, Cortes C, Castellanos-Juarez V, Martinez P, Herrera-Estrella A (2007) The MAP kinase TVK1 regulates conidiation, hydrophobicity and the expression of genes encoding cell wall proteins in the fungus *Trichoderma virens*. Microbiology 153:2137–2147

Montero M, Sanz L, Rey M, Monte E, Llobell A (2005) BGN163, a novel acidic beta-1-6-glucanase from mycoparasitic fungus *Trichoderma harzianum*. CECT 2413 FEBS J 272:3441–3448

Montero-Barrientos M, Cardoza RE, Gutierrez S, Monte E, Hermosa R (2007) The heterologous overexpression of hsp23, a small heat-shock protein gene from *Trichoderma virens*, confers thermotolerance to *T harzianum*. Curr Genet 52:45–53

Mora A, Earle ED (2001) Combination of *Trichoderma harzianum* endochitinase and a membrane-affecting fungicide on control of *Alternaria* leaf spot in transgenic broccoli plants. Appl Microbiol Biotechnol 55:306–310

Mukherjee M, Horwitz BA, Sherkhane PD, Hadar R, Mukherjee PK (2006) A secondary metabolite biosynthesis cluster in *Trichoderma virens*: evidence from analysis of genes underexpressed in a mutant defective in morphogenesis and antibiotic production. Curr Genet 50:193–202

Mukherjee M, Mukherjee PK, Kale SP (2007) cAMP signaling is involved in growth, germination, mycoparasitism and secondary metabolism in *Trichoderma virens*. Microbiology 153:1734–1742

Mukherjee PK, Haware MP, Jayanthi S (1995a) Preliminary investigations in integrated biocontrol of *Botrytis* gray mold of chickpea. Ind Phytotpath 48:141–149

Mukherjee PK, Mukhopadhyay AN, Sarmah D, Shreshtha SM (1995b) Comparative antagonistic properties of *Gliocladium virens* and *Trichoderma harzianum* on *Sclerotium rolfsii* and *Rhizoctonia solani* – its relevance to understanding the mechanisms of biocontrol. J Phytopathol 143:275–279

Mukherjee PK, Latha J, Hadar R, Horwitz BA (2003) TmkA, a mitogen activated protein kinase of *Trichoderma virens*, is involved in biocontrol properties and repression of conidiation in the dark. Eukaryot Cell 2:446–455

Mukherjee PK, Latha J, Hadar R, Horwitz A (2004) Role of two G-protein alpha subunits, TgaA and TgaB, in the antagonism of *Trichoderma virens* against plant pathogens. Appl Environ Microbiol 70:542–549

Mukherjee PK, Hadar R, Pardovitz-Kedmi E, Trushina N, Horwitz BA (2006) *MRSP1*, encoding a novel *Trichoderma* secreted protein, is negatively regulated by MAPK. Biochem Biophys Res Commun 350:716–722

Mukhopadhyay AN, Mukherjee PK (1996) Fungi as fungicides. Int J Trop Plant Dis 14:1–17

Mukhopadhyay AN, Shrestha SM, Mukherjee PK (1992) Biological seed treatment for control of soilborne plant pathogens FAO. Plant Prot Bull 40:21–30

Noël A, Levasseur C, Le V-Q, Séguin A (2005) Enhanced resistance to fungal pathogens in forest trees by genetic transformation of black spruce and hybrid poplar with a *Trichoderma harzianum* endochitinase gene. Physiol Mol Plant Pathol 67:92–99

Olmedo-Monfil V, Mendoza-Mendoza A, Gomez I, Cortes C, Hererra-Estrella A (2002) Multiple environmental signals determine the transcriptional activation of the mycoparasitism related gene prb1 in *Trichoderma atroviride*. Mol Genet Genom 267:703–712

Olson HA, Benson DM (2007) Induced systemic resistance and the role of binucleate *Rhizoctonia* and *Trichoderma hamatum* 382 in biocontrol of *Botrytis blight* in geranium. Biol Control 42:233–241

Peterbauer C, Brunner K, Mach RL, Kubicek CP (2002a) Identification of the N-acetyl-D-glucosamine-inducible element in the promoter of the *Trichoderma atroviride* bag1 gene encoding N-acetyl-glucosamidase. Mol Genet Genomics 267:162–170

Peterbauer C, Litscher D, Kubicek CP (2002b) The *Trichoderma atroviride* seb1 (stress response element binding) gene encodes an AGGGG-binding protein which is involved in the response to high osmolarity stress. Mol Genet Genomics 268:223–231

Pozo MJ, Baek JM, Garcia JM, Kenerley CM (2004) Functional analysis of tvsp1, a serine protease-encoding gene in the biocontrol agent *Trichoderma virens* Fungal Genet Biol 41:336–348

Reithner B, Brunner K, Schumacher R, Peissl I, Seidl V, Krska R, Zeilinger S (2005) The G protein alpha ?subunit Tga1 of *Trichoderma atroviride* is involved in chitinase formation and differential production of antifungal metabolites Fungal Genet Biol 42:749–760

Reithner B, Brunner K, Schumacher R, Stoppacher N, Pucher M, Brunner K, Zeilinger S (2007) Signaling via the Trichoderma atroviride mitogen-activated protein kinase Tmk1 differentially affects mycoparasitism and plant protection. Fungal Genet Biol 44:1123–1133; doi:101016/jfgb200704001

Rocha-Ramirez V, Omero C, Chet I, Horwitz BA, Herrera-Estrella A (2002) *Trichoderma atroviride* G-protein alpha subunit gene *tga1* is involved in mycoparasitic coiling and conidiation. Eukaryot Cell 1:594–605

Sakuno E, Yabe K, Hamasaki T, Nakajima H (2000) A new inhibitor of 5′-hydroxyaverantin dehydrogenase, an enzyme involved in aflatoxin biosynthesis, from *Trichoderma hamatum* J Nat Prod 63:1677–1678

Samac DA, Tesfaye M, Dornbusch M, Saruul P, Temple SJ (2004) A comparison of constitutive promoters for expression of transgenes in alfalfa (*Medicago sativa*). Transgenic Res 13:349–361

Sanz L, Montero M, Redondo J, Llobell A, Monte E (2005) Expression of an alpha-1,3-glucanase during mycoparasitic interaction of *Trichoderma asperellum*. FEBS J 272:493–499

Seidl V, Huemer B, Seiboth B, Kubicek CP (2005) A complete survey of *Trichoderma* chitinases reveals three distinct subgroups of family 18 chitinases. FEBS J 272:5923–5939

Seidl V, Marchetti M, Schandl R, Allmaier G, Kubicek CP (2006a) Epl1, the major secreted protein of *Hypocrea atroviridis* on glucose, is a member of a strongly conserved protein family comprising plant defense response elicitors. FEBS J 273:4346–4359

Seidl V, Schmoll M, Scherm B, Balmas V, Seiboth B, Migheli Q, Kubicek CP (2006b) Antagonism of *Pythium* blight of zucchini by *Hypocrea jecorina* does not require cellulase gene expression but is improved by carbon catabolite derepression. FEMS Microbiol Lett 257:145–151

Shakeri J, Foster HA (2007) Proteolytic activity and antibiotic production by *Trichoderma harzianum* in relation to pathogenicity to insects. Enz Micro Technol 40(4):961–968

Shanmugam V, Sriram S, Babu S, Nandakumar R, Raguchander T, Balasubramanian P, Samiyappan R (2001) Purification and characterization of an extracellular alpha-glucosidase protein from *Trichoderma viride* which degrades a phytotoxin associated with sheath blight disease in rice. J Appl Microbiol 90:320–329

Shoresh M, Gal-On A, Leibman D, Chet I (2006) Characterization of a mitogen-activated protein kinase gene from cucumber required for *Trichoderma*-conferred plant resistance. Plant Physiol 142:1169–1179

Siddiqui IA, Shaukat SS (2004) *Trichoderma harzianum* enhances the production of nematicidal compounds in vitro and improves biocontrol of *Meloidogyne javanica* by *Pseudomonas fluorescens* in tomato. Lett Appl Microbiol 38:169–175

Sivasithamparam K, Ghisalbarti E (1998) Secondary metabolism. In: Harman GE, Kubicek CP (eds) *Trichoderma* and *Gliocladium*, vol 1. Basic biology, taxonomy and genetics Taylor and Francis London, UK, pp 139–191

Suarez MB, Sanz L, Chamorro MI, Rey M, Gonzalez FJ, Llobell A, Monte E (2005) Proteomic analysis of secreted proteins from *Trichoderma harzianum*. Identification of a fungal cell wall-induced aspartic protease. Fungal Genet Biol 42:924–934

Viterbo A, Chet I (2006) TasHyd1, a new hydrophobin gene from the biocontrol agent *Trichoderma asperellum* is involved in plant root colonization. Mol Plant Path 7:249–258

Viterbo A, Montero M, Ramot O, Friesem D, Monte E, Llobell A, Chet I (2002a) Expression regulation of the endochitinase chit36 from *Trichoderma asperellum* (*T harzianum* T203). Curr Genet 42:114–122

Viterbo A, Ramot O, Chemir I, Chet I (2002b) Significance of lytic enzymes from *Trichoderma* spp in the biocontrol of fungal plant pathogens. Anton Leewenhoek 81:549–556

Viterbo A, Harel M, Horwitz BA, Chet I, Mukherjee PK (2005) *Trichoderma* mitogen-activated protein kinase signaling is involved in induction of plant systemic resistance. Appl Environ Microbiol 71:6241–6246

Vizcaino JA, Sanz L, Cardoza RE, Monte E, Gutierrez S (2005) Detection of putative peptide synthetase genes in *Trichoderma* species: application of this method to the cloning of a gene from *T harzianum* CECT 2413. FEMS Microbiol Lett 244:139–148

Wiest A, Grzegorski D, Xu BW, Goulard C, Rebuffat S, Ebbole DJ, Bodo B, Kenerley C (2002) Identification of peptaibols from *Trichoderma virens* and cloning of a peptaibol synthetase. J Biol Chem 277:20862–20868

Wilhite SE, Lumsden RD, Straney DC (1994) Mutational analysis of gliotoxin production by the biocontrol fungus *Gliocladium virens* in relation to suppression of Pythium damping-off Phytopathology 84:816–821

Wilhite SE, Lumsden RD, Straney DC (2001) Peptide synthetase gene in *Trichoderma virens*. Appl Environ Microbiol 67:5055–5062

Woo SL, Donzelli B, Scala F, Mach R, Harman GE, Kubicek CP, Del Sorbo G, Lorito M (1999) Disruption of the ech42 (endochitinase-encoding) gene affects biocontrol activity in *Trichoderma harzianum* P1. Mol Plant Microbe Interact 12:419–429

Xu JR (2000) Map kinases in fungal pathogens Fungal Genet Biol 31:137–152

Yates IE, Meredith F, Smart W, Bacon CW, Jaworski AJ (1999) *Trichoderma viride* suppresses fumonisin B1 production by *Fusarium moniliforme*. J Food Prot 62:1326–1332

Yedidia I, Benhamou N, Chet I (1999) Induction of defense responses in cucumber plants (Cucumis sativus L) by the biocontrol agent *Trichoderma harzianum*. Appl Environ Microbiol 65:1061–1070

Yedidia I, Benhamou N, Kapulnik Y, Chet I (2000) Induction and accumulation of PR proteins activity during early stages of root colonization by the mycoparasite *Trichoderma harzianum* strain T-203. Plant Physiol Biochem 38:863–873

Yedidia I, Shoresh M, Kerem Z, Benhamou N, Kapulnik Y, Chet I (2003) Concomitant induction of systemic resistance to *Pseudomonas syringae* pv *lachrymans* in cucumber by *Trichoderma asperellum* (T-203) and accumulation of phytoalexins. Appl Env Microbiol 69:7343–7353

Zeilinger S, Galhaup C, Payer K, Woo SL, Mach RL, Fekete C, Lorito M, Kubicek CP (1999) Chitinase gene expression during mycoparasitic interaction of *Trichoderma harzianum* with its host. Fungal Genet Biol 26:131–140

Zeilinger S, Reithner B, Scala V, Piessl I, Lorito M, Mach R (2005) Signal transduction by Tga3, a novel G protein alpha subunit of *Trichoderma atroviride*. Appl Environ Microbiol 71:1591–1597

Zhou X, Xu S, Liu L, Chen J (2007) Degradation of cyanide by *Trichoderma* mutants constructed by restriction enzyme mediated integration (REMI). Bio Res Technol 98:2958–2962

Part III
Coexistence Between Molecules

Chapter 11
Quorum Sensing in Bacteria-Plant Interactions

Kristien Braeken, Ruth Daniels, Maxime Ndayizeye, Jos Vanderleyden, and Jan Michiels(✉)

11.1 Introduction

Multicellular organisms rely on an accurate communication between individual cells to coordinate many aspects of physiology and development. Prokaryotic organisms, although unicellular, also express certain traits only when a critical number of bacteria has been reached. Here, the individual bacterium benefits from joint multicellular behaviour to survive, compete and persist in nature, or to colonize a particular host. Therefore, they have to communicate with each other. Fuqua et al. (1994) introduced the term "quorum sensing (QS)" to describe the process where bacterial communication is used to monitor population density and to change bacterial gene expression and behaviour accordingly (Fuqua et al. 2001; von Bodman et al. 2003a). Essentially, QS is based on production of low-mass signalling molecules, the extracellular concentration of which is related to the population density of the producing organisms. These signalling molecules can be sensed by the bacterial cells and this allows the population to initiate a concerted action once a critical concentration ("quorum") has been reached (Whitehead et al. 2001). A wide range of (potential) low-mass signalling molecules have been identified. These include peptide-based signals in various Gram-positive organisms and the N-acyl homoserine lactone (AHL) signals found in many Gram-negative bacteria (Proteobacteria) (Fuqua et al. 2001; Whitehead et al. 2001) as well as many other signal molecules (for an overview see Visick and Fuqua 2005). However, Redfield (2002) suggested that in some cases quorum sensing might be a side effect of cells monitoring their diffusion environment instead of communicating. By this means, cells can regulate the secretion of effectors to minimize losses to extracellular diffusion. Most QS-regulated processes in plant-associated bacteria are mediated

J. Michiels
Centre of Microbial and Plant Genetics, Katholieke Universiteit Leuven,
Kasteelpark Arenberg 20, B-3001 Leuven, Belgium
e-mail: jan.michiels@biw.kuleuven.be

C.S. Nautiyal, P. Dion (eds.) *Molecular Mechanisms of Plant
and Microbe Coexistence.* Soil Biology 15, DOI: 10.1007/978-3-540-75575-3
© Springer-Verlag Berlin Heidelberg 2008

by AHL (*N*-acyl homoserine lactone (HSL))-based QS systems, which is the main focus of this chapter.

11.2   The Paradigm of AHL Quorum Sensing: The *lux* System

The first QS system described is that of the marine bacterium, *Vibrio fisheri*, which produces light when colonizing the light organs of the squid *Eyprymna scolopes*. The *V. fisheri* QS system involves two major components: *luxI*, the AHL synthase-encoding gene, and the transcriptional activator encoded by *luxR*. At low cell densities, low levels of LuxI inside the bacterial cell are responsible for production of *N*-(3-oxo-hexanoyl)-L-HSL (3-oxo-C_6-HSL), a signal molecule moving freely across bacterial membranes. Once a critical concentration of this signal molecule has been reached (corresponding to a "quorum" of bacteria), it can bind and herewith activate LuxR inside the cell. Activated LuxR is thought to bind a 20-bp element of dyad symmetry, called *lux*-box, which results in transcriptional activation of the *luxICDABEG* genes, leading to increasing production of light and of 3-oxo-C_6-HSL. Therefore the process was originally called autoinduction and this QS system was thought to be unique for marine vibrios (reviewed by Fuqua et al. 2001; Whitehead et al. 2001).

However, identification of AHL-based systems in other bacteria during the past 20 years proved this phenomenon is widespread among Proteobacteria with more than 50 species now recognized to produce AHLs (Fuqua et al. 2001). In both pathogenic and beneficial plant-associated bacteria, a large number of AHL-based QS systems were identified and shown to affect processes such as swarming, biofilm formation, conjugal plasmid transfer, stress survival and synthesis of colonization and virulence factors such as surfactants, exopolysaccharides (EPS), antibiotics and extracellular enzymes. An overview of the diverse phenomena regulated in representative groups of these bacteria (symbiotic rhizobia, *Agrobacterium* sp., *Erwinia* sp. and plant-associated pseudomonads) is presented in Table 11.1.

The structure of the AHLs discovered vary in the size of the acyl chains with lengths from 4 to 18 carbon atoms being identified so far (Whitehead et al. 2001; Marketon et al. 2002). Variability also exists in the third carbon position of the acyl chain, where a hydrogen, hydroxyl or oxo substitution can be found. Furthermore, unsaturated chains have been identified (von Bodman et al. 2003a). The produced AHL is released into the environment, either by passive diffusion, as observed for 3-oxo-C_6-HSL in *V. fisheri* (Kaplan and Greenberg 1985), or by a combination of diffusion and active transport for AHLs with longer acyl-side chains as described for 3-oxo-C_{12}-HSL in *Pseudomonas aeruginosa*, where the *mexAB-oprM* operon, a member of a large family of antibiotic transporters, encodes a specific efflux pump involved in active transport of 3-oxo-C_{12}-HSL (Pearson et al. 1999). It is not yet known if the release of the very long-chain AHLs produced by some plant-associated

Table 11.1 AHL production and associated phenomena in plant-associated bacteria

Bacterium	AHLs	Gene loci involved	Associated phenomena	Additional regulatory compounds	Refs
Agrobacterium tumefaciens	3-oxo-C_8-HSL	TraI/TraR	Ti Plasmid transfer	TraM (TraR antiactivator)	1–8, 67
				TrlR (forming of non-productive TraR-TrlR heterodimer) for some octopine-type Ti plasmids)	
				AttM lactonase controlled by Rel_{Atum} succinic semialdehyde and GABA of wounded plants stimulates expression *attM*	
Agrobacterium vitis	NYD[a]	AvhR	Diminished AHL production, Grape necrosis, HR[b] response on tobacco	/	9
		AviR	Grape necrosis, HR response	/	10
	Long chain AHLs	AvsI/AvsR	Grape necrosis, HR response	/	11
Pseudomonas aureofaciens	C_6-HSL	PhzI/PhzR	Phenazine production, rhizosphere colonization, protease production	/.	12–14
	NYD	CsaI/CsaR	Exoprotease production, Wheat rhizosphere colonization, cell surface properties (CsaR)	Regulated by GacAS. In conjunction with PhzRI	14
P. putida IsoF	3-oxo-C_{12}-HSL, 3-oxo-C_{10}-HSL, 3-oxo-C_8-HSL, 3-oxo-C_6-HSL	PpuI/PpuR	Biofilm structural development	*rsaL* homologue between *ppuR* and *ppuI*	15–16
P. putida WCS358	Idem as IsoF DKPs[c]	PpuI/PpuR	/	*ppuI* expression is negatively regulated by *RsaL*, GacA positively affects *ppuI* expression. QS and *rpoS* affect each other	17–18

(continued)

Table 11.1 (continued)

Bacterium	AHLs	Gene loci involved	Associated phenomena	Additional regulatory compounds	Refs
P. putida PCL1445	NYD	PpuI/PpuR	Production of cyclic lipopeptides (putisolvin I and II); Biofilm formation	RsaL functions as a negative regulator of QS system	19
P. syringae pv. syringae	3-oxo-C_6-HSL	AhlI/AhlR	Diminished EPS (alginate) production, Epiphytic fitness/disease development, Hydrogen peroxide susceptibility, Motility	Indirect control of QS by AefR and GacA	20–21
P. syringae pv. tomato DC3000	NYD	PsyI/PsyR	/	PsrA controls AHL level through AefR and RpoS	68
P. chlororaphis PCL1391	C_6-HSL	PhzI/PhzR	Activation of *phz* (phenazine-1-carboxamide) biosynthetic operon	Regulated by a cascade involving GacAS, RpoS and PsrA, *phzRI* expression negatively affected by fusaric acid	22–25
	C_4-HSL, C_6-HSL	NYD	Involved in phenazine-1-carboxamide synthesis	Suppressed by PsrA	23
P. fluorescens 2–79	3-OH-C_6-HSL 3-OH-C_7-HSL 3-OH-C_8-HSL 3-OH-C10-HSL, C_6-HSL, C_8-HSL	PhzI/PhzR	Phenazine-1-carboxylate synthesis	/	26
P. fluorescens NCIMB10586	NYD	MupI/MupR	Mupirocin (pseudomonic acid) biosynthesis	/	27
P. fluorescens 2P4	3-oxo-C_6-HSL 3-oxo-C_8-HSL	PcoI/PcoR	Biofilm formation, Colonization on wheat rhizosphere and biocontrol ability	/	28
P. fluorescens CHA0	Non-AHL QS signal (NYD)	Unknown	Linked with biocontrol via the GacAS-Small RNAs (Rsm) cascade	Signal molecule induces Small RNAs. Perception needs functional GacS	29

Pseudomonas sp. M18	C_4-HSL, C_6-HSL	RhlI/RhlR	Repression of pyoluteorin biosynthesis, Stationary phase survival	/	65
P. corrugata CFBP5454	C_6-HSL, C_8-HSL, 3-oxo-C_6-HSL,	PcoI/PcoR	PcoR is involved in swarming, Tobacco HR response and tomato pith necrosis	/	66
Pantoea stewartii subsp *stewartii*	3-oxo-C_6-HSL 3-oxo-C_8-HSL	EsaI/EsaR	EPS production, Biofilm formation, Adhesion, Xylem dissemination, Pathogenicity	EsaR directly represses the *rcsA* gene encoding an essential coactivator for the RcsA/RcsB-mediated transcriptional activation of the *cps* genes (EPS production)	30–33
Erwinia carotovora subsp *carotovora (Ecc)* ATCC39048	3-oxo-C_6-HSL, C_6-HSL	CarI (=ExpI)[d]/ CarR	Regulation of carbapenem biosynthetic genes	/	34
Ecc Ecc71 *Ecc* SCRI193	**3-oxo-C_6-HSL**[e]	ExpI/ExpR	ExpI:Regulation of plant cell wall-degrading enzymes production and Hrp (type III) secretion system in different *Ecc* strains	Main ExpI AHL (3-oxo-C_6-HSL or 3-oxo-C_8-HSL) prevents corresponding ExpR-*rsmA* binding and ExpR-mediated activation of *rsmA* transcription	34–37
Ecc SCC3193 *Ecc* EC153	3-oxo-C_6-HSL, **3-oxo-C_8-HSL**[e]	ExpI/ ExpR1/ ExpR2	Both ExpR1 and ExpR2 cooperate in regulation of plant virulence factor production via RsmA	In Ecc SCC3193, ExpR1 senses the HSL produced by the cognate ExpI, while ExpR2 has broader specifity.	37
E. carotovora subsp. *betavasculorum* strain 168	3-oxo-C_6-HSL	EcbI/EcbR	Antibiotic synthesis and pectate lyase activity	/	35,38
Erwinia carotovora subsp. *atroseptica*	3-oxo-C_6-HSL	ExpI/ExpR VirR(=ExpR2)	Exoenzyme production, Virulence	VirR represses virulence genes at low cell density	39
Erwinia chrysanthemi	3-oxo-C_6-HSL, C_5-HSL	ExpI/ExpR	No clear effect on pectate lyase production, ExpR binds *pel* (pectate lyase) promoters	/	40

(continued)

Table 11.1 (continued)

Bacterium	AHLs	Gene loci involved	Associated phenomena	Additional regulatory compounds	Refs
R. leguminosarum bv. *viciae*	3-OH-$C_{14:1}$-HSL	CinI/CinR	Growth inhibition (CinI)	CinIR is atop of the QS regulatory cascade	41
	3-OH-C_8-HSL, C_6-HSL, C_7-HSL, C_8,-HSL	RaiI/RaiR	Unknown	/	42
	C_6-HSL, C_8-HSL	RhiI/RhiR	Induction of *rhi*ABC genes (involved in nodulation of pea and vetch)	/	43–44
	3-oxo-C_8-HSL C_8-HSL	TraI/TraR/BisR	Conjugal plasmid transfer, growth inhibition (BisR, TraR)	*TraM* (TraR anti-activator)	45–46
Rhizobium etli CFN42	3-oxo-C_8-HSL	TraI/TraR/CinR	Conjugal plasmid transfer	CinR (p42a) responds to an additional signal, *traM* remains unexpressed	47
	3-OH-C_8-HSL[g]	RaiI/RaiR[f]	Unknown	/	48
	NYD	CinI/CinR$_{ch}$[f]	Unknown	/	48
R. etli CNPAF512	3-OH-slc-HSL[h]	CinI/CinR	Nitrogen fixation, growth, symbiosome development, Swarming	Rel$_{Ret}$ upregulates QS systems	49–51
	Short chain AHLs	RaiI/RaiR	Nodulation		51–52
Rhizobium sp NGR234	3-oxo-C_8-HSL	TraI/TraR	Conjugal plasmid transfer Growth inhibition (TraR)	/	53
	Long-chain AHL	NYD			
Bradyrhizobium sp.	Bradyoxetin	Unknown/NswB	Affecting *nod* gene expression	/	54–56
Sinorhizobium meliloti RM1021 and 8530 (ExpR⁺)	C_{12}-HSL, 3-oxo-C_{14}-HSL, $C_{16:1}$-HSL, 3-oxo-$C_{16:1}$-HSL, C_{18}-HSL, C_{16}-HSL, 3-oxo-C_{16}-HSL	SinI/SinR	EPS II production, Motility, Delay in nodule initiation	Most of the regulation of SinI is through ExpR	57–60, 61
		ExpR	ExpR is inactivated in Rm1021 by an IS[j]		
	short chain AHLs	*mel*[i]	/	/	57

Organism	AHL	LuxI/LuxR homologues	Function	Other	Ref.
S. meliloti Rm41	C_{12} to C_{18}-HSL	SinI/SinR ExpR	Regulation of EPS II synthesis	Mutation of $GroEL_C$ leads to reduced AHLs levels	62
	3-oxo-C_8-HSL. C_8-HSL, 3-OH-C_8-HSL	TraI/TraR	Conjugal plasmid transfer	TraM (TraR anti-activator)	62
	Other short chain AHLs	*mel* [i]	/	/	57
Mesorhizobium thianshanense	NYD	MrtI/MrtR	Root hair attachment, Nodulation deficient	/	63
M. huakuii	C_8-HSL	Unknown	Reduced biofilm formation upon overexpression of *A. tumefaciens* TraR	/	64

[a] NYD: not yet determined; [b] HR: hypersensitive response; [c] non-AHL QS molecules: DKP diketopiperazines identified in *P. putida* WCS358; [d] CarI is also called ExpI; [e] in bold is the only or main HSL produced by ExpI in these *Ecc* strains Class I strains: 3-oxo C_8-HSL, Class II strains: 3-oxo-C_6-HSL; [f] homologues identified in genome sequence; [g] based on homology with *R. leguminosarum*; [h] slc: saturated long chain, [i] putative system, homologue of *hdtS*, [j] IS: insertion sequence

(1) Piper et al. 1993, (2) Hwang et al. 1994, (3) Fuqua and Winans 1994, (4) Fuqua et al. 1995, (5) Chai et al. 2001, (6) Zhang et al. 2002, (7) Zhang et al. 2004, (8) Chevrot et al. 2006, (9) Hao et al. 2005, (10) Zheng et al. 2003, (11) Hao and Burr 2006, (12) Wood et al. 1997; (13) Chancey et al. 1999, (14) Zhang and Pierson 2001, (15) Steidle et al. 2002, (16) Arevalo-ferro et al. 2005, (17) Bertani and Venturi 2004, (18) Degrassi et al. 2002, (19) Dubern et al. 2006, (20) Quinones et al. 2004, (21) Quinones et al. 2005, (22) Chin-A-Woeng et al. 2001, (23) Chin-A-Woeng et al. 2005, (24) Girard et al. 2006, (25)Van Rij et al. 2005, (26) Khan et al. 2005, (27) El-Sayed et al. 2001, (28) Wei and Zhang 2006, (29) Kay et al. 2005, (30) Beck von Bodman and Farrand 1995, (31) von Bodman et al. 1998, (32) Koutsoudis et al. 2006, (33) Minogue et al. 2005, (34) Whitehead et al. 2001, (35) Chatterjee et al. 2005, (36) Cui et al. 2005, (37) Sjöblom et al. 2006, (38) Costa and Lopez, 1997, (39) Burr et al. 2006, (40) Nasser et al. 1998, (41) Lithgow et al. 2000, (42) Wisniewsky-Dyé et al. 2002, (43) Cubo et al. 1992, (44) Rodelas et al. 1999, (45) Wilkinson et al. 2002, (46) Danino et al. 2003, (47) Tun-Garrido et al. 2003, (48) Gonzalez et al. 2006, (49) Daniels et al. 2002, (50) Daniels et al. 2004, (51) Mor·s et al. 2005, (52) Rosemeyer et al. 1998, (53) He et al. 2003, (54) Loh et al., 2002, (55) Loh et al., 2002a, (56) Loh and Stacey, 2003, (57) Marketon et al., 2002, (58) Marketon et al. 2003, (59) Hoang et al. 2004, (60) Gao et al. 2005, (61) Pellock et al. 2002, (62) Marketon and Gonzalez, 2002, (63) Zheng et al. 2006, (64) Wang et al. 2004, (65) Yan et al. 2007, (66) Licciardello et al. 2007, (67) Wang et al. 2006, (68) Chatterjee et al. 2007

bacteria is also assisted by efflux pumps. Further research in *P. aeruginosa* revealed that the MexGHI-OpmD pump is also essential for proper cell-cell communication as mutation of the genes encoding the efflux proteins resulted in the inability to produce 3-oxo-C_{12}-HSL and PQS, *Pseudomonas* quinolone signal, another component of the quorum sensing network in *P. aeruginosa* (Aendekerk et al. 2005). This effect is probably due to intracellular accumulation of a toxic PQS precursor and demonstrates the importance of proper functioning of efflux systems for QS at least in *P. aeruginosa*.

11.3 Molecular Mechanisms of AHL Production and Detection

11.3.1 AHL Production

Most of the organisms producing AHL were shown to possess one or more *luxI*-type genes encoding LuxI homologous proteins that catalyze AHL formation. The reaction involves linking and lactonizing the methionine moiety from *S*-adenosyl-methionine (SAM) to particular fatty acyl chains carried primarily on the acyl carrier protein (ACP) (Moré et al. 1996; Schaefer et al. 1996; Parsek et al. 1999). LuxI-type proteins are about 200 amino acids long and are most conserved in the amino-terminal portion, whereas the carboxy terminus is more divergent suggesting a role in recognition of the acyl chain (Fuqua et al. 2001). Recently, the structures of EsaI and LasI were determined (Watson et al. 2002; Gould et al. 2004). This revealed that the acyl binding is indeed determined by specific residues in the C-terminal part. Different sizes of hydrophobic side chains of amino acids in this binding pocket contribute to the structure of the closed binding pocket in EsaI, producing 3-oxo-C_6-HSL and minor amounts of 3-oxo-C_8-HSL, vs a tunnel like structure in LasI producing 3-oxo-C_{12}-HSL. The latter structure places theoretically no restriction on the length of the acyl-chain that can be bound. How LasI selects longer acyl-ACPs relative to shorter, more prevalent, ones is not known. Determination of a co-crystal structure of acyl-ACP and AHL-synthase might resolve this question (Gould et al. 2004). Besides residues limiting the length of the acyl chain in the binding pocket, the presence or absence of a Ser/Thr at position 140 in EsaI was shown to constitute the basis for the C_3 substitution of the acyl chain (Watson et al. 2002).

Besides the LuxI-type of AHL synthases, a different class of AHL synthases was described in *Vibrio* species: LuxM/AinS. The precursors for AinS seem to be similar to those of LuxI-type of AHL synthases although both octanoyl-CoA and octanoyl-ACP could serve as acyl donors. Whether the same is true for LuxM remains to be determined (Hanzelka et al. 1999; Fuqua et al. 2001). Finally, in *Pseudomonas fluorescens* 113, a third potential AHL synthase was described (Laue et al. 2000). A gene, named *hdtS*, was identified and this locus directs the synthesis of a protein of approximately 33 kDa, capable of synthesizing 3-OH-$C_{14:1}$-HSL, C_{10}-HSL and C_6-HSL in *E. coli* (Laue et al. 2000). Further research revealed that HdtS is the primary lysophosphatidic acid (LPA)-acyltransferase in *P. fluorescens* 113, normally responsible for the production of phosphatidic acid, a crucial phospholipid intermediate in cell membrane biosynthesis by acyl chain transfer to LPA. These authors also failed to detect AHL production in *E. coli* after transfer of *hdtS*, so at present it is not clear whether HdtS is involved in AHL production (Cullinane et al. 2005).

11.3.2 AHL Detection

AHL levels influence gene expression through their interaction with LuxR-type transcriptional regulators. Biochemical and genetic studies of a number of LuxR homologues have revealed that they are two-domain proteins (Fuqua et al. 2001). The N-terminal domain binds to a specific AHL and mediates oligomerization, while the C-terminal domain contains a helix-turn-helix DNA binding region (Fuqua et al. 2001). Although no membrane spanning elements are present in *V. fisheri*, it has been proposed that LuxR contacts the inner side of the cytoplasmic membrane through amphipathic interactions which is in line with the observation that monomeric TraR in absence of AHLs cofractionates with the membrane fraction in *A. tumefaciens* (Qin et al. 2000). However, AHL-binding results in structural changes of TraR and shifts the equilibrium towards stable dimer formation and release of TraR complexes in the cytoplasm (Qin et al. 2000). It is proposed that the AHL serves as a scaffold for folding and stabilizes the DNA binding conformation of the activator TraR, and that the lack of AHL enhances the proteolysis of the TraR protein (Zhu and Winans 2001). The resolution of the crystal structure for TraR in complex with its cognate signal, 3-oxo-C_8-HSL, and its target *lux* box-like sequence proved that functional TraR is a dimer and that the AHL is entirely buried within its binding pocket (Vannini et al. 2002; Zhang RG et al. 2002). The N-terminal AHL-binding domain of TraR is sufficient for 3-oxo-C_8-HSL binding and dimerization, as TraR fragments containing only this domain are able to form inactive heterodimers with full-length protein (Luo et al. 2003). This might constitute a higher level of regulation in *A. tumefaciens* as a natural deletion allele, TrlR has been described (Chai et al. 2001). In addition, the TraM protein, although it shares no homology with TraR, exerts its function through formation of inactive heterodimers (Hwang et al. 1999; Vannini et al. 2004).

Although homologues, diversity in the mechanism of activation seems to occur between LuxR-type proteins. While CarR of *Erwinia carotovora* also binds its autoinducer, CarR exists as a preformed dimer and autoinducer binding causes the dimers to form higher-order multimers (Welch et al. 2000). LuxR-type activator proteins usually require *cis*-acting DNA elements, referred to as *lux* box homologues (Fuqua et al. 2001). However, several reports mention regulated genes for which there are no obvious defined *lux*-type boxes. Once bound to the DNA sequence, LuxR-type proteins facilitate the binding of RNAP, via specific residues contacting RNAP. Direct interaction between N-terminal domains of TraR and the α subunit of RNAP has also been described (White and Winans 2005).

While most characterized members of LuxR-type proteins are activators of transcription as described above, a few LuxR-type proteins, such as EsaR and $ExpR_{Ecc}$, act as repressors. These homologues recognize and bind to a DNA binding site, which is positioned in a way that it blocks the transcriptional activity of the RNA polymerase (von Bodman et al. 1998; Andersson et al. 2000). In contrast to the stabilizing effect of AHL binding on the activating LuxR-type proteins, AHL-binding to EsaR promotes structural changes that result in reduced DNA binding potential (Minogue et al. 2002). Whether these conformational changes also render AHL-EsaR sensitive to proteolysis, remains to be studied. A study of von Bodman et al. (2003b) revealed that, although the AHL responsiveness of both proteins is the opposite of that shown by most LuxR family members, EsaR and $ExpR_{Ecc}$ have preserved the ability to interact with RNA polymerase. Indeed, when expression from a typical activator-type sequence (*luxI* promoter) was measured, EsaR and ExpR could bind and activate transcription, although to a lower level as LuxR because of their lower affinity for the *lux* box. In contrast to LuxR, EsaR and ExpR bind and activate expression in absence of AHL and activation is abolished upon addition of the corresponding AHLs.

Despite the overall homology of LuxI/LuxR homologues between related strains, the AHLs produced and recognized might be strain-specific as is illustrated by the ExpI/ExpR homologues found in closely related *Erwinia* strains. In these strains, ExpR is inactivated specifically by the main AHL produced by the corresponding ExpI (Class I strains: 3-oxo-C_8-HSL; Class II strains: 3-oxo-C_6-HSL; see also Table 11.1) (Chatterjee et al. 2005). However, recent research revealed that the situation is even more complex as several *Erwinia* strains possess two ExpR homologues, with ExpR1 reacting with the strains cognate AHL while the ExpR2 has a much broader specificity and might respond to signals from other strains or species (Sjöblom et al. 2006; Burr et al. 2006).

Finally, AHLs might have other roles besides their function as signal molecules. Kaufmann et al. (2005) demonstrated that *N*-(3-oxododecanoyl)-HSL and its nonenzymatically formed tetramic acid degradation product 3-(1-hydroxydecylidene)-5-(2-hydroxyethyl)pyrrolidine-2,4-dione function as antibacterial agents. The latter product was shown to bind iron with comparable affinity to known bacterial siderophores, which might play a role in the observed bactericidal activity of the molecule. Daniels et al. (2006) described a role for long-chain AHLs as biosurfactants during swarming. These molecules were shown to possess significant surface activity and to induce liquid flows, known as Marangoni flows, as a result of gradients in surface tension at biologically relevant concentrations. As a high population density is most likely needed to obtain a sufficient concentration of AHL biosurfactant in the extracellular environment, the link with other quorum sensing-regulated phenomena is logical.

11.4 The Complexity of QS: QS Networks, Interspecies Crosstalk, Quorum Quenching, QS Mimics and Host Responses

11.4.1 QS Networks

During the last decade, extensive studies on QS revealed the complexity and variety in molecular arrangements that enable communication between bacterial cells (reviewed by Waters and Bassler 2005). One aspect of this is that bacteria often possess multiple QS systems either functioning in parallel or in a hierarchic mode. The best studied example of QS systems operating in parallel is *V. harveyi*, possessing three parallel systems. Inputs of all these signals are integrated at the level of LuxU phosphorylation and finally results in bioluminescence via the LuxO response regulator (reviewed by Waters and Bassler 2005). The prototype of hierarchical QS systems is the pathogen *P. aeruginosa* with the LasIR QS system atop of the RhlIR system. The *P. aeruginosa* quinolone signal, 2-heptyl-3-hydroxy-4-quinolone (PQS), adds a further level of complexity to the QS network, as it provides an additional link between the *las* and *rhl* systems. In addition, the quorum sensing cascade of *P. aeruginosa* is subjected to regulation by a number of additional regulatory factors (reviewed by: Whitehead et al. 2001; Daniels et al. 2004; Venturi 2006).

Within the plant-associated bacteria, the most complex QS network has been described in *Rhizobium leguminosarum* bv. *viciae* with the chromosomal *cinIR* system atop of a cascade involving *raiIR* (pIJ9001), *rhiIR* and *traIR-bisR* (pSYM) as depicted in Fig. 11.1 (Rodelas et al. 1999; Lithgow et al. 2000; Wisniewsky-Dyé et al. 2002, Wilkinson et al. 2002). Most strains of *Rhizobium leguminosarum* were known to produce a low-molecular weight component that was referred to as small bacteriocin (Hirsch 1979; Wijffelman et al. 1983) because it results in growth inhibition of a sensitive *R. leguminosarum* strain. Later on, this molecule was characterized structurally as $3\text{-OH-C}_{14:1}\text{-HSL}$ (Schripsema et al. 1996). Independently, Gray et al. (1996) identified this molecule as an inducer of the *rhiABC* operon, regulated by the LuxR-type regulator RhiR. The *rhiABC* genes are located on the symbiotic plasmid pRL1JI and, although their function is unknown, *rhiA* was shown to be highly expressed in bacteroids (Dibb et al. 1984). Furthermore, *rhiABC* expression is repressed by flavonoids (Economou et al. 1989) and is involved in efficient nodulation of pea and vetch (Cubo et al. 1992). Induction of *rhiABC* expression by $3\text{-OH-C}_{14:1}\text{-HSL}$ was found to be dependent on RhiR (Gray et al. 1996). Later work revealed the presence of an AHL synthase encoding gene, *rhiI* upstream of *rhiABC* (Rodelas et al. 1999). RhiI produces $C_6\text{-HSL}$ and $C_8\text{-HSL}$, both of which activate *rhiR*-dependent induction of *rhiABC*. This led to the hypothesis that $3\text{-OH-C}_{14:1}\text{-HSL}$ is an upstream regulatory molecule positively influencing production of other AHLs (Rodelas et al. 1999). The discovery of the *cin* locus, containing CinI responsible for $3\text{-OH-C}_{14:1}\text{-HSL}$ synthesis, confirmed this hypothesis (Lithgow et al. 2000) and further work confirmed the role of CinIR as a master

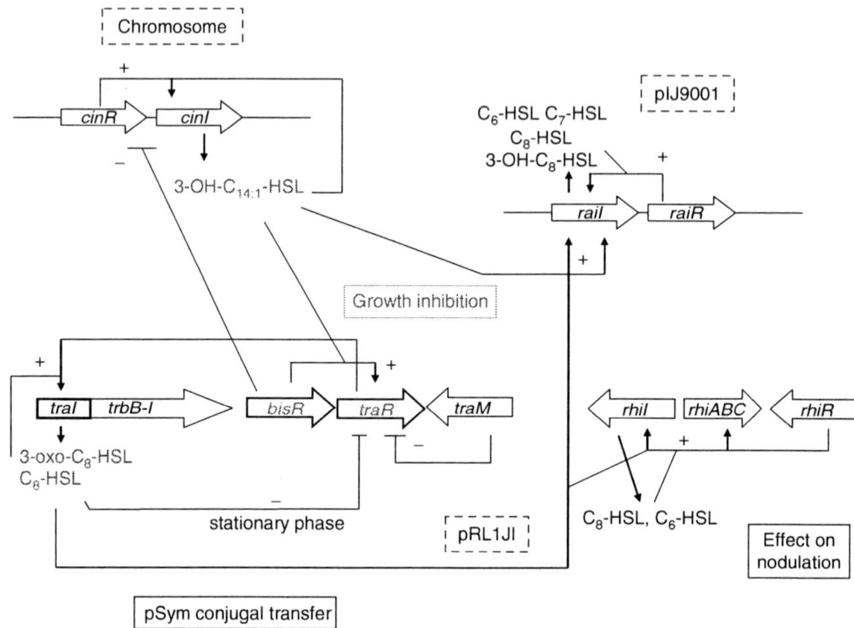

Fig. 11.1 QS network in *R. leguminosarum,* Four identified QS systems in *R. leguminosarum* strain 34. The chromosomally located *cinRI* system, producing 3-OH-C$_{14:1}$, is on top of the regulatory cascade. The dual relationship between *cinRI* and the *traI-bisR-traR* locus enables pRL1JI-carrying donor strains to switch on transfer genes only when 3-OH-C$_{14:1}$-HSL producing acceptor strains are in the close environment. The 3-oxo-C$_8$-HSL produced by TraI also influences expression of the *rhi* and *rai* locus situated on plasmid pRL1JI and pIJ9001 respectively. Effects of 3-OH-C$_{14:1}$ on the *rhi* locus are probably indirect via TraI-produced QS signal molecules. Loci involved in the growth inhibition phenomenon are *indicated in gray*. + and − indicate activation or repression of the genes at the end of the arrow/line

regulator of the QS systems in *R. leguminosarum,* including the later on described pIJ9001-located *raiIR* locus and the pRL1JI-located locus (*traI, bisR, traR*), highly homologous to the plasmid transfer region of the *A. tumefaciens* Ti-plasmid (Lithgow et al. 2001; Wisniewski-Dyé et al. 2002; Wilkinson et al. 2002). Together, these systems form a complex intertwined network.

The plasmid-borne *rai* system is highly similar to the *rai* system of *R. etli* CNPAF512 (Rosemeyer et al. 1998). RaiI mainly synthesizes 3-OH-C$_8$-HSL and minor amounts of C$_6$-HSL, C$_7$-HSL and C$_8$-HSL. The *raiI* gene is upregulated by RaiR and 3-OH-C8-HSL, but also influenced by 3-OH-C$_{14:1}$-HSL and 3-oxo-C$_8$-HSL (Wisniewski-Dyé et al. 2002). The source of this 3-oxo-C$_8$-HSL was identified as TraI, located on the symbiotic plasmid. Together with the downstream located *bisR* and *traR* genes, *traI* is mainly involved in regulation of the conjugal plasmid transfer genes. Besides 3-oxo-C$_8$-HSL, TraI also produces small amounts of C$_8$-HSL. The expression of *traR* is induced by BisR in the presence of low concentrations of the CinI-made 3-OH-C$_{14:1}$-HSL. BisR also represses *cinI* expression in donor strains carrying pSym. Afterwards, TraR activates expression of *traI* in cooperation with the TraI-made signal molecules. Downstream of *traR*, a *traM*-homologue was found corresponding to the organization in *A. tumefaciens*. TraM was proven to reduce premature expression of the *traI-trb* operon, probably because TraM titrates TraR at low expression levels of traR. In the stationary phase, TraI-made QS signal molecules exert a negative effect on *traR* expression (Wilkinson et al. 2002; Danino et al. 2003).

Although the complex relationships between the QS systems are now becoming apparent, the functions of these networks during symbiosis are less clear. No clear symbiotic phenotypes were associated with the RaiIR or CinIR QS systems. It was suggested they could play a role in environmental adaptation not readily observed in standard laboratory tests of growth and nodulation (Lithgow et al. 2000; Wisniewski-Dyé et al. 2002). The *traIR bisR* system was shown to be involved in conjugal transfer of the pSym, and the relationship with the *cin* system enables donor strains to induce plasmid transfer specifically when pRL1JI-deficient strains, producing 3-OH-C$_{14:1}$-HSL, are in close proximity (Wilkinson et al. 2002; Danino et al. 2003). Involvement of quorum sensing in conjugal plasmid transfer was also described for *Rhizobium* NGR234, *Rhizobium etli* CFN42and *Sinorhizobium meliloti* Rm41 (He et al. 2003; Marketon and Gonzalez 2002; Tun-Garrido et al. 2003). Furthermore, 3-OH-C$_{14:1}$-HSL in *R. leguminosarum* is associated with growth inhibition by converting exponential growing cells into stationary phase cells, arresting further growth even though cell densities remain low (Gray et al. 1996). However, this effect requires the presence of pRL1JI. Moreover, addition of 3-OH-C$_{14:1}$-HSL could rescue starvation survival in certain *R. leguminosarum* strains that entered the stationary phase at low cell density, although additional components of spent medium are required to observe this in *R. leguminosarum* 8401/pRL1JI and no effect was observed in *R. leguminosarum* 8401 (Thorne and Williams 1999; Lithgow et al. 2000). These data correlate well with the work of Wilkinson et al. (2002), who demonstrated that the *bisR* and *traR* loci on the symbiotic plasmid pRL1JI, in addition to the TraI-made signal molecules 3-oxo-C$_8$-HSL and C$_8$-HSL, are required for the growth inhibition phenomenon. Most likely, high level induction of TraR by the 3-OH-C$_{14:1}$-HSL-BisR complex causes growth effects in the presence of 3-oxo-C$_8$-HSL and C$_8$-HSL as TraR in conjunction with one of these AHLs affects additional genes in the bacterium. Probably these genes are also located elsewhere in the genome, although they have not yet been characterized (Wilkinson et al. 2002).

The only reported systematic investigation for QS-regulated genes so far in *R. leguminosarum* species, is a proteomic analysis of QS-regulated genes in another strain, *R. leguminosarum* bv. *viciae* UPM791, by a quorum quenching approach. This strain harbours four native plasmids and the proteomic analysis revealed that only a modest fraction of the proteins was affected during the quorum quenching approach. Moreover, the number of regulated genes identified also depended on the presence of pSym and another endogenous plasmid (Cantero et al. 2006). The three main quorum-induced polypeptides appeared to be isoforms of the RhiA protein, although the origin and role of these modifications are presently unknown (Cantero et al. 2006).

Finally, a relatively new topic in the QS regulatory cascades is the relationship with small RNAs-mediated-gene regulation. Basically, these small RNAs (RsmB in *E. carotovora*; RsmX, RsmY and RsmZ in *P. fluorescens*) function by sequestering an RNA-binding protein RsmA involved in repression of secondary metabolism in *E. carotovora* or synthesis of extracellular secondary metabolites in *P. fluorescens* (von Bodman et al. 2003a; Kay et al. 2005). RsmA expression in certain *E. carotovora* species was shown to be affected by ExpR (Cui et al. 2005; Sjöblom et al. 2006). In *P. fluorescens*, the small RNAs contribute to the fine-tuning of the GacS/A controlled population-density dependent regulation (Kay et al. 2005) (see also Table 11.1; reviewed by Bejerano-Sagie and Xavier 2007).

11.4.2 Interspecies Crosstalk

The observation that many QS signal molecules are produced by multiple bacterial species suggests that these molecules function in intraspecies as well as in interspecies communication (see also Fig. 11.2). This is most extensively studied for the LuxS-produced AI-2 signal, as this pathway occurs in more than 55 species of both Gram-positive and Gram-negative bacteria (reviewed by Waters and Bassler 2005).

Many bacteria inhabiting the rhizosphere and plant surfaces produce AHL QS signals. Members of the genus *Rhizobium* show the greatest diversity, with some producing only one and others producing as many as nine detectable putative signals (Cha et al. 1998; Marketon and González 2002). Moreover, TLC analysis revealed that many of these species have AHLs in common such as, e.g. 3-oxo-C_8-HSL which is found in many rhizobia, *A. tumefaciens*, *Erwinia carotovora* pv. *atroseptica* and others. Also, several isolates mainly from the rhizobia, produce extremely nonpolar compounds indicative of very long acyl side-chains. Communication between these species, either synergistically or competitively, may therefore play an important role in the dynamics of these microbial communities. In this respect, Sjöblom et al. (2006) reported that ExpR2 of *Erwinia carotovora* subsp. *carotovora* is able to interact with non-cognate AHLs produced by other bacteria and that this can affect gene expression. Evidence of crosstalk arose earlier from a study showing that extracts of *P. aeruginosa* can induce QS-regulated virulence factor production in *Burkholderia cepacia* (McKenney et al. 1995) whereas extracts of QS mutants failed to do this. Also, 8% of bacterial isolates from wheat root surfaces stimulated QS-regulated phenazine synthesis in *P. aureofaciens* strain 30–84 when co-inoculated and growing in situ on the root surface (Pierson et al. 1998). The use of Gfp-based monitoring strains, that allow in situ visualization of AHL-mediated communication between individual cells in the tomato rhizosphere, confirmed that ca. 12% provoked a positive signal with one or more of the monitor strains.

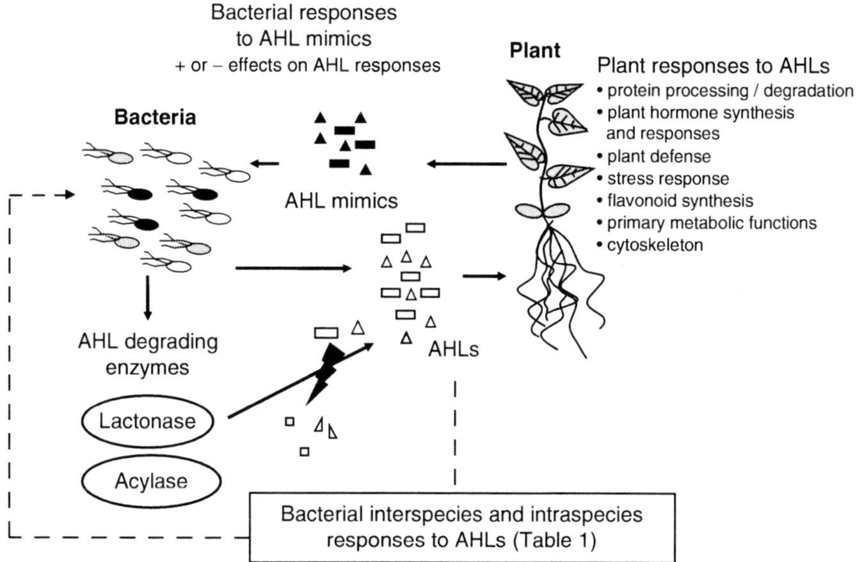

Fig. 11.2 AHL-mediated responses in the plant-bacterium interaction. A pool of AHL signals (*white triangles and rectangles*) produced by different Gram-negative bacteria is present in the neighborhood of plants. These AHLs can be detected by the bacteria present and affect a range of processes in the bacterial community (Table 11.1). Besides AHLs production, a number of bacteria are also shown to interfere with the outcome of AHL signaling by the production of enzymes that degrade the AHLs. Plants also respond to the bacterial AHLs signals, as has mainly been studied for *M. truncatula* by Mathesius et al. (2003). More importantly, different plants were found to participate in the signaling cascade by the production of AHL mimics (*black rectangles and triangles*) that can positively or negatively affect bacterial QS, possibly via effects on bacterial AHL synthesis or secretion

Moreover, this study showed that AHLs are capable of diffusing over relatively long distances in the rhizosphere (Steidle et al. 2001). In addition to the positive interaction between the species described above, numerous reports have demonstrated that interference with QS-mediated signal molecules also frequently occurs not only between bacteria but also between bacteria and higher plants (see below). Finally, the discovery of a LuxR homologue, SdiA, in *Salmonella enterica* serovar Typhimurium and *E. coli*, without the presence of a gene homologous to AHL synthases, further complicates the communication network as these bacteria may be able to eavesdrop on other microbes communication without producing these molecules themselves (Ahmer et al. 1998; Michael et al. 2001). Recent work revealed that a number of AHL-type molecules can induce conformational changes upon binding SdiA, thereby releasing the SdiA protein in a soluble form. The fact that a number of AHL can bind SdiA is consistent with its postulated biological function as a detector of the presence of other species of bacteria (Yao et al. 2006).

11.4.3   Quorum Quenching, QS Mimics and Host Response

Many *Bacillus* species arc now shown to secrete an AHL lactonase, encoded by *aiiA* homologous genes, that is non specific with regard to the AHL side chain and this enables them to interfere with AHL-based QS between other bacteria (Dong et al. 2002). Recent resolution of the crystal structure the *Bacillus thuringiensis* lactonase in complex with AHL reveals that it is a metalloenzyme containing two zinc ions involved in catalysis (Liu et al. 2005; Kim et al. 2005). In addition, species have been identified that degrade AHL to use the breakdown products as carbon or nitrogen source. In *Variovorax paradoxus* and probably also *Ralstonia* sp., this is mediated by an AHL-acylase, encoded by *aiiD* or homologous genes (Leadbetter and Greenberg 2000; Lin et al. 2003). Enzymatic AHL-degrading activities seem to be much more widespread and have now been described in many other species including *Arthrobacter* and *Klebsiella* (Park et al. 2003; d'Angelo-Picard et al. 2005; Yang et al. 2005). Interestingly, some bacteria use these mechanisms to degrade their own AHLs. These include *A. tumefaciens*, encoding AttM, an AHL lactonase, whose expression is upregulated at the stationary phase and results in a sharp decline of 3-oxo-C_8-HSL levels, necessary for Ti-plasmid conjugal transfer (Zhang HB et al. 2002). Upregulation depends on a functional rel_{Atum} gene (Zhang HB et al. 2004). Recently, Chevrot et al. (2006) showed that the activity of the lactonase is also influenced by the level of GABA, a non-protein amino acid whose concentration increases rapidly in wounded plant tissues. GABA stimulated the inactivation of 3-oxo-C_8-HSL by inducing the expression of the *attKLM* operon, of which only *attM* is functionally characterized and found to encode the AHL lactonase (Zhang HB et al. 2002). Further research revealed that mutation of the *aldH* gene, encoding a succinic semialdehyde dehydrogenase (an enzyme of the GABA degradation pathway) which is involved in the conversion of succinic semialdehyde (SSA) to succinic acid, also results in early expression of *attM*. SSA was shown to bind the AttJ repressor (Wang et al. 2006). Also, *P. aeruginosa* was shown to encode an AiiD-type acylase, PvdQ, that specifically degrades long-acyl but not short-acyl-HSLs. This enzyme was shown to be sufficient although not necessary for AHL utilization (Huang et al. 2003). Recently, a second gene, *quiP* (for **q**uorum signal **u**tilization and **i**nactivation **p**rotein), was discovered that probably encodes the main enzyme responsible for the observed AHL acylase activity as *quiP* mutants are defective for growth in a culture containing C_{10}-HSL as the sole carbon and nitrogen source (Huang et al. 2006). Further research is required to determine how *P. aeruginosa* balances expression of QS systems and its AHL acylase activities to avoid futile cycling. Finally, Pierson and coworkers also found that a substantial fraction of bacterial isolates negatively influenced phenazine production in *P. aureofaciens* strain 30–84. The negatively acting signals from all strains tested were not extractable by nonpolar solvents in contrast to other described QS inhibitors (see below). Further characterization of the compound from one strain revealed that it is heat stable and protease resistant making it unlikely that an enzyme degrading AHLs is involved (Morello et al. 2004).

Besides bacteria, plants are also able to interfere with or to mimic QS signalling between bacteria. The first reported AHL mimic was discovered in the red algae *Delisea pulchra*, which produces a halogenated furanone bearing structural similarity to AHLs and specifically inhibiting swarming behaviour in *Serratia liquefaciens* (Givskov et al. 1996). The furanones probably exert their action by binding LuxR-type proteins in a non-agonist fashion, thereby accelerating LuxR decay (Manefield et al. 2002). More effects of furanones on QS and QS-regulated phenotypes were recently reviewed by Shiner et al. (2005). A second compound interfering with QS are diketopiperazines, a family of cyclic dipeptides found in the supernatant of numerous bacterial species. Holden et al. (1999) reported that these compounds can modulate QS in several species by acting as AHL antagonists of some LuxR-based systems and as agonists in others. However, concentrations necessary to activate biosensors are high compared to those of natural AHLs and of the furanones that antagonize swarming in *S. liquefaciens*. Many higher plant species, such as pea, rice, soybean, tomato, crown vetch, and *Medicago truncatula*, secrete substances that mimic AHL signals and affect QS behaviour in bacteria. Both stimulatory and inhibitory effects have been described (Teplitski et al. 2000; Gao et al. 2003; reviewed by Bauer and Mathesius 2004). In *M. truncatula*, secretion of mimics depended on the developmental age of the seedlings and secretion of some compounds possibly also depends on prior exposure of the plant to bacteria (Gao et al. 2003). Most of the compounds partition into organic solvents in a different way compared to AHLs, suggesting they likely are novel compounds that interfere with QS in bacteria, although the exact structures remain to be identified (Bauer and Mathesius 2004). In the unicellular alga *Chlamydomonas reinhardtii*, ethyl acetate extracts of culture supernatants contained more than a dozen chemically separable but unidentified substances capable of specifically stimulating the LasR or CepR AHL bacterial QS reporter strains but not other tested LuxR homologues including LuxR itself. Interestingly, in *S. meliloti*, one of these highly purified *Chlamydomonas* compounds stimulating the LasR reporter had both stimulatory and inhibitory effects on the accumulation of proteins that were altered in response to the bacterium's own AHL signals (Teplitski et al. 2004). Furthermore, Keshavan et al. (2005) identified L-canavanine, an arginine analogue, as one of the compounds produced by seed exudates of the *S. meliloti* host plant alfalfa that interferes with QS in certain reporter strains and with QS-regulated *exp* gene expression in *S. meliloti*. This provides evidence that plants can effectively influence and even disrupt bacterial QS by secretion of mimics. Mathesius et al. (2003) also reported on the effects of AHLs on plant gene expression by determining the effect of exposure of *M. truncatula* roots to 3-oxo-C_{12}-HSL or 3-oxo-$C_{16:1}$-HSL. The abundance of over 150 proteins was changed, although the response depended on the concentration and identity of the AHLs, suggesting that plants can differentiate between QS signals from different bacteria. In addition, AHL treatment affected metabolites secreted by the roots, including the AHL mimics (Mathesius et al. 2003; Teplitski et al. 2004).

11.5 Conclusions

Research over the past decades showed that many plant-associated bacteria, from both pathogenic and beneficial species, use QS to regulate specific traits, some of these being important in the interaction with other bacteria or the host plant. Often these QS systems are part of complex regulatory networks that have only begun to be unraveled. Although many bacteria are now found to possess AHL-based LuxR/LuxI homologous systems, the situation is far more complex as other types of QS systems and signaling molecules have been described. In addition, despite sequence homology, function and regulation of the QS systems may be adapted at species or even strain level and much more research is required to unravel the roles they play in the microbial communities and during bacteria-plant interaction. Moreover, besides their signaling role, additional biological functions of AHL and/or their degradation products have also been reported, including biosurfactant activity, antimicrobial activity and a role as siderophore. The fact that host plants are able to respond or interrupt bacterial QS further illustrates that AHL signaling is an important factor in determining the outcome of plant-bacteria interaction. Moreover, an increasing number of bacteria is described that can degrade AHLs, sometimes produced by the bacterium itself. Studying the dynamics of AHL production and degradation and their effects on microbial communities and plant-interaction will help to fully understand the role of QS in plant-microbe interaction and may reveal further control points for manipulation of these interactions.

References

Aendekerk S, Diggle SP, Song Z, Hoiby N, Cornelis P, Williams P, Camara M (2005) The MexGHI-OpmD multidrug efflux pump controls growth, antibiotic susceptibility and virulence in *Pseudomonas aeruginosa* via 4-quinolone-dependent cell-to-cell communication. Microbiology 151:1113–1125

Ahmer BM, van Reeuwijk J, Timmers CD, Valentine PJ, Heffron F (1998) *Salmonella typhimurium* encodes an SdiA homolog, a putative quorum sensor of the LuxR family, that regulates genes on the virulence plasmid. J Bacteriol 180:1185–1193

Andersson RA, Eriksson AR, Heikinheimo R, Mae A, Pirhonen M, Koiv V, Hyytiainen H, Tuikkala A, Palva ET (2000) Quorum sensing in the plant pathogen *Erwinia carotovora* subsp. *carotovora*: the role of expR(Ecc). Mol Plant Microbe Interact 13:384–393

Arevalo-Ferro C, Reil G, Gorg A, Eberl L, Riedel K (2005) Biofilm formation of *Pseudomonas putida* IsoF: the role of quorum sensing as assessed by proteomics. Syst Appl Microbiol 28:87–114

Bauer WD, Mathesius U (2004) Plant responses to bacterial quorum sensing signals. Curr Opin Plant Biol 7:429–433

Beck von Bodman S, Farrand SK (1995) Capsular polysaccharide biosynthesis and pathogenicity in *Erwinia stewartii* require induction by an *N*-acylhomoserine lactone autoinducer. J Bacteriol 177:5000–5008

Bejerani-Sagie M, Xavier KB (2007) The role of small RNAs in quorum sensing. Curr Opin Microbiol 10:189–198

Bertani I, Venturi V (2004) Regulation of the *N*-acyl homoserine lactone-dependent quorum-sensing system in rhizosphere *Pseudomonas putida* WCS358 and cross-talk with the stationary-phase RpoS sigma factor and the global regulator GacA. Appl Environ Microbiol 70:5493–5502

Burr T, Barnard AM, Corbett MJ, Pemberton CL, Simpson NJ, Salmond GP (2006) Identification of the central quorum sensing regulator of virulence in the enteric phytopathogen, *Erwinia carotovora*: the VirR repressor. Mol Microbiol 59:113–125

Cantero L, Palacios JM, Ruiz-Argueso T, Imperial J (2006) Proteomic analysis of quorum sensing in *Rhizobium leguminosarum* biovar *viciae* UPM791. Proteomics 6:S97–S106

Cha C, Gao P, Chen YC, Shaw PD, Farrand SK (1998) Production of acyl-homoserine lactone quorum-sensing signals by gram-negative plant-associated bacteria. Mol Plant Microbe Interact 11:1119–1129

Chai Y, Zhu J, Winans SC (2001) TrlR, a defective TraR-like protein of *Agrobacterium tumefaciens*, blocks TraR function in vitro by forming inactive TrlR:TraR dimers. Mol Microbiol 40:414–421

Chancey ST, Wood DW, Pierson LS III (1999) Two-component transcriptional regulation of *N*-acyl-homoserine lactone production in *Pseudomonas aureofaciens*. Appl Environ Microbiol 65:2294–2299

Chatterjee A, Cui Y, Hasegawa H, Leigh N, Dixit V, Chatterjee AK (2005) Comparative analysis of two classes of quorum-sensing signaling systems that control production of extracellular proteins and secondary metabolites in *Erwinia carotovora* subspecies. J Bacteriol 187:8026–8038

Chatterjee A, Cui Y, Hasegawa H, Chatterjee AK (2007) PsrA, the *Pseudomonas* sigma regulator, controls regulators of epiphytic fitness, quorum-sensing signals, and plant interactions in *Pseudomonas syringae* pv. tomato strain DC3000. Appl Environ Microbiol 73:3684–3694

Chevrot R, Rosen R, Haudecoeur E, Cirou A, Shelp BJ, Ron E, Faure D (2006) GABA controls the level of quorum-sensing signal in *Agrobacterium tumefaciens*. Proc Natl Acad Sci (USA) 103:7460–7464

Chin-A-Woeng TF, van den Broek D, de Voer G, van der Drift KM, Tuinman S, Thomas-Oates JE, Lugtenberg BJ, Bloemberg GV (2001) Phenazine-1-carboxamide production in the biocontrol strain *Pseudomonas chlororaphis* PCL1391 is regulated by multiple factors secreted into the growth medium. Mol Plant Microbe Interact 14:969–979

Chin-A-Woeng TF, van den Broek D, Lugtenberg BJ, Bloemberg GV (2005) The *Pseudomonas chlororaphis* PCL1391 sigma regulator *psrA* represses the production of the antifungal metabolite phenazine-1-carboxamide. Mol Plant Microbe Interact 18:244–253

Costa JM, Loper JE (1997) EcbI and EcbR: homologs of LuxI and LuxR affecting antibiotic and exoenzyme production by *Erwinia carotovora* subsp. *betavasculorum*. Can J Microbiol 43:1164–1171

Cubo MT, Economou A, Murphy G, Johnston AW, Downie JA (1992) Molecular characterization and regulation of the rhizosphere-expressed genes *rhiABCR* that can influence nodulation by *Rhizobium leguminosarum* biovar *viciae*. J Bacteriol 174:4026–4035

Cui Y, Chatterjee A, Hasegawa H, Dixit V, Leigh N, Chatterjee AK (2005) ExpR, a LuxR homolog of *Erwinia carotovora* subsp. *carotovora*, activates transcription of *rsmA*, which specifies a global regulatory RNA-binding protein. J Bacteriol 187:4792–4803

Cullinane M, Baysse C, Morrissey JP, O'Gara F (2005) Identification of two lysophosphatidic acid acyltransferase genes with overlapping function in *Pseudomonas fluorescens*. Microbiology 151:3071–3080

d'Angelo-Picard C, Faure D, Penot I, Dessaux Y (2005) Diversity of *N*-acyl homoserine lactone-producing and -degrading bacteria in soil and tobacco rhizosphere. Environ Microbiol 7:1796–1808

Daniels R, De Vos DE, Desair J, Raedschelders G, Luyten E, Rosemeyer V, Verreth C, Schoeters E, Vanderleyden J, Michiels J (2002) The *cin* quorum sensing locus of *Rhizobium etli* CNPAF512 affects growth and symbiotic nitrogen fixation. J Biol Chem 277:462–468

Daniels R, Vanderleyden J, Michiels J (2004) Quorum sensing and swarming migration in bacteria. FEMS Microbiol Rev 28:261–289

Daniels R, Reynaert S, Hoekstra H, Verreth C, Janssens J, Braeken K, Fauvart M, Beullens S, Heusdens C, Lambrichts I, De Vos DE, Vanderleyden J, Vermant J, Michiels J (2006) Quorum signal molecules as biosurfactants affecting swarming in *Rhizobium etli*. Proc Natl Acad Sci (USA) 103:14965–14970

Danino VE, Wilkinson A, Edwards A, Downie JA (2003) Recipient-induced transfer of the symbiotic plasmid pRL1JI in *Rhizobium leguminosarum* bv. *viciae* is regulated by a quorum-sensing relay. Mol Microbiol 50:511–525

Degrassi G, Aguilar C, Bosco M, Zahariev S, Pongor S, Venturi V (2002) Plant growth-promoting *Pseudomonas putida* WCS358 produces and secretes four cyclic dipeptides: cross-talk with quorum sensing bacterial sensors. Curr Microbiol 45:250–254

Dibb NJ, Downie JA, Brewin NJ (1984) Identification of a rhizosphere protein encoded by the symbiotic plasmid of *Rhizobium leguminosarum*. J Bacteriol 158:621–627

Dong YH, Gusti AR, Zhang Q, Xu JL, Zhang LH (2002) Identification of quorum-quenching *N*-acyl homoserine lactonases from *Bacillus* species. Appl Environ Microbiol 68:1754–1759

Dubern JF, Lugtenberg BJ, Bloemberg GV (2006) The *ppuI-rsaL-ppuR* quorum-sensing system regulates biofilm formation of *Pseudomonas putida* PCL1445 by controlling biosynthesis of the cyclic lipopeptides putisolvins I and II. J Bacteriol 188:2898–2906

Economou A, Hawkins FKL, Downie JA, Johnson AWB (1989) Transcription of *rhiA*, a gene on a *Rhizobium leguminosarum* bv. *viciae* Sym plasmid, requires *rhiR* and is expressed by flavonoids that induce nod genes. Mol Microbiol 3:87–93

El-Sayed AK, Hothersall J, Thomas CM (2001) Quorum-sensing-dependent regulation of biosynthesis of the polyketide antibiotic mupirocin in *Pseudomonas fluorescens* NCIMB 10586. Microbiology 147:2127–2139

Fuqua WC, Winans SC (1994) A LuxR-LuxI type regulatory system activates *Agrobacterium* Ti plasmid conjugal transfer in the presence of a plant tumor metabolite. J Bacteriol 176:2796–2806

Fuqua WC, Winans SC, Greenberg EP (1994) Quorum sensing in bacteria: the LuxR-LuxI family of cell density-responsive transcriptional regulators. J Bacteriol 176:269–275

Fuqua C, Burbea M, Winans SC (1995) Activity of the *Agrobacterium* Ti plasmid conjugal transfer regulator TraR is inhibited by the product of the *traM* gene. J Bacteriol 177:1367–1373

Fuqua C, Parsek MR, Greenberg EP (2001) Regulation of gene expression by cell-to-cell communication: acyl-homoserine lactone quorum sensing. Annu Rev Genet 35:439–468

Gao M, Teplitski M, Robinson JB, Bauer WD (2003) Production of substances by *Medicago truncatula* that affect bacterial quorum sensing. Mol Plant Microbe Interact 16:827–834

Gao M, Chen H, Eberhard A, Gronquist MR, Robinson JB, Rolfe BG, Bauer WD (2005) *sinI-* and *expR*-dependent quorum sensing in *Sinorhizobium meliloti*. J Bacteriol 187:7931–7944

Girard G, van Rij ET, Lugtenberg BJ, Bloemberg GV (2006) Regulatory roles of *psrA* and *rpoS* in phenazine-1-carboxamide synthesis by *Pseudomonas chlororaphis* PCL1391. Microbiology 152:43–58

Givskov M, de Nys R, Manefield M, Gram L, Maximilien R, Eberl L, Molin S, Steinberg PD, Kjelleberg S (1996) Eukaryotic interference with homoserine lactone-mediated prokaryotic signalling. J Bacteriol 178:6618–6622

Gonzalez V, Santamaria RI, Bustos P, Hernandez-Gonzalez I, Medrano-Soto A, Moreno-Hagelsieb G, Janga SC, Ramirez MA, Jimenez-Jacinto V, Collado-Vides J, Davila G (2006) The partitioned *Rhizobium etli* genome: genetic and metabolic redundancy in seven interacting replicons. Proc Natl Acad Sci (USA) 103:3834–3839

Gould TA, Schweizer HP, Churchill ME (2004) Structure of the *Pseudomonas aeruginosa* acyl-homoserine lactone synthase LasI. Mol Microbiol 53:1135–1146

Gray KM, Pearson JP, Downie JA, Boboye BE, Greenberg EP (1996) Cell-to-cell signaling in the symbiotic nitrogen-fixing bacterium *Rhizobium leguminosarum*: autoinduction of a stationary phase and rhizosphere-expressed genes. J Bacteriol 178:372–376

Hanzelka BL, Parsek MR, Val DL, Dunlap PV, Cronan JE Jr, Greenberg EP (1999) Acylhomoserine lactone synthase activity of the *Vibrio fischeri* AinS protein. J Bacteriol 181:5766–5770

Hao G, Burr TJ (2006) Regulation of long-chain *N*-acyl-homoserine lactones in *Agrobacterium vitis*. J Bacteriol 188:2173–2183

Hao G, Zhang H, Zheng D, Burr TJ (2005) *luxR* homolog *avhR* in *Agrobacterium vitis* affects the development of a grape-specific necrosis and a tobacco hypersensitive response. J Bacteriol 187:185–192

He X, Chang W, Pierce DL, Seib LO, Wagner J, Fuqua C (2003) Quorum sensing in *Rhizobium* sp. strain NGR234 regulates conjugal transfer (*tra*) gene expression and influences growth rate. J Bacteriol 185:809–822

Hirsch PR (1979) Plasmid-determined bacteriocin production by *Rhizobium leguminosarum*. J Gen Microbiol 113:219–228

Hoang HH, Becker A, Gonzalez JE (2004) The LuxR homolog ExpR, in combination with the Sin quorum sensing system, plays a central role in *Sinorhizobium meliloti* gene expression. J Bacteriol 186:5460–5472

Holden MT, Ram Chhabra S, de Nys R, Stead P, Bainton NJ, Hill PJ, Manefield M, Kumar N, Labatte M, England D, Rice S, Givskov M, Salmond GP, Stewart GS, Bycroft BW, Kjelleberg S, Williams P (1999) Quorum-sensing cross talk: isolation and chemical characterization of cyclic dipeptides from *Pseudomonas aeruginosa* and other Gram-negative bacteria. Mol Microbiol 33:1254–1266

Huang JJ, Han JI, Zhang LH, Leadbetter JR (2003) Utilization of acyl-homoserine lactone quorum signals for growth by a soil pseudomonad and *Pseudomonas aeruginosa* PAO1. Appl Environ Microbiol 69:5941–5949

Huang JJ, Petersen A, Whiteley M, Leadbetter JR (2006) Identification of QuiP, the product of gene PA1032, as the second acyl-homoserine lactone acylase of *Pseudomonas aeruginosa* PAO1. Appl Environ Microbiol 72:1190–1197

Hwang I, Li PL, Zhang L, Piper KR, Cook DM, Tate ME, Farrand SK (1994) TraI, a LuxI homologue, is responsible for production of conjugation factor, the Ti plasmid *N*-acylhomoserine lactone autoinducer. Proc Natl Acad Sci (USA) 91:4639–4643

Hwang I, Smyth AJ, Luo ZQ, Farrand SK (1999) Modulating quorum sensing by antiactivation: TraM interacts with TraR to inhibit activation of Ti plasmid conjugal transfer genes. Mol Microbiol 34:282–294

Kaplan HB, Greenberg EP (1985) Diffusion of autoinducer is involved in regulation of the *Vibrio fisheri* luminescence system. J Bacteriol 163:1210–1214

Kaufmann GF, Sartorio R, Lee SH, Rogers CJ, Meijler MM, Moss JA, Clapham B, Brogan AP, Dickerson TJ, Janda KD (2005) Revisiting quorum sensing: discovery of additional chemical and biological functions for 3-oxo-*N*-acylhomoserine lactones. Proc Natl Acad Sci (USA) 102:309–314

Kay E, Dubuis C, Haas D (2005) Three small RNAs jointly ensure secondary metabolism and biocontrol in *Pseudomonas fluorescens* CHA0. Proc Natl Acad Sci (USA) 102:17136–17141

Keshavan ND, Chowdhary PK, Haines DC, Gonzalez JE (2005) L Canavanine made by *Medicago sativa* interferes with quorum sensing in *Sinorhizobium meliloti*. J Bacteriol 187:8427–8436

Khan SR, Mavrodi DV, Jog GJ, Suga H, Thomashow LS, Farrand SK (2005) Activation of the *phz* operon of *Pseudomonas fluorescens* 2–79 requires the LuxR homolog PhzR, *N*-(3-OH-Hexanoyl)-L-homoserine lactone produced by the LuxI homolog PhzI, and a cis-acting *phz* box. J Bacteriol 187:6517–6527

Kim MH, Choi WC, Kang HO, Lee JS, Kang BS, Kim KJ, Derewenda ZS, Oh TK, Lee CH, Lee JK. (2005) The molecular structure and catalytic mechanism of a quorum-quenching *N*-acyl-L-homoserine lactone hydrolase. Proc Natl Acad Sci (USA) 102:17606–17611

Koutsoudis MD, Tsaltas D, Minogue TD, von Bodman SB (2006) Quorum-sensing regulation governs bacterial adhesion, biofilm development, and host colonization in *Pantoea stewartii* subspecies *stewartii*. Proc Natl Acad Sci (USA) 103:5983–5988

Laue BE, Jiang Y, Chhabra SR, Jacob S, Stewart GS, Hardman A, Downie JA, O'Gara F, Williams P (2000) The biocontrol strain *Pseudomonas fluorescens* F113 produces the *Rhizobium* small bacteriocin, *N*-(3-hydroxy-7-*cis*-tetradecenoyl) homoserine lactone, via HdtS, a putative novel *N*-acylhomoserine lactone synthase. Microbiology 146:2469–2480

Leadbetter JR, Greenberg EP (2000) Metabolism of acyl-homoserine lactone quorum-sensing signals by *Variovorax paradoxus*. J Bacteriol 182:6921–6926

Licciardello G, Bertani I, Steindler L, Bella P, Venturi V, Catara V (2007) *Pseudomonas corrugata* contains a conserved *N*-acyl homoserine lactone quorum sensing system; its role in tomato pathogenicity and tobacco hypersensitivity response. FEMS Microbiol Ecol [Epub ahead of print]

Lin YH, Xu JL, Hu J, Wang LH, Ong SL, Leadbetter JR, Zhang LH (2003) Acyl-homoserine lactone acylase from *Ralstonia* strain XJ12B represents a novel and potent class of quorum-quenching enzymes. Mol Microbiol 47:849–860

Lithgow JK, Wilkinson A, Hardman A, Rodelas B, Wisniewski-Dye F, Williams P, Downie JA (2000) The regulatory locus *cinRI* in *Rhizobium leguminosarum* controls a network of quorum-sensing loci. Mol Microbiol 37:81–97

Lithgow JK, Danino VE, Jones J, Downie JA (2001) Analysis of *N*-acyl homoserine-lactone quorum-sensing molecules made by different strains and biovars of *Rhizobium leguminosarum* containing different symbiotic plasmids. Plant Soil 232:3–12

Liu D, Lepore BW, Petsko GA, Thomas PW, Stone EM, Fast W, Ringe D (2005) Three-dimensional structure of the quorum-quenching *N*-acyl homoserine lactone hydrolase from *Bacillus thuringiensis*. Proc Natl Acad Sci (USA) 102:11882–11887

Loh J, Stacey G (2003) Nodulation gene regulation in *Bradyrhizobium japonicum*: a unique integration of global regulatory circuits. Appl Environ Microbiol 69:10–17

Loh J, Carlson RW, York WS, Stacey G (2002a) Bradyoxetin, a unique chemical signal involved in symbiotic gene regulation. Proc Natl Acad Sci (USA) 99:14446–14451

Loh J, Lohar DP, Andersen B, Stacey G (2002b) A two-component regulator mediates population-density-dependent expression of the *Bradyrhizobium japonicum* nodulation genes. J Bacteriol 184:1759–1766

Luo ZQ, Smyth AJ, Gao P, Qin Y, Farrand SK (2003) Mutational analysis of TraR. Correlating function with molecular structure of a quorum-sensing transcriptional activator. J Biol Chem 278:13173–13182

Manefield M, Rasmussen TB, Henzter M, Andersen JB, Steinberg P, Kjelleberg S, Givskov M (2002) Halogenated furanones inhibit quorum sensing through accelerated LuxR turnover. Microbiology 148:1119–1127

Marketon MM, González JE (2002) Identification of two quorum-sensing systems in *Sinorhizobium meliloti*. J Bacteriol 184:3466–3475

Marketon MM, Gronquist MR, Eberhard A, González JE (2002) Characterization of the *Sinorhizobium meliloti sinR/sinI* locus and the production of novel *N*-acyl homoserine lactones. J Bacteriol 184:5686–5695

Marketon MM, Glenn SA, Eberhard A, González JE (2003) Quorum sensing controls exopolysaccharide production in *Sinorhizobium meliloti*. J Bacteriol 185:325–333

Mathesius U, Mulders S, Gao M, Teplitski M, Caetano-Anolles G, Rolfe BG, Bauer WD (2003) Extensive and specific responses of a eukaryote to bacterial quorum-sensing signals. Proc Natl Acad Sci (USA) 100:1444–1449

McKenney D, Brown KE, Allison DG (1995) Influence of *Pseudomonas aeruginosa* exoproducts on virulence factor production in *Burkholderia cepacia*: evidence of interspecies communication. J Bacteriol 177:6989–6992

Michael B, Smith JN, Swift S, Heffron F, Ahmer BM (2001) SdiA of *Salmonella enterica* is a LuxR homolog that detects mixed microbial communities. J Bacteriol 183:5733–5742

Minogue TD, Wehland-von Trebra M, Bernhard F, von Bodman SB (2002) The autoregulatory role of EsaR, a quorum-sensing regulator in *Pantoea stewartii* ssp. *stewartii*: evidence for a repressor function. Mol Microbiol 44:1625–1635

Minogue TD, Carlier AL, Koutsoudis MD, von Bodman SB (2005) The cell density-dependent expression of stewartan exopolysaccharide in *Pantoea stewartii* ssp. *stewartii* is a function of EsaR-mediated repression of the *rcsA* gene. Mol Microbiol 56:189–203

Moré MI, Finger LD, Stryker JL, Fuqua C, Eberhard A, Winans SC (1996) Enzymatic synthesis of a quorum-sensing autoinducer through use of defined substrates. Science 272:1655–1658

Morello JE, Pierson EA, Pierson LS III (2004) Negative cross-communication among wheat rhizosphere bacteria: effect on antibiotic production by the biological control bacterium *Pseudomonas aureofaciens* 30–84. Appl Environ Microbiol 70:3103–3109

Moris M, Braeken K, Schoeters E, Verreth C, Beullens S, Vanderleyden J, Michiels J (2005) Effective symbiosis between *Rhizobium etli* and *Phaseolus vulgaris* requires the alarmone ppGpp. J Bacteriol 187:5460–5469

Nasser W, Bouillant ML, Salmond G, Reverchon S (1998) Characterization of the *Erwinia chrysanthemi expI-expR* locus directing the synthesis of two N-acyl-homoserine lactone signal molecules. Mol Microbiol 29:1391–1405

Park SY, Lee SJ, Oh TK, Oh JW, Koo BT, Yum DY, Lee JK (2003) AhlD, an N-acylhomoserine lactonase in *Arthrobacter* sp., and predicted homologues in other bacteria. Microbiology 149:1541–1550

Parsek MR, Val DL, Hanzelka BL, Cronan JE Jr, Greenberg EP (1999) Acyl homoserine-lactone quorum-sensing signal generation. Proc Natl Acad Sci (USA) 96:4360–4365

Pearson JP, Van Delden C, Iglewski BH (1999) Active efflux and diffusion are involved in transport of *Pseudomonas aeruginosa* cell-to-cell signals. J Bacteriol 181:1203–1210

Pellock BJ, Teplitski M, Boinay RP, Bauer WD, Walker GC (2002) A LuxR homolog controls production of symbiotically active extracellular polysaccharide II by *Sinorhizobium meliloti*. J Bacteriol 184:5067–5076

Pierson EA, Wood DW, Cannon JA, Blachere FM (1998) Interpopulation signaling via N-acyl-homoserine lactones among bacteria in the wheat rhizosphere. Mol Plant Microbe Interact 11:1078–1084

Piper KR, von Bodman SB, Farrand SK (1993) Conjugation factor of *Agrobacterium tumefaciens* regulates Ti plasmid transfer by autoinduction. Nature 362:448–450

Qin Y, Luo ZQ, Smyth AJ, Gao P, Beck von Bodman S, Farrand SK (2000) Quorum-sensing signal binding results in dimerization of TraR and its release from membranes into the cytoplasm. EMBO J 19:5212–5221

Quinones B, Pujol CJ, Lindow SE (2004) Regulation of AHL production and its contribution to epiphytic fitness in *Pseudomonas syringae*. Mol Plant Microbe Interact 17:521–531

Quinones B, Dulla G, Lindow SE (2005) Quorum sensing regulates exopolysaccharide production, motility, and virulence in *Pseudomonas syringae*. Mol Plant Microbe Interact 18:682–693

Redfield RJ (2002) Is quorum sensing a side effect of diffusion sensing? Trends Microbiol 10:365–370

Rodelas B, Lithgow JK, Wisniewski-Dye F, Hardman A, Wilkinson A, Economou A, Williams P, Downie JA (1999) Analysis of quorum-sensing-dependent control of rhizosphere-expressed (*rhi*) genes in *Rhizobium leguminosarum* bv. *viciae*. J Bacteriol 181:3816 3823

Rosemeyer V, Michiels J, Verreth C, Vanderleyden J (1998) *luxI*- and *luxR*-homologous genes of *Rhizobium etli* CNPAF512 contribute to synthesis of autoinducer molecules and nodulation of *Phaseolus vulgaris*. J Bacteriol 180:815–821

Schaefer AL, Val DL, Hanzelka BL, Cronan JE Jr, Greenberg EP (1996) Generation of cell-to-cell signals in quorum sensing: acyl homoserine lactone synthase activity of a purified *Vibrio fischeri* LuxI protein. Proc Natl Acad Sci (USA) 93:9505–9509

Schripsema J, de Rudder KE, van Vliet TB, Lankhorst PP, de Vroom E, Kijne JW, van Brussel AA (1996) Bacteriocin small of *Rhizobium leguminosarum* belongs to the class of N-acyl-L-homoserine lactone molecules, known as autoinducers and as quorum sensing co-transcription factors. J Bacteriol 178:366–371

Shiner EK, Rumbaugh KP, Williams SC (2005) Inter-kingdom signaling: deciphering the language of acyl homoserine lactones. FEMS Microbiol Rev 29:935–947

Sjöblom S, Brader G, Koch G, Palva ET (2006) Cooperation of two distinct ExpR regulators controls quorum sensing specificity and virulence in the plant pathogen *Erwinia carotovora*. Mol Microbiol 60:1474–1489

Steidle A, Sigl K, Schuhegger R, Ihring A, Schmid M, Gantner S, Stoffels M, Riedel K, Givskov M, Hartmann A, Langebartels C, Eberl L (2001) Visualization of *N*-acylhomoserine lactone-mediated cell-cell communication between bacteria colonizing the tomato rhizosphere. Appl Environ Microbiol 67:5761–5770

Steidle A, Allesen-Holm M, Riedel K, Berg G, Givskov M, Molin S, Eberl L (2002) Identification and characterization of an *N*-acylhomoserine lactone-dependent quorum-sensing system in *Pseudomonas putida* strain IsoF. Appl Environ Microbiol 68:6371–6382

Teplitski M, Robinson JB, Bauer WD (2000) Plants secrete substances that mimic bacterial *N*-acyl homoserine lactone signal activities and affect population density-dependent behaviors in associated bacteria. Mol Plant Microbe Interact 13:637–648

Teplitski M, Chen H, Rajamani S, Gao M, Merighi M, Sayre RT, Robinson JB, Rolfe BG, Bauer WD (2004) *Chlamydomonas reinhardtii* secretes compounds that mimic bacterial signals and interfere with quorum sensing regulation in bacteria. Plant Physiol 134:137–146

Thorne SH, Williams HD (1999) Cell density-dependent starvation survival of *Rhizobium legumi-nosarum* bv. *phaseoli:* identification of the role of an *N*-acyl homoserine lactone in adaptation to stationary-phase survival. J Bacteriol. 181:981–990

Tun-Garrido C, Bustos P, Gonzalez V, Brom S (2003) Conjugative transfer of p42a from *Rhizobium etli* CFN42, which is required for mobilization of the symbiotic plasmid, is regulated by quorum sensing. J Bacteriol 185:1681–1692

Vannini A, Volpari C, Gargioli C, Muraglia E, Cortese R, De Francesco R, Neddermann P, Marco SD (2002) The crystal structure of the quorum sensing protein TraR bound to its autoinducer and target DNA. EMBO J 21:4393–4401

Vannini A, Volpari C, Di Marco S (2004) Crystal structure of the quorum-sensing protein TraM and its interaction with the transcriptional regulator TraR. J Biol Chem 279:24291–24296

van Rij ET, Girard G, Lugtenberg BJ, Bloemberg GV (2005) Influence of fusaric acid on phena-zine-1-carboxamide synthesis and gene expression of *Pseudomonas chlororaphis* strain PCL1391. Microbiology 151:2805–2814

Venturi V (2006) Regulation of quorum sensing in *Pseudomonas*. FEMS Microbiol Rev 30:274–291

Visick KL, Fuqua C (2005) Decoding microbial chatter: cell-cell communication in bacteria. J Bacteriol 187:5507–5519

von Bodman SB, Majerczak DR, Coplin DL (1998) A negative regulator mediates quorum-sensing control of exopolysaccharide production in *Pantoea stewartii* subsp. *stewartii*. Proc Natl Acad Sci (USA) 95:7687–7692

von Bodman SB, Bauer WD, Coplin LD (2003a) Quorum sensing in plant-pathogenic bacteria. Annu Rev Phytopathol 41:455–482

von Bodman SB, Ball JK, Faini MA, Herrera CM, Minogue TD, Urbanowski ML, Stevens AM (2003b) The quorum sensing negative regulators EsaR and ExpR$_{Ecc}$, homologues within the LuxR family, retain the ability to function as activators of transcription. J Bacteriol 185:7001–7007

Wang H, Zhong Z, Cai T, Li S, Zhu J (2004) Heterologous overexpression of quorum-sensing regulators to study cell-density-dependent phenotypes in a symbiotic plant bacterium *Mesorhizobium huakuii*. Arch Microbiol 182:520–525

Wang C, Zhang HB, Wang LH, Zhang LH (2006) Succinic semialdehyde couples stress response to quorum-sensing signal decay in *Agrobacterium tumefaciens*. Mol Microbiol 62:45–56

Waters CM, Bassler BL (2005) Quorum sensing: cell-to-cell communication in bacteria. Annu Rev Cell Dev Biol 21:319–346

Watson WT, Minogue TD, Val DL, von Bodman SB, Churchill ME (2002) Structural basis and specificity of acyl-homoserine lactone signal production in bacterial quorum sensing. Mol Cell 9:685–694

Wei HL, Zhang LQ (2006) Quorum-sensing system influences root colonization and biological control ability in *Pseudomonas fluorescens* 2P24. Antonie Van Leeuwenhoek 89:267–280

Welch M, Todd DE, Whitehead NA, McGowan SJ, Bycroft BW, Salmond GP (2000) *N*-Acyl homoserine lactone binding to the CarR receptor determines quorum-sensing specificity in *Erwinia*. EMBO J 19:631–641

White CE, Winans SC (2005) Identification of amino acid residues of the *Agrobacterium tumefaciens* quorum-sensing regulator TraR that are critical for positive control of transcription. Mol Microbiol 55:1473–1486

Whitehead NA, Barnard AM, Slater H, Simpson NJ, Salmond GP (2001) Quorum-sensing in Gram-negative bacteria. FEMS Microbiol Rev 25:365–404

Wijffelman CA, Pees E, van Brussel AA, Hooykaas PJJ (1983) Repression of small bacteriocin excretion in *Rhizobium leguminosarum* and *Rhizobium trifolii* by transmissible plasmids. Mol Gen Genet 192:171–176

Wilkinson A, Danino V, Wisniewski-Dye F, Lithgow JK, Downie JA (2002) *N*-Acyl-homoserine lactone inhibition of rhizobial growth is mediated by two quorum-sensing genes that regulate plasmid transfer. J Bacteriol 184:4510–4519

Wisniewski-Dyé F, Jones J, Chhabra SR, Downie JA (2002) *raiIR* genes are part of a quorum-sensing network controlled by *cinI* and *cinR* in *Rhizobium leguminosarum*. J Bacteriol 184:1597–1606

Wood DW, Gong F, Daykin MM, Williams P, Pierson LS III (1997) *N*-Acyl-homoserine lactone-mediated regulation of phenazine gene expression by *Pseudomonas aureofaciens* 30–84 in the wheat rhizosphere. J Bacteriol 179:7663–7670

Yan A, Huang X, Liu H, Dong D, Zhang D, Zhang X, Xu Y (2007) An *rhl*-like quorum-sensing system negatively regulates pyoluteorin production in *Pseudomonas* sp. M18. Microbiology 153:16–28

Yang WW, Han JI, Leadbetter JR (2005). Utilization of homoserine lactone as a sole source of carbon and energy by soil *Arthrobacter* and *Burkholderia* species. Arch Microbiol 10:1–8

Yao Y, Martinez-Yamout MA, Dickerson TJ, Brogan AP, Wright PE, Dyson HJ (2006) Structure of the *Escherichia coli* quorum sensing protein SdiA: activation of the folding switch by acyl homoserine lactones. J Mol Biol 355:262–273

Zhang HB, Wang LH, Zhang LH (2002a) Genetic control of quorum-sensing signal turnover in *Agrobacterium tumefaciens*. Proc Natl Acad Sci (USA) 99:4638–4643

Zhang HB, Wang C, Zhang LH (2004) The quormone degradation system of *Agrobacterium tumefaciens* is regulated by starvation signal and stress alarmone (p)ppGpp. Mol Microbiol 52:1389–1401

Zhang RG, Pappas T, Brace JL, Miller PC, Oulmassov T, Molyneaux JM, Anderson JC, Bashkin JK, Winans SC, Joachimiak A (2002b) Structure of a bacterial quorum-sensing transcription factor complexed with pheromone and DNA. Nature 417:971–974

Zhang Z, Pierson LS III (2001) A second quorum-sensing system regulates cell surface properties but not phenazine antibiotic production in *Pseudomonas aureofaciens*. Appl Environ Microbiol 67:4305–4315

Zheng D, Zhang H, Carle S, Hao G, Holden MR, Burr TJ (2003) A *luxR* homolog, *aviR*, in *Agrobacterium vitis* is associated with induction of necrosis on grape and a hypersensitive response on tobacco. Mol Plant Microbe Interact 16:650–658

Zheng H, Zhong Z, Lai X, Chen WX, Li S, Zhu J (2006) A LuxR/LuxI-type quorum-sensing system in a plant bacterium, *Mesorhizobium tianshanense*, controls symbiotic nodulation. J Bacteriol 188:1943–1949

Zhu J, Winans SC (2001) The quorum-sensing transcriptional regulator TraR requires its cognate signaling ligand for protein folding, protease resistance, and dimerization. Proc Natl Acad Sci (USA) 98:1507–1512

Chapter 12
Signals in the Underground: Microbial Signaling and Plant Productivity

Fazli Mabood, Woo Jin Jung, and Donald L. Smith(✉)

12.1 Introduction

Green plants are the vehicle by which virtually all energy enters the terrestrial biosphere. The rhizosphere is the first place where other organisms have a chronic access to this energy source and therefore an area of intense biological activity. Plants exude about 40% of photosynthates into the rhizosphere which makes it energy rich (Lynch and Whipps 1991). Due to the availability of substrates for metabolism, the rhizosphere is able to support large populations of microbes such as bacteria, actinomyctes, fungi, protozoa, algae, and viruses, etc. These rhizosphere inhabiting microorganisms compete for water, nutrients and space and sometimes improve their competitiveness by developing an intimate association with the plant. The rhizosphere can also be a battlefield among these microorganisms, with continuous competition and hostility among its inhabitants. The victors sometimes affect plant growth and development. However, it is becoming increasingly clear that establishment of the association between rhizobacteria and plants depends upon an exchange of signal molecules and it is during this initial signal exchange that plants and bacteria sometimes accept or reject each other.

PGPR, first defined by Joseph W. Kloepper, are soil bacteria that are able to colonize plant roots and promote plant growth and development (Kloepper and Schroth 1978). These include intracellular PGPR (iPGPR) and extracellular PGPR (ePGPR) that promote plant growth and development via diverse mechanisms (Gray and Smith 2005). Inoculants of iPGPR (rhizobia – *Rhizobium, Bradyrhizoum, Sinorhizobium*) are currently used as commercial products to enhance nodulation and nitrogen fixation by legumes (Vessey 2003). The use of ePGPR in commercial inoculants, to promote plant growth and development, has also reached a reasonable degree of sophistication (Haas and Defago 2005; Vessey 2003).

D.L. Smith
Plant Science Department, McGill University/Macdonald Campus,
21,111 Lakeshore Road, Ste. Anne de Bellevue, Quebec, Canada H9X 3V9
e-mail: donald.smith@mcgill.ca

C.S. Nautiyal, P. Dion (eds.) *Molecular Mechanisms of Plant and Microbe Coexistence.* Soil Biology 15, DOI: 10.1007/978-3-540-75575-3
© Springer-Verlag Berlin Heidelberg 2008

Rhizobacteria and plants communicate using a variety of signaling molecules that can determine the nature of associations between plants and the respective PGPR strain. In this section we will describe the best studied signaling interactions that occur between rhizobacteria (iPGPR and ePGPR) and plants, which occur in the beginning of rhizobacteria-plant associations, associations that lead to enhanced plant growth and development. Specifically, we elaborate signaling interactions that ultimately promote plant growth and development by: 1) signaling in symbiotic plant-bacteria associations, 2) signaling in free living rhizobacteria-plant associations, and 3) signaling in bacteria-bacteria interactions.

12.2   Peace Talks in the Underground: Plant Microbe Signaling in *Rhizobium*-Legume Symbiosis

Legume plants are unique in that they are able to enter into symbiosis with soil bacteria of the genera *Rhizobium, Bradyrhizobium, Azorhizobium, Mesorhizobium* and *Sinorhizobium*, collectively known as rhizobia. A successful interaction between a legume plant and the appropriate rhizobia leads to formation of a new plant organ, the nodule, which are generally formed on roots and bacteria reside inside them, in the form of bacteroids, and fix atmospheric dinitrogen into ammonia (Perret et al. 2000). The interaction between the two symbiotic partners is a highly specialized process and involves 'sophisticated molecular passwords' released by both partners; this signal exchange plays a role in host specificity. The first step in host specificity is the release of specific signal molecules, mostly flavonoids and isoflavonoids that induce the transcription of bacterial nodulation genes leading to the biosynthesis and secretion of lipo-chitooligosaccharides (LCOs), so called Nod factors, by rhizobia. The second step in host specificity lies in the structure of Nod factors, which act as bacteria-to-plant signal molecules. The correct Nod factors elicit responses in the host plant that lead to infection and formation of a nitrogen fixing nodule.

In this section we focus on the initial rhizosphere signaling that occurs between rhizobia and legume plants. Specifically we describe the various plant-to-bacteria signal molecules (isoflavonoids and others), their ability to induce the transcription of bacterial nodulation genes, and the biosynthesis of bacteria-to-plant signal molecules (Nod factors).

12.2.1   *The Legume Signals: Biosynthesis and Function of the Nodulation Gene Inducers*

(Iso)flavonoids are the best understood group of *nod* gene inducer molecules in rhizobia-legume symbioses, and their role in *nod* gene induction has been widely studied (Mabood et al. 2006a). Besides (iso)flavonoid *nod* gene inducers, various

Table 12.1 Compounds produced by different legumes that induce nodulation genes in their respective rhizobial partner

Plant source	Compound	Reference
Alfalfa	Luteolin, Chrysoeriol,	Peters et al. (1986); Hartwig et al. (1990)
	4,4′-Dihydroxy-2′-methoxychalcone	Maxwell et al. (1989)
	Liquiritigenin, 4′, 7-Dihydroxyflavone,	Maxwell et al. (1989)
	4,4′-dihydroxy-2′-methoxychalcone	Maxwell et al. (1989)
	Formononetin-7-O-(6″-O-malonylglycoside)	Dakora et al. (1993a)
	Stachydrine, Trigonelline (Non flavonoids)	Phillips et al. (1992, 1995)
Common bean	Delphinidin, Kaempferol, Malvidin	Hungria et al. (1991a)
	Myricetin, Petunidin, Quercetin	Hungria et al. (1991a)
	Eriodictyol, Genistein, Naringenin	Hungria et al. (1991b)
	Daidzein, Coumestrol	Dakora et al. (1993b)
Bambara groundnut	Genistein, Daidzein, Coumestrol	Dakora and Muofhe (1996)
Clover (white)	4′,7-Dihydroxyflavone geraldone	Redmond et al. (1986)
	4′-Hydroxy-7-methoxyflavone	
Cowpea	Genistein, Daidzein, Coumestrol	Dakora (2000)
Lupin	Erythronic and Tetronic acid (non-flavonoids)	Gagnon and Ibrahim (1998)
Pea	Apigenin, Eriodictyol	Firmin et al. (1986)
	Hesperatin, Naringenin, Luetolin	Begum et al. (2001)
Soybean	Genistein, Daidzein, Coumestrol	Kosslak et al. (1987)
	Genistein-7-O-glucoside	Smit et al. (1992)
	Genistein-7-O-(6″-O-malonylglucoside)	Smit et al. (1992)
	Daidzein-7-O-(6″-O-malonylglucoside)	Smit et al. (1992)
	Isoliquiritigenin	Kape et al. (1992)
	Jasmonates (JA, MeJA) (non flavonoid)	Mabood and Smith (2005); Mabood et al. (2006b)
Vetch	3,5,7,3′-Tetrahydroxy-4′-methoxyflavanone	Zaat et al. (1989)
	7,3′-Dihydroxy-4′-methoxyflavanone	Zaat et al. (1989)
	4,2′,4′-Trihydroxychalcone	Recourt et al. (1991)
	4,4′-Dihydroxy-2′-methoxychalcone	Recourt et al. (1991)
	Naringenin	Recourt et al. (1991)
	Liquiritigenin	Recourt et al. (1991)
	7,4′-Dihydroxy-3′-methoxyflavanone	Recourt et al. (1991)
	5,7,4′-Trihydroxy-3-methoxyflavanone	Recourt et al. (1991)
	5,7,3′-Trihydroxy-4′-methoxyflavanone	Recourt et al. (1991)

non-flavonoid inducers have also been identified (Table 12.1). These belong to the following chemical groups: betaines (stachydrine and trigonelline) (Phillips et al. 1992), aldonic acids (erythronic and tetronic acids) (Gagnon and Ibrahim 1998), and jasmonates (jasmonic acid and its derivative methyl jasmonate – MeJA) (Rosas et al. 1998; Mabood and Smith 2005).

12.2.1.1 Flavonoids

Flavonoids are phenolic compounds that are widely distributed in vascular plants. More than 4000 flavonoids have been identified in higher plants; these have diverse physiological and ecological functions (Debeaujon et al. 2000; Mathesius et al. 1998; Perret et al. 2000; Stafford 1990). The biosynthesis of flavonoids starts with phenylalanine, a product of the shikimic acid pathway, which is then converted into *trans*-cinnamic acid via the action of phenylalanine ammonialyase (PAL) enzyme. Through a series of reactions *trans*-cinnamic acid is converted into flavonoids and isoflavonoids. Isoflavonoids are limited to the legume family and the enzyme iso-flavone synthase plays a major role in the biosynthesis of isoflavonoids (Yu et al. 2000; Broughton et al. 2003; Mabood et al. 2006a; Perret et al. 2000). Flavonoid storage occurs in plant cell vacuoles, in either glycosylated or malonylated forms.

(Iso)flavonoids can act as signal molecules in rhizobia-legume symbioses. They at as chemoattractants to rhizobia (Currier and Strobel 1976; Gitte et al. 1978; Gaworzewska and Carlile 1982; Caetano-Anolles et al. 1988) and are able to induce the transcription of *nod* genes of the rhizobial partner. It is well documented that specific (iso)flavonoids act as signals and induce *nod* genes in the correct rhizobia. At this time we do not fully understand how this specificity occurs; however, it is clear that the interaction between (iso)flavonoid signal compounds and rhizobial NodD proteins is part of this specificity. For example, genistein (an isoflavonoid) is a potent inducer of *Bradyrhizobium japonicum*, which nodulates soybean (*Glycine max*) (Kosslak et al. 1987), while it inhibits the nodulation genes of *Rhizobium leguminosarum* bv. viceae, which nodulates pea (*Pisum sativum*) (Firmin et al. 1986). Hesperetin and naringenin (flavanones) are the strongest inducers of *Rhizobium leguminosarum* bv. viceae (Firmin et al. 1986; Begum et al. 2001). Similarly, genistein inhibits the *nod* genes of *Sinorhizobium meliloti*, which nodulates alfalfa (Hirsch et al. 2001), while these genes are induced by luteolin (a flavone) (Hartwig et al. 1990). Different legumes produce different flavonoids and these act as specific signals for specific rhizobia. A list of the *nod* gene inducing flavonoids produced by legume plants is presented in Table 12.1.

12.2.1.2 Other *nod* Gene Inducers

Other groups of compounds have also been shown to act in rhizobial *nod* gene induction. These are jasmonates (jasmonic acid – JA, and methyl jasmonate – MeJA), betaines (stachydrine and trigonelline) and aldonic acids (erythronic and tetronic acids). However, their ability to induce nodulation genes is not well investigated. These non-flavonoid compounds are structurally different from each other and are biosynthesized via pathways different from phenyl propanoid path-way used to synthesize flavonoid inducer molecules.

Jasmonates are fatty acid derivatives and are ubiquitous in plants. They are bio-synthesized from linolenic acid via the octadecanoid pathway. In the octadecanoid pathway, linolenic acid is converted into 13-hydroperoxylinolenic acid through the action of lipoxygenase (LOX), which, through a series of reactions, is then converted into JA and MeJA (Vick and Zimmerman 1984; Creelman and Mullet 1997). Jasmonates have several functions in plant growth and development (Corbineau et al. 1988; Creelman and Mullet 1997; Gundlach et al. 1992; Kramell et al. 1995). Jasmonates are synthesized in large quantities by germinating soybean seedlings (Creelman et al. 1992) and are able to induce nodulation genes of *B. japonicum* (Mabood and Smith 2005). When applied to bacterial cultures of *B. japonicum,* jasmonates are able to induce Nod factor (lipo chitooligosaccharides – LCOs) production in the culture medium (Mabood et al. 2006b). Pre-incubation of *B. japonicum* with jasmonates (JA and MeJA) also promoted soybean nodulation, nitrogen fixation and plant growth under greenhouse conditions, at both optimal and suboptimal root zone temperatures (Mabood and Smith 2005). Suboptimal root zone temperature has been shown to inhibit soybean nodulation, nitrogen fixation and plant growth (Zhang and Smith 1994). When *B. japonicum* was incubated with MeJA alone or together with genistein (an isoflavone), it enhanced soybean nodulation, nitrogen fixation (Mabood et al. 2006c), plant growth and yield (Mabood et al. 2006d) under short season field conditions in southeastern Quebec, Canada.

Trigonelline and stachydrine are found in higher plants and are biosynthesized from aspartic acid and ornithine, respectively (Phillips 2000). Their biosynthesis is generally related to osmotic stress (Jones et al. 1986). Alfalfa seeds produce trigonelline and stachydrine which activate the nodulation genes of *S. meliloti* by interacting with the NodD2 protein (Phillips et al. 1992, 1995). The two aldonic acids, erythronic and tetronic acids, are exuded by lupin (*Lupinus albus*) roots and induce the nodulation genes of *S. meliloti* and *Rhizobium fredii* (Gagnon and Ibrahim 1998).

12.2.1.3 Industrial Applications

Plant-to-bacteria signal molecules are now used commercially to promote legume nodulation and nitrogen fixation. For instance, genistein and daidzein, the *nod* gene inducers of *B. japonicum*, are used in commercial inoculants under the name SoyaSignal. Leibovitch et al. (2001) have shown that SoyaSignal technology was effective in promoting soybean yield under field conditions over six years in eastern Canada and the northern United States. SoyaSignal is either applied directly to seeds or in soil furrows, when adequate indigenous bradyrhizobial populations are present, to promote soybean nodulation and nitrogen fixation (Smith and Zhang 1999).

12.2.2 The Bacterial Signals: Biosynthesis of Nod Factors

As described, plant-to-bacteria signals are able to induce the transcription of the bacterial *nod* genes. Coordinated expression of the bacterial *nod* genes (*nod, noe, nol*) results in biosynthesis of Nod factors, structurally known as lipo-chitooligosaccharides

(LCOs) (Stacey et al. 1995). LCOs are key signal molecules that play a major role in the recognition of the micro-symbiont and the early stages of nodule organogenesis (Mabood et al. 2006a). This was first demonstrated in 1990 when the chemical structure of Nod factor and its activities were first identified by Lerouge et al. (1990), who isolated a Nod factor from *S. meliloti* culture filtrate. Since then, researchers around the globe, investigating the rhizobia-legume symbiosis, have characterized a wide range of Nod factors from other rhizobia (see reviews by Hungria and Stacey 1997; Hanin et al. 1999; D'Haeze and Holsters 2002).

A Nod-factor consists of a backbone of 3–5 β-1,4 linked *N*-acetyl-D-glucosamine units which is *N*-acylated at the terminal non-reducing end, leading to the designation lipo-chitooligosaccharides (LCOs) (Fig. 12.1). This basic Nod factor structure is synthesized by the bacterial *nodA*, *nodB* and *nodC* gene products. The first stage of Nod factor biosynthesis is catalyzed by the NodC (chitin synthase), thus producing chitooligosaccharide (Geremia et al. 1994; Spaink et al. 1994). The elongation of the oligosaccharide chain by NodC occurs at the nonreducing end (Kamst et al. 1997, 1999; Mergaert et al. 1995). The NodB (*N*-deacetylase) removes the acetyl residue from the non-reducing end of the chitooligosaccharide oligomer (John et al. 1993; Spaink et al. 1994). NodA (acyl transferase) adds a fatty acyl group to the non-reducing end of the oligosaccharide backbone at the C-2 position. The acyl group varies in the number of carbon atoms (16–20) and degree of unsaturation (0–4) (Atkinson et al. 1994; Rohrig et al. 1994; Debelle et al. 1996). The genes coding for the enzymes that are responsible for the production of the chitin oligomeric backbone (*nodA, nodB,* and *nodC*), known as common *nod* genes, and their regulatory gene (*nodD*) are conserved in all rhizobia species (Perret et al. 2000).

The basic structure of the Nod factor is modified in a species-specific manner, so that they vary in the length of the chitin backbone and at both reducing and non-reducing ends, resulting in a substantial diversity of Nod factors (Fig. 12.1). Modifications that occur at the reducing end include methyl fucose, fucose, sulfate, acetyl and arabinose groups, while modifications at the non-reducing end include carbamoyl or acetyl groups, besides *N*-acylation (common to all Nod factors) (Fig. 12.1). The genes responsible for the modification of the *N*-acetylglucosamine

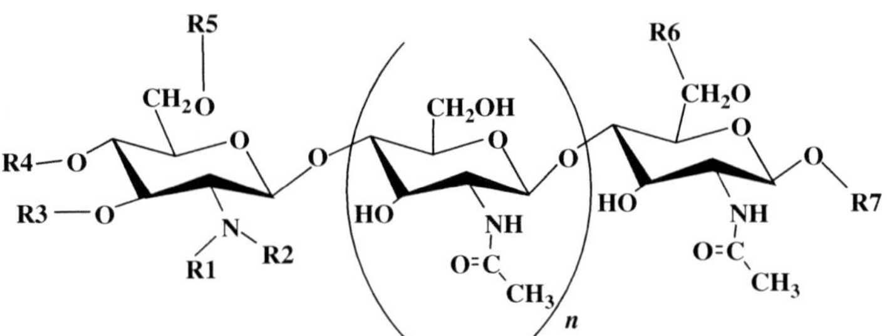

Fig. 12.1 The structure of Nod factors. The basic structure of Nod factors is comprised of GlcNAc units carrying different substitutions at both reducing and non-reducing ends

backbone are known as 'host specific *nod* genes' since they are not common to all rhizobia and they play a role in determining host specificity. Modifications to the terminal sugar residues also play a part in defining host specificity of Nod factors. For instance, *B. japonicum*, a microsymbiont of soybean, produces a Nod factor with a methyl-fucose group at the reducing end that is encoded by the 'host-specific' *nodZ* gene (Lopez-Lara et al. 1996). The methyl-fucosylation is essential to interactions with the host legume, soybean. *Sinorhizobium meliloti*, a microsymbiont of alfalfa, produces a Nod factor with a sulfate group that is encoded by the 'host specific' *nodH* and *Q* genes (Denarie et al. 1996). *Sinorhizobium meliloti* strains with the appropriate mutations in these genes produce non-sulfated Nod factors and are unable to infect alfalfa, their normal host legume (Denarie et al. 1996). Nod factors are active at very low concentrations. At submicromolar concentrations, they induce many physiological changes in legumes such as root hair deformation (Fig. 12.2), calcium spiking, changes in hormone levels, gene expression and are also able to initiate nodule organogenesis in legumes plant roots (Dénarié and Cullimore 1993; D'Haeze and Holsters 2002; Goedhart et al. 2003).

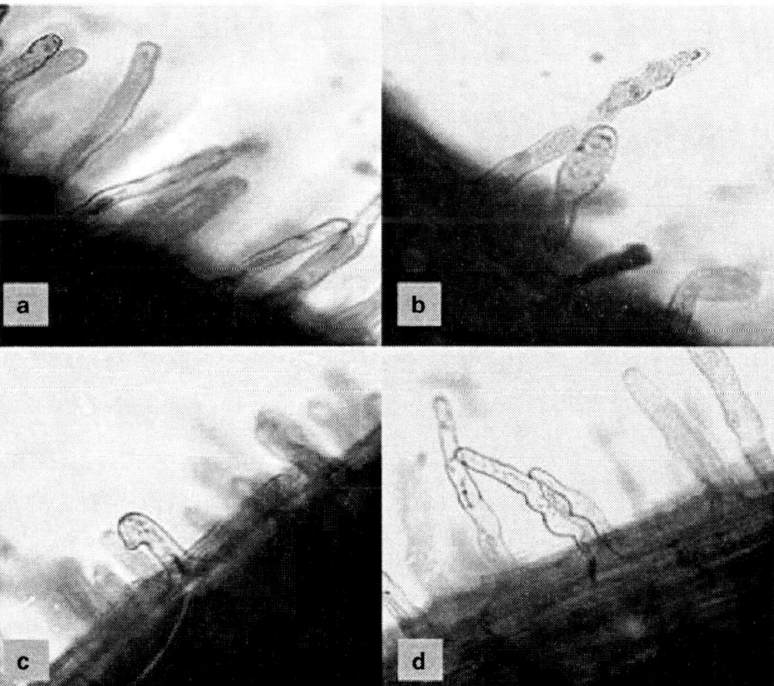

Fig. 12.2 Root hair deformation types of soybean root segments treated with purified Nod factor (Nod Bj-V (C$_{18:1}$, MeFuc)) from *Bradyrhiozbium japonicum*. **a** Untreated root hairs (control). **b** Root hair bulging. **c** Root hair curling, the so-called shepherds crook. **d** Root hair wiggling

Recent data suggests that non-legume plants are also able to perceive Nod factors. In cell cultures of the non-legume species, carrot mutants and Norway spruce, Nod factors are able to promote cell division and restore or promote somatic embryo formation (De Jong et al. 1993; Dyachok et al. 2000, 2002). Purified Nod factor (Nod Bj-V $C_{18:1}$, MeFuc) from *Bradyrhizobium japonicum* promoted soybean (a host) and corn (a non-host) root growth when applied in a hydroponic solution (Souleimanov et al. 2002). Prithiviraj et al. (2003) also reported that, at submicromolar concentrations, purified Nod factors from *B. japonicum* promoted seed germination and early seedling growth and development of diverse crop plants including legumes and non-legumes.

12.2.2.1 Industrial Applications of Nod Factors

The application of Nod factors in commercial inoculants to promote nodulation and plant growth is a very new technology. Nitragin Inc. has recently introduced a new product, Optimize, which includes *B. japonicum* cells together with LCOs (Smith 2005). Interestingly, McIver (2005) has recently reported the release of a new LCO based product by Agribiotics Inc. that is able to stimulate growth in a wide range of non-legume crops.

12.3 PGPR Signals that Promote Plant Growth and Development

PGPR employ complex mechanisms to promote plant growth and development. Important among them are biofertilization, biocontrol (production of antimicrobial compounds, lytic enzymes and induction of plant defense responses), phytostimulation and production of volatile organic compounds.

12.3.1 Biofertilization: Living Fertilizers in the Rhizosphere

A biofertilizer is defined as "a substance which contains living microorganisms which, when applied to seed, plant surfaces, or soil, colonizes the rhizosphere or the interior of the plant and promotes growth by increasing the supply or availability of primary nutrients to the host plant" (Vessey 2003). Some PGPR promote plant growth and development by acting as biofertilizers. PGPR that act as biofertilizers improve the nutrient status of plants via the following mechanisms: 1) biological N_2 fixation, 2) increasing nutrient availability in the rhizosphere, and 3) enhancing symbioses as helper bacteria (Vessey 2003).

Nitrogen fixing rhizobia are the most widely used bacterial biofertilizers. Rhizobia (including the genera *Allorhizobium, Azorhizobium, Bradyrhizobium,*

Mesorhizobium, Rhizobium, and *Sinorhizobium*) are able to enter into symbiosis with their host legume plants and are able to fix atmospheric dinitrogen in highly specialized structures known as nodules. The nitrogen fixed inside the nodules is provided to the plants, while the plants provide photosynthetically fixed carbon to the rhizobia residing inside the nodules. However, this legume-rhizobia symbiosis is highly host specific and interaction between homologous partners leads to the formation of nodules on the host plants. Host specificity is defined by "passwords", in the form of bacterial and plant signals (Stacey et al. 1995; Spaink 2000), as described above. Rhizobial inoculants are increasingly used as biofertilizers in commercial inoculants (Vessey 2003). Given the large quantities of fossil fuels used to produce nitrogen fertilizers and steep rise in fossil fuel prices over the last two years, expansion of biofertilizer use is likely to continue.

Diazotrophic bacteria that live and fix nitrogen outside of formal symbioses are referred to as free-living nitrogen-fixing bacteria. Important among them are *Azospirillum, Acetobacter, Herbaspirillum, Azoarcus* and *Azotobacter* (Steenhoudt and Vanderleyden 2000). They are important in agricultural systems since they are able to fix nitrogen in association with non-legume plants (Boddey et al. 1991).

PGPR that act as biofertilizers also promote plant growth and development through other mechanisms. Phosphate solubilizing bacteria (PSB) (Fig. 12.3) produce phosphatases that help in mineralization of organic phosphorus, while others release organic acids that help in phosphate solubilization (Kim et al. 1998; Rodriguez and Fraga 1999). Some PGPR produce siderophores that play an important role in iron availability to plants (Bloemberg and Lugtenberg 2001). Iron is an essential micronutrient required for plant growth and development and is relatively

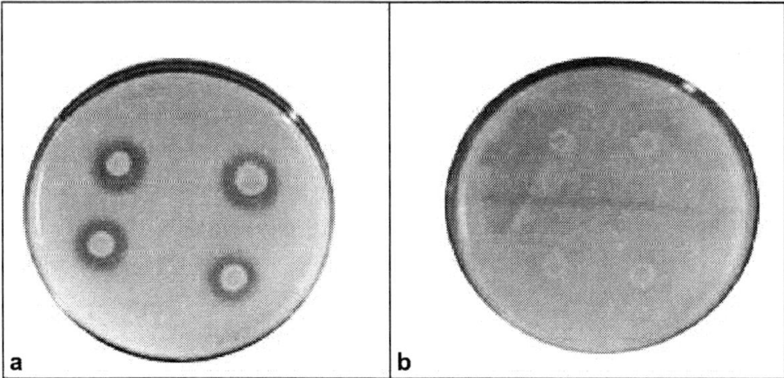

Fig. 12.3 Phosphate solubilization by a rhizobacterial isolate in the authors' laboratory. Bacterial culture (10 µL, OD_{600} = 1.6) was spotted onto medium containing tricalcium phosphate and photograph taken after 48 h of incubation at 28±1 °C. **a** Solubilization of insoluble phosphate by rhizobacterial isolate (S-14) as indicated by clear halo formation around the area of bacterial growth on medium containing tricalcium phosphate. **b** Rhizobacterial isolate (S-18) not capable of solubilizing tricalcium phosphate as evident by lack of clear halo formation around bacterial growth

unavailable to plants because of formation of insoluble ferric hydroxide complexes
in the presence of oxygen (Guerinot and Yi 1994).

12.3.2   Biocontrol: Warfare in the Underground – Signaling
in Hostile Associations

12.3.2.1   Production of Antibiotics: Powerful Microbial Tools
of Mass Destruction

Some rhizobacteria produce antimicrobial compounds that kill a diverse range of
other microorganisms (Figs. 12.4 and 12.5). In the past two decades there has been
considerable research work on the role of antibiosis as a biocontol mechanism
employed by rhizobacteria (Whipps 2001). Fluorescent pseudomonads produce a
variety of antibiotic substances, such as amphisin, 2,4-diacetylphloroglucinol
(DAPG), hydrogen cyanide, oomycin A, phenazine, pyoluteorin, pyrrolnitrin,
tensin, tropolone, and cyclic lipopeptides (Defago 1993; de Souza et al. 2003;
Nielson et al. 2002; Nielson and Sorensen 2003; Raaijmakers et al. 2002). Besides
pseudomonads, *Bacillus*, *Streptomyces*, and *Stenotrophomonas* spp. also produce
antibiotics, such as oligomycin A, kanosamine, zwittermicin A, and xanthobaccin
(Hashidoko et al. 1999; Kim BS et al. 1999; Milner et al. 1995, 1996; Nakayama
et al. 1999).

The synthesis of antibiotics is dependent on several factors, such as major and
minor minerals, type of carbon source, pH and growth temperature and availability
of trace elements (Bender et al. 1999b; Duffy and Defago 1997, 1999, 2000;
Georgakopoulos et al. 1994; Keel et al. 1989; Milner et al. 1995, 1996; Ownley

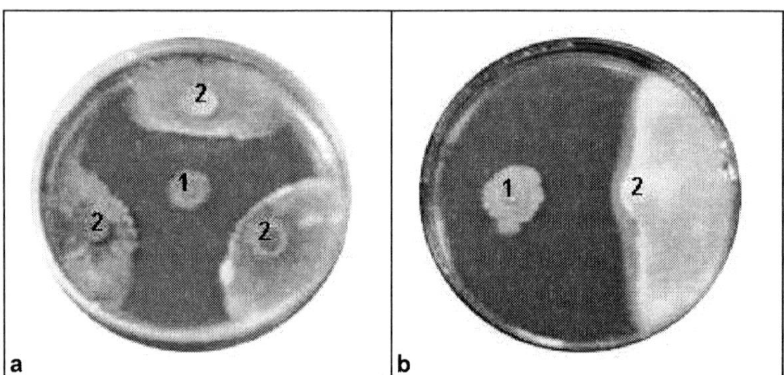

Fig. 12.4 Antifungal activity of rhizobacterial isolates from the authors' laboratory. In-vitro
antifungal activity of a rhizobacterial isolate (5–22) against **a** *Rhizoctonia solani* (the fungus that
causes damping-off disease of soybean) (1: rhizobacterial isolate 5–22, 2: *Rhizoctonia solani* GG-
4), and **b** *Phytopthora infestans* (the fungus that causes the late blight disease of tomato and
potato) (1: rhizobacterial isolate 5–22, 2: *Phytopthora infestans*)

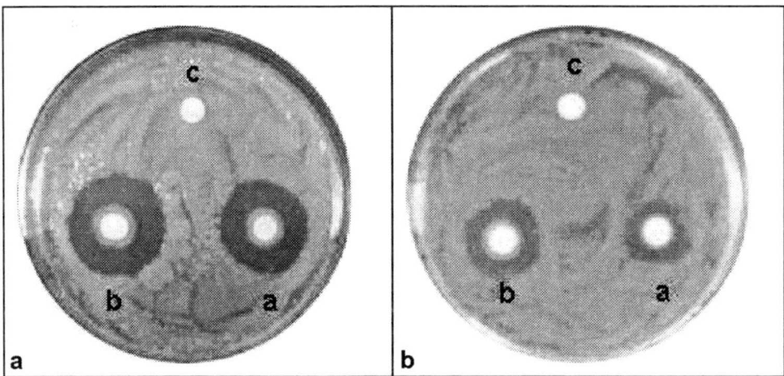

Fig. 12.5 Antibacterial activity of rhizobacterial isolates from the authors' laboratory. **a** Antibacterial activity of a rhizosphere isolate (8–11) against *Clavibacter michiganensis* ssp. michiganensis (causing bacterial wilt of tomato). *c*: control (sterilized dd H_2O), *a*: 5 µL, and *b*: 10 µL of isolate 8–11 (OD_{600} = 2.04). **b** Antibacterial activity of a rhizosphere isolate (S-14) against *Pseudomonas syringae* pv. tomato DC 3000 (causing bacterial speck of tomato). *c*: control (sterilized dd H_2O), *a*: 5 µL, and *b*: 10 µL of isolate S-14 (OD_{600} = 2.07)

et al. 1992, 2003). Nutrient availability can have an effect on the type of antimicrobial compound biosynthesized by a particular bacterial strain. For instance, in *P. fluorescens* CHA0, when glucose is added as a carbon source to the media biosynthesis of DAPG is stimulated, while pyoluteorin biosynthesis is repressed. However, in the absence of glucose pyoluteorin is the dominant antimicrobial compound produced by this strain (Duffy and Defago 1999). Similarly the age of the host plant also affects production of antimicrobial compounds by the colonizing rhizobacteria; root exudates from young plants do not induce DAPG production, while it is induced by the root exudates of older plants (Picard et al. 2000).

12.3.2.2 Production of Antimicrobial Peptides (Bacteriocins): Discrimination Among Friends and Foes

Some rhizobacteria produce antimicrobial peptides known as bacteriocins which kill closely related bacteria, facilitating a fratricidal struggle for nutrients and niche space. Bacteriocins are small peptides produced by bacteria that show bactericidal or static effects against bacteria closely related to the producer strain (Jack et al. 1995). Bacteriocins produced by Gram-positive bacteria have been studied quite extensively and have received a great deal of interest, since they can be used as food preservatives, to inhibit pathogenic bacteria. For example, Nisin, a widely used bacteriocin produced by lactic acid bacteria, is an effective food biopreservative (Hansen 1994).

Bateriocins are getting a great deal of attention due to their efficacy in the biological control of food spoilage and of pathogenic organisms (Delves-Broughton 1990; Parret and Mot 2002). For example, Kamoun et al. (2005) have shown that *B. thuringiensis* subsp. kurstaki strain BUPM4 produces a bacteriocin known as Bacthuricin (BF4) that can kill closely related bacteria in the rhizosphere. Recently, Gray et al. (2006a) showed that *Bacillus thuringiensis* NEB 17 produces a bacteriocin, Thuricin 17 (T17), which shows antimicrobial activity against a broad range of closely related bacterial species. Interestingly, the two bacteriocins, T17 and BF4, share similarity at the N terminus sequence and have similar molecular weights (T17 – 3162 Da and BF4 – 3160 Da) (Gray et al. 2006a,b; Kamoun et al. 2005). We have recently shown that these two bacteriocins show similar spectra of antimicrobial activities and thus have structural and functional similarities (Jung et al. 2008).

12.3.2.3 Production of Extracellular Lytic Enzymes: Interactions with Pathogenic Microorganisms

Although rhizobacteria and actinomycetes used in biocontrol produce antibiotics to suppress bacterial and fungal pathogens, some also produce cell wall degrading enzymes that exert antifungal activities and contribute to the biocontol activity of these rhizobacteria. The cell wall degrading enzymes are chitinases, glucanases, cellulases, and proteases that cause lysis and degradation of the fungal cell walls. Selected actinomycete isolates were used for the control of *Phytophthora fragariae* var. rubi, which causes raspberry root rot. These isolates produced β-1,3-, β-1,4-, and β-1,6-glucanases that are able to hydrolyze glucans from *Phytophthora* cell walls, thus causing lysis of *Phytophthora* cells (Valois et al. 1996). Application of chitinase producing *Enterobacter agglomerans* (Chernin et al. 1995), *Bacillus cereus* (Pleban et al. 1997) and *Paenibacillus illinoisensis* (Jung et al. 2003) decreased *Rhizoctonia solani* disease incidence in cotton and cucumber. Chitinase producing *P. illinoisensis* has also been shown to improve biocontrol of *Phytopthora* blight of pepper (*Capsicum annuum* L.) caused by *Phytopthora capsici* (Jung et al. 2005).

Extracellular proteases produced by *Stenotrophomonas maltophila* W81 are involved in the biocontrol of *Pythium ultimum* in the sugar beet rhizosphere (Dunne et al. 1997). Selected strains of *Bacillus subtilis*, *Erwinia herbicola*, *Serratia plymuthica*, and an actinomycete showed antifungal activity against the grapevine dieback fungus *Eutypa lata* by producing fungal hydrolases such as chitinases, proteases and cellulases (Schmidt et al. 2001). El-Tarabily et al. (1996) also found biocontrol activity by cellulase-producing *Micromonospora carbonacea* against *Phytophthora cinnamomi*, the causal organism of the root rot of *Banksia grandis*.

The ability of rhizobacteria to produce lytic enzymes is considered of crucial importance in the biological control of fungal plant diseases. For instance, *Paenibacillus* sp. 300 and *Streptomyces* sp. 385 produce chitinase and β-1,3-glucanase, which play an important role in the biological control of *Fusarium* wilt of cucumber (*Cucumis sativus*) caused by *Fusarium oxysporum* f. sp. *cucumerinum* (Singh et al. 1999).

Pseudomonas cepacia produces β-1,3 glucanase, which lyse fungal cell walls and decrease the incidence of diseases caused by *Rhizoctonia solani, Sclerotium rolfsii* and *Pythium ultimum* (Fridlender et al. 1993).

12.3.2.4 Production of Extracellular Lytic Enzymes: Interaction with Beneficial Microorganisms

Interactions with the Mycorrhizal Symbiosis

The symbiotic association of beneficial fungi with the roots of plants is of considerable importance in agriculture. Mycorrhizal fungi are able to increase the absorption surface of the infected plants and improve water (Subramanian et al. 1995) and uptake of nutrients such as phosphorus (P) (Li et al. 1991; Ortas et al. 1996), copper (Cu) (Li et al. 1991), zinc (Zn) (Burkert and Robson 1994) and cadmium (Cd) (Guo et al. 1996) by plants. Mycorrhizal fungi also improve nitrogen nutrition (N) of plants by taking up nitrogen from NH_4^+-N mineral fertilizers (Ames et al. 1983; Johansen et al. 1993) or senescing roots of distant plants and transporting it to other plants (Hamel et al. 1991a,b). Chitinolytic bacteria may have negative effect on the development of mycorrhizae. Lysis of mycorrhizal hyphae in the external rhizosphere has been observed and may be due to the mycolytic activity of chitinolytic bacteria producing chitinases outside the rhizosphere zone, that do not repress chitinase induction by rhizobacteria (de Boer and van Veen 2001). Root exudates contain amino acids and sugars that suppress chitinase production by chitinolytic bacteria, indicating that mycolytic activity occurs only under starvation conditions (de Boer and van Veen 2001).

Interactions in *Rhizobium*-Legume Symbiosis

During the initial events of legume-rhizobia interactions, there is a two way signaling between the macro- and the micro-symbionts (described in more detail above). In order to invade the host plants successfully, rhizobia must produce Nod factors, which are lipo-chitooligosaccharides (LCOs), in response to plant produced *nod* gene inducing molecules (such as isoflavonoids). Nod factors show similarity to chitin in that the core structure of the Nod factor is composed of *N*-acetylglucosamine (GlcNAc). Some soil bacteria produce chitinases that degrade (hydrolyse) chitin. Although chitinolytic bacteria may be used as potential biocontrol agents, there are still some questions about their application to the rhizosphere soil as biocontrol agents. They might also cause problems in the rhizosphere by disrupting inter-organismal signaling in rhizobia-legume symbiosis, a crucial step in the early stages of nitrogen fixing symbiosis. We found that the soil bacteria *Paenibacillus illinoisensis* and *Bacillus thuringiensis* subsp. *pakistani* produce chitinases that are able to cleave Nod factors. In vitro treatment of Nod factors from *B. japonicum* (Nod BjV ($C_{18:1}$, MeFuc) with crude chitinases from *P. illinoisensis* and *B. thuringiensis* subsp. *pakistani* led to a substantial degradation of Nod factor molecules (Jung et al. 2006). The negative effect of chitinolytic bacteria on inter-organismal signal exchange in the nitrogen fixing symbiosis could lead to reduced nodulation.

12.3.3 Biocontrol: Alliances in the Underground – Microbial Signals Enhancing Resistance of Plants to Phytopathogens

12.3.3.1 Systemic Acquired Resistance (SAR)

The role of SA as a potent inducer of SAR and pathogenesis-related (PR) proteins is well documented (Enyedi et al. 1992; Gaffney et al. 1993; Ward et al. 1991; Ryals et al. 1996). It has been shown that under iron-limited conditions, some rhizobacteria are able to produce SA, which may trigger the SAR pathway, thereby inducing systemic resistance in plants (De Meyer and Hofte 1997; Maurhofer et al. 1994; Buysens et al. 1996; Leeman et al. 1996; Press et al. 1997). However this is uncertain for some strains. Systemic induction was still observed by *Pseudomonas fluorescens* WCS417 under conditions when the rhizobacteria did not produce SA on the roots (Leeman et al. 1996). Also, SA deficient mutants of *Serratia marcescens* 90–166 were able to induce systemic resistance in plants the same way as the wild-type strain (Press et al. 1997), suggesting that SA production is not required for induction of ISR by some strains. However, SA seems to be the major signal causing induction of SAR in plants by some rhizobacteria. *Pseudomonas aeruginosa* 7NSK2 produces SA that activates SAR in bean plants (De Meyer et al. 1999). Studies with SA-deficient mutants of *P. aeruginosa* 7NSK2 have shown that this strain loses its ability to induce resistance to *Botrytis cinerea* in bean (De Meyer and Hofte 1997). Introduction of SA biosynthetic genes into *P. fluorescens* strain P3, incapable of producing SA, made this strain capable of producing SA and induced resistance to tobacco necrosis virus in tobacco plants (Maurhofer et al. 1998).

12.3.3.2 Induced Systemic Resistance (ISR)

PGPR inoculated onto seeds or seedling roots may trigger another mechanism, known as induced systemic resistance (ISR), to suppress pathogenic microorgan-isms and arrest disease development in the host plant (van Loon et al. 1998; Kloepper et al. 1999). As a result plants are able to resist subsequent pathogen infections (van Loon 1997). Stimulation of ISR by benign root colonizing PGPR is considered different from systemic acquired resistance (SAR). SAR involves a sys-temic activation of defense responses to necrotizing pathogens in distal parts of plants and involves hypersensitive response and expression of several pathogenesis-related (PR) genes (Ryals et al. 1996); however, ISR, as induced by PGPR, does not provoke a visible hypersensitive response (Wei et al. 1991). Although both SAR and ISR enhance plant defense responses, these defense responses show differences in signal transduction pathways and the in-planta molecules that are involved the signal transduction cascades. In order to make a clear distinction between these two defense responses, the term ISR is applied to resistance responses induced by rhizobacteria while SAR is applied to resistance responses induced by pathogenic microorganisms (Bakker et al. 2003). Salicylic acid plays a central role in the SAR pathway and studies have shown that there is a strong relationship between SAR and accumulation of SA in plants (Sticher et al. 1997). However, ISR is not SA dependent and involves jasmonic acid (JA) and ethylene (ET) in a signal transduction cascade leading to improved disease resistance (Pieterse et al. 1998). Interestingly, studies have shown that these two defense pathways are induced independently and that SA- and JA-mediated defense pathways show antagonistic activity; induction of the SA pathway inhibits the JA pathway and vice versa (Dong 1998; Felton et al. 1999; Niki et al. 1998).

While there has been considerable research into the molecular mechanisms of ISR (induced by rhizobacteria) and SAR (induced by pathogenic microorganisms) in the past decade, the microbial factors inducing each still need to be identified. It was recently suggested that some rhizobacteria produce volatile organic compounds that are able to induce ISR in *Arabidopsis* plants (Ryu et al. 2004). Research into microbial signals and plant receptors responsible for ISR and SAR responses will unravel the complexities of systemic resistance in plants and will enable us to use bacterial factors to induce plants, for better biocontrol activities.

12.3.4 Phytostimulation: Constructive Communication – Microbial Production of Plant Growth Promoting Compounds

Other PGPR signals that provoke plant growth promotion are the better known plant hormones. It has been shown that some PGPR are able to produce phytohormones such as auxins, cytokinins, gibberellins and ethylene (Steenhoudt and Vanderleyden 2000; Yanni et al. 2001; Strzelczyk and Pokojska-Burdziej 1984; Kucey 1988; Patten and Glick 1996; Persello-Cartieaux et al. 2003; Weingart and Volksch 1997; Bender et al. 1999a), as well as salicylic acid (De Meyer and Hofte 1997; Parker 2003; Bloemberg and Lugtenberg 2001) and have a role in controlling the biosynthesis of ethylene (Persello-Cartieaux et al. 2003).

The production of phytohormones by PGPR is of considerable value in agriculture since these hormones directly affect the physiology of crop plants (for instance root growth, root hair formation and ion uptake) thereby affecting plant growth and development. For instance, it has been shown that *Azospirillum spp.* produce auxins, cytokinins and gibberellins that enhance root growth and development, thereby enhancing nutrient uptake and increasing biomass and yield (Steenhoudt and Vanderleyden 2000). Phytohormone production by rhizobacteria, especially indole-3-acetic acid (IAA), is an important source of plant growth promotion (Bashan and Levanov 1990; Okon and Labandera-Gonzales 1994; Frankenberger and Arshad 1995). A range of *Bacillus* species have been reported to produce IAA (Chanway and Nelson 1990; Selvadurai et al. 1991). Some rhizobacteria produce indole analogs such as indole-3-ethanol (TOL), which also play an important role in plant growth promotion (Frankenberger and Arshad 1995; Chanway and Nelson 1990).

Some rhizobacteria, for instance *Bacillus cereus* (Selvadurai et al. 1991) and *Paenibacillus polymyxa* (Lebuhn et al. 1997), produce indole compounds, different from IAA, and have been reported to modulate plant growth and development. Plants can easily take up TOL and convert it into IAA by plant TOL-oxidase and O_2 (Sandberg 1984). Thus the stimulatory effect of TOL may be due to the conversion of TOL into IAA inside plant roots.

PGPR are also known to produce cytokinins, hormones important in the control of cell division, chloroplast development and bud formation (Arshad and Frankenberger 1991; Serdyuk et al. 1995). For instance, *Rhizobium leguminosarum* strains promote early plant growth and development of canola and lettuce due to their ability to produce phytohormones such as IAA and cytokinins (Noel et al. 1996). Similarly *Methylobacterium* spp. can increase soybean seed germination and this effect may be mimicked by medium containing cytokinins (Holland 1994).

Rhizobacteria can also control ethylene biosynthesis in-planta. Auxin producing rhizobacteria are able to cause an increase in the activity level of the enzyme 1-aminocyclopropane-1-carboxylate (ACC) synthase, an enzyme that is involved in the biosynthesis of ethylene in plants (Xie et al. 1996). Plants exude ACC into the rhizosphere and this is hydrolyzed by rhizobacteria that produce ACC deaminase (Campbell and Thomson 1996; Shah et al. 1998). As a result, plants continue to exude ACC into the rhizosphere, in an attempt to maintain hormonal balance; thus the rhizobacteria are able to control biosynthesis of ethylene in plant roots (Persello-Cartieaux et al. 2003).

PGPR that produce salicylates are able to induce systemic acquired resistance (SAR) in colonized plants. The effects of SA producing rhizobacteria on plants have been described in Sect. 12.3.3.

12.3.5 *Volatile Signals from PGPR: A Scent of Victory – Role in Plant Growth Promotion and Systemic Resistance*

It is well documented that plants communicate with each other using air borne volatile signal molecules that induce resistance in plants. Examples of these are methyl salicylate, a SA derivative and a product of the phenylpropanoid pathway, methyl jasmonate, a JA derivate and a product of the octadecanoid pathway, and ethylene, a gaseous plant hormone, are potent activators of plant defense responses. These compounds are released, at very low concentrations, by damaged plants; they act as signals to neighbouring plants (Baldwin et al. 2006; Dong 1998; Farmer and Ryan 1990; Shulaev et al. 1997). However, it was not clear until recently that rhizosphere bacteria are also able to produce volatile compounds able to play an important role in plant growth and development.

Recently it has been demonstrated that rhizobacteria also produce a blend of air borne volatile organic compounds (VOCs) that are known to regulate plant growth and development (Ryu et al. 2003) and induce systemic resistance in *Arabidopsis* (Ryu et al. 2004). The PGPR strains *Bacillus subtilis* GB03 and *Bacillus amyloliquefaciens* IN937a release two volatile compounds, 3-hydroxy-2-butanone (acetoin) and 2,3-butanediol (2,3-B), that promote growth and development of *Arabidopsis*. PGPR strains that did not produce these VOCs did not show growth promotion activities. The plant growth promoting effect of these VOCs was comparable with those of commercial acetoin and 2,3-butanediol (Ryu et al. 2003). When *Arabidopsis* seedlings were exposed to bacterial VOCs (acetoin and 2,3-B) from *B. subtilis* and *B. amyloliquefaciens* for only four days the ISR pathway was induced, resulting in symptoms of disease caused by *Erwinia carotovora* subsp. carotovora (Ryu et al. 2004).

12.4 Conclusions and Future Prospects

In the past two decades there has been significant research progress regarding the molecular mechanisms leading to beneficial plant microbe interactions and how they ultimately promote plant growth and development. As a result, scientists have been successful in harnessing plant-microbe signaling into commercial formulations. Significant among them are the rhizobia based inoculants used as biofertilizers and other PGPR inoculants that promote plant growth and development directly or indirectly.

It seems that the market potential for bio-based products that enhance plant growth and development and promote biocontrol of plant diseases is increasing (Haas and Defago 2005). Conventional and organic growers are taking more interest in bio-based products (Rzewnicki 2000) since rhizobacteria based products have already been successfully applied in the greenhouse industry, and there seems to be considerable potential for further growth in this area (Paulitz and Belanger 2001).

Although PGPR have been used successfully for biocontrol of plant diseases under laboratory and greenhouse conditions (Paulitz and Belanger 2001), results in the field have not always shown promise (Nelson 2004). This may be due to the controlled environment and high frequency of fungal diseases under greenhouse conditions (Paulitz and Belanger 2001). Field conditions are highly variable and knowledge of biotic and abiotic factors influencing the survival, proliferation and performance of PGPR may lead to the development of better PGPR inoculant formulations and PGPR based products (Bowen and Rovira 1999; McSpadden Gardener and Fravel 2002).

Other challenges lie in PGPR product formulation, such as scale up of production method to the level of industrial fermenters, shelf life of the product, methods

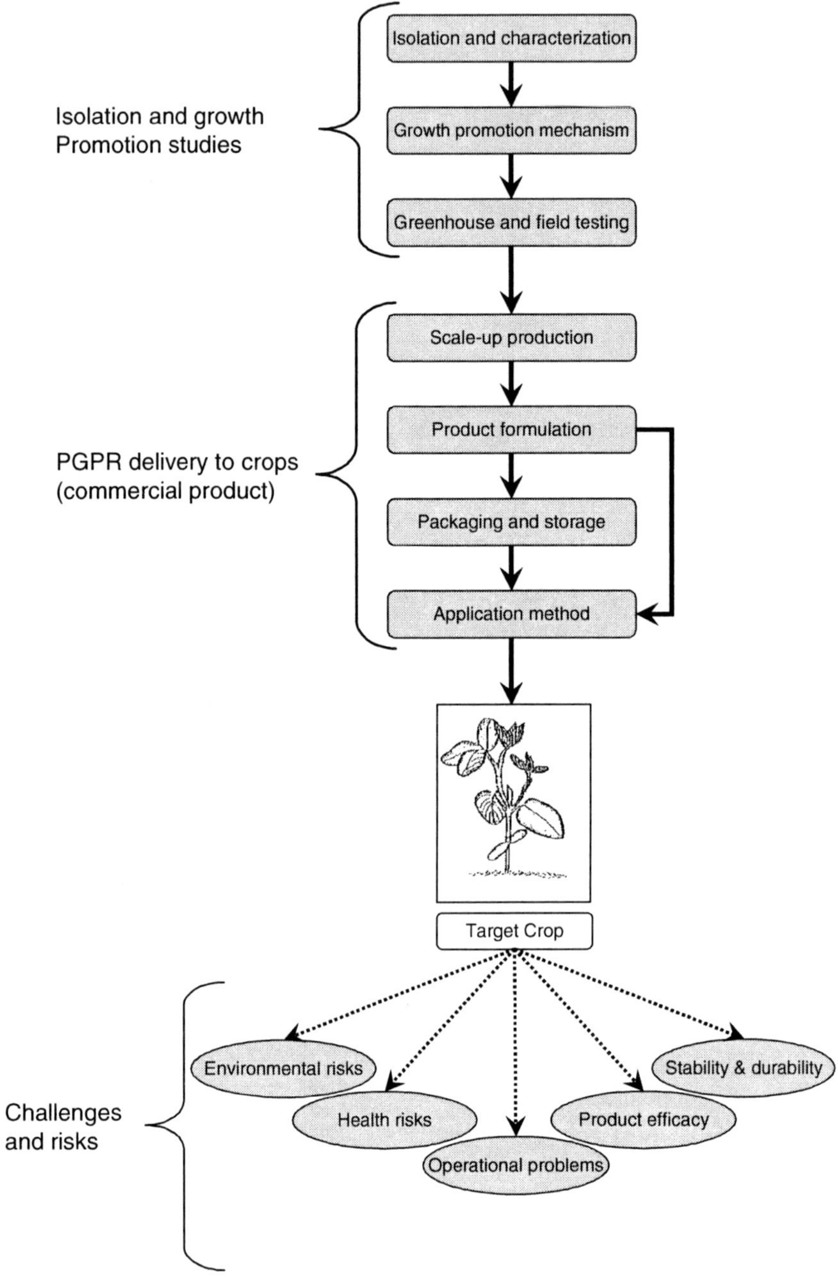

Fig. 12.6 A schematic diagram showing different stages of PGPR research, the development of appropriate delivery method to target plants and the challenges and risks associated with the PGPR-based product

and ease of application, cost of the PGPR product and possible compatibility with chemical and other seed treatments. Before a PGPR based product is delivered, health and safety testing (effect on non-target organisms including human beings, toxicity, pathogenicity and allerginicity) and environmental issues (persistence in the environmental and potential gene transfer to other organisms) must be addressed (Fravel et al. 1999; Mathre et al. 1999; McSpadden Gardener and Fravel 2002). A thorough market study of the PGPR based products is also of crucial importance before a product is registered and commercialized (Fig. 12.6).

References

Ames RN, Reid CPP, Porter LK, Cambardella C (1983) Hyphal uptake and transport of nitrogen from two ^{15}N-labelled sources by *Glomus mosseae*, a vesicular arbuscular mycorrhizal fungus. New Phytol 95:381–396

Arshad M, Frankenberger WT (1991) Microbial production of plant hormones. Plant Soil 133:1–8

Atkinson EM, Palcic MM, Hindsgaul O, Long SR (1994) Biosynthesis of *Rhizobium meliloti* lipooligosaccharide Nod factors: NodA is required for an *N*-acyltransferase activity. Proc Natl Acad Sci USA 91:8418–8422

Bakker PAHM, Ran LX, Pieterse CMJ, van Loon LC (2003) Understanding the involvement of rhizobacteria mediated induction of systemic resistance in biocontrol of plant diseases. Can J Plant Pathol 25:5–9

Baldwin IT, Halitschke R, Paschold A, von Dahl CC, Presto CA (2006) Volatile signaling in plant-plant interactions: "Talking Trees" in the genomics era. Science 311:812–815

Bashan Y, Levanoy H (1990) Current status of *Azospirillum* inoculation technology: *Azospirillum* as a challenge for agriculture. Can J Microbiol 36:591–608

Begum AA, Leibovitch S, Migner P, Zhang F (2001) Specific flavonoids induced *nod* gene expression and pre-activated *nod* genes of *Rhizobium leguminosarum* increased pea (*Pisum sativum* L.) and lentil (*Lens culinaris* L.) nodulation in controlled growth chamber environments. J Exp Bot 52:1537–1543

Bender CL, Alarcon-Chaidez F, Gross DC (1999a) *Pseudomonas syringae* phytotoxins: mode of action, regulation, and biosynthesis by peptide and polyketide synthesases. Microbiol Mol Biol Rev 63:266–292

Bender C, Rangaswamy V, Loper J (1999b) Polyketide production by plant-associated pseudomonads. Annu Rev Phytopathol 37:175–196

Bloemberg GV, Lugtenberg BJJ (2001) Molecular basis of plant growth promotion and biocontrol by rhizobacteria. Curr Opin Plant Biol 4:343–350

Boddey RM, Urquiaga S, Reis V, Dobereiner J (1991) Biological nitrogen fixation associated with sugar cane. Plant Soil 137:111–117

Bowen GD, Rovira AD (1999) The rhizosphere and its management to improve plant growth. Adv Agron 66:1–102

Broughton WJ, Zhang F, Perret X, Staehelin C (2003) Signals exchanged between legumes and *Rhizobium*: agricultural uses and perspectives. Plant Soil 252:129–137

Burkert B, Robson A (1994) ^{65}Zn uptake in subterranean clover (*Trifolium subterraneum* L.) by three vesicular-arbuscular mycorrhizal fungi in a root-free sandy soil. Soil Biol Biochem 26:1117–1124

Buysens S, Heungens K, Poppe J, Hofte M (1996) Involvement of pyochelin and pyoverdin in suppression of Pythium-induced damping-off of tomato by *Pseudomonas aeruginosa* 7NSK2. Appl Environ Microbiol 62:865–871

Caetano-Anolles G, Crist-Estes DK, Bauer WD (1988) Chemotaxis of *Rhizobium meliloti* to the plant flavone luteolin requires functional nodulation genes. J Bacteriol 170:3164–3169

Campbell BGM, Thomson JA (1996) 1-Aminocyclopropane-1-carboxylate deaminase genes from *Pseudomonas* strains. FEMS Microbiol Lett 138:207–210

Chanway CP, Nelson LM (1990) Field and laboratory studies of *Triticum aestivum* L. inoculated with co-existent growth-promoting *Bacillus* strains. Soil Biol Biochem 22:789–795

Chernin L, Ismailov Z, Haran S, Chet I (1995) Chitinolytic *Enterobacter agglomerans* antagonistic to fungal plant pathogens. App Env Microbiol 61:1720–1726

Corbineau F, Rudnicki RM, Come D (1988) The effects of methyl jasmonate on sunflower (*Helianthus annuus* L.) seed germination and seedling development. Plant Growth Regul 7:157–169

Creelman RA, Mullet JE (1997) Biosynthesis and action of jasmonate in plants. Annu Rev Plant Physiol Plant Mol Biol 48:355–381

Creelman RA, Tierney ML, Mullet JE (1992) Jasmonic acid/methyl jasmonate accumulate in wounded soybean hypocotyls and modulate wound gene expression. Proc Natl Acad Sci USA 89:4938–4941

Currier WW, Strobel GA (1976) Chemotaxis of Rhizobium species to plant root exudates. Plant Physiol 57:802–823

D'Haeze W, Holsters M (2002) Nod factor structures, responses, and perception during initiation of nodule development. Glycobiology 12:79R–105R

Dakora FD (2000) Commonality of root nodulation signals and nitrogen assimilation in tropical grain legumes belonging to the tribe Phaseoleae. Aust J Plant Physiol 27:885–892

Dakora FD, Muofhe ML (1996) Molecular signals involved in nodulation of the African Bambara groundnut (*Vigna subterranea* L.). In: Proceedings of the International Bambara groundnut Symposium, University of Nottingham, UK, July 23–25, pp 171–179

Dakora FD, Joseph CM, Phillips DA (1993a) Alfalfa (*Medicago sativa* L.) root exudates contain isoflavonoids in the presence of *Rhizobium meliloti*. Plant Physiol 101:819–824

Dakora FD, Joseph CM, Phillips DA (1993b) Common bean root exudates contain elevated levels of daidzein and coumestrol in response to *Rhizobium* inoculation. Mol Plant Microbe Interact 6:665–668

de Boer V, van Veen JA (2001) Are chitinolytic rhizosphere bacteria really beneficial to plants? In: Jeger MJ, Spence NJ (eds) Biotic interactions in plant-pathogen associations. CABI Publishing, New York, USA

De Jong AJ, Heidstra K, Spaink HP, Hartog MV, Meijer EA, Hendriks T, Lo Schiavo F, Terzi M, Bisseling T, Van Kammen A, De Vries SC (1993) *Rhizobium* lipooligosaccharides rescue a carrot somatic embryo mutant. Plant Cell 5:615–620

De Meyer G, Hofte M (1997) Salicylic acid produced by the rhizobacterium *Pseudomonas aeruginosa* 7NSK2 induces resistance to leaf infection by *Botrytis cinerea* on bean. Phytopathology 87:588–593

De Meyer G, Capieau K, Audenaert K, Buchala A, Metraux J-P, Hofte M (1999) Nanogram amounts of salicylic acid produced by the rhizobacterium *Pseudomonas aeruginosa* 7NSK2 activate the systemic acquired resistance pathway in Bean. Mol Plant-Microbe Interact 12:450–458

de Souza JT, de Boer M, de Waard P, van Beek TA, Raaijmakers JM (2003) Biochemical, genetic, and zoo sporicidal properties of cyclic lipopeptide surfactants produced by *Pseudomonas fluorescens*. Appl Environ Microbiol 69:7161–7172

Debeaujon FL, Leon-Kloosterziel IKM, Koornneef M (2000) Influence of the testa on seed dormancy, germination, and longevity in Arabidopsis. Plant Physiol 122:403–413

Debelle F, Plazanet C, Roche P, Pujol C, Savagnac A, Rosenberg C, Prome J-C, Denarie J (1996) The NodA proteins of *Rhizobium meliloti* and *Rhizobium tropici* specify the *N*-acylation of Nod factors by different fatty acids. Mol Microbiol 22:303–314

Defago G (1993) 2,4-Diacetylphloroglucinol, a promising compound in biocontrol. Plant Pathol 42:311–312

Delves-Broughton J (1990) Nisin and its application as a food preservative. J Soc Dairy Tech 43:73–76

Denarie J, Cullimore JV (1993) Lipo-oligosaccharide nodulation factors: a new class of signalling molecules mediating recognition and morphogenesis. Cell 74:951–954

Denarie J, Debelle F, Prome JC (1996) Rhizobium lipo-chitooligosaccharide nodulation factors: signaling molecules mediating recognition and morphogenesis. Ann Rev Biochem 65:503–535

Dong X (1998) SA, JA, ethylene, and disease resistance in plants. Curr Opin Plant Biol 1:316–323

Duffy BK, Defago G (1997) Zinc improves biocontrol of Fusarium crown and root rot of tomato by *Pseudomonas fluorescens* and represses the production of pathogen metabolites inhibitory to bacterial antibiotic biosynthesis. Phytopathology 87:1250–1257

Duffy BK, Defago G (1999) Environmental factors modulating antibiotic and siderophore biosynthesis by *Pseudomonas fluorescens* biocontrol strains. Appl Environ Microbiol 65:2429–2438

Duffy BK, Defago G (2000) Controlling instability in gacS-gacA regulatory genes during inoculum production of *Pseudomonas fluorescens* bicontrol strains. Appl Environ Microbiol 66:3142–3150

Dunne C, Crowley JJ, Moenne-Loccoz Y, Dowling DN, de Bruijn FJ, O'Gara F (1997) Biological control of *Pythium ultimum* by *Stenotrophomonas maltophilia* W81 is mediated by an extracellular proteolytic activity. Microbiology 143:3921–3931

Dyachok JV, Tobin AE, Price NPJ, Von Arnold S (2000) Rhizobial Nod factors stimulate somatic embryo development in *Picea abies*. Plant Cell Rep 19:290–297

Dyachok JV, Wiweger M, Kenne L, Von Arnold S (2002) Endogenous Nod-factor-like signal molecules promote early somatic embryo development in Norway Spruce. Plant Physiol 128:523–533

El-Tarabily KA, Sykes ML, Kurtbke ID, Hardy GEStJ, Barbosa AM, Dekker RFH (1996) Synergistic effects of a cellulase-producing *Micromonospora carbonacea* and an antibiotic-producing *Streptomyces violascens* on the suppression of *Phytophthora cinnamomi* root rot of *Banksia grandis*. Can J Bot 74:618–624

Enyedi AJ, Yalpani N, Silverman P, Raskin I (1992) Localization, conjugation, and function of salicylic acid in tobacco during the hypersensitive reaction to tobacco mosaic virus. Proc Natl Acad Sci USA 89:2480–2484

Farmer EE, Ryan CA (1990) Interplant communication: airborne methyl jasmonate induces synthesis of proteinase inhibitors in plant leaves. Proc Natl Acad Sci USA 87:7713–7716

Felton GW, Korth KL, Bi JL, Wesley SV, Huhman DV, Mathews MC, Murphy JB, Lamb C, Dixon RA (1999) Inverse relationship between systemic resistance of plants to microorganisms and to insect herbivory. Curr Biol 9:317–320

Firmin JL, Wilson KE, Rossen L, Johnston AWB (1986) Flavonoid activation of nodulation genes in *Rhizobium* reversed by other compounds present in plants. Nature 324:90–92

Frankenberger WT Jr, Arshad M (1995) Phytohormones in soils. Marcel Dekker, New York

Fravel DR, Rhodes DJ, Larkin RP (1999) Production and commercialization of biocontrol products. In: Albajes R, Gullino ML, van Lenteren JC, Elad Y (eds) Integrated pest and disease management in greenhouse crops. Kluwer Academia Publishers, Dordrecht, pp 365–376

Fridlender M, Inbar J, Chet I (1993) Biological control of soilborne plant pathogens by a β-1,3-glucanase-producing *Pseudomonas cepacia*. Soil Biol Biochem 25:1211–1221

Gaffney T, Friedrich L, Vernooij B, Negrotto D, Nye G, Uknes S, Ward E, Kessmann H, Ryals J (1993) Requirement of salicylic acid for the induction of systemic acquired resistance. Science 261:754–756

Gagnon H, Ibrahim RK (1998) Aldonic acids: a novel family of *nod* gene inducers of *Mesorhizobium loti*, *Rhizobium lupini*, and *Sinorhizobium meliloti*. Mol Plant-Microbe Interact 11:988–998

Gaworzewska ET, Carlile MJ (1982) Positive chemotaxis of *Rhizobiuim leguminosarum* and other bacteria towards exudates from legumes and other plants. J Gen Microbiol 128:1179–1188

Georgakopoulos DG, Hendson M, Panopoulos NJ, Schroth MN (1994) Cloning of a phenazine biosynthetic locus of *Pseudomonas aureofaciens* PGS12 and analysis of its expression in vitro with the ice nucleation reporter gene. App Environ Microbiol 60:2931–2938

Geremia RA, Mergaert P, Geelen D, Van Montagu M, Holsters M (1994) The NodC protein of *Azorhizobium caulinodans* is an *N*-acetylglucosaminyltransferase Proc Natl Acad Sci USA 91:2669–2673

Gitte RR, Vittal Rai P, Patil RB (1978) Chemotaxis of Rhizobium species towards root exudates of *Cicer arietinium*. Plant Soil 50:553–566

Goedhart J, Bono JJ, Bisseling T, Gadella TW Jr (2003) Identical accumulation and immobilization of sulfated and non-sulfated Nod factors in host and non-host root hair cell walls. Mol Plant–Microbe Interact 16:884–892

Gray EJ, Smith DL (2005) Intracellular and extracellular PGPR: commonalities and distinctions in the plant-bacterium signaling processes. Soil Biol Biochem 37:395–412

Gray EJ, Lee KD, Souleimanov A, Di Falco MR, Zhou X, Ly A, Charles TC, Smith DL (2006a) A novel bacteriocin, thurici 17, produced by PGPR strain *Bacillus thuringiensis* NEB17: isolation and classification. J Appl Microbiol 100:545–554

Gray EJ, Di Falco M, Souleimanov A, Smith DL (2006b) Proteomic analysis of the bacteriocin thuricin 17 produced by *Bacillus thuringiensis* NEB17. FEMS Microbiol Lett 255:27–32

Guerinot ML, Yi Y (1994) Iron: nutritious, noxious, and not readily available. Plant Physiol 104:815–820

Gundlach H, Muller MJ, Kutchan TM, Zenk MH (1992) Jasmonic acid is a signal transducer in elicitor-induced plant cell cultures. Proc Natl Acad Sci USA 89:2389–2393

Guo Y, George E, Marschner H (1996) Contribution of an arbuscular mycorrhizal fungus to the uptake of cadmium and nickel in bean and maize plants. Plant Soil 184:195–205

Haas D, Defago G (2005) Biological control of soil-borne pathogens by fluorescent pseudomonads. Nat Rev Microbiol 3:307–319

Hamel C, Barrantes-Cartin U, Furlan V, Smith DL (1991a) Endo-mycorrhizal fungi in nitrogen transfer from soybean to maize. Plant Soil 138:33–40

Hamel C, Nesser C, Barrantes-Cartin U, Smith DL (1991b) Endo-mycorrhizal fungal species mediate ^{15}N transfer from soybean to maize in non fumigated soil. Plant Soil 138:41–47

Hanin M, Jabbouri S, Broughton WJ, Fellay R, Quesada-Vincens D (1999) Molecular aspects of host-specific nodulation. In: Stacey G, Keen NT (eds) Plant microbe interactions, vol 4. APS Press, St Paul, Minnesota

Hansen JN (1994) Nisin as a model food preservative. Crit Rev Food Sci Nutr 4:69–93

Hartwig UA, Maxwell CA, Joseph CM, Phillips DA (1990) Chrysoeriol and luteolin released from alfalfa seeds induce nod genes in *Rhizobium meliloti*. Plant Physiol 92:116–122

Hashidoko Y, Nakayama T, Homma Y, Tahara S (1999) Structure elucidation of xanthobaccin A, a new antibiotic produced from *Stenotrophomonas* sp. strain SB-K88. Tetrahedron Lett 40:2957–2960

Hirsch AM, Lum MR, Downie JA (2001) What makes the rhizobia-legume symbiosis so special? Plant Physiol 127:1484–1492

Holland MA (1994) PPFMs and other covert contaminants: is there more to plant physiology than just plant? Ann Rev Plant Physiol Plant Mol Biol 45:197–209

Hungria M, Stacey G (1997) Molecular signal exchanges between host plants and rhizobia: basic aspects and potential application in agriculture. Soil Biol Biochem 29:819–830

Hungria M, Joseph CM, Phillips DA (1991a). Anthocyanidins and flavonols, major *nod* gene inducers from seeds of a black-seeded common bean (Phaseolus vulgaris L.). Plant Physiol 97:751–758

Hungria M, Joseph CM, Phillips DA (1991b) *Rhizobium nod* gene inducers exuded naturally from roots of common bean (*Phaseolus vulgaris* L.). Plant Physiol 97:759–764

Jack WE, Tagg JR, Ray B (1995) Bacteriocins of Gram-positive bacteria. Microbiol Rev 59:171–200

Johansen A, Jakobsen I, Jensen ES (1993) External hyphae of vesicular-arbuscular mycorrhizal fungi associated with *Trifolium subterraneum* L. 3. Hyphal transport of 32P and 15N. New Phytol 124:61–68

John M, Rohrig H, Schmidt J, Wieneke U, Schell J (1993) *Rhizobium* NodB protein involved in nodulation signal synthesis is a chitooligosaccharide deacetylase. Proc Natl Acad Sci USA 90:625–629

Jones GP, Naidu BP, Starr RK, Paleg LG (1986) Estimates of solutes accumulating in plants by ^{1}H nuclear magnetic resonance spectroscopy. Aust J Plant Physiol 13:649–658

Jung WJ, An KN, Jin YL, Park RD, Lim KT, Kim KY, Kim TH (2003) Biological control of damping-off caused by *Rhizoctonia solani* using chitinase producing *Paenibacillus illinoisensis* KJA-424. Soil Biol Biochem 35:1261–1264

Jung WJ, Jin YL, Kim KY, Park RD, Kim TH (2005) Changes in pathogenesis-related proteins in pepper plants with regard to biological control of phytopthora blight with *Paenibacillus illinoisensis*. Biocontrol 50:165–178

Jung WJ, Mabood F, Souleimanov A, Park RD, Smith DL (2006) Chitinases produced by *Paenibacillus illinoisensis* and *Bacillus thuringiensis* subsp. *pakistani* degrade Nod factor from *Bradyrhizobium japonicum*. Microbiol Res (in press) doi:10.1016/j.micres.2006.06.013

Jung WJ, Mabood F, Souleimanov A, Zhou X, Jaoua S, Kamoun F, Smith DL (2008) Stability and antibacterial activity of bacteriocins produced by *Bacillus thuringiensis* and *Bacillus thuringiensis* ssp. Kurstaki. J Microbiol Biotechnol (In Press)

Kamoun F, Mejdoub H, Aouissaoui H, Reinbolt J, Hammami A, Jaoua S (2005) Purification, amino acid sequence and characterization of Bacthuricin F4, a new bacteriocin produced by *Bacillus thuringiensis*. J Appl Microbiol 98:881–888

Kamst E, Pilling J, Raamsdonk LM, Lugtenberg BJJ, Spaink HP (1997) *Rhizobium* nodulation protein NodC is an important determinant of chitin oligosaccharide chain length in Nod factor biosynthesis. J Bacteriol 179:2103–2108

Kamst E, Bakkers J, Quaedvlieg NE, Pilling J, Kijne JW, Lugtenberg BJJ, Spaink HP (1999) Chitin oligosaccharide synthesis by rhizobia and zebrafish embryos starts by glycosyl transfer to O4 of the reducing-terminal residue. Biochemistry 38:4045–4052

Kape R, Parniske M, Brandt S, Werner D (1992) Isoliquiritigenin, a strong nod gene- and glyceollin resistance-inducing flavonoid from soybean root exudate. App Env Microbiol 58:1705–1710

Keel C, Voisard C, Berling CH, Kahr G, Défago G (1989) Iron sufficiency, a prerequisite for the suppression of tobacco black root rot by *Pseudomonas fluorescens* strain CHA0 under gnotobiotic conditions. Phytopathology 79:584–589

Kim BS, Moon SS, Hwang BK (1999) Isolation, identification and antifungal activity of a macrolide antibiotic, oligomycin A, produced by *Streptomyces libani*. Can J Bot 77:850–858

Kim KY, Jordan D, McDonald GA (1998) Effect of phosphatesolubilizing bacteria and vesicular–arbuscular mycorrhizae on tomato growth and soil microbial activity. Biol Fertil Soil 26:79–87

Kloepper JW, Schroth MN (1978) Plant growth-promoting rhizobacteria on radishes. In: Proceedings of the 4th International Conference on Plant Pathogenic Bacteria, vol 2, Station de Pathologie Vegetale et Phytobacteriologie INRA Angers France, pp 879–882

Kloepper JW, Rodriguez-Kabana R, Zehnder GW, Murphy J, Sikora E, Fernandez C (1999) Plant root-bacterial interactions in biological control of soilborne diseases and potential extension to systemic and foliar diseases. Aust J Plant Pathol 28:27–33

Kosslak RM, Bookland R, Berkei J, Paaren HE, Appelbaum ER (1987) Induction of *Bradyrhizobium japonicum* common *nod* genes by isoflavones isolated from *Glycine max*. Proc Natl Acad Sci USA 84:7428–7432

Kramell R, Altzorn R, Schneider G, Miersch O, Bruckner C, Schmidt J, Sembdner G, Parthier B (1995) Occurrence and identification of jasmonic acid and its amino acid conjugates induced by osmotic stress in barley leaf tissue. J Plant Growth Regul 14:29–36

Kucey RMN (1988) Plant growth-altering effects of *Azospirillum brasilense* and *Bacillus* C-11–25 on two wheat cultivars. J App Bacteriol 64:187–196

Lebuhn M, Heulin T, Hartmann A (1997) Production of auxin and other indolic and phenolic compounds by *Paenibacillus polymyxa* strains isolated from different proximity to plant roots. FEMS Microbiol Ecol 22:325–334

Leeman M, den Ouden FM, van Pelt JA, Dirkx FPM, Steijl H, Bakker PAHM, Schippers B (1996) Iron availability affects induction of systemic resistance to Fusarium wilt of radish by *Pseudomonas fluorescens*. Phytopathology 86:149–155

Leibovitch S, Migner P, Zhang F, Smith DL (2001) Evaluation of the effect of SoyaSignal technology on soybean yield [*Glycine max* (L.) Merr.] under field conditions over 6 years in eastern Canada and the northern United States. J Agron Crop Sci 187:281–292

Lerouge P, Roche P, Faucher C, Maillet F, Trucet G, Prome J-C, Denarie J (1990) Symbiotic host-specificity of *Rhizobium meliloti* is determined by a sulphated and acylated glucosamine oligosaccharide. Nature 344:781–784

Li X-L, Marschner H, George E (1991) Acquisition of phosphorus and copper by VA-mycorrhizal hyphae and root-to-shoot transport in white clover. Plant Soil 136:49–57

Lopez-Lara IM, Bloktip L, Quinto C, Garcia ML, Stacey G, Bloemberg GV, Lamers GEM, Lugtenberg BJJ, Thomasoates JE, Spaink HP (1996) NodZ of *Bradyrhizobium* extends the nodulation host range of *Rhizobium* by adding a fucosyl residue to nodulation signals. Mol Microbiol 21:397–408

Lynch JM, Whipps JM (1991) Substrate flow in the rhizosphere. In: Keister DL, Cregan B (eds) The rhizosphere and plant growth. Beltsville Symposium in Agricultural Research 14. Kluwer, Dordrecht The Netherlands, pp 15–24

Mabood F, Smith DL (2005) Pre-incubation of *Bradyrhizobium japonicum* with jasmonates accelerate nodulation and nitrogen fixation in soybean (*Glycine max*) at optimal and sub-optimal root zone temperatures. Physiol Plant 125:311–323

Mabood F, Gray EJ, Lee KD, Supanjani, Smith DL (2006a) Exploiting inter-organismal chemical communication for improved inoculants. Can J Plant Sci 86:951–966

Mabood F, Souleimanov A, Khan W, Smith DL (2006b) Jasmonates induce Nod factor production by *Bradyrhizobium japonicum*. Plant Physiol Biochem 44:759–765

Mabood F, Zhou X, Smith DL (2006c) *Bradyrhizobium japonicum* preincubated with methyl jasmonate increases soybean nodulation and nitrogen fixation. Agron J 98:289–294

Mabood F, Zhou X, Lee KD, Smith DL (2006d) Methyl jasmonate, alone or in combination with genistein, and *Bradyrhizobium japonicum* increases soybean (*Glycine max* L.) plant dry matter production and grain yield under short season conditions. Field Crops Res 95:412–419

Mathesius U, Bayliss C, Weinman JJ, Schlaman HRM, Spaink HP, Rolfe BG, McCully ME, Djordjevic MA (1998) Flavonoids synthesized in cortical cells during nodule initiation are early developmental markers in white clover. Mol Plant Microbe Interact 11:1223–1232

Mathre DE, Cook RJ, Callan NW (1999) From discovery to use. Traversing the world of commercializing biocontrol agents for plant disease control. Plant Dis 83:972–983

Maurhofer M, Hase C, Meuwly P, Metraux J-P, Defago G (1994) Induction of systemic resistance of tobacco to tobacco necrosis virus by the root-colonizing *Pseudomonas fluorescens* strain CHA0: influence of the gacA gene and of pyoverdine production. Phytopathology 84:139–146

Maurhofer M, Reimmann C, Schmidli-Sacherer P, Heeb S, Haas D, Defago G (1998) Salicylic acid biosynthetic genes expressed in *Pseudomonas fluorescens* P3 improve the induction of systemic resistance in tobacco against tobacco necrosis virus. Phytopathology 88:678–684

Maxwell CA, Hartwig UA, Joseph CM, Phillips DA (1989) A chalcone and two related flavonoids released from alfalfa roots induce *nod* genes of *Rhizobium meliloti*. Plant Physiol 91:842–847

McIver H (2005) The use of legume/rhizobial signals for yield enhancement in non-leguminous crops. Inoculant forum 2005, Saskatoon, Saskatchewan, March 17–18

McSpadden Gardener BB, Fravel DR (2002) Biological control of plant pathogens: research, commercialization, and application in the USA. Online. Plant Health Progr doi:10.1094/PHP-2002-0510-01-RV[http://www.apsnet.org/online/feature/biocontrol/top.html]

Mergaert P, D'Haeze W, Geelen D, Prome D, van Montagu M, Geremia R, Prome J-C, Holsters M (1995) Biosynthesis of *Azorhizobium caulinodans* Nod factor: study of the activity of the NodABC proteins by expression of the genes in *Escherichia coli*. J Biol Chem 270:29217–29223

Milner JL, Raffel SJ, Lethbridge BJ, Handelsman J (1995) Culture conditions that influence accumulation of zwittermicin A by *Bacillus cereus* UW85. Appl Microbiol Biotechnol 43:685–691

Milner JL, Silo-Suh L, Lee JC, He H, Clardy J, Handelsman J (1996) Production of kanosamine by *Bacillus cereus* UW85. Appl Environ Microbiol 62:3061–3065

Nakayama T, Homma Y, Hashidoko Y, Mizutani J, Tahara S (1999) Possible role of xanthobaccins produced by *Stenotrophomonas* sp. strain SB-K88 in suppression of sugar beet damping-off disease. Appl Environ Microbiol 65:4334–4339

Nelson LM (2004) Plant growth promoting rhizobacteria (PGPR): prospects for new inoculants. Online. Crop Manag doi:10.1094/CM-2004-0301-05-RV[http://www.plantmanagementnetwork.org/pub/cm/review/2004/rhizobacteria/]

Nielsen TH, Sorensen J (2003) Production of cyclic lipopeptides by *Pseudomonas fluorescens* strains in bulk soil and in the sugar beet rhizosphere. Appl Environ Microbiol 69:861–868

Nielsen TH, Sorensen D, Tobiasen C, Andersen JB, Christeophersen C, Givskov M, Sorensen J (2002) Antibiotic and biosurfactant properties of cyclic lipopeptides produced by fluorescent *Pseudomonas* spp. from the sugar beet rhizosphere. Appl Environ Microbiol 68:3416–3423

Niki T, Mitsuhara I, Seo S, Ohtsubo N, Ohashi Y (1998) Antagonistic effect of salicylic acid and jasmonic acid on the expression of pathogenesis-related (PR) protein genes in wounded mature tobacco leaves. Plant Cell Physiol 39:500–507

Noel TC, Sheng C, Yost CK, Pharis RP, Hynes MF (1996) *Rhizobium leguminosarum* as a plant growth-promoting rhizobacterium: direct growth promotion of canola and lettuce. Can J Microbiol 42:279–283

Okon Y, Labandera-Gonzales CA (1994) Agronomic applications of *Azospirillum*: an evaluation of 20 years worldwide field inoculation. Soil Biol Biochem 26:1591–1601

Ortas I, Harris PJ, Rowell DL (1996) Enhanced uptake of phosphorus by mycorrhizal sorghum plants as influenced by forms of nitrogen. Plant Soil 184:255–264

Ownley BH, Weller DM, Thomashow LS (1992) Influence of in situ and in vitro pH on suppression of *Gaeumannomyces graminis* var. tritici by *Pseudomonas fluorescens* 2–79. Phytopathology 82:178–184

Ownley BH, Duffy BK, Weller DM (2003) Identification and manipulation of soil properties to improve the biological control performance of phenazine-producing *Pseudomonas fluorescens*. Appl Environ Microbiol 69:3333–3343

Parker JE (2003) Plant recognition of microbial patterns. Trends Plant Sci 8.245–247

Parret AHA, Mot RD (2002) Bacteria killing their own kind: novel bacteriocins of Pseudomonas and other γ proteobacteria. Trend Microbiol 10:107–112

Patten CL, Glick BR (1996) Bacterial biosynthesis of indole-3-acetic acid. Can J Microbiol 42:207–220

Paulitz TC, Belanger RB (2001) Biological control in greenhouse systems. Ann Rev Phytopathol 39:103–133

Perret X, Staehelin C, Broughton WJ (2000) Molecular basis of symbiotic promiscuity. Microbiol Mol Biol Rev 64:180–201

Persello-Cartieaux F, Nussaume L, Robaglia C (2003) Tales from the underground: molecular plant–rhizobacteria interactions. Plant Cell Environ 26:189–199

Peters NK, Frost JW, Long SR (1986) A plant flavone, luteolin, induces expression of *Rhizobium meliloti* nodulation genes. Science 233:977–980

Phillips DA (2000) Biosynthesis and release of rhizobial nodulation gene inducers by legumes. In: Triplett EW (ed) Prokaryotic nitrogen fixation: a model system for the analysis of a biological process. Horizon Science Press, Wymondham Norfolk England, pp 349–364

Phillips DA, Joseph CM, Maxwell CA (1992) Trigonelline and stachydrine released from alfalfa seeds activate NodD2 protein in *Rhizobium meliloti*. Plant Physiol 99:1526–1531

Phillips DA, Wery J, Joseph CM, Jones AD, Teubers LR (1995) Release of flavonoids and betaines from seeds of seven *Medicago* species. Crop Sci 35:805–808

Picard C, Di Cello F, Ventura M, Fani R, Guckert A (2000) Frequency and biodiversity of 2,4-diacetylphloroglucinol-producing bacteria isolated from the maize rhizosphere at different stages of plant growth. Appl Environ Microbiol 66:948–955

Pieterse CMJ, Van Wees SCM, Van Pelt JA, Knoester M, Laan R, Gerrits H, Weisbeek PJ, van Loon LC (1998) A novel signaling pathway controlling induced systemic resistance in Arabidopsis. Plant Cell 10:1571–1580

Pleban S, Chernin L, Chet I (1997) Chitinolytic activity of an endophytic strain of *Bacillus cereus*. Lett App Microbiol 25:284–288

Press CM, Wilson M, Tuzun S, Kloepper JW (1997) Salicylic acid produced by *Serratia marcescens* 90–166 is not the primary determinant of induced systemic resistance in cucumber or tobacco. Mol Plant Microbe Interact 10:761–768

Prithiviraj B, Zhou X, Souleimanov A, Khan WM, Smith DL (2003) A host-specific bacteria-to-plant signal molecule (Nod factor) enhances germination and early growth of diverse crop plants. Planta 21:437–445

Raaijmakers JM, Vlami M, de Souza JT (2002) Antibiotic production by bacterial biocontrol agents. Antonie Leeuwenhoek 81:537–547

Recourt K, Schripsema J, Kijn JW, van Brussel AAN, Lugtenberg BJJ (1991) Inoculation of *Vicia sativa* subsp. *nigra* roots with *Rhizobium leguminosarum* biovar *viciae* results in release of *nod* gene activating flavanones and chalcones. Plant Mol Biol 16:841–852

Redmond JW, Batley M, Djordjevic MA, Innes RW, Kuempel PL, Rolfe BG (1986) Flavones induce expression of nodulation genes in *Rhizobium*. Nature 323:632–635

Rodriguez H, Fraga R (1999) Phosphate solubilizing bacteria and their role in plant growth promotion. Biotechnol Adv 17:319–339

Rohrig H, Schmidt J, Wieneke U, Kondorosi E, Barlier I, Schell J, John M (1994) Biosynthesis of lipooligosaccharide nodulation factors-*Rhizobium* NodA protein is involved in N-acylation of the chitooligosaccharide backbone. Proc Natl Acad Sci USA 91:3122–3126

Rosas S, Soria R, Correa N, Abdala G (1998) Jasmonic acid stimulates the expression of *nod* genes in Rhizobium. Plant Mol Biol 38:1161–1168

Ryals JA, Neuenschwander UH, Willits MG, Molina A, Steiner HY, Hunt MD (1996) Systemic acquired resistance. Plant Cell 8:1809–1819

Ryu C-M, Farag MA, Hu C-H, Reddy MS, Wei H-X, Pare PW, Kloepper JW (2003) Bacterial volatiles promote growth in Arabidopsis. Proc Natl Acad Sci USA 100:4927–4932

Ryu C-M, Farag MA, Hu C-H, Reddy MS, Kloepper JW, Pare PW (2004) Bacterial volatiles induce systemic resistance in Arabidopsis. Plant Physiol. 134:1017–1026

Rzewnicki P (2000) Ohio organic producers: final survey results. Online. Ohio State University, Extension Research Bulletin Special Circular 174[http://ohioline.osu.edu/sc174/index.html]

Sandberg G (1984) Biosynthesis and metabolism of indole-3-ethanol and indole-3-acetic acid by *Pinus sylvestris* L. needles. Planta 161:398–403

Schmidt CS, Lorenz D, Wolf GA (2001) Biological control of the grapevine dieback fungus *Eutypa lata* I: screening of bacterial antagonists. J Phytopathol 149:427–435

Selvadurai EL, Brown AE, Hamilton JTG (1991) Production of indole-3-acetic acid analogues by strains of *Bacillus cereus* in relation to their influence on seedling development. Soil Biol Biochem 23:401–403

Serdyuk OP, Smolygina LD, Muzafarov EN, Adanin VM, Arinbasarov MU (1995) 4-Hydroxyphenethyl alcohol, a new cytokinin-like substance from the phototrophic purple bacterium *Rhodospirillum rubrum* 1R. FEBS Lett 365:10–12

Shah S, Li J, Moffatt BA, Glick BR (1998) Isolation and characterization of ACC deaminase genes from two different plant growth-promoting rhizobacteria. Can J Microbiol 44:833–843

Shulaev V, Silverman P, Raskin I (1997) Airborne signalling by methyl salicylate in plant pathogen resistance. Nature 385:718–721

Singh PP, Shin YC, Park CS, Chung YR (1999) Biological control of *Fusarium Wilt* of cucumber by chitinolytic bacteria. Phytopathology 89:92–99

Smit G, Puvanesarajah V, Carlson RW, Barbour WM, Stacey G (1992) *Bradyrhizobium japonicum nodD1* can be specifically induced by soybean flavonoids that do not induce the *nodYABCSUIJ* operon. J Biol Chem 267:310–318

Smith DL, Zhang F (1999) Composition for enhancing grain yield and protein yield of legumes grown under environmental conditions that inhibit or delay nodulation thereof. US Patent 5922316

Smith S (2005) New patented growth promoter technology to enhance early season soybean development & grain yield. Inoculant Forum 2005, Saskatoon, Saskatchewan, March 17–18

Soulemainov A, Prithiviraj B, Smith DL (2002) The major Nod factor of *Bradyrhizobium japonicum* promotes early growth of soybean and corn. J Exp Bot 53:1929–1934

Spaink HP (2000) Root nodulation and infection factors produced by rhizobial bacteria. Ann Rev Microbiol 54:257–288

Spaink HP, Wijfjes AHM, van der Drift KMGM, Haverkamp J, Thomas-Oates JE, Lugtenberg BJJ (1994) Structural identification of metabolites produced by the NodB and NodC proteins of *Rhizobium leguminosarum*. Mol Microbiol 13:821–831

Stacey G, Sanjuan J, Luka S, Dockendorff T, Carlson RW (1995) Signal exchange in the *Bradyrhizobium*-soybean symbiosis. Soil Biol Biochem 27:473–483

Stafford H (1990) Flavonoid metabolism. CRC Press, Boca Raton, FL

Steenhoudt O, Vanderleyden J (2000) *Azospirillum*, a free-living nitrogen-fixing bacterium closely associated with grasses: genetic, biochemical and ecological aspects. FEMS Microbiol Rev 24:487–506

Sticher L, Mauch-Mani B, Metraux JP (1997) Systemic acquired resistance. Annu Rev Phytopathol 35:235–270

Strzelczyk E, Pokojska-Burdziej A (1984) Production of auxins and gibberellin-like substances by mycorrhizal fungi, bacteria and actinomycetes isolated from soil and the mycorrhizosphere of pine (*Pinus silvestris* L.). Plant Soil 81:185–194

Subramanian KS, Charest C, Dwyer LM, Hamilton RI (1995) Arbuscular mycorrhizae and water relations in maize under drought stress at tasseling. New Phytol 129:643–650

Valois D, Fayad K, Barbasubiye T, Garon M, Dery C, Brzezinski R, Beaulieu C (1996) Glucanolytic actinomycetes antagonistic to *Phytophthora fragariae* var. *rubi*, the causal agent of raspberry root rot. App Env Microbiol 62:1630–1635

van Loon LC (1997) Induced resistance in plants and the role of pathogenesis-related proteins. Eur J Plant Pathol 103:753–765

van Loon LC, Bakker PAHM, Pierterse CMJ (1998) Systemic resistance induced by rhizosphere bacteria. Annu. Rev. Phytopathol 36:453–483

Vessey JK (2003) Plant gowth promoting rhizobacteria as biofertilizers. Plant Soil 255:571–586

Vick BA, Zimmerman DC (1984) Biosynthesis of jasmonic acid by several plant species. Plant Physiol 75:458–461

Ward ER, Uknes SJ, Williams SC, Dincher SS, Wiederhold DL, Alexander DC, Ahl-Goy P, Metraux J-P, Ryals JA (1991) Coordinate gene activity in response to agents that induce systemic acquired resistance. Plant Cell 3:1085–1094

Wei G, Kloepper JW, Tuzun S (1991) Induction of systemic resistance of cucumber to *Colletotrichum orbiculare* by select strains of plant growth promoting rhizobacteria. Phytopathology 81:1508–1512

Weingart H, Volksch B (1997) Ethylene production by *Pseudomonas syringae* pathovars in vitro and in planta. App Env Microbiol 63:156–161

Whipps JM (2001) Microbial interactions and biocontrol in the rhizosphere. J Exp Bot 52:487–511

Xie H, Pasternak JJ, Glick BR (1996) Isolation and characterization of mutants of the plant growth-promoting rhizobacterium *Pseudomonas putida* GR12-2 that overproduce indoleacetic acid. Curr Microbiol 32:67–71

Yanni YG, Rizk RY, Abd El-Fattah FK, Squartini A, Corich V, Giacomini A, De Bruijn F, Rademaker J, Maya-Flores J, Ostrom P, Vega-Hernandez M, Hollingsworth RI, Martinez-Molina

E, Mateos P, Velazquez E, Wopereis J, Triplett E, Umali-Garcia M, Anarna JA, Rolfe BG, Ladha JK, Hill J, Mujoo R, Ng PK, Dazzo FB (2001) The beneficial plant growth-promoting association of *Rhizobium leguminosarum* bv. trifolii with rice roots. Aust J Plant Physiol 28:845–870

Yu O, Jung W, Shi J, Croes RA, Fader GM, McGonigle B, Odell JT (2000) Production of the isoflavones genistein and daidzein in non-legume dicot and monocot tissues. Plant Physiol 124:781–794

Zaat SAJ, Schripsema J, Wijffelman CA, van Brussel AAN, Lugtenberg BJJ (1989) Analysis of the major inducers of the *Rhizobium nodA* promoter from *Vicia sativa* root exudate and their activity with different *nodD* genes. Plant Mol Biol 13:175–188

Zhang F, Smith DL (1994) Effects of low root zone temperatures on the early stages of symbiosis establishment between soybean [*Glycine max* (L.) Merr.] and *Bradyrhizobium japonicum*. J Exp Bot 279:1467–1473

Chapter 13
Protein-Protein Interactions in Plant Virus Movement and Pathogenicity

Joachim F. Uhrig(✉) and Stuart A. MacFarlane

13.1   Introduction

Viruses generally consist of a rather small number of molecular components including a few proteins, sometimes a membranous envelope, and an RNA or DNA genome encompassing a very limited set of coding sequences. For every step in the viral life cycles such as replication of the viral genomes, transcription/translation of viral gene products, intra- and intercellular movement and virus assembly and transmission, viruses make use of the biosynthetic and regulatory capacities of the host cells.

The rigid nature of plant cell walls and the lack of a cardiovascular system prevent plant viruses, in contrast to animal viruses, from spreading within the infected organism by cell lysis or budding and subsequent passive transport in a liquid circulation system. Similarly, the lack of a circulating antibody system has brought about RNA silencing mechanisms as the major antiviral defence strategy applied by plant cells. Due to these two main differences in viral infections between plant and animal systems, plant viruses have evolved specialized proteins and protein functions. On the one hand, so-called movement proteins utilize plant host structures and mechanisms to facilitate intra cellular, inter-cellular and long-distance transport; on the other, silencing suppressors serve to counteract and escape the host plant antiviral defence machineries.

Virus movement and the suppression of silencing cannot be separated unambiguously. Historically, a number of different plant viral proteins have been termed "movement proteins" based on the observation that defects in the respective genes affect cell-to-cell movement and limit systemic spread. In some cases it is now becoming clear that the genuine function of some of these movement proteins might rather be suppression of silencing. While it is obvious that the suppression of silencing is the prerequisite for the establishment of an infection of a cell during the

J.F. Uhrig
University of Cologne, Department of Botany III, Gyrhofstr. 15, D-50931 Cologne, Germany
e-mail: Joachim.Uhrig@uni-koeln.de

C.S. Nautiyal, P. Dion (eds.) *Molecular Mechanisms of Plant
and Microbe Coexistence.* Soil Biology 15, DOI: 10.1007/978-3-540-75575-3
© Springer-Verlag Berlin Heidelberg 2008

movement process, there is evidence that many viral proteins are multifunctional, and that the functions in movement and suppression of silencing may be separable (Bayne et al. 2005). Advances in research in both fields in recent years not only have greatly improved our understanding of the molecular basis of plant viral infection cycles, but also have revealed insight into as yet poorly understood cellular processes. Long-distance transport of RNA and/or proteins, and micro RNA-based control mechanisms are emerging as ancient fundamental regulatory processes underlying plant development and the processing of environmental signals (Lough and Lucas 2006; Murchison and Hannon 2004).

The appropriation of cellular functions and regulatory machineries often is accomplished by physical interaction between viral and host proteins facilitating a redirection of the host protein's function to serve a function in the viral life cycle. The investigation of host proteins interacting with viral proteins has proven to be a very promising approach to dissect the molecular basis of viral infections and to understand how viruses integrate in the complex structural and regulatory networks controlling plant growth and development.

13.2 Cell-to-Cell Movement of Plant Viruses

Plant viral movement proteins facilitate cell-to-cell and long-distance transport of viral structures and thereby allow the virus to establish a systemic infection in the host plant. In general, movement of viruses in plants does not involve extra-cellular stages, but rather occurs via plasmodesmata, the specialized, plant-specific intercellular structures connecting the cytoplasms of adjacent cells. Historically, plasmodesmata were regarded as simple channels allowing passive trafficking of low-molecular weight growth regulators and nutrients. An emerging picture now is that plasmodesmata are highly complex structures that regulate the selective trafficking of macromolecules (Lucas and Lee 2004).

Recent progress, in the analysis of cell-to-cell and long-distance transport processes in plants – stimulated by the investigation of virus movement in plants – revealed a complex communication network based on the transport of signalling molecules. The trafficking of proteins, RNA or ribonucleotide-protein complexes might be a fundamental means of plants to control development and to communicate and respond to environmental signals (Lough and Lucas 2006; Lucas and Lee 2004). A growing body of evidence supports the notion that proteins that move through plasmodesmata, so-called non-cell-autonomous proteins (NCAPs), can contribute to patterning and the establishment of cell fate in plant tissues (Gallagher and Benfey 2005; Lough and Lucas 2006; Lucas and Lee 2004). In fact, the immunological relatedness of CmPP16, an NCAP isolated from *Cucurbita maxima* phloem sap, and the movement protein of *Red clover necrotic mosaic virus* was interpreted as an indication of a potential common evolutionary origin of viral movement proteins and NCAPs (Xoconostle-Cazares et al. 1999). The observation that viral movement proteins compete with endogenous NCAPs for the plasmodesmata trafficking machinery supports the notion that plant viruses have evolved movement proteins to "hitch-hike" this NCAP trafficking pathway (Lucas 2006).

The mechanisms and molecular machineries involved in the inter-cellular and long-distance movement of NCAPs, viruses or other macromolecules are only beginning to be understood, but research in this field has been greatly stimulated by the investigation of the molecular interactions of viral movement proteins, especially by the identification of plant host proteins interacting with viral movement proteins (Boevink and Oparka 2005; see Table 13.1).

Table 13.1 Plant proteins interacting with viral movement proteins

Virus	Movement protein	Biological function(s)	Interacting host proteins	References for interaction
CLCV	NSP	Virus movement, nuclear shuttling	AtNSI1 acetyltransferase	Carvalho and Lazarowitz (2004); McGarry et al. (2003)
TGMV TCrLYV	NSP	Virus movement, nuclear shuttling	LeNIK, GmNIK, NIK1, NIK2, NIK3 LRR receptor-like kinases	Fontes et al. (2004); Mariano et al. (2004)
TMV	MP, p30	Virus movement	KELP, MBF1, transcriptional co-activators; NtRIO kinase PME MPB2C Calreticulin PAPK kinase	Chen MH et al. (2000, 2005); Dorokhov et al. (1999); Kragler et al. (2003); Lee et al. (2005); Matsushita et al. (2001, 2002); Yoshioka et al. (2004)
TuMV	VPg	RNA replication and translation, virus movement, virulence factor	PVIP PHD finge cysteine-rich protein	Dunoyer et al. (2004)
ToMV	CP	Coat protein, virus movement	IP-L	Li et al. (2005)
TCV	p8	Virus movement	Atp8	Lin and Heaton (2001)
PPV	CI	Replication, virus movement	Photosystem I PSI-K protein	Jimenez et al. (2006)
TSWV	NSm	Virus movement	AtA39, NtDNAJ_M541 At4/1	Soellick et al. (2000); von Bargen et al. (2001)
PMTV	TGB2	Virus movement	TIP1, TIP2, TIP3; RME-8 J-domain protein	Fridborg et al. (2003); Haupt et al. (2005)
TBSV	P22	Virus movement	HD-ZIP	Desvoyes et al. (2002)
GFLV	MP	Virus movement	KNOLLE syntaxin	Laporte et al. (2003)
CPMV	60 K	Virus movement	VAP27-1/2 SNARE	Carette et al. (2002)
CaMV	MP	Virus movement	MP17 rab-acceptor	Huang et al. (2001)

Intercellular movement of macromolecules involves as the first step the intracellular translocation of the proteins, RNAs or nucleo-protein complexes to the cell periphery and the sites of the plasmodesmata. In contrast to other translocation or localization events such as transport into chloroplasts or mitochondria, or targeting to the endoplasmic reticulum (ER) and the secretory pathway, no recognizable conserved sequence patterns or structural elements have been identified so far, targeting proteins to plasmodesmata. There is now increasing evidence that different pathways can direct proteins to the plasmodesmata, involving cytoskeletal elements and/or the endomembrane system. Both ways have recently been shown to be employed by viruses as well.

The second step is the translocation of the macromolecules through the plasmodesmata. Still, only very few structural components of plasmodesmata have been identified up to date. Therefore, the mechanism and the selectivity of plasmodesmatal transport are largely unknown. However, some of the identified host proteins interacting with viral movement proteins allow developing hypotheses on the molecular mechanisms involved in this step of viral movement.

Long-distance transport of viral structures usually occurs via the plant vasculature. This is probably the least understood step in plant viral movement and clear molecular data about the mechanisms involved are not available at present.

13.3 The Role of the Cytoskeleton

In order to move from the site of replication to the plasmodesmata at the cell periphery, viruses make use of the transporting capacities of the plant cytoskeleton. More than ten years ago plant viral movement proteins had already been demonstrated to co-localize with and bind to both, microtubules and actin filaments (Heinlein et al. 1995; McLean et al. 1995). However, the exact roles of these two cytoskeletal elements in the intracellular translocation and the intercellular transport of viral structures are just emerging. The most extensively investigated plant viral movement protein is the P30 protein from TMV. Fusions of P30 with the green fluorescent protein (GFP) have been analyzed with respect to intracellular localization, colocalization with cytoskeletal elements, and dynamics during infection and movement processes (Boyko et al. 2000; Epel et al. 1996; Heinlein et al. 1995; Mas and Beachy 2000; Padgett et al. 1996; Reichel et al. 1999). A conserved amino acid sequence motif has been identified within the sequences of tobamoviral movement proteins resembling a region in tubulin that was proposed earlier to mediate lateral contacts between microtubules (Boyko et al. 2000). This finding indicates that tobamoviral movement proteins bind to microtubules by mimicking tubulin interaction and assembly surfaces, thereby presumably competitively displacing γ-tubulin, and probably make use of microtubule polymerization to drive the transport process (Boyko et al. 2000; Wick 2000).

The TMV movement protein has furthermore been shown to interact with MPB2C, a previously uncharacterized microtubule associated plant protein (Kragler et al. 2003).

In accordance with the idea of microtubule-based intra- or intercellular transport, At4/1, a protein with homologies to myosin/kinesin motor proteins, has been found to interact with NSm, the movement protein of *Tomato Spotted Wilt Virus* (TSWV), a virus with a negative/ambisense ssRNA genome (von Bargen et al. 2001).

However, the role of microtubules in viral movement is still not quite clear, and it is very likely that different viruses utilize different cellular structures and mechanisms for intra- and inter-cellular movement. In the case of the TMV movement protein it has been shown that MPB2C is not required for movement but is necessary for microtubule association (Curin et al. 2007). In fact, MBP2C seems to be a negative effector of intracellular movement, and there is evidence that microtubules are dispensable for inter-cellular movement (Kragler et al. 2003). These and other recent findings indicate that microtubules and microtubule-associated factors might be involved in degradation of movement proteins, and that a transport function necessary for translocation of movement proteins and viral structures to the cell periphery and the plasmodesmata might rather be provided by the microfilaments of the actin cytoskeleton in connection with the endomembrane system (Boevink and Oparka 2005; Liu et al. 2005).

So far, no direct physical interaction between viral movement proteins and actin filaments has been demonstrated. However, co-localization and application of actin-destabilizing drugs indicate an association of for example the TMV movement protein with actin filaments (McLean et al. 1995). Similarly, an association of the movement proteins TGB2 and TGB3 of Potato mop-top virus (PMTV) have been shown to co-align with the actin cytoskeleton and, upon application of the actin-depolymerizing drug latrunculin, subcellular localization of these two proteins was changed and movement was abolished (Haupt et al. 2005).

Beet yellows virus requires five viral proteins for movement, including a homolog of a class of eukaryotic heat shock proteins of approximately 70kDa, Hsp70h (Peremyslov et al. 1999). Recently this protein has been found to form motile granules that are associated with actin microfilaments, and translocation to plasmodesmata was dependent on an intact actin cytoskeleton (Prokhnevsky et al. 2005). The involvement of Hsp70 proteins in viral movement is interesting because this class of proteins has been shown to be involved not only in protein folding but also in protein translocation processes (Young et al. 2003). The function of Hsp70 proteins is dependent on co-chaperones such as proteins of the DnaJ family, which are required for stable binding of Hsp70 to their substrates. Intriguingly, several plant viral movement proteins have been shown to interact with proteins containing a J-domain conserved in DnaJ proteins (Haupt et al. 2005; Soellick et al. 2000; von Bargen et al. 2001). These findings indicate a potentially general role of Hsp70 in viral movement either in connection with transport along the cytoskeleton or in the process of partial unfolding of viral structures that may be required for passage through plasmodesmata.

In addition to the direct recruitment of the host cell's transport machineries there are indications that hint at an indirect way of trafficking by "hitch-hiking" a protein that itself is transported into the neighbouring cell via the plasmodesmata. Recently, an interaction between the TBSV movement protein P22 and a plant homeodomain leucin zipper (HD-ZIP) protein has been identified (Desvoyes et al. 2002). One of the first non cell-autonomous proteins identified in plants is the homeodomain protein KNOTTED1, that has been shown to move between cells via plasmodesmata (Lucas et al. 1995). Therefore, the proposition in the case of the TBSV-P22 protein is that by way of interacting with the HD-ZIP protein it may be transported to and possibly passaged through the plasmodesmata (Desvoyes et al. 2002). So far, the mechanism of how homeodomain proteins move to and traffic through plasmodesmata is not known. However, recently a microtubule-associated protein has been identified interacting with and regulating the intra-cellular localization of homeodomain proteins of the three amino acid loop extension (TALE) class (Hackbusch et al. 2005). Whether this intriguing novel protein family might be a candidate for a component of intra- or intercellular transport machineries remains to be investigated.

13.4 Involvement of the Endomembrane System in Virus Movement

Plamodesmata are plasma membrane-lined inter-cellular channels that establish continuity of the cytoplasms of adjacent plant cells. Although the molecular structure and composition of plasmodesmata is still obscure, it is well established that parts of the endomembrane system, the so-called desmotubules, extend through these channels from one cell to the other.

An emerging picture is that both inter-cellular trafficking of endogenous signalling molecules like NCAPs, and virus movement might proceed with involvement of the endomembrane transport system (Boevink and Oparka 2005). Specific targeting of some cellular proteins to the plasmodesmata via a Golgi-dependent pathway has been revealed recently (Sagi et al. 2005).

There is a growing body of evidence that many steps in plant viral life cycles may be connected with cellular endomembrane systems. Virus replication often is associated with the ER, and a number of viral proteins facilitating movement similarly locate to the ER.

The TMV P30 movement protein and the TMV 126/183-kDa protein, both required for TMV movement, have been shown to locate to the ER (Heinlein et al. 1998). Similarly, an association of the movement protein (p6) of beet yellow closterovirus with the ER has been observed (Huang and Zhang 1999; Peremyslov et al. 2004).

Several host proteins involved in the secretion pathway or membrane trafficking events have been identified recently as interaction partners of plant viral movement proteins, supporting the view that the dynamic endomembrane system plays a central role in viral movement processes.

Movement of Grapevine fanleaf virus (GFLV) virions proceeds through tubules, formed by the virus-encoded movement protein, that penetrate modified plasmodesmata. Tubule formation is dependent on a functional secretory pathway, and the GFLV movement protein physically interacts with KNOLLE, a syntaxin involved in cytokinesis (Laporte et al. 2003). Syntaxins are membrane proteins belonging to the SNARE family that regulates vesicle targeting and fusion (Chen YA and Scheller 2001). Interestingly, the Cowpea mosaic virus 60 K movement protein interacts with VAP33/SNARE (Carette et al. 2002), indicating that Syntaxin/SNARE-mediated vesicle fusion, targeting or trafficking is a mechanism for intracellular translocation that different viruses take advantage of.

In addition to interacting with the syntaxin KNOLLE, the movement protein of GFLV has been shown to colocalize with calreticulin-containing foci (Laporte et al. 2003). This again might be a common theme in movement processes of different plant viruses, because recently a direct interaction between the TMV movement protein p30 and calreticulin has been described (Chen MH et al. 2005). Calreticulin is a ubiquitous calcium-binding chaperone involved in integrin-mediated cell adhesion in animals (Coppolino et al. 1997). In plants, calreticulin has been shown to localize to the ER and to plasmodesmata. The interaction with p30, co-localization in vivo and the observation that over-expression of calreticulin affects p30 localization and TMV movement have led to the conclusion that calreticulin might be functionally involved in viral movement (Chen MH et al. 2005). Indirect support for the idea of an involvement of calreticulin in plant viral movement comes from the finding that one of the two movement proteins of turnip crinkle virus (TCV) interacts with a protein designated Atp8, containing so-called RGD motifs (Lin and Heaton 2001). These motifs are known to function as cell-attachment sequences that are recognized by integrins which in turn may be bound by calreticulin (D'Souza et al. 1991). Direct or indirect binding of movement proteins to calreticulin might serve for anchoring viral movement complexes to peripheral attachment sites which frequently are associated with plasmodesmata (Boevink and Oparka 2005).

MPI7, an *Arabidopsis thaliana* protein interacting with the cauliflower mosaic virus (CaMV) movement protein (MP) provides more, albeit indirect, evidence for a possible involvement of membrane or vesicle trafficking in viral movement processes. Direct interaction between MP and MPI7 in yeast and in planta has been demonstrated by yeast two-hybrid analyses and fluorescence resonance energy transfer (FRET) (Huang et al. 2001). Biological significance and a potential important role of this interaction in viral infectivity were inferred from the observation that two amino acid exchanges in the MP that were shown previously to abolish infectivity likewise disrupted the interaction. Furthermore, an infectious second-site mutant that differed from the non-infective mutant by only a single amino acid restored the interaction in the yeast two-hybrid system (Huang et al. 2001). MPI7 has homologies to a class of mammalian Rab acceptor proteins. Rab proteins, in turn, are small ras-like GTP binding proteins now known to function in both constitutive and regulated exocytosis, as well as in endocytosis and transcytosis. Proteins of the Rab class tether incoming vesicles to the correct target organelle contributing to the specificity of membrane trafficking and the proper flow of cargo within the cell (Zerial and McBride 2001).

Potyviruses have three overlapping genes encoding the so-called "triple-gene block" (TGB) proteins that function together to promote virus cell-to-cell movement. TGBp2 of poa semilatent virus, as well as TGBp2 and TGB3p3 of potato virus X (PVX) and potato mop top virus (PMTV) have recently been shown to locate to the ER (Haupt et al. 2005; Solovyev et al. 2000). PMTV TGBp3, and similarly a number of different viral movement proteins contain a conserved amino acid motif that was shown to be essential for correct targeting of the proteins (Haupt et al. 2005; Laporte et al. 2003). In animals, similar amino acid motifs have been shown to be recognized by clathrin-coated vesicle adaptors at the Golgi and the plasma membrane. Moreover, the movement proteins of Potato mop-top virus physically interact with a conserved RMA-8 family of J-domain proteins essential for endocytic trafficking (Haupt et al. 2005).

Thus, these recent data strongly support the idea that intracellular movement of plant viruses proceeds via hitch-hiking the host's membrane and vesicle trafficking pathways.

13.5   Modification of and Passage Through Plasmodesmata

A characteristic generally associated with plant viral movement proteins is their ability to increase the size exclusion limits (SEL) of plasmodesmata. Plasmodesmata connecting mesophyll cells usually have an SEL of approximately 60 kDa, i.e. they are permeable to molecules of up to 60 kDa. Plasmodesmal SEL may change depending on cell type and developmental stage, and in mature tissue, for example, SEL may be as low as 1 kDa or less (Imlau et al. 1999). Although numerous viral movement proteins have been shown to possess the ability to increase this SEL, the exact mechanism of how this is accomplished is largely unknown. Here again, the identification of host factors interacting with plant viral movement proteins might be helpful in understanding the molecular basis of plasmodesmal dynamics. The deposition of callose to close plasmodesmata during wound responses and defence reactions is supposed to have a function in antiviral defence by blocking systemic viral spread (Beffa and Meins 1996). It is therefore an intriguing finding that the PVX TGB2 protein interacts with three ankyrin repeat-containing proteins, TIP1, TIP2 and TIP3, that in turn interact with beta-1,3-glucanase (Fridborg et al. 2003). Beta-1,3-glucanase is a callose-degrading enzyme that thereby may regulate plasmodesmal SEL, suggesting that a potential strategy of PVX to gate plasmodesmata is to accelerate callose degradation.

The TMV movement protein has been shown to interact with a pectin methylesterase (PME) (Chen MH et al. 2000; Dorokhov et al. 1999). While the functional implication favoured by the authors was that the main function of this interaction might be to recruit TMV movement protein to the cell periphery, more recently, it has been speculated that on the contrary, the movement protein might recruit the activity of PME to loosen the cell wall surrounding plasmodesmata to increase the SEL (Boevink and Oparka 2005).

Long-discussed regulatory steps in the passage through plasmodesmata include protein phosphorylation by plasmodesmata-associated kinases and an influence of local Ca^{2+} concentrations (Citovsky et al. 1993; Roberts and Oparka 2003). The recent finding of an interaction between the TMV movement protein and calreticulin might provide a link to calcium-dependent regulatory processes (see above). Furthermore, PAPK, a plasmodemata localized member of the casein kinase I family has been shown recently to specifically recognize and phosphorylate the TMV movement protein and a number of endogenous NCAPs, a modification that has been shown previously to be important for TMV movement protein function (Citovsky et al. 1993; Lee et al. 2005).

13.6 Movement and Pathogenicity

The identification of a particular plant virus protein as a movement protein is based often on observations of loss-of-function, using infectious cDNA clones of viruses, whereby mutation of the gene encoding the protein leads to an inability of the virus to move out of the initial infected cell (i.e. debilitated in local or cell-to-cell move-ment) or from the inoculated leaf into other leaves (i.e. debilitated in systemic or long-distance movement). Gain-of-function approaches include co-infection studies where various viruses were shown to be able to enhance the movement of other viruses, with the complementation leading to local movement of the dependent virus in an otherwise non-host plant (Malyshenko et al. 1989). In another approach, movement-viable viruses have been constructed in which the movement gene of one virus has been directly replaced with that of a different virus (Dejong and Ahlquist 1992; Ryabov et al. 1999). It is also possible to complement the movement of otherwise defective (mutated) viruses by inoculating them to transgenic plants that themselves express the virus movement protein (Kaplan et al. 1995). Approaches such as these have also been used widely to assign a pathogenicity function to particular virus proteins, where pathogenicity may signify an increase in virus replication/accumulation as well as stimulation of symptom production, both in terms of intensity and of distribution throughout the plant (Brigneti et al. 1998; Liu et al. 2002; Yelina et al. 2002). Further investigation of the mechanism of action of different plant virus pathogenicity proteins has revealed an association between the ability of the virus to overcome host defence responses, specifically RNA silencing (also known in plants as post-transcriptional gene silencing) and the involvement of virus proteins in the process of virus movement (see Table 13.2).

13.7 Antiviral Defence by RNA Silencing

The term RNA silencing refers to an enzymatic process occurring in plant (and other organisms) cells where RNA molecules are targeted in a sequence-specific manner for cleavage and further degradation (Brodersen and Voinnet 2006). The initiator for RNA silencing is double-stranded (ds) RNA, which in the case of

Table 13.2 Plant proteins interacting with viral silencing suppressors

Virus	Suppressor	Biological function(s)	Interacting host proteins	References for interaction
TEV PVY	HC-Pro	Systemic movement, transmission by aphids, genome amplification	Rgs-CaM HIP1, HIP1	Anandalakshmi et al. (2000); Guo D et al. (2003)
CMV TAV	2b	Systemic and cell-to-cell movement, Pathogenicity	LytB, karyo-pherin α, TLP1	Ham et al. (1999); Kim et al. (2005); Wang et al. (2004b)
TBSV	p19	Pathogenicity, cell-to-cell and systemic movement	ALY (Hin19)	Park et al. (2004; Uhrig et al. (2004)
PVX	p25	Cell-to-cell movement, egress from veins in systemic leaves, RNA helicase	TIP1, TIP2, TIP3	Fridborg et al. (2003)
BWYV CABYV PLRV	P0	Symptom production, virus accumulation	SKP1, SKP2	Pazhouhandeh et al. (2006)
TCV	CP	Capsid formation, virus movement	TIP	Ren et al. (2000)
ToMV TMV	126 kDa	RNA-dependent RNA polymerase, virus movement	PAP1/IAA26, AAA AtPase, 33 K subunit of photosystem II, P58[IPK]	Abbink et al. (2002; Bilgin et al. (2003); Padmanabhan et al. (2005)
ACMV TGMV TYLCV BCTV	AC2, AL2, C2, L2	Pathogenicity, activation of virus gene expression	SNF1 kinase, ADK	Hao et al. (2003); Wang et al. (2003)

viruses may be provided as an intermediate of replication or by base pairing of regions of the single-stranded genomic or messenger RNA (Molnar et al. 2005). The initiator dsRNA molecules are cleaved into 20–26 nt ds small interfering (si)RNAs by RNaseIII-domain-containing Dicer proteins, or Dicer-like (DCL) proteins in plants. *Arabidopsis thaliana* encodes four DCL proteins, which have distinct but overlapping roles in the processing of dsRNA from various sources. The ds siRNAs are unwound and one strand is incorporated in the RNA-induced silencing complex (RISC) which contains, among others, Argonaut (AGO) proteins. The single-stranded siRNA in RISC base pairs with its complementary target RNA, which is then cleaved by the AGO component. The RNA silencing system is efficiently triggered by virus RNA, and can effectively damp down if not completely prevent virus infection. This action is associated with the phenomenon of recovery in which plants fight off an initially strong infection, e.g. with nepoviruses or tobraviruses, and reach a state where they contain extremely low levels of virus and are protected from further infection by the same or a very similar virus (Ratcliff et al. 1997). The introduction of techniques such as the transient expression of virus and reporter proteins in plants using infiltration with *Agrobacterium tumefaciens* cultures, the production of transgenic plants expressing virus proteins, and the creation of hybrid viruses expressing genes from different sources have led to the understanding that the pathogenicity proteins of many viruses interfere in some way or other with RNA silencing, and are now commonly referred to as silencing suppressor proteins (Palukaitis and MacFarlane 2006; Voinnet 2005a).

13.8 Plant Proteins Interacting with Viral Silencing Suppressors

Many of the virus encoded silencing suppressor proteins had previously been shown to be involved in virus movement (see Table 13.2). Further studies revealed that some proteins interfere with local RNA silencing, whereas others prevent silencing in systemic infected leaves. In addition, grafting experiments have shown that some of the proteins interfere with the initiation of silencing whereas others interrupt the movement of a silencing signal that is necessary for propagation of the silenced state (Roth et al. 2004; Voinnet 2005b). Information on the precise mechanism of action of most silencing suppressors is not available. However, recent studies have shown that dsRNA-binding, including siRNA-binding may be common to many different suppressors (Lakatos et al. 2006; Merai et al. 2006). In fact the crystal structure of two suppressors, one complexed with siRNA, have been obtained (Vargason et al. 2003; Ye and Patel 2005).

Mainly by applying the yeast two-hybrid system, a number of host proteins have been identified that bind to various silencing suppressor proteins. However, while the interactors of viral movement proteins have led to some conclusive hypotheses on the mechanisms of viral movement (see above), the functions of most of the proteins interacting with viral silencing suppressors are unknown. Therefore, there is still no general picture of how these cellular targets of silencing suppressors actually integrate in the complex network the RNA silencing process.

One of the most studied suppressor proteins is the helper component-proteinase (HC-Pro) that is encoded by potyviruses such as TEV and PVY. This protein is multifunctional, being required for virus transmission by aphids, viral polyprotein processing and systemic movement (Maia et al. 1996). HC-Pro also is a strong silencing suppressor that binds to ds siRNAs (Lakatos et al. 2006), interferes with 3 methylation of another class of small RNAs, miRNAs (Yu et al. 2006), and inhibits the ribonuclease activity of the 20 S proteasome (Ballut et al. 2005). How many of these functions are directly related to suppression of silencing is not known. However, the introduction into HC-Pro of mutations that affected long-distance movement and genome amplification also inhibited silencing suppression activity, whereas mutations that inactivated the proteolytic activity of HC-Pro had no effect on silencing suppression (Kasschau and Carrington 2001). The TEV

HC-Pro was found to interact with a calmodulin related protein called rgsCaM (regulator of gene silencing-calmodulin-like protein) whose expression was upregulated by HC-Pro (Anandalakshmi et al. 2000). Over-expression of this protein in plants also led to suppression of silencing, suggesting that it might be an endogenous suppressor and that the calcium-signaling pathway might play a role in silencing. Two other proteins, HIP1, a RING-finger protein, and HIP2, with no identifiable functional motifs, bind to HC-Pro in yeast, although the significance of these interactions is not known (Guo D et al. 2003).

Mutations in the gene encoding the P19 protein, present in tombusviruses such as TBSV and CymRSV, affect cell-to-cell and systemic movement of the virus in a host-specific manner, as well as symptom production (Scholthof et al. 1995a,b; Turina et al. 2003). The p19 protein is a very strong silencing suppressor that binds siRNAs in vitro and in vivo and was suggested to function solely by sequestration of these molecules without the involvement of any host proteins (Lakatos et al. 2004). Nevertheless, yeast two-hybrid experiments revealed that P19 interacts with members of the ALY family of RNA-binding proteins, which in animals are involved in export of RNAs from the nucleus (Park et al. 2004; Uhrig et al. 2004). In plants, expression of P19 leads to re-localization of two of the four ALY proteins from the nucleus to the cytoplasm. By contrast, the two ALY proteins that remain in the nucleus themselves sequester the P19 protein in the nucleus (Canto et al. 2006). This relocalisation of P19 inhibits its activity as a silencing suppressor. Whether the suppression activity of P19 is directly responsible for its influence on virus movement is not clear, although mutations in P19 that affected silencing suppression activity also affected virus movement and interaction of P19 with ALY (Chu et al. 2000; Uhrig et al. 2004).

Mutation of the gene encoding the 2b protein of CMV does not affect virus replication in protoplasts (Soards et al. 2002) but does affect the degree of movement of the virus in tobacco and cucumber (Ding et al. 1996). The 2b protein functions as a silencing suppressor, which differs in its activity to HC-Pro and TBSV P19 as it interferes with the long range spread of the silencing signal away from the point of initiation (Guo HS and Ding 2002). In doing so the 2b protein enters the cell nucleus, where it also reduces methylation of DNA sequences. Mutation of sequences necessary for nuclear localisation of the 2b protein affects its ability to suppress silencing and to promote a pathogenic synergistic interaction with ZYMV (Wang et al. 2004b). Yeast two-hybrid studies have identified a prokaryotic LytB homologue from tobacco that interacted with the CMV 2b protein in yeast (Ham et al. 1999), a karyopherin α protein from *Arabidopsis* that is likely involved in nuclear import of the 2b protein (Wang et al. 2004a), and a tobacco thaumatin-like protein (TLP1) whose expression is upregulated by CMV infection (Kim et al. 2005). In this latter example TLP1 also interacted in yeast with the CMV movement and capsid proteins.

The CP (P38) of TCV suppresses local silencing and prevents the accumulation of siRNAs (Qu et al. 2003). A 25 amino acid region at the N-terminus of the protein, that is sequestered inside assembled virus capsids, was shown to be important for suppression activity as well as for interaction with the TIP, a transcription factor from *Arabidopsis thaliana* (Ren et al. 2000). Furthermore, the CP:TIP interaction is required for a hypersensitive resistance response in Arabidopsis. However, single amino acid mutations in the N-terminal regions can separate the TIP-binding and suppression activities of the CP, suggesting that TIP may not be involved in the silencing pathway (Choi et al. 2004). Mutations were introduced into P38 that retained suppressor function but abolished encapsidation. Movement of TCV did not require encapsidation but did require P38-mediated silencing suppression (Deleris et al. 2006).

Poleroviruses, which include PLRV, BWYV and CABYV, are aphid transmitted viruses that accumulate only within the phloem system of plants. This tissue limitation is likely to be due in part to their lack of a particular movement function, as co-infection with an umbravirus, PEMV-2, or with PVX expressing the movement protein (ORF4) of PEMV-2, enabled PLRV to move out of the phloem into the mesophyll tissue (Ryabov et al. 2001). The P0 protein of poleroviruses is a silencing suppressor, and mutation of the gene encoding this protein greatly reduces or abolishes accumulation of viral RNA (Pfeffer et al. 2002; Sadowy et al. 2001). The polerovirus P0 protein interacts via an F-Box-like motif with AtSKP1 and AtSKP2 (ubiquitin E3 ligases), and mutation of the F-box motif in P0 prevented interaction with SKP1/SKP2 and inhibited the silencing suppression activity of P0 (Pazhouhandeh et al. 2006). Knock-down of SKP1 in *Nicotiana benthamiana* by virus-induced gene silencing (VIGS) made these plants resistant to PLRV infection. These results suggest that P0 might function as an F-box protein potentially directing ubiquitination and degradation by the 26 S proteasome of an essential component of the host posttranscriptional gene silencing machinery.

The 126 K protein of TMV, and its homologue in other tobamoviruses, is a component of the viral replicase. It contains motifs associated with methyltransferase and RNA helicase proteins, and has a role in cell-to-cell movement of the virus (Hirashima and Watanabe 2001). The 126 K protein is also a suppressor of RNA silencing (Kubota et al. 2003). Three different yeast two-hybrid studies have isolated different host proteins that interact with the helicase domain of this protein. In the first study, interaction was found with the tobacco AAA ATPase and with the 33 K subunit protein of the oxygen-evolving photosystem II complex (Abbink et al. 2002). Silencing by VIGS in *Nicotiana benthamiana* of the ATPase gene decreased TMV accumulation twofold and also reduced accumulation of PVX and AMV. Silencing of the 33 K subunit gene led to a tenfold increase in TMV accumulation as well as an enhancement of PVX and AMV accumulation. In the second study, the TMV helicase domain interacted with the *Arabidopsis* AUX/IAA protein PAP1/IAA26 which is a putative regulator of plant auxin genes involved in plant development (Padmanabhan et al. 2005). Silencing of PAP1 induced symptoms similar to those seen during virus infection, and infection of plants with TMV prevented the normal accumulation of PAP1 in the nucleus which led to a disruption in the expression pattern of auxin-stimulated genes. The third study used a yeast three-hybrid approach to identify proteins that complexed with the TMV helicase domain and the tobacco N gene, a TMV-resistance gene (Bilgin et al. 2003). This identified P58[IPK], which inhibits cell death mediated by a double-stranded RNA-activated protein kinase (PKR), that in animals is part of the interferon response to virus infection. Plants in which the P58[IPK] gene was silenced or mutated were hypersusceptible to TMV and tobacco etch virus resulting in plant death upon infection with these viruses, suggesting that the virus does not inhibit P58[IPK] activity but in some way modulates it to prevent the induction of cell death.

Geminiviruses, which are comprised of single-stranded DNA rather than RNA also encode a silencing suppressor protein (Voinnet et al. 1999). The suppressor protein of ACMV is the AC2 protein, which is a transcriptional activator protein involved in CP expression. The homologous protein from TGMV is called the AL2 protein, and the homologue from TYLCV is called the C2 protein (Dong et al. 2003). Transgenic plants expressing AL2 or the positional homologue L2 from BCTV are more susceptible to these viruses and to TMV, an unrelated RNA virus (Sunter et al. 2001). AL2 and L2 interact in plants with SNF1 kinase which controls the activity of a range of metabolic pathway transcriptional activators and repressors in response to nutritional and environmental stress (Hao et al. 2003). Overexpression of SNF1 causes enhanced resistance to geminivirus infection, and the AL2 and L2 proteins bind SNF1 to inhibit its kinase activity in vitro and in vivo (in yeast). In a further yeast two-hybrid screen TGMV AL2 and BCT L2 proteins were also shown to interact with adenosine 5 phosphotransferase (ADK) (Wang et al. 2003). The viral proteins inactivate ADK in vitro and in vivo, as also occurs in transgenic plant expressing these proteins or in plants infected with geminiviruses. SNF1 is activated by 5 AMP; therefore, these observations indicate that global regulation of metabolism by SNF1 might be part of antiviral defences, and the inactivation of ADK and SNF1 by the geminivirus proteins might represent a dual strategy to counter this defense.

13.9 Conclusion

The molecular investigation of plant viruses has stimulated and promoted the recent advance in understanding of two fundamental cellular processes underlying the coordination of developmental programs and the response to both environmental cues and pathogen challenge: RNA silencing, as a part of the general microRNA pathways regulating gene expression, and the long-distance communication networks in plants based on the intercellular trafficking of signalling macromolecules. Protein interaction studies using viral movement proteins and silencing suppressors have now identified a large number of plant proteins providing a valuable source of information about the molecular basis of these two processes. Interactors of silencing suppressors indicate functional links between RNA silencing and calcium signalling, nuclear shuttling, global regulation of metabolism, auxin action and protein degradation. While these protein interaction data still not connect to give a complete picture of the RNA silencing process, a more conclusive idea of the mechanisms of intercellular movement through plasmodesmata is emerging, involving the participation of the cytoskeleton and the endomembrane system, as well as the action of chaperones and cell wall-modifying enzymes.

Abbreviations: ACMV, African cassava mosaic virus; AMV, Alfalfa mosaic virus; BCTV, Beet curly top virus; BSMV, Barley stripe mosaic virus; BWYV, Beet western yellows virus; BYSV, Beet yellow stunt virus; BYV, Beet yellows virus; CABYV, Cucurbit aphid-borne yellows virus; CTV, Citrus tristeza virus; CMV, Cucumber mosaic virus; CPMV, Cowpea mosaic virus; CymRSV, Cymbidium ringspot virus; PCV, Peanut clump virus; PEMV-2, Pea enation mosaic virus 2; PLRV, Potato leafroll virus; PoLV, Pothos latent virus; PSLV, Poa semilatent virus; PVX, Potato virus X; PVY, Potato virus Y; RDV, Rice dwarf virus, RHBV, Rice hoja blanca virus; RYMV, Rice yellow mottle virus; SBWMV, Soilborne wheat mosaic virus, TAV, Tomato aspermy virus; TBSV, Tomato bushy stunt virus; TCV, Turnip crinkle virus; TEV, Tobacco etch virus; TGMV, Tomato golden mosaic virus; TMV, Tobacco mosaic virus; ToMV, Tomato mosaic virus; TRV, Tobacco rattle virus; TSWV, Tomato spotted wilt virus; TYLCV, Tomato yellow leaf curl virus; TYMV, Turnip yellow mosaic virus; ZYMV, Zucchini yellow mosaic virus

References

Abbink TE, Peart JR, Mos TN, Baulcombe DC, Bol JF, Linthorst HJ (2002) Silencing of a gene encoding a protein component of the oxygen-evolving complex of photosystem II enhances virus replication in plants. Virology 295:307–319

Anandalakshmi R, Marathe R, Ge X, Herr JM Jr, Mau C, Mallory A, Pruss G, Bowman L, Vance VB (2000) A calmodulin-related protein that suppresses posttranscriptional gene silencing in plants. Science 290:142–144

Ballut L, Drucker M, Pugniere M, Cambon F, Blanc S, Roquet F, Candresse T, Schmid HP, Nicolas P, Le Gall O, Badaoui S (2005) HcPro, a multifunctional protein encoded by a plant RNA virus, targets the 20S proteasome and affects its enzymic activities. J Gen Virol 86:2595–2603

Bayne EH, Rakitina DV, Morozov SY, Baulcombe DC (2005) Cell-to-cell movement of potato potexvirus X is dependent on suppression of RNA silencing. Plant J 44:471–482

Beffa R, Meins F Jr (1996) Pathogenesis-related functions of plant beta-1,3-glucanases investigated by antisense transformation–a review. Gene 179:97–103

Bilgin DD, Liu Y, Schiff M, Dinesh-Kumar SP (2003) P58(IPK), a plant ortholog of double-stranded RNA-dependent protein kinase PKR inhibitor, functions in viral pathogenesis. Dev Cell 4:651–661

Boevink P, Oparka KJ (2005) Virus-host interactions during movement processes. Plant Physiol 138:1815–1821

Boyko V, Ferralli J, Ashby J, Schellenbaum P, Heinlein M (2000) Function of microtubules in intercellular transport of plant virus RNA. Nat Cell Biol 2:826–832

Brigneti G, Voinnet O, Li WX, Ji LH, Ding SW, Baulcombe DC (1998) Viral pathogenicity determinants are suppressors of transgene silencing in *Nicotiana benthamiana*. Embo J 17:6739–6746

Brodersen P, Voinnet O (2006) The diversity of RNA silencing pathways in plants. Trends Genet 22:268–280

Canto T, Uhrig JF, Swanson M, Wright KM, MacFarlane SA (2006) Translocation of tomato bushy stunt virus P19 protein into the nucleus by ALY proteins compromises its silencing suppressor activity. J Virol 80:9064–9072

Carette JE, Verver J, Martens J, van Kampen T, Wellink J, van Kammen A (2002) Characterization of plant proteins that interact with cowpea mosaic virus '60K' protein in the yeast two-hybrid system. J Gen Virol 83:885–893

Carvalho MF, Lazarowitz SG (2004) Interaction of the movement protein NSP and the Arabidopsis acetyltransferase AtNSI is necessary for Cabbage leaf curl geminivirus infection and pathogenicity. J Virol 78:11161–11171

Chen MH, Sheng J, Hind G, Handa AK, Citovsky V (2000) Interaction between the tobacco mosaic virus movement protein and host cell pectin methylesterases is required for viral cell-to-cell movement. Embo J 19:913–920

Chen MH, Tian GW, Gafni Y, Citovsky V (2005) Effects of calreticulin on viral cell-to-cell movement. Plant Physiol 138:1866–1876

Chen YA, Scheller RH (2001) SNARE-mediated membrane fusion. Nat Rev Mol Cell Biol 2:98–106

Choi CW, Qu F, Ren T, Ye XH, Morris TJ (2004) RNA silencing-suppressor function of Turnip crinkle virus coat protein cannot be attributed to its interaction with the Arabidopsis protein TIP. J Gen Virol 85:3415–3420

Chu M, Desvoyes B, Turina M, Noad R, Scholthof HB (2000) Genetic dissection of tomato bushy stunt virus p19-protein-mediated host-dependent symptom induction and systemic invasion. Virology 266:79–87

Citovsky V, McLean BG, Zupan JR, Zambryski P (1993) Phosphorylation of tobacco mosaic virus cell-to-cell movement protein by a developmentally regulated plant cell wall-associated protein kinase. Genes Dev 7:904–910

Coppolino MG, Woodside MJ, Demaurex N, Grinstein S, St-Arnaud R, Dedhar S (1997) Calreticulin is essential for integrin-mediated calcium signalling and cell adhesion. Nature 386:843–847

Curin M, Ojangu EL, Trutnyeva K, Ilau B, Truve E, Waigmann E (2007) MPB2C, a microtubule-associated plant factor, is required for microtubular accumulation of tobacco mosaic virus movement protein in plants. Plant Physiol 143:801–811

Dejong W, Ahlquist P (1992) A hybrid plant Rna virus made by transferring the noncapsid movement protein from a rod-shaped to an icosahedral virus is competent for systemic infection. Proc Nat Acad Sci USA 89:6808–6812

Deleris A, Gallego-Bartolome J, Bao J, Kasschau KD, Carrington JC, Voinnet O (2006) Hierarchical action and inhibition of plant dicer-like proteins in antiviral defense. Science 313:68–71

Desvoyes B, Faure-Rabasse S, Chen MH, Park JW, Scholthof HB (2002) A novel plant homeodomain protein interacts in a functionally relevant manner with a virus movement protein. Plant Physiol 129:1521–1532

Ding SW, Shi BJ, Li WX, Symons RH (1996) An interspecies hybrid RNA virus is significantly more virulent than either parental virus. Proc Nat Acad Sci USA 93:7470–7474

Dong XL, van Wezel R, Stanley J, Hong YG (2003) Functional characterization of the nuclear localization signal for a suppressor of posttranscriptional gene silencing. J Virol 77:7026–7033

Dorokhov YL, Makinen K, Frolova OY, Merits A, Saarinen J, Kalkkinen N, Atabekov JG, Saarma M (1999) A novel function for a ubiquitous plant enzyme pectin methylesterase: the host-cell receptor for the tobacco mosaic virus movement protein. FEBS Lett 461:223–228

D'Souza SE, Ginsberg MH, Plow EF (1991) Arginyl-glycyl-aspartic acid (RGD): a cell adhesion motif. Trends Biochem Sci 16:246–250

Dunoyer P, Thomas C, Harrison S, Revers F, Maule A (2004) A cysteine-rich plant protein potentiates Potyvirus movement through an interaction with the virus genome-linked protein VPg. J Virol 78:2301–2309

Epel BL, Padgett HS, Heinlein M, Beachy RN (1996) Plant virus movement protein dynamics probed with a GFP-protein fusion. Gene 173:75–79

Fontes EP, Santos AA, Luz DF, Waclawovsky AJ, Chory J (2004) The geminivirus nuclear shuttle protein is a virulence factor that suppresses transmembrane receptor kinase activity. Genes Dev 18:2545–2556

Fridborg I, Grainger J, Page A, Coleman M, Findlay K, Angell S (2003) TIP, a novel host factor linking callose degradation with the cell-to-cell movement of Potato virus X. Mol Plant Microbe Interact 16:132–140

Gallagher KL, Benfey PN (2005) Not just another hole in the wall: understanding intercellular protein trafficking. Genes Dev 19:189–195

Guo D, Spetz C, Saarma M, Valkonen JP (2003) Two potato proteins, including a novel RING finger protein (HIP1), interact with the potyviral multifunctional protein HCpro. Mol Plant Microbe Interact 16:405–410

Guo HS, Ding SW (2002) A viral protein inhibits the long range signaling activity of the gene silencing signal. Embo J 21:398–407

Hackbusch J, Richter K, Muller J, Salamini F, Uhrig JF (2005) A central role of Arabidopsis thaliana ovate family proteins in networking and subcellular localization of 3-aa loop extension homeodomain proteins. Proc Natl Acad Sci USA 102:4908–4912

Ham BK, Lee TH, You JS, Nam YW, Kim JK, Paek KH (1999) Isolation of a putative tobacco host factor interacting with cucumber mosaic virus-encoded 2b protein by yeast two-hybrid screening. Mol Cells 9:548–555

Hao L, Wang H, Sunter G, Bisaro DM (2003) Geminivirus AL2 and L2 proteins interact with and inactivate SNF1 kinase. Plant Cell 15:1034–1048

Haupt S, Cowan GH, Ziegler A, Roberts AG, Oparka KJ, Torrance L (2005) Two plant-viral movement proteins traffic in the endocytic recycling pathway. Plant Cell 17:164–181

Heinlein M, Epel BL, Padgett HS, Beachy RN (1995) Interaction of tobamovirus movement proteins with the plant cytoskeleton. Science 270:1983–1985

Heinlein M, Padgett HS, Gens JS, Pickard BG, Casper SJ, Epel BL, Beachy RN (1998) Changing patterns of localization of the tobacco mosaic virus movement protein and replicase to the endoplasmic reticulum and microtubules during infection. Plant Cell 10:1107–1120

Hirashima K, Watanabe Y (2001) Tobamovirus replicase coding region is involved in cell-to-cell movement. J Virol 75:8831–8836

Huang Z, Zhang L (1999) Association of the movement protein of alfalfa mosaic virus with the endoplasmic reticulum and its trafficking in epidermal cells of onion bulb scales. Mol Plant Microbe Interact 12:680–690

Huang Z, Andrianov VM, Han Y, Howell SH (2001) Identification of arabidopsis proteins that interact with the cauliflower mosaic virus (CaMV) movement protein. Plant Mol Biol 47:663–675

Imlau A, Truernit E, Sauer N (1999) Cell-to-cell and long-distance trafficking of the green fluorescent protein in the phloem and symplastic unloading of the protein into sink tissues. Plant Cell 11:309–322

Jimenez I, Lopez L, Alamillo JM, Valli A, Garcia JA (2006) Identification of a plum pox virus CI-interacting protein from chloroplast that has a negative effect in virus infection. Mol Plant Microbe Interact 19:350–358

Kaplan IB, Shintaku MH, Zhang L, Marsh LE, Palukaitis P (1995) Complementation of virus movement in transgenic tobacco expressing cucumber mosaic-virus 3a gene. Virology 209:188–199

Kasschau KD, Carrington JC (2001) Long-distance movement and replication maintenance functions correlate with silencing suppression activity of potyviral HC-Pro. Virology 285:71–81

Kim MJ, Ham BK, Kim HR, Lee IJ, Kim YJ, Ryu KH, Park YI, Paek KH (2005) In vitro and in planta interaction evidence between Nicotiana tabacum thaumatin-like protein 1 (TLP1) and Cucumber mosaic virus proteins. Plant Mol Biol 59:981–994

Kragler F, Curin M, Trutnyeva K, Gansch A, Waigmann E (2003) MPB2C, a microtubule-associated plant protein binds to and interferes with cell-to-cell transport of tobacco mosaic virus movement protein. Plant Physiol 132:1870–1883

Kubota K, Tsuda S, Tamai A, Meshi T (2003) Tomato mosaic virus replication protein suppresses virus-targeted posttranscriptional gene silencing. J Virol 77:11016–11026

Lakatos L, Szittya G, Silhavy D, Burgyan J (2004) Molecular mechanism of RNA silencing suppression mediated by p19 protein of tombusviruses. Embo J 23:876–884

Lakatos L, Csorba T, Pantaleo V, Chapman EJ, Carrington JC, Liu YP, Dolja VV, Calvino LF, Lopez-Moya JJ, Burgyan J (2006) Small RNA binding is a common strategy to suppress RNA silencing by several viral suppressors. Embo J

Laporte C, Vetter G, Loudes AM, Robinson DG, Hillmer S, Stussi-Garaud C, Ritzenthaler C (2003) Involvement of the secretory pathway and the cytoskeleton in intracellular targeting and tubule assembly of Grapevine fanleaf virus movement protein in tobacco BY-2 cells. Plant Cell 15:2058–2075

Lee JY, Taoka K, Yoo BC, Ben-Nissan G, Kim DJ, Lucas WJ (2005) Plasmodesmal-associated protein kinase in tobacco and Arabidopsis recognizes a subset of non-cell-autonomous proteins. Plant Cell 17:2817–2831

Li Y, Wu MY, Song HH, Hu X, Qiu BS (2005) Identification of a tobacco protein interacting with tomato mosaic virus coat protein and facilitating long-distance movement of virus. Arch Virol 150:1993–2008

Lin B, Heaton LA (2001) An Arabidopsis thaliana protein interacts with a movement protein of Turnip crinkle virus in yeast cells and in vitro. J Gen Virol 82:1245–1251

Liu H, Reavy B, Swanson M, MacFarlane SA (2002) Functional replacement of the Tobacco rattle virus cysteine-rich protein by pathogenicity proteins from unrelated plant viruses. Virology 298:232–239

Liu JZ, Blancaflor EB, Nelson RS (2005) The tobacco mosaic virus 126-kilodalton protein, a constituent of the virus replication complex, alone or within the complex aligns with and traffics along microfilaments. Plant Physiol 138:1853–1865

Lough TJ, Lucas WJ (2006) Integrative plant biology: tole of phloem long-distance macromolecular trafficking. Annu Rev Plant Biol 57:203–232

Lucas WJ (2006) Plant viral movement proteins: agents for cell-to-cell trafficking of viral genomes. Virology 344:169–184

Lucas WJ, Lee JY (2004) Plasmodesmata as a supracellular control network in plants. Nat Rev Mol Cell Biol 5:712–726

Lucas WJ, Bouche-Pillon S, Jackson DP, Nguyen L, Baker L, Ding B, Hake S (1995) Selective trafficking of KNOTTED1 homeodomain protein and its mRNA through plasmodesmata. Science 270:1980–1983

Maia IG, Haenni AL, Bernardi F (1996) Potyviral HC-Pro: a multifunctional protein. J Gen Virol 77:1335–1341

Malyshenko SI, Kondakova OA, Taliansky ME, Atabekov JG (1989) Plant-virus transport function – complementation by helper viruses is non-specific. J Gen Virol 70:2751–2757

Mariano AC, Andrade MO, Santos AA, Carolino SM, Oliveira ML, Baracat-Pereira MC, Brommonshenkel SH, Fontes EP (2004) Identification of a novel receptor-like protein kinase that interacts with a geminivirus nuclear shuttle protein. Virology 318:24–31

Mas P, Beachy RN (2000) Role of microtubules in the intracellular distribution of tobacco mosaic virus movement protein. Proc Natl Acad Sci USA 97:12345–12349

Matsushita Y, Deguchi M, Youda M, Nishiguchi M, Nyunoya H (2001) The tomato mosaic tobamovirus movement protein interacts with a putative transcriptional coactivator KELP. Mol Cells 12:57–66

Matsushita Y, Miyakawa O, Deguchi M, Nishiguchi M, Nyunoya H (2002) Cloning of a tobacco cDNA coding for a putative transcriptional coactivator MBF1 that interacts with the tomato mosaic virus movement protein. J Exp Bot 53:1531–1532

McGarry RC, Barron YD, Carvalho MF, Hill JE, Gold D, Cheung E, Kraus WL, Lazarowitz SG (2003) A novel Arabidopsis acetyltransferase interacts with the geminivirus movement protein NSP. Plant Cell 15:1605–1618

McLean BG, Zupan J, Zambryski PC (1995) Tobacco mosaic virus movement protein associates with the cytoskeleton in tobacco cells. Plant Cell 7:2101–2114

Merai Z, Kerenyi Z, Kertesz S, Magna M, Lakatos L, Silhavy D (2006) Double-stranded RNA binding may be a general plant RNA viral strategy to suppress RNA silencing. J Virol 80:5747–5756

Molnar A, Csorba T, Lakatos U, Varallyay E, Lacomme C, Burgyan J (2005) Plant virus-derived small interfering RNAs originate predominantly from highly structured single-stranded viral RNAs. J Virol 79:7812–7818

Murchison EP, Hannon GJ (2004) miRNAs on the move: miRNA biogenesis and the RNAi machinery. Curr Opin Cell Biol 16:223–229

Padgett HS, Epel BL, Kahn TW, Heinlein M, Watanabe Y, Beachy RN (1996) Distribution of tobamovirus movement protein in infected cells and implications for cell-to-cell spread of infection. Plant J 10:1079–1088

Padmanabhan MS, Goregaoker SP, Golem S, Shiferaw H, Culver JN (2005) Interaction of the tobacco mosaic virus replicase protein with the Aux/IAA protein PAP1/IAA26 is associated with disease development. J Virol 79:2549–2558

Palukaitis P, MacFarlane SA (2006) Viral counter-defense molecules. Springer

Park JW, Faure-Rabasse S, Robinson MA, Desvoyes B, Scholthof HB (2004) The multifunctional plant viral suppressor of gene silencing P19 interacts with itself and an RNA binding host protein. Virology 323:49–58

Pazhouhandeh M, Dieterle M, Marrocco K, Lechner E, Berry B, Brault V, Hemmer O, Kretsch T, Richards KE, Genschik P, Ziegler-Graff V (2006) F-box-like domain in the polerovirus protein P0 is required for silencing suppressor function. Proc Natl Acad Sci USA 103:1994–1999

Peremyslov VV, Hagiwara Y, Dolja VV (1999) HSP70 homolog functions in cell-to-cell movement of a plant virus. Proc Natl Acad Sci USA 96:14771–14776

Peremyslov VV, Pan YW, Dolja VV (2004) Movement protein of a closterovirus is a type III integral transmembrane protein localized to the endoplasmic reticulum. J Virol 78:3704–3709

Pfeffer S, Dunoyer P, Heim F, Richards KE, Jonard G, Ziegler-Graff V (2002) P0 of beet western yellows virus is a suppressor of posttranscriptional gene silencing. J Virol 76:6815–6824

Prokhnevsky AI, Peremyslov VV, Dolja VV (2005) Actin cytoskeleton is involved in targeting of a viral Hsp70 homolog to the cell periphery. J Virol 79:14421–14428

Qu F, Ren T, Morris TJ (2003) The coat protein of turnip crinkle virus suppresses posttranscriptional gene silencing at an early initiation step. J Virol 77:511–522

Ratcliff F, Harrison BD, Baulcombe DC (1997) A similarity between viral defense and gene silencing in plants. Science 276:1558–1560

Reichel C, Mas P, Beachy RN (1999) The role of the ER and cytoskeleton in plant viral trafficking. Trends Plant Sci 4:458–462

Ren T, Qu F, Morris TJ (2000) HRT gene function requires interaction between a NAC protein and viral capsid protein to confer resistance to turnip crinkle virus. Plant Cell 12:1917–1926

Roberts AG, Oparka KJ (2003) Plasmodesmata and the control of symplastic transport. Plant Cell Environ 26:103–124

Roth BM, Pruss GJ, Vance VB (2004) Plant viral suppressors of RNA silencing. Virus Res 102:97–108

Ryabov EV, Robinson DJ, Taliansky ME (1999) A plant virus-encoded protein facilitates long-distance movement of heterologous viral RNA. Proc Nat Acad Sci USAmerica 96:1212–1217

Ryabov EV, Fraser G, Mayo MA, Barker H, Taliansky M (2001) Umbravirus gene expression helps Potato leafroll virus to invade mesophyll tissues and to be transmitted mechanically between plants. Virology 286:363–372

Sadowy E, Maasen A, Juszczuk M, David C, Zagorski-Ostoja W, Gronenborn B, Hulanicka MD (2001) The ORF0 product of Potato leafroll virus is indispensable for virus accumulation. J Gen Virol 82:1529–1532

Sagi G, Katz A, Guenoune-Gelbart D, Epel BL (2005) Class 1 reversibly glycosylated polypeptides are plasmodesmal-associated proteins delivered to plasmodesmata via the golgi apparatus. Plant Cell 17:1788–1800

Scholthof HB, Scholthof KBG, Jackson AO (1995a) Identification of tomato bushy stunt virus host-specific symptom determinants by expression of individual genes from a potato-virus-X vector. Plant Cell 7:1157–1172

Scholthof HB, Scholthof KBG, Kikkert M, Jackson AO (1995b) Tomato bushy stunt virus spread is regulated by 2 nested genes that function in cell-to-cell movement and host-dependent systemic invasion. Virology 213:425–438

Soards AJ, Murphy AM, Palukaltis P, Carr JP (2002) Virulence and differential local and systemic spread of Cucumber mosaic virus in tobacco are affected by the CMV 2b protein. Mol Plant-Microbe Interact 15:647–653

Soellick T, Uhrig JF, Bucher GL, Kellmann JW, Schreier PH (2000) The movement protein NSm of tomato spotted wilt tospovirus (TSWV): RNA binding, interaction with the TSWV N protein, and identification of interacting plant proteins. Proc Natl Acad Sci USA 97:2373–2378

Solovyev AG, Stroganova TA, Zamyatnin AA Jr, Fedorkin ON, Schiemann J, Morozov SY (2000) Subcellular sorting of small membrane-associated triple gene block proteins: TGBp3-assisted targeting of TGBp2. Virology 269:113–127

Sunter G, Sunter JL, Bisaro DM (2001) Plants expressing tomato golden mosaic virus AL2 or beet curly top virus L2 transgenes show enhanced susceptibility to infection by DNA and RNA viruses. Virology 285:59–70

Turina M, Omarov R, Murphy JF, Bazaldua-Hernandez C, Desvoyes B, Scholthof HB (2003) A newly identified role for Tomato bushy stunt virus P19 in short distance spread. Mol Plant Pathol 4:67–72

Uhrig JF, Canto T, Marshall D, MacFarlane SA (2004) Relocalization of nuclear ALY proteins to the cytoplasm by the tomato bushy stunt virus P19 pathogenicity protein. Plant Physiol 135:2411–2423

Vargason JM, Szittya G, Burgyan J, Hall TMT (2003) Size selective recognition of siRNA by an RNA silencing suppressor. Cell 115:799–811

Voinnet O (2005a) Induction and suppression of RNA silencing: Insights from viral infections. Nature Rev Genet 6:206-U1

Voinnet O (2005b) Non-cell autonomous RNA silencing. FEBS Lett 579:5858–5871

Voinnet O, Pinto YM, Baulcombe DC (1999) Suppression of gene silencing: a general strategy used by diverse DNA and RNA viruses of plants. Proc Nat Acad Sci USA 96:14147–14152

von Bargen S, Salchert K, Paape M, Piechulla B, Kellmann JW (2001) Interactions between the tomato spotted wilt virus movement protein and plant proteins showing homologies to myosin, kinesin and DnaJ-like chaperones. Plant Physiol Biochem 39:1083–1093

Wang H, Hao L, Shung CY, Sunter G, Bisaro DM (2003) Adenosine kinase is inactivated by geminivirus AL2 and L2 proteins. Plant Cell 15:3020–3032

Wang Y, Tzfira T, Gaba V, Citovsky V, Palukaitis P, Gal-On A (2004a) Functional analysis of the Cucumber mosaic virus 2b protein: pathogenicity and nuclear localization. J Gen Virol 85:3135–3147

Wang YZ, Tzfira T, Gaba V, Citovsky V, Palukaitis P, Gal-On A (2004b) Functional analysis of the Cucumber mosaic virus 2b protein: pathogenicity and nuclear localization. J Gen Virol 85:3135–3147

Wick S (2000) Plant microtubules meet their MAPs and mimics. Nat Cell Biol 2:E204–206

Xoconostle-Cazares B, Xiang Y, Ruiz-Medrano R, Wang HL, Monzer J, Yoo BC, McFarland KC, Franceschi VR, Lucas WJ (1999) Plant paralog to viral movement protein that potentiates transport of mRNA into the phloem. Science 283:94–98

Ye KQ, Patel DJ (2005) RNA silencing suppressor p21 of beet yellows virus forms an RNA binding octameric ring structure. Structure 13:1375–1384

Yelina NE, Savenkov EI, Solovyev AG, Morozov SY, Valkonen JPT (2002) Long-distance movement, virulence, and RNA silencing suppression controlled by a single protein in hordei- and potyviruses: Complementary functions between virus families. J Virol 76:12981–12991

Yoshioka K, Matsushita Y, Kasahara M, Konagaya K, Nyunoya H (2004) Interaction of tomato mosaic virus movement protein with tobacco RIO kinase. Mol Cells 17:223–229

Young JC, Barral JM, Ulrich Hartl F (2003) More than folding: localized functions of cytosolic chaperones. Trends Biochem Sci 28:541–547

Yu B, Chapman EJ, Yang Z, Carrington JC, Chen X (2006) Transgenically expressed viral RNA silencing suppressors interfere with microRNA methylation in Arabidopsis. FEBS Lett 580:3117–3120

Zerial M, McBride H (2001) Rab proteins as membrane organizers. Nat Rev Mol Cell Biol 2:107–117

Chapter 14
Effects of Root Exudates in Microbial Diversity and Activity in Rhizosphere Soils

P. Nannipieri(✉), J. Ascher, M.T. Ceccherini, L. Landi, G. Pietramellara, G. Renella, and F. Valori

14.1 Introduction

The rhizosphere is the soil volume at the root-soil interface that is under the influence of the plant roots and the term was introduced by Hiltner in 1904 (Brimecombe et al. 2001). Microbial population in the rhizosphere has continuous access to a flow of low and high molecular weight organic substrates derived from roots. This continuous flow of organic compounds may affect together with specific physio-chemical and biological conditions microbial activity and community structure of the rhizosphere soil (Sorensen 1997; Brimecombe et al. 2001). Current techniques still lack the adequate sensitivity and resolution for data collection at the micro-scale, and the question 'How important are various soil processes acting at different scales for ecological function?' is therefore challenging to answer. The nano-scale secondary ion mass spectrometer (NanoSIMS) represents the latest generation of ion microprobes, which link high-resolution microscopy with isotopic analysis. Recently Herrmann et al. (2007) have described the principles of NanoSIMS and discusses the potential of this tool to contribute to the field of biogeochemistry and soil ecology.

Both microbial activity and microbial diversity of the rhizosphere have been extensively studied as testimonies by numerous chapters and books (Keister and Creagen 1991; Lynch 1990a; Pinton et al. 2001, 2007; Waisel et al. 1991). This interest depends on the important effects that microorganisms inhabiting the rhizosphere have on plant activity. Both beneficial and detrimental interactions occur between microorganisms and plants (Lynch 1990b); among the former symbiotic dinitrogen fixation, association with mycorrhizae, biocontrol against pathogens and production of plant growth promoting compounds by beneficial rhizobacteria have

P. Nannipieri
Dipartimento della Scienza del Suolo e Nutrizione della Pianta, Universita' degli Studi di Firenze, 50144 Firenze, Italy
e-mail: paolo.nannipieri@unifi.it

C.S. Nautiyal, P. Dion (eds.) *Molecular Mechanisms of Plant and Microbe Coexistence.* Soil Biology 15, DOI: 10.1007/978-3-540-75575-3
© Springer-Verlag Berlin Heidelberg 2008

been the most studied (Brimecombe et al. 2001). Detrimental interactions are due to the presence of plant pathogens and deleterious rhizobacteria which inhibit plant growth without causing disease symptoms (Brimecombe et al. 2001). However, despite the wealth of information of the effects of rhizodeposition on microbial activity and microbial diversity, there is still considerable debate on the underlying mechanisms and the extent of the relative effects.

This review discusses the state-of-the-art of microbial activity and microbial diversity in the rhizosphere soil. Since it is not possible to prepare an exhaustive review as the complexity and vastness of the treated matter exceeds the limits of a single chapter, we have operated a selection of related topics. In the case of microbial diversity we have mainly discussed the recent advances obtained by using molecular techniques, which allow the detection of unculturable microorganisms. Among the various parameters used to determine microbial activity we have focused the discussion on soil respiration and enzyme activities because the former is strictly linked to organic C mineralization, and thus to oxidation of root exudates, and the latter represent specific reactions involved in the release of plant and microbial available nutrients. The difficulties of interpreting of community-level physiological profiles are also examined. Given the major effects of rhizodeposition on composition and activities of microbial communities inhabiting rhizosphere soil, the initial focus will be on the classification, collection, functions, and factors affecting root exudates.

14.2 Rhizodeposition: Classification, Quantification and Effects on Biotic Processes of the Rhizosphere Soil

Through rhizodeposition, roots introduce into the soil water soluble exudates, polymers such as carbohydrates and proteins, lysates and cell walls, whole cells, whole roots and gases such as CO_2 and ethylene (Morgan and Whipps 2001). Although most root products are C compounds, the rhizodeposited products include ions, sometimes O_2 and even water. Excretions and secretions have a perceived functional role whereas diffusates and root debris do not (Uren 2007). Excretions are the product of internal metabolism such as respiration while secretions are deemed to facilitate external processes such as nutrient acquisition. Both excretion and secretion require energy and some exudates may act as either. For example, protons derived from CO_2 production in respiration are deemed excretions while those derived from an organic acid involved in nutrient acquisition are deemed secretions. Root products differ, not only in their function, but also on the basis of their (a) chemical properties such as composition, solubility, volatility and molecular weight and (b) site of origin. The chemical properties determine in turn the biological activities of these root products and their behaviour in soils; thus the persistence in soil depends on chemical properties, particularly sorption and biodegradability of the root compounds. However, it is important to emphasize, first, that biological activity of some compounds of rhizodeposition such as phytohormones, exoenzymes, phytoalexins, allelochemicals, and phytotoxins has been studied primarily in solution cultures or under axenic conditions, whereas it is well established that the behaviour of these compounds depends on their survival in the soil (Uren 2007) and, second, that usually specific roles or functions have been assigned to single compounds released from roots (Uren 2007), but it is likely that rhizosphere soil processes are the result of combined effects of more than one root exudate. In addition to attributing specific functions to a root exudate, it is important to calculate the relative amount needed to carry out the specific function. For example, Uren (2007) has calculated that the amount of ascorbic acid released from wheat plants to reduce and dissolve sufficient Mn oxide in the rhizosphere to give the observed Mn concentration in the mature plant should be so high as to be unrealistic. Thus root exudates in addition to resorcinol should be involved in the reduction of Mn.

Amounts and composition of root exudates depend on plant genotype, plant growth stage and environmental conditions such as CO_2, light, pH, temperature, moisture and nutrients (Grayston et al. 1996; Neumann and Römheld 2001). It is well established that plants devote a large proportion of the C fixed to root exudation (0–40% of the total net C assimilated by plants) (Lynch and Whipps 1991). According to Uren (2007), the fixed C committed to roots by the plant is generally divided equally between root tissue and root products. Among the root products, 60% are used in root respiration and the remainder (10% of the net fixed C) is released as border cells, root debris, diffusates and secretions, with the latter being the minor components (Darrah 1996; Lynch and Whipps 1991; Whipps 1990). Of course, caution is required in generalizing these estimates because sick or stressed plants can involve a larger commitment than healthy plants (Farrar et al. 2003).

Root border cells represent root cap cells that are separated from the root apex during root growth (Hawes and Lin 1990; Hawes et al. 2003). In the rhizosphere soil of maize plants these cells remain active among root hairs, secreting mucilage, for up to three weeks after their separation from the root (Foster et al. 1983; McCully 1989, 1995). Then, sooner or later, these cells die and are lysed with release of their content in the external milieu; in this way they contribute to the C transfer from roots to soil (Uren and Reisenauer 1988).

14.3 Methodology for Collecting Root Exudates and Studying the Rhizosphere Effect

Systems used to collect the overall root exudation products such as those based on collection of water soluble root exudates by immersion of the roots into aerated trap solutions (usually solutions with Ca^{2+} so as to stabilize the membrane) yield no information to allow distinguishing the part of the root that has produced the root exudates or determining spatial variability in root exudation (Neumann and Römheld 2001). Thus, these methods do not enable one to determine which section of the root is the source of exudation. It is advisable to avoid long-term exposure of roots to solutions with low ionic strength, as these can stimulate exudation (Jones and Darrah 1993; Prikryl and Vancura 1980). However, the main inconvenience of these systems is the lack of impedance of solid growth media that stimulate the root exudation (Neumann and Römheld 2001). This limit may be overcome by collecting exudates of plants growing in solid media such as sand or vermiculite and then percolating the media with the trap solution (Johnson et al. 1996) or letting roots grow directly on filter papers (Neumann et al. 1999), resin foil (Kamh et al. 1999) or micro suction cups (Göttlein et al. 1996). The drawback of these methods is the possibility that certain exudates might be adsorbed by the solid medium.

Under field conditions rhizosphere soil samples are usually taken by removing the soil attached to roots and the effect of plant on soil properties is studied by comparing the behaviour of rhizosphere soil with that of the bulk soil. However, this approach presents several drawbacks. In particular, bulk soil can contain root hairs (Norvel and Cary 1992). Furthermore, it is not possible to control the various variables (plant physiology, root age, root section, temperature, moisture, etc.) affecting the root exudation and thus the rhizosphere soil. Microbial community structure assessed by phospholipid fatty acid analysis (PLFA) was more affected by plant species than soil moisture. Community level physiological profiles (CLPP), in terms of diversity of substrate utilization and average well colour development (AWCD), were affected by plant species and soil moisture (Chen et al. 2007).

The study of the rhizosphere is a very complex task and several microcosms have been created to study the dynamics and the interaction between soil, microorganisms and plants in the rhizosphere. Most of these systems are based on the physical separation of rhizosphere and the adjacent bulk soil by porous membranes (Kuchenbuch and Jungk 1982) and may allow sampling soil at different distances from the rhizoplane. These systems are enclosed in so-called rhizoboxes, delimiting vertical or horizontal nets. In some systems the nets are localised in the bottom of the box and the soil below the roots is considered rhizosphere soil (roots mate approach) (Kuchenbuch and Jungk 1982). In the slit system (Hinsinger and Gilkes 1997) the division is made with a 0.2-μm membrane and only the hair roots may grow in a thin layer of soil. In the rhizobox of Li et al. (1991), membranes with different pore diameters can allow penetration of root hairs and hyphae in the soil compartment. The rhizobox set up by Wenzel et al. (2001) allows monitoring of the pH, redox potential and soil moisture during the experiment, through the use of proper microsensors; it also allows the monitoring of root development, distribution and morphology without involving destructive sampling. In spite of the advantages of this rhizobox, which is amenable to dynamic measurements, Wenzel et al. (2001) underlined the importance of combining studies using their rhizobox with field measurements of rhizosphere processes, since any experimental approach based on an in vitro system implies deviations from the indigenous soil-plant system.

Model root systems (MRS) have been used to discriminate between the effect of the various root exudates. In a simple device reported by Badalucco and Kuikman (2001), the soil is pressed to a precise density (1.4 g cm^3) into a plastic ring standing on a Petri dish covered with aluminium foil. The top of the soil is covered with a cellulose paper filter (Whatman 41), which can be wetted with different solutions of root exudates (MRE, model root exudates). This system produces a gradient of the root exudate at increasing distance from the filter paper (Falchini et al. 2003) and it is possible to sample soil slices at different distances from the filter paper, which simulates the rhizoplane.

When working with model compounds, these should be incorporated in amounts reflecting the daily carbon input to the rhizosphere, ranging usually between 50 and 100 µg C g^{-1} soil (Trofymow et al. 1987; Iijima et al. 2000). Therefore it is unrealistic to apply several mg C g^{-1} of soil (Badalucco and Kuikman 2001; Baudoin et al. 2003).

14.4 Effects of Root Exudates on Microbial Activity of Rhizosphere Soils

Microbial activity of rhizosphere soil varies among different plant species, probably because these differ in root exudates composition (Van der Krift et al. 2001; Warembourg et al. 2003). Plants with a high concentration of root solutes and a rapid growth should stimulate a high rhizosphere microbial activity. This hypothesis was evaluated by Valè et al. (2005) by comparing the rhizosphere microbial activity (in vitro mineralization of a small amount of ^{14}C-glucose) and bacterial abundance (expressed as the number of CFU, or colony-forming units) between six herbaceous species grown in the greenhouse and differing in plant biomass and root C concentrations (including soluble and insoluble C). The microbial activity was positively correlated with root soluble C concentration and shoots biomass and negatively correlated with concentration of insoluble C in roots.

14.4.1 Competition Between Plants and Microorganisms for Soil Nutrients

Since it stimulates microbial growth, rhizodeposition might actually decrease the availability of mineral nutrients in the region of the rhizosphere where it is released. Information on the topology and timing of release of specific compounds from different root districts has been obtained using whole cell biosensors (Jasper et al. 2001; Casavant et al. 2003; see this volume, Chap. 9). However, mineral nutrient immobilization following rhizodeposition does not impair plant growth because the apical regions of the root (i.e. the zone extending from the root hair zone to the root apex) extract most of those nutrients that are available for uptake before extensive rhizospheric colonization by saprophytic microorganisms occurs. If this hypothesis is true, microbial growth should be limited.

On the other hand, microbial growth itself becomes limited following depletion of mineral nutrients in the rhizodeposition zone. This effect was observed in the maize rhizosphere (Merckx et al. 1987). Similarly it was found that microbial respiration was not limited in the rhizosphere of winter wheat by available C

(Cheng et al. 1996) but probably by some other nutrients. Among the major nutrients, N may be the main limiting nutrient for microbial growth due to the high C to N ratio of root products (Marschner 1995); indeed, rhizodeposition by maize plants increased N immobilization but also microbial denitrification due to the presence of easily available organic C (Qian et al. 1997).

Some factors complicate the study of the competition between roots and microorganisms for nutrients. Among these are the fact that: 1) lateral roots explore a soil different from that explored by the main roots; 2) nutrients immobilized by microbial assimilation are likely to be recycled following the death and degradation of microbes and microfauna; 3) root exudation of soluble organic compounds correspond to a net release between efflux and influx; 4) the uptake of any nutrient by plants and microorganisms depends on the form of this nutrient in the rhizosphere soil.

The apices of lateral roots must grow through the rhizosphere of the superior axis from which they originate, and thus they may experience the effects related to the type of exudates produced by the main axis and the exudate using microbial population (Uren 2007).

Nutrients immobilized by microbial assimilation are likely to be recycled following the death and degradation of microbes and microfauna, but the timing and location of such events in relation to the nutrient-absorbing regions of the root are difficult to investigate because of the difficulties in simulating these processes under laboratory conditions. However, Mary et al. (1993) measured the recycling of C and N during the decomposition of root mucilage, glucose and roots by simply mixing low concentrations of these substrates with soil.

The net release between efflux and influx (Darrah and Roose 2001) can be influenced by microorganisms (Phillips et al. 2004). For example, 2,4-diacetylphloroglucinol (DAPG) produced by *Pseudomonas* can block the amino acid influx whereas the fungal product zearalenone can increase amino acids efflux in different plants (alfalfa, *Medicago sativa*, maize, *Zea mays*, and wheat, *Triticum aestivum*). Studies based on the use of the reporter bacteria have highlighted that DAPG is likely to mediate the chemical communication among different populations of rhizobacteria (Maurhofer et al. 2004; see this volume, Chap. 9).

Generally both plants and microorganisms prefer ammonium to nitrate but this is not always so (Badalucco and Kuikman 2001). For example, tomatoes prefer nitrate, white spruce ammonium, and some arctic sedges amino acids (Badalucco and Kuikman 2001). In the rhizosphere soil of pine (*Pine ponderosa*) the plant accounted for 70% of the total NO_3^- consumption whereas it only accounted for 30% of the total NH_4^+ consumption (Norton and Firestone 1996). However, the presence of roots reduced NH_4^+ consumption by both nitrifiers and heterotrophs, and the nitrifiers competed with heterotrophs for NH_4^+, transforming it to nitrate and thus keeping inorganic N in a form available to roots. Using two bacterial biosensors for detecting NO_3^- availability, De Angelis et al. (2005) reported significantly lower NO_3^- availability in the rhizosphere of wild oat than in bulk soil and the competition between roots and the whole-cell bioreporters could be attenuated by soil amendment with NO_3^-.

14.4.2 Soil Respiration

Soil respiration has often been used as an index of microbial activity (Nannipieri et al. 1990). The respiration of rhizosphere soil is greater than that of the bulk soil not located around organic debris, because in the former case CO_2 can originate not only from microbial respiration of soil organic C but also from root respiration and microbial decomposition of rhizodeposition. Raich and Mora (2005) reported that root and rhizosphere soil respiration of an annual crop accounted for 27–30% of the overall soil respiration. Separation between root respiration and CO_2 evolution from rhizosphere soil is methodologically difficult. An interesting approach was followed by Cheng et al. (1993), who saturated soil with unlabelled glucose before the ^{14}C pulse-labelling of plant shoots so as to eliminate the use of labelled substrates released from roots by soil microorganisms. Root respiration and microbial respiration of rhizosphere soil accounted for 40% and 60% of the overall respiration, respectively. Kuzyakov (2002b) suggested that microbial respiration accounted for 50–60% of the total plant-induced respiration. Usually, microbial respiration in the rhizosphere soil is highly dependent on climatic conditions, nutrient availability and root exudation, which is itself controlled by the rate of photosynthesis during light periods (Kim and Verma 1992).

Another interesting approach was followed by Rochette and Flanagan (1997), who grew C_4 (*Zea mays* L) plants on soil developed under C_3 plants (mixture of grass and legumes) to measure the isotope ratio ($\delta^{13}C$) of CO_2 evolved from soil under corn or from soil kept free of vegetation. The $\delta^{13}C$ values of CO_2 from the control soil were significantly lower than those of CO_2 from the corn plots and this allowed to estimate the CO_2 of the rhizosphere soil, which ranged 18–25% of crop net photosynthesis and 24–35% of crop net CO_2 assimilation during most of the growing season.

Recently, CO_2-C evolution from decomposition of root exudates has been studied in systems simulating the rhizosphere zone, by monitoring the mineralization of single synthetic low molecular weight organic compounds commonly present in root exudates (Kozdroj and van Elsas 2000; Badalucco and Kuikman 2001; Baudoin et al. 2003; Falchini et al. 2003; Landi et al. 2005). Thus, Falchini et al. (2003) monitored the diffusion of ^{14}C-labelled glucose, oxalic acid, or glutamic acid into soil from a filter placed on the surface of a sandy loam soil. Glutamate showed a higher mineralization than glucose during the first three days, whereas the mineralization of oxalic acid showed a three-day lag phase. Both glutamate and glucose addition caused a positive priming effect. Hamer and Marschner (2005) reported that fructose and alanine induced a stronger priming effect in forest soil when compared with other root exudates such as oxalic acid and catechol. Oxalic acid induced both negative and positive priming effects whereas catechol always reduced mineralization of soil organic matter. It was not possible to predict the occurrence and magnitude of the priming effect from the chemical and physical soil properties, but it was observed that the priming effect was more pronounced in forest soils containing low biodegradable organic carbon (Hamer and Marschner

2005). Li and Yagi (2004) observed that C inputs by rice (*Oryza sativa*) grown under elevated CO_2 retarded the mineralization of organic matter in the 0–5-cm surface layer of a paddy soil. Several hypotheses have been proposed to interpret positive priming effects. According to Fontaine et al. (2003), addition of easily available organic C stimulates the growth of r-strategists and the successive growth of k-strategists is responsible of the degradation of recalcitrant organic matter. Another hypothesis explains the positive priming effect as due to the increase in the turnover of native microbial biomass (Chander and Joergensen 2001; De Nobili et al. 2001) whereas Kuzyakov et al. (2000) suggested that the activation of soil microorganisms by the addition of the easily available organic C increases enzyme synthesis with higher degradation of soil organic matter. The real and apparent priming effects caused by the addition of ^{15}N labelled fertilizers have been discussed by Jenkinson et al. (1985).

14.4.3 Nutrient Dynamics and Functional Aspects of Rhizosphere Soil

It is well established that root exudation of easily available organic compounds affects nutrient dynamics through the turnover and mineralization of organic compounds (Grayston et al. 1996; Hamilton and Frank 2001; Kuzyakov and Cheng 2001; Kuzyakov 2002a). In this regard the most studied plant nutrients are nitrogen and phosphorus.

The assimilation of root-derived C stimulates microbial N immobilization because the average C/N ratio of rhizodeposition is higher than the C/N ratio of soil microflora (Badalucco and Kuikman 2001). This may result in a temporary reduction of available N to the plant. Only under conditions of low N availability may the rhizosphere be a region of excess C supply where N concentration limits microbial growth (Merckx et al. 1987). Norton and Firestone (1996) suggested that N immobilization rates of rhizosphere soil of *Pinus ponderosa* seedling were limited by NH_4^+ rather than by C availability. The microbial N immobilization promoted by the microbial assimilation of root-derived C can be counterbalanced by the protozoan stimulation of the N mineralization; root exudation promotes bacterial growth with mineralization of organic N and then bacteria are grazed by protozoa with release of NH_4^+ due to the higher C/N ratio of the protozoa than bacteria (Clarholm 1985; Kuikman et al. 1990; Liljeroth et al. 1990, 1994). The higher N mineralization in the rhizosphere soil has been indirectly demonstrated by the higher protease and histidinase activities, both involved in N mineralization processes, in the rhizosphere than bulk soil (Badalucco et al. 1996). It has been postulated that nematodes are the primary consumers of bacteria in the rhizosphere (Griffiths 1990) and they also release ammonia when grazing bacteria because the C/N ratio of the former is higher than that of the latter (Badalucco and Kuikman 2001). According to Jones et al. (2005) the NH_4^+ release as the result of grazing of bacteria by protozoa, nematodes and invertebrates and the slower turnover times of roots compared to microorganisms (Hodge et al. 2000) counteract the effects due to the fact that microorganisms are superior competitors than plants for both inorganic and organic N sources.

It has been shown that root-derived C stimulates N immobilization-mineralization turnover and denitrification in a greenhouse experiment based on growing maize plants, monitoring ^{13}C natural abundance and ^{15}N added as $^{15}NH_4{}^{15}NO_3$ fertilizer (Qian et al. 1997). The N mineralization-immobilization turnover (MIT) is very important in regulating the amount of N available to plants and it is based on the transformation of organic N to $NH_4{}^+$ and the opposite reaction. The alternative pathway is the so-called "direct route" in which microorganisms take up simple organic molecules, such as amino acids, and once these amino acids are inside the microbial cells they are deaminated and the surplus $NH_4{}^+$ is released into the extracellular soil environment (Barraclough 1997). Amino acid uptake by rhizobacteria has been proven by using reporter bacteria (Jaeger et al. 1999; Espinosa-Urgel and Ramos 2001; see this volume, Chap. 9).

The amount of available N supplied as fertilizer can affect the type of reactions in the rhizosphere soil. By growing barley plants for 46 days in a sandy loam soil in a cabinet with a ^{14}C-labeled atmosphere and applying either high (169.1 mg N Kg^{-1} wet soil) or low (34.2 mg N Kg^{-1} wet soil) amounts of fertilizer N as ^{15}N labeled ammonium sulphate, it was observed that the proportion of ^{14}C translocated below ground was slightly higher in the high-N than in the low-N treatment and the decomposition of organic matter was reduced in the high-N treatment (Zagal et al. 1993). In contrast, soil microorganisms in the low-N treatment preferred C from soil organic matter over root derived C.

From the global change perspective, increase of atmospheric CO_2 and land cover transformation are among the major impacts caused by human activities. Pinay et al. (2007) determined the effects of two years atmospheric CO_2 enrichment on soil potential respiration (SIR), denitrification (DEA) and nitrification (NEA) activities. Soil microbial activities measured by SIR, DEA and NEA were not sensitive to an increase of atmospheric CO_2 but some of these functions were significantly altered by the type of plant cover, i.e. annuals vs perennials. The relative changes of these microbial activities induced by annual and perennial plants was inversely related to the density and the diversity of the corresponding functional bacterial groups, i.e. change in nitrification (NEA)> change in denitrification (DEA)> change in respiration (SIR). In other words, the functional community with the least diversity and density, i.e. nitrification, was the most affected by the plant cover type and these changes remained after the rain event. In contrast, the respiration process, under the control of a wide diversity and density of microorganisms, did not present any significant change. Denitrification presented an intermediate pattern with significant rate differences after the two year experiment, but functionally converged two months after the rain event.

Increases in bacterial activity were observed in both rhizosphere and bulk soil by Christensen and Christensen (1994) when N was present in limiting concentrations. In addition, Söderberg and Bååth (2004) observed that addition of $NH_4{}^+$ but not $NO_3{}^+$ decreased bacterial activity (as determined by thymidine and leucine incorporation) of rhizosphere soil of barley seedlings but not bacterial activity of the bulk soil because plant NH_4^+ uptake decreased the pH of rhizosphere soil due to secretion of H^+ by roots. However, a change in the composition of root exudates in response to the different N source cannot be excluded.

The functional aspects of the microflora inhabiting the rhizosphere soil has been monitored by the community-level physiological profiles (CLPP) generated with sole-carbon-source-utilization tests from BIOLOG (Garland and Mills 1991). In spite of the fact that three closely related legumes, i.e. alfalfa (*Medicago sativa*), common bean (*Phaseolus vulgaris*) and clover (*Trifolium pratense*), showed differences in the composition of the rhizosphere communities as assessed by PCR-SSCP (single strand conformation polymorphism) analysis of 16S rRNA genes, the overall analysis by CLPP indicated that the metabolic potential of all rhizosphere samples was similar (Miethling et al. 2003). Treatment of soil for 14 days with artificial root exudate (glucose, fructose, sucrose, citric acid, lactic acid, succinic acid, alanine, serine and glutamic acid) solution at a rate of $100\,\mu g$ C g^{-1} day^{-1}, to simulate a daily carbon input to soil by root, markedly changed the BIOLOG oxidation pattern (Baudoin et al. 2003). However, the addition of the exogenous substrates was not followed by an increased oxidation of these compounds in the BIOLOG plates and this does not support the hypothesis that BIOLOG oxidation patterns can be used as an index of substrate availability in situ. It is well established that the CLPP presents several drawbacks. In particular, it is culture-dependent, it does not determine the contribution of fungi and it does not maintain the composition of the microbial communities constant during the incubation (Nannipieri et al. 2003). Degens and Harris (1997) overcame these limitations by measuring the utilization patterns of various substrates by soil microbial communities using short-term responses of soil treated with amino acids, carboxylic acids, carbohydrates and organic polymers.

14.4.4   *Enzyme Activity in the Rhizosphere Soil*

Enzyme activity is generally higher in rhizosphere than in bulk soil, as a result of a greater microbial activity sustained by root exudates or due to the release of enzymes from roots (Badalucco and Kuikman 2001). The overall enzyme activity of the rhizosphere soil can depend on enzymes localized in root cells, root remains, microbial cells, microbial cell debris, microfaunal cells and the related cell debris, free extracellular enzymes or enzymes adsorbed or inglobated in soil particles. Ultracytochemical techniques have been used with electron microscopy to localize enzymes in electron-transparent components of soil such as microbial and root extracellular polysaccharides, fragments of cell walls and microbial membranes but these techniques cannot be applied in regions of soil with naturally electron-dense particles such as minerals (Ladd et al. 1996). Thus, acid phosphatase has been detected in roots, mycorrhizae, soil microbial cells and fragments of microbial membranes as small as $7\times20\,nm$.

Soil microbes release extracellular enzymes useful for the initial degradation of high molecular weight substrates such as cellulose, chitin and lignin, and mineralise organic compounds to mineral N, P, S and other elements. Enzymes attached to the outer surface of microbial cells, the ectoenzymes, can also carry out the hydrolysis of high-molecular weight substrates (Burns 1982; Nannipieri 1994). In addition to extracellular enzymes, active intracellular enzymes can also be released after cell lysis and remain active in the extracellular soil environment insofar as they do not require cofactors for their activity, extracellular pH and temperature are not denaturing and abiotic inactivation or proteolytic degradation does not occur (Nannipieri 1994). Sorption by soil colloids may protect an enzyme from microbial degradation or chemical hydrolysis and the enzyme can retain its activity if it is not denaturated and its active site is available to the substrates (Nannipieri 1994).

Most extracellular enzymes have a low mobility in soil due to their molecular size and charge characteristics, and thus any secreted enzyme must operate close to the point of secretion and its substrate must be able to diffuse towards it. Acid phosphatase was secreted in response to P deficiency stress by epidermal cells of the main tip roots of white lupin and was present in the cell walls and intercellular spaces of lateral roots (Wasaki et al. 1997). Such apoplastic phosphatase is protected against microbial degradation and cannot be adsorbed by soil colloids, but is effective only when soluble organophosphates, normally present in the soil solution, diffuse in the apoplastic space (Seeling and Jungk 1996). The role of phosphatases in the rhizosphere remains uncertain (Tinker and Nye 2000).

Tarafdar and Jungk (1987) carried out a very interesting study on the relationship between enzyme activity of soil and nutrient cycling in the rhizosphere. They sampled a silt loam soil at different distances from the rhizoplane of either clover (*Trifolium alexandrinum*, 10 days old) or wheat (*Triticum aestivum*, 15 days old) and found that the total P and organic P contents decreased in the rhizosphere soil, whereas the inorganic P content increased in the vicinity of the rhizoplane. Such an increase was probably due to the increase of both acid and alkaline phosphatase activities in the rhizosphere soil and it paralleled the increase in both fungal and bacterial counts, suggesting a probable microbial origin of both enzymes in the rhizosphere soil. Both phosphatase activities increased with plant age, probably as the result of the increase in microbial biomass and/or the increase in total root surface. It has been speculated that plants do not need to secrete phosphatases because the phosphatase activity (mostly of microbial origin) in the rhizosphere soil is generally sufficient to ensure sufficient available P (Tarafdar and Jungk 1987; Tarafdar and Marschner 1994). *Bacillus amyloliquefaciens* FZB45, a plant-growth-promoting rhizobacterium, stimulated growth of maize seedlings under phosphate limitation in the presence of phytate whereas a phytase-negative mutant strain FZB45/M2 did not stimulate plant growth (Idriss et al. 2002). However, the plant origin of phosphatase as of any enzyme of the rhizosphere soil cannot be excluded because plant-borne enzyme can be released in the rhizosphere (Tarafdar and Jungk 1987). Indeed, transgenic *Nicotiana tabacum* (tobacco) or *Arabidopsis thaliana*, which expressed constitutively β-propeller phytase from *Bacillus subtilis* (*168phyA*), secreted extracellular phytase in much higher amounts than the respective wild-type plants and used sodium phytate as the sole P source (Lung et al. 2005). Similarly, transgenic *Arabidopsis thaliana* with phytase gene (*phyA*) from *Aspergillus niger* was capable of taking up P from a range of organic phosphorus substrates added to agar under sterile conditions (Richardson et al. 2001). However, transgenic *Trifolium subterraneum* L constitutively expressing a phytase gene (*phyA*) from *Aspergillus niger* was capable of exuding phytase and taking up more P than wild-type plant when grown in agar with phytate, but it was not successful when it was grown in soil (George et al. 2004), probably because plant-exuded phytase was adsorbed by soil colloids and/or degraded by soil protease (George et al. 2005).

In a soil-plant (wheat) microcosm, bacterial numbers, protozoan numbers, histidinase and casein hydrolysing activity were monitored after 21 and 33 days of plant growth (Badalucco et al. 1996). Microbial numbers and enzyme activities were higher in the rhizosphere than in the bulk soil; the closer to the soil-root interface, the higher the numbers and the enzyme activities (Badalucco et al. 1996). It was hypothesised that bacteria were the main source of histidinase, whereas protease activity was suggested to be produced by bacteria, protozoa and root hairs.

Using the model rhizosphere system described above, Renella et al. (2005) reported that different root exudates were mineralized to different extents and had different stimulatory effects on microbial growth and on hydrolase activities, mostly localized in the rhizosphere zone. In particular, the rapid increase in the alkaline phosphatase activity could be considered as an indirect evidence of the important role of rhizobacteria in the synthesis of this enzyme in the rhizosphere (Tarafdar and Jungk 1987).

Measurements of enzyme activities have been used to study the effect of transgenic plants on soil metabolism. Both dehydrogenase and alkaline phosphatase activities of soil sampled from transgenic alfalfa, regardless of association with recombinant nitrogen-fixing soil *Sinorhizobium meliloti*, were significantly lower than those of soil sampled from parental alfalfa (Donegan et al. 1999).

Enzyme activities of rhizosphere soil have been measured to assess the perturbation resulting from the introduction of genetically modified microorganisms in the ecosystem (Naseby and Lynch 1998). The inoculation of wheat seeds with a genetically modified strain of *Pseudomonas fluorescens* increased urease and chitobiosidase activities of rhizosphere soil at 0–20 cm depth and decreased alkaline phosphatase but not acid phosphatase activity (Naseby and Lynch 1997). The reduction in alkaline phosphatase activity was attributed to a displacement of the rhizosphere communities producing the enzyme. Opposite changes in the measured enzyme activities were observed when inoculation of wheat seeds with the genetically modified *P. fluorescens* was carried out in the presence of a mixture of urea, chitin and glycerophosphate (Naseby and Lynch 1997).

P. fluorescens F113, which naturally produces the antifungal 2,4-diacetylphloroglucinol (DAPG) and is marked with a *lacZY* gene cassette, increased alkaline phosphatase, phosphodiesterase and arylsulfatase activities of pea rhizosphere whereas the other inocula reduced enzyme activities compared to the control (without bacterial inoculum) (Naseby and Lynch 1998). It was suggested that increases in enzyme activities were caused by the production of DAPG, which decreased the available inorganic phosphate and sulphate in the rhizosphere being the synthesis of these enzymes controlled by these nutrients (Naseby et al. 1998). However, an opposite trend was found for acid phosphatase activity, which is mostly of plant origin, contrarily to the primarily microbially-determined alkaline phosphatase activity. Therefore, acid phosphatase activity is more dependent upon the nutritional status of the plant. The presence of the F113 strain was associated with low β-galactosidase, β-glucosidase, *N*-acetylglucosaminidase activities and probably this behaviour depended on the increase in available C. On the other hand, no effects on enzyme activities were observed when *Pseudomonas fluorescens* F113 was present in the rhizosphere of field-grown sugar beet (Naseby et al. 1998). It was concluded that the impact of various genetically modified *Pseudomonas* on the rhizosphere populations and functions depended on the nature of the genetic modification (Naseby and Lynch 1998).

The potentialities of enzymes produced by rhizosphere microorganisms, including genetically modified microorganisms, in bioremediation and biocontrol of pests and diseases have been discussed by Naseby and Lynch (2002).

The main problem in interpreting the meaning of enzyme activities in soil are, first, that the current enzyme assays measure the potential rather than the real enzyme activity because the conditions of incubation assays are based on optimal pH and temperature values, optimal substrate concentrations, presence of a buffer and shaking of soil slurries; of course the conditions for enzymes in situ are much different from those used in the assay (Burns 1982; Nannipieri et al. 2002; Gianfreda and Ruggiero 2006) and, second, that the current enzyme assays do not distinguish among different enzymes contributing to the measured total enzyme activity (Burns 1982; Nannipieri 1994; Nannipieri et al. 2002; Gianfreda and Ruggiero 2006). It has been suggested that enzymes can be present in soil in different locations, as intracellular enzymes in active, resting, and dead cells as well as in cell debris and as extracellular enzymes in the soil solution, adsorbed by inorganic colloids or associated in various ways with humic molecules (Nannipieri et al. 2002). It would be important to determine the intracellular enzyme activity of active microbial cells so as to obtain meaningful information on the microbial functional diversity (Nannipieri et al. 2002). Several methods have been proposed to distinguish the extracellular stabilized enzyme activity (activity due to enzyme adsorbed or englobated in soil colloids) from intracellular enzyme activity but all of these have disadvantages (Nannipieri et al. 2002). As discussed above, the situation is more complex in the rhizosphere than in bulk soil, due to the presence of active and still intact root cells detached from the roots, of mycorrhizal cells strictly linked to roots and active bacterial, fungal and faunal cells. All these cells present a broad array of active enzymes.

The source of active enzyme in soil by using the molecular techniques will be discussed in this volume (see Chap. 9).

14.5 Microbial Diversity in the Rhizosphere

Rhizosphere microorganisms are classified on the basis of the interactions with plants, as they can have negative (e.g. phytopathogenic), positive (e.g. plant growth promotion, symbiosis), or neutral (e.g. no benefits) effects on plants (Brimecombe et al. 2001). Symbiotic and pathogenic microorganisms have been well characterized by cultural or direct methods, because of the possible applications of such work in plant protection or crop production, whereas the role of neutral microorganisms in the rhizosphere has been generally neglected. Recently, however, the role of 'neutral' rhizosphere microorganisms in plant nutrition has been re-evaluated (Hirsch et al. 2003; Talbot 2003).

The number of rhizosphere-colonizing microbes has been determined by plate counts or the Most Probable Number (Bakken 1997; Brimecombe et al. 2001; Johnsen et al. 2001). Soil treated with artificial root exudate (glucose, fructose, sucrose, citric acid, lactic acid, succinic acid, alanine, serine and glutamic acid) solution at a rate of $100\,\mu g$ C g^{-1} day^{-1}, to simulate a daily carbon input to soil by root, markedly increased bacterial counts (Baudoin et al. 2003). Variations of the C/N ratio of the solution added to soil had no effect on the bacterial numbers. It is well established that culture-dependent methods only detect 1–10% of the microorganisms inhabiting the soil (Torsvik et al. 1996). Molecular techniques based on the extraction, purification and characterization of nucleic acids from soil, the BIOLOG technique and phospholipid fatty acid analysis (PLFA) have provided alternative methods for analyses of the microbial diversity in the rhizosphere (Lynch et al. 2004). The limits of the BIOLOG technique have been already discussed. The phospholipid fatty acid (PLFA) technique, which is based on the extraction, fractionation, methylation and chromatography of the phospholipid component of soil lipids, can only be used to estimate gross changes in community structure (Zelles 1999). It is possible to identify species by fatty acid analysis using standard cultural-based media and a suitable database. Molecular methods based on the extraction, purification and characterization of nucleic acids are generally used to study both culturable and unculturable microorganisms in soil. There are a broad variety of these methods for low, intermediate and high resolution analysis (Johnsen et al. 2001; Lynch et al. 2004). Generally, these methods cannot provide the resolution of microbial diversity where it is necessary to identify key microbial species at the community level or to elucidate their role in the ecosystem. These limitations can be overcome to some extent by rRNA gene analysis for microbial diversity studies (see this volume, Chap. 9).

Low-resolution techniques, such as the determination of base distribution in community DNA and the rate at which denatured, single-stranded DNA reanneals, give an estimate of the total genetic diversity (Torsvik and Øvreas 2002; Lynch et al. 2004). Most of the intermediate-resolution techniques are based on comparative analysis of conserved genes such as those coding for ribosomal RNA (rRNA), the so-called rDNA (Johnsen et al. 2001; Lynch et al. 2004). They usually involve polymerase chain reaction (PCR) amplification of rRNA genes from soil DNA samples, combined with fingerprinting techniques, such as denaturing gradient gel electrophoresis (DGGE), temperature gradient gel electrophoresis (TGGE), terminal restriction fragment length polymorphisms (T-RFLP), amplified rDNA restriction analysis (ARDRA), cloning and sequencing (Johnsen et al. 2001; Torsvik and Øvreas 2002; Lynch et al. 2004). The DGGE (which separates chemically-denatured PCR products of the same size but of different sequences) and T-RFLP (which distinguishes between PCR products by recognizing only the terminal fragment of restriction digestion), have been generally used to study rhizosphere–microbe interactions.

By using DGGE, it was revealed that different plants support different bacterial (Marschner et al. 2002), fungal (Gomes 2003) and archaeal (Nicol et al. 2003) communities and that the structure of microbial communities was affected by root architecture, plant age and various perturbations (Marschner et al. 2002; Nicol et al. 2003). The effect of plant roots on the composition of archaeal communities has been confirmed by PCR single stranded conformation polymorphism (PCR-SSCP) with significant differences in the composition of the Crenarchaeota populations between the rhizosphere soil of different plant species and their respective bulk soil (Sliwinski and Goodman 2004). The effect of root architecture on the composition of bacterial communities from the rhizosphere of grassland species has also been shown by T-RFLP analyses (Kuske et al. 2002). Marschner et al. (2001) found that the composition of bacterial rhizosphere community from three plant species (chickpea, rape and Sudan grass), as determined by PCR-DGGE of 16 S rDNA, depended on the complex interaction between soil type, plant species and root zone location. Bacterial diversity was higher in mature root zones than at the root tips in the sand and clay soils but not in the loamy sand soil. They also showed that N fertilization had no significant effect on the composition of bacterial community of the rhizosphere soil whereas both fertilization and soil type influenced plant growth. Stark et al. (2007) recently demonstrated that the addition of green manures improved soil biology by increasing microbial biomass and activity irrespective of management history, that no direct relationship existed among microbial structure, enzyme activity and N mineralization, and that microbial community structure (by PCR-DGGE) was more strongly influenced by inherent soil and environmental factors than by short-term management practices.

Plate counts, but not TGGE fingerprintings, showed an effect of root age on bacterial community structure of rhizosphere soil of *Zea mays* (Gomes et al. 2001). Bacterial community composition on the cluster roots and in the rhizosphere soil, determined by DGGE, differed among three species of *Banksia* (*B. attenuata* R. Brown, *B. ilicifolia* R.Brown and *B. menziesii* R.Brown), and depended on sampling times and cluster root age, as young, mature and senescing roots were distinguished (Marschner et al. 2005). No changes were observed in both acid and alkaline phosphatases, whereas both β-glucosidase and protease activities increased with time. The three species differed in asparaginase activity. Smalla et al. (2001) showed a plant-dependent shift in the relative abundance of bacterial populations by comparing DGGE-fingerprinting of 16 S rDNA fragments amplified by PCR from soil or rhizosphere. This shift was more pronounced in the second than in the first year. DNA sequencing showed that most of the dominant bands from the rhizo-sphere patterns corresponded to Gram-positive bacteria. The study concerned rhizosphere communities of field grown strawberry (*Fragaria ananassa* Duch), oilseed rape (*Brassica napus* L.) and potato (*Solanum tuberosum* L.), those three plants being host to the pathogenic fungus *Verticillium dahlae*. Both the abundance and composition of *V. dahlae* bacterial antagonists were plant species dependent (Berg et al. 2002). While most studies to date have focused on a single functional gene, analysis of a more complex suite of genes would enable us to better address the role of the community structure in controlling various processes in soil. In the recent years there has been a growing interest in genes and transcripts coding for metabolic enzymes. Besides questions addressing redundancy and diversity, more and more attention is given on the abundance of specific DNA and mRNA in the different habitats. Sharma et al. (2007) have recently reviewed several PCR tech-niques that are suitable for quantification of functional genes and transcripts such as most probable number (MPN)-PCR, competitive PCR and real-time PCR. These new quality of data is of high relevance to improve mathematic models of turnover processes.

The T-RFLP analysis showed that composition of eubacterial community of rhizosphere of conventionally managed continuous corn and organically managed corn was similar to that of soil light fraction, which includes plant debris of soil, but differed respect to that of heavy fraction, which includes the mineral particles and associated humic matter (Blackwood and Paul 2003). Nunan et al. (2005) have studied the rhizoplane bacterial community rather than the rhizosphere communi-ties, after hypothesising that plant effects on microbial community should be more pronounced on the rhizoplane than in the rhizosphere soil. Neither T-RFLP nor DGGE fingerprints of PCR-amplified 16rDNA did not show any effect of grassland plant species on the bacterial community of rhizoplane.

Neither DGGE nor T-RFLP analyses provide information on key microbial spe-cies nor elucidate their role in the rhizosphere if cloning and sequencing are not carried out. In addition, it should be stressed that the selected method for extraction and purification of nucleic acid from rhizosphere soil can markedly affect observa-tions on bacterial community structure (Niemi et al. 2001).

It was found that there was a higher diversity of *amo*A genes (Briones et al. 2003) and *nif*H genes (Cocking 2003) in rhizospheres of rice cultivars and non-legumes than in bulk soils, respectively. The higher diversity of these two genes encoding key functions in N cycling might suggest that through rhizodeposition, plants select functional groups rather than taxonomic groups of microbes.

Many factors (root architecture, root age, perturbation, stability of soil micro-flora, etc.) can interfere with the recognition of the effects of plant species on the composition of microbial communities inhabiting rhizosphere soil. In addition, soil microflora appears very stable, since changes due to perturbations are transitory (Nannipieri et al. 2003). An ingenious approach for studying the effects of plant species on composition of microflora by eliminating the problem of the presence of a stable microbial community was carried out by Bardgett and Walker (2004), who studied the effect of colonizer plant species on microbial growth and composition on recently deglaciated terrain in south-east Alaska by analysing PLFA. Bacterial biomass was increased by *Rhacomitrium, Alnus* and *Equisetum* and fungal biomass was increased by *Rhacomitrium* and *Alnus* with respect to bare soil.

The relative importance of specific plant properties vs soil characteristics in determining the composition of bacterial communities of the rhizosphere soil was examined in an innovative experiment, in which *Carex arenaria*, a non-mycorrhizal plant species, was chosen so as to eliminate the confounding factor represented by different levels of mycorrhizal colonization; this plant was grown in 10 different sites with soils presenting different properties (De Ridder-Duine et al. 2005). Bacterial diversity of rhizosphere and bulk soil was analysed by DGGE. It was observed that the diversity of a particular rhizosphere community was more similar to that of the bulk soil community of the same site rather than to that of rhizosphere communities from other sites.

Better insights on the effects of plant root in modifying the structure of soil microbial communities can be obtained by studies in which the rhizosphere effect is simulated by adding specific compounds occurring in root exudates. Both oxalic and glutamic acid changed the DGGE profiles of soil bacterial communities, causing the appearance of few extra-bands in the 0–2-mm soil layer of the model root system (Falchini et al. 2003). Microbial diversity, as determined by ribosomal intergenic spacer analysis (RISA), was changed when soil was treated with a mixture of root exudate compounds (glucose, fructose, saccharose, citric acid, lactic acid, succinic acid, alanine, serine and glutamic acid) at a rate of $100\,\mu g$ C g^{-1} day^{-1} for 14 days whereas the C/N ratio of the added solution had no effect (Baudoin et al. 2003).

In recent years molecular tools have been developed to analyze the structures of the rhizosphere-associated fungal communities from several crops (Gomes et al. 2003; Kowalchuk 1999; Smit et al. 1999), and also the function and possible role of the observed fungal diversity associated with plant roots, especially their antagonistic potential (Kowalchuk et al. 1997; Vandenkoornhuyse et al. 2003). Gomes et al. (2003) showed a rhizosphere effect of two maize cultivars grown in tropical soils on fungal communities analysed by DGGE of 18 S rDNA amplified by an universal primer. Plant growth development had an effect, whereas no difference was observed between fungal communities of the rhizospheres of the two cultivars. Cloning and sequencing of the dominant bands showed a dominance of members of Ascomycetes and Pleosporales families in young maize plants and a dominance of Ascomycetes and Basidiomycetous yeast in the rhizosphere of senescent plants.

14.6   Effect of Transgenic Plants on Microbial Diversity in the Rhizosphere Soil

Few studies have been conducted to investigate the effect of transgenic plants on soil microbial communities in spite of the several thousands field releases of transgenic crop plants (Kowalchuk et al. 2003; Lynch et al. 2004). Two possible effects can occur in the rhizosphere soil. With the first, bacterial population inhabiting the rhizosphere soil can capture and stably integrate transgenic plant DNA. In this case it may be risky the acquisition of antibiotic resistance genes, generally used as markers in transgenic crops, because it may change the composition of microbial communities. With the second, both composition and activity of microbial communities of rhizosphere soil can be changed as a consequence of altered root exudation or root morphology in transgenic plants. For example, in the case of transgenic modifications made to improve the plant resistance against microbial pathogens, the composition of rhizosphere microbial communities should be monitored. Indeed, the introduced resistance trait is based on the release of transgenic products such as cell-wall-attacking enzymes or molecules like T4-lysozyme, chitinase or cecropine, which can affect not only bacterial and fungal pathogens but also non-pathogenic microorganisms. Lynch et al. (2004) suggested that a requirement to an accurate monitoring of the effect of transgenic plants on soil microbial diversity is the careful collection of baseline data so as to take natural variations into consideration. In addition, Kowalchuk et al. (2003) have recommended to study effects of transgenic plants on soil microflora in small-scale field experiments since greenhouse conditions can markedly differ from field conditions. The parent variety of the transformed crop should also be included in the experimental design. For example, Dunfield and Germida (2001) observed that the microbial rhizosphere community of the transgenic glyphosate-tolerant oilseed, as determined by fatty acid methyl ester (FAME), was different from those of the three glufosinate ammonium-tolerant oilseed varieties. However, since the parental non-transgenic variety was not compared with the transgenic glyphosate-tolerant oilseed it was not possible to determine if the observed changes were due to the transgenic modification or not (Lynch et al. 2004). Both rhizosphere samples and bulk soil were sampled six times in a two-year, multi-site field study involving a transgenic canicola variety and a conventional variety and the composition of microbial communities as affected by the use of transgenic plants was investigated by community-level physiological profiles, fatty acid methyl ester profiles and terminal amplified ribosomal DNA (Dunfield and Germida 2003). Changes in the composition of microbial communities associated with the introduction of the transgenic variety were temporary and did not persist throughout.

A transgenic T4-lysozyme-expressing potato released detectable amount of T4 lysozyme (De Vries et al. 1999) with bactericidal activity at the rhizoplane (Ahrenholtz et al. 2000). However, the composition of bacterial communities of rhizosphere of both transgenic and a transgenic control without the T4 lysozyme gene, monitored by the BIOLOG approach, fatty acid analysis, DGGE or cloning and sequencing, were influenced by sampling period, plant developmental stage and field site and not by the T4 lysozyme expression (Heuer and Smalla 1997; Heuer et al. 1999; Muyzer and Smalla 1998).

Lynch et al. (2004) have suggested that relevant effects of transgenic plants on composition of microbial communities inhabiting soil should be more important than those due to season and field site.

14.7 Conclusions

Microbial activity, as determined by CO_2 evolution of rhizosphere soil, is often positively correlated with root soluble C concentrations and negatively correlated with non-soluble C. Enzyme activities are also generally higher in the rhizosphere than bulk soil, probably due to higher microbial sources but the currently used enzyme assays do not make it possible to distinguish the respective contributions of microbial, plant and faunal enzymes to the overall measured enzyme activity. Studies on microbial (mostly bacterial) diversity seem to indicate that plant species do not always affect the composition of microbial communities inhabiting the rhizosphere soil, because other factors, such as root architecture, root age, perturbations, stability of soil microflora, etc, also exert a significant influence. Growth in elevated CO_2 may, however, affect decomposition by changing the amount and dynamics of litter fall by modifying litter quality through changes in plant community composition; and by altering the soil environment and its biological activity (by increase of soil water, C input to soil, rhizosphere activity, etc.) with consequent priming of the decomposition of old stable organic matter. These indirect effects can be tested only by long-term studies on litter decomposition in forests exposed to elevated CO_2, but the current literature comprises results only from short-term incubations (Hyvönen et al. 2007).

Techniques used to determine microbial diversity such as those based on DNA fingerprinting have the potential to reveal genetic diversity but say nothing about the expression of these genes (see this volume, Chap. 9). Therefore, a higher microbial diversity that would be promoted by a higher flux of available C in the rhizosphere, as compared to bulk soil, would not necessarily imply a consequent higher diversity of functional genes. A possible objective of further studies would be to relate the utilization of plant C by microbial species to a particular functional role in the ecosystem (see this volume, Chap. 9). However because of current limitations on our understanding with respect to acclimation of the physiological processes, the climatic constraints, and feedbacks among these processes – particularly those acting at the biome scale – projections of C-sink strengths beyond a few decades are highly uncertain.

Interpretation of experiments examining in situ responses of soil microorganisms should be made with caution as incubation studies represent model systems under optimum conditions that rarely occur in the field. However, assessing soil properties under constant conditions allows variables such as soil moisture levels, temperature, microbial-plant interactions and soil type, to be studied individually (Stark et al. 2007). Due to limitations of the currently used methods, studies on microbial processes in the rhizosphere soil should be based on combined measurements of microbial diversity and microbial activity (Fig. 14.1). In this context, an interesting approach was followed by Kourtev et al. (2003), who worked with two exotic plant species, a Japanese barberry (*Berberis thubergii*, D.C., a hardy shrub forming tickets of multi-stemmed plants) and a Japanese stilt grass (*Microstegium vimineum*, Camus, a C4 annual plant), and with a native under-story plant

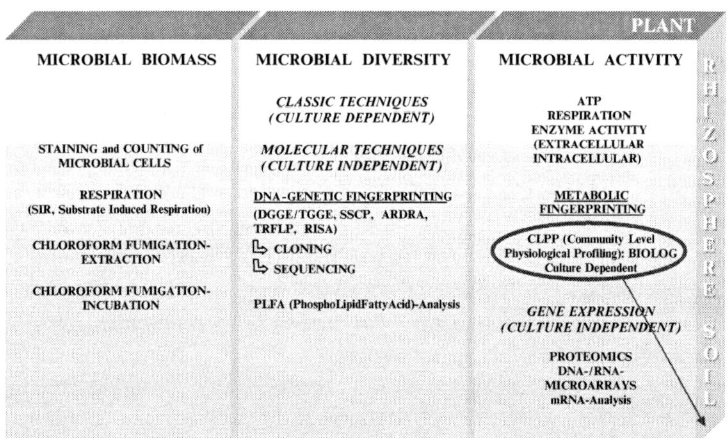

Fig. 14.1 Recommended multidisciplinary approach combining studies on microbial diversity, activity and biomass, to assess the rhizosphere soil microflora and its interactions with plants

(*Vaccinium* spp, a blueberry). The two exotic and the native species were grown in the same soil, and their effect on both microbial diversity and microbial activity was examined. After a three-month incubation period, soils planted with the exotic species differed in PLFA profiles, enzyme activities and SIR profile as compared with the initial soil or with the soil under the native under-story plant. Endocellulase, aminopeptidase, alkaline phosphatase and phenol oxidase activities were higher in stilt grass soil than in barberry or blueberry soils, whereas cellulase, acid phosphatase, peroxidase activity decreased in all planted soils.

There is need for better integration between plant physiology and molecular biology with soil chemistry, physics, and microbial and mesofaunal ecology. Working in isolation can still advance the field; however, the biggest advances will be made when scientific fields are integrated. It has been assumed that each compound released by roots has a specific role or function, but the reality is that very few proposed effects are established (Uren 2007). Probably the reality involves combination of more than one single root exudate compounds and future research should be directed at quantifying the significance of root exudates under realistic plant-soil systems.

References

Ahrenholtz I, Harms K, de Vries J, Wackernagel W (2000) Increased killing of *Bacillus subtilis* on hair roots of transgenic T4 lysozyme-producing potatoes. Appl Environ Microbiol 66:1862–1865

Badalucco L, Kuikman PJ (2001) Mineralization and immobilization in the rhizosphere. In: Pinton R, Varanini Z, Nannipieri P (eds) The rhizosphere. Biochemistry and organic substances at the soil-plant interface. Marcel Dekker, New York, pp 141–196

Badalucco L, Kuikman PJ, Nannipieri P (1996) Protease and deaminase activities in wheat rhizosphere and their relation to bacterial and protozoan populations. Biol Fertil Soils 23:99–104

Bakken LR (1997) Culturable and nonculturable bacteria in soil. In: van Elsas JD, Trevors JT, Wellington EMH (eds) Modern soil microbiology. Marcel Dekker, New York, pp 47–61

Bardgett RD, Walker LR (2004) Impact of coloniser plant species on the development of decomposer microbial communities following deglaciation. Soil Biol Biochem 36:555–559

Barraclough D (1997) The direct or MIT route for nitrogen immobilization: a ^{15}N mirror image study with leucine and glycine. Soil Biol Biochem 29:101–108

Baudoin E, Benizri E, Guckert A (2003) Impact of artificial root exudates on the bacterial community structure in bulk soil and maize rhizosphere. Soil Biol Biochem 35:1183–1192

Berg G, Roskot N, Steidle A, Eberl L, Zock A, Smalla K (2002) Plant-dependent genotypic and phenotypic diversity of antagonistic rhizobacteria isolated from different *Verticillium* host plants. Appl Environ Microbiol 68:3328–3338

Blackwood CB, Paul EA (2003) Eubacterial community structure and population size within the soil light fraction, rhizosphere, and heavy fraction of several agricultural systems. Soil Biol Biochem 35:1245–1255

Brimecombe MJ, De Lelj FA, Lynch JM (2001) The rhizosphere. The effect of root exudates on rhizosphere microbial populations. In: Pinton R, Varanini Z, Nannipieri P (eds) The rhizosphere. Biochemistry and organic substances at the soil-plant interface. Marcel Dekker, New York, pp 95–140

Briones AM, Satoshi O, Yoshiaki U, Niels-Birger R, Wolfgang R, Hidetoshi O (2003) Ammonia-oxidising bacteria on root biofilms and their possible contribution to N use efficiency of different rice cultivars. Plant Soil 250:335–348

Burns RG (1982) Enzyme activity in soil: location and a possible role in microbial ecology. Soil Biol Biochem 14:423–427

Casavant NC, Thompson D, Beattle GA, Phillips GJ, Halverson LJ (2003) Use of a site-specific recombination-based biosensor for detecting bioavailable toluene and related compounds on roots. Environ Microbiol 5:238–249

Chander K, Joergensen RG (2001) Decomposition of ^{14}C glucose in two soils with different amounts of heavy metal contamination. Soil Biol Biochem 33:1811–1816

Chen MM, Zhu YG, Su YH, Chen BD, Fu BJ, Marschner P (2007) Effects of soil moisture and plant interactions on the soil microbial community structure. Eur J Soil Biol 43:31–38

Cheng W, Coleman DC, Carrol CR, Hoffman CA (1993). *In situ* measurement of root respiration and soluble C concentrations in the rhizosphere. Soil Biol Biochem 25:1189–1196

Cheng W, Zhang Q, Coleman D (1996) Is available carbon limiting microbial respiration in the rhizosphere? Soil Biol Biochem 28:1283–1288

Christensen H, Christensen S (1994) ^{3}H-thymidine incorporation of rhizosphere bacteria influenced by plant N-status. Plant Soil 162:113–116

Clarholm M (1985) Interactions of bacteria, protozoa, and plants leading to mineralization of soil nitrogen . Soil Biol Biochem 17:181–187

Cocking EC (2003) Endophytic colonisation of plant roots by nitrogen-fixing bacteria. Plant Soil 252:169–175

Darrah PR (1996) Rhizodeposition under ambient and elevated CO_2 levels. Plant Soil 187:265–275

Darrah PR, Roose T (2001) Modeling the rhizosphre. In: Pinton R, Varanini Z, Nannipieri P (eds) The rhizosphere. Biochemistry and organic substances at the soil-plant interface. Marcel Dekker, New York, pp 327–372

De Angelis KM, Ji P, Firestone MK, Lindow SE (2005) Two novel bacterial biosensors for detection of nitrate availability in the rhizosphere. Appl Environ Microbiol 71:8537–8547

De Nobili M, Contin M, Mondini C, Brookes PC (2001) Soil microbial biomass is triggered into activity by trace amounts of substrate. Soil Biol Biochem 33:1163–1170

De Ridder-Duine AS, Kowalchuk GA, Klein Gunnewiek PJA, Smant W, van een JA, de Boer W (2005) Rhizosphere bacterial community composition in natural stands of *Carex arenaria* (sand sedge) is determined by bulk soil community composition. Soil Biol Biochem 37:349–357

De Vries J, Harms K, Broer Mahn A, During K, Wachernagel W (1999) The bacteriolyitc activity in transgenic potatoes expressing a chimeric T4 lysozyme gene and the effect of T4 lysozyme on soil- and phytopathogenic bacteria Syst Appl Microbiol 22:280–286

Degens BP, Harris JA (1997) Development of a physiological approach to measuring the catabolic diversity of soil microbial communities. Soil Biol Biochem 29:1309–1320

Donegan KK, Seidler RJ, Doyle JD, Porteous LA (1999) A field study with genetically engineered alfalfa inoculated with recombinant *Sinorhizobium meliloti*: effects on the soil ecosystem. J Appl Ecol 36:920–936

Dunfield KE, Germida J (2001) Diversity of microbial communities in the rhizosphere and root interior of field-grown genetically modified *Brassica napus*. FEMS Microbiol Ecol 38:1–9

Dunfield KE, Germida J (2003) Seasonal changes in the rhizosphere microbial communities associated with field-grown genetically modified canola (Brassica napus) Appl Environ Microbiol 69:7310–7318

Espinosa-Urgel M, Ramos JL (2001) Expression of a Pseudomonas putida involved in lysine metabolism is induced in the rhizosphere. Appl Environ Microbiol 67:5219–5224

Falchini L, Naumova N, Kuikman PJ, Bloem J, Nannipieri P (2003) CO_2 evolution and denaturing gradient gel electrophoresis profiles of bacterial communities in soil following addition of low molecular weight substrates to simulate root exudation. Soil Biol Biochem 36:775–782

Farrar J, Haes D, Jones D, Lindow S (2003) How roots control the flux of carbon to the rhizosphere Ecology 84:827–837

Fontaine S, Mariotti A, Abbadie L (2003) The priming effect of organic matter: a question of microbial competition? Soil Biol Biochem 35:837–843

Foster RC, Rovira AD, Cock TW (1983) Ultrastructure of the root-soil interface, American Phytopathological Society, St. Paul, MN, USA

Garland JL, Mills AL (1991) Classification and characterisation of heterotrophic microbial communities on the basis of patterns of community-level-sole-carbon-source-utilization. Appl Environ Microbiol 57:2351–2359

George TS, Richardson AE, Hadobas PA, Simpson RJ (2004) Characterization of transgenic *Trifolium subterraneum* L which expresses *phyA* and release extracellular phytase: growth and P nutrition in laboratory media and soil. Plant Cell Environ 27:1351–1361

George TS, Richardson AE, Simpson RJ (2005) Behaviour of plant-derived extracellular phytase upon addition to soil. Soil Biol Biochem 37:977–988

Gianfreda L, Ruggiero P (2006) Enzyme activities in soil. In: Nannipieri P, Smalla K (eds) Nucleic acids and proteins in soil. Springer, Berlin Heidelberg New York, pp 257–311

Gomes NCM (2003) Dynamics of fungal communities in bulk and maize rhizosphere soil in the tropics. Appl Environ Microbiol 69:3758–3766

Gomes NCM, Heuer H, Schonfeld J, Costa R, Hagler-Mendoca L, Smalla K (2001) Bacterial diversity of the rhizosphere of maize (*Zea mays*) grown in tropical soil studied by temperature gradient gel electrophoresis. Plant Soil 233:167–180

Gomes NCM, Fagbola O, Costa R, Rumjanek NG, Buchner A, Mendonc L, Hagler A, Smalla K (2003) Dynamics of fungal communities in bulk and maize rhizosphere soil in the tropics. Appl Environ Microbiol 69:3758–3766

Göttlein A, Hell U, Blasek R (1996) A system for microscale tensiometry and lysimetry. Geoderma 69:147–156

Grayston SJ, Vaughan D, Jones D (1996) Rhizosphere carbon flow in trees, in comparison with annual plants: the importance of root exudation and its impact on microbial activity and nutrient availability. Appl Soil Ecol 5:29–56

Griffiths BS (1990) A comparison of microbial-feeding nematodes and protozoa in the rhizosphere of different plants. Biol Fertil Soils 9:83–88

Hamer U, Marschner B (2005) Priming effects in different soil types induced by fructose, alanine, oxalic acid and catechol additions. Soil Biol Biochem 37:445–454

Hamilton EW, Frank DA (2001) Can plants stimulate soil microbes and their own nutrient supply? Evidence from a grazing tolerant grass. Ecology 82:2397–2402

Hawes MC, Lin HJ (1990) Correlation of pectolytic enzyme activity with the programmed release of cells from root caps of pea (*Pisum sativum*). Plant Physiol 94:1855–1859

Hawes MC, Bengough G, Cassab G, Ponce G (2003) Root caps and rhizosphere. J Plant Growth Regul 21:352–367

Herrmann AM, Ritz K, Nunan N, Clode PL, Pett-Ridge J, Kilburn MR, Murphy DV, O'Donnell AG, Stockdale EA (2007) Nano-scale secondary ion mass spectrometry — a new analytical tool in biogeochemistry and soil ecology. Soil Biol Biochem 39:1835–1850

Heuer H, Smalla K (1997) Evaluation of community level catabolic profiling using BIOLOG GN microplates to study microbial community changes in potato phyllosphere. J Microbiol Methods 30:49–61

Heuer H, Hartung K, Wieland G, Kramer I, Smalla K (1999) Polynucleotide probes that target a hypervariable region of 16 S rRNA genes to identify bacterial isolates corresponding to bands of community fingerprints. Appl Environ Microbiol 65:1045–1049

Hinsinger P, Gilkes RJ (1997) Dissolution of phosphate rock in the rhizosphere of five plant species grown in an acid, P-fixing mineral substrate. Geoderma 75:231–249

Hirsch AM, Bauer WD, Bird DM, Cullimore J, Tyler B, Yoder JI (2003) Molecular signal and receptors: controlling rhizosphere interactions between plants and other organisms. Ecology 84:858–868

Hodge A, Robinson D, Fitter AH (2000) Are microorganisms more effective than plants at competing for nitrogen? Trends Plant Sci 5:304–308

Hyvönen R, Göran IÅ, Linder S, Persson TM, Cotrufo F, Ekblad A, Freeman M, Grelle A, Janssens JA, Jarvis PG, Kellomäki S, Lindroth A, Loustau D, Lundmark T, Norby RJ, Oren R, Pilegaard K, Ryan MG, Sigurdsson BD, Strömgren M, van Oijen M, Wallin G (2007) The likely impact of elevated [CO_2], nitrogen deposition, increased temperature and management on carbon sequestration in temperate and boreal forest ecosystems: a literature review. New Phytol 173:463–480

Idriss EI, Makarewicz O, Farouk A, Rosner K, Greiner R, Bochow H, Ritcher T, Borriss R (2002) Extracellular phytase activity of *Bacillus amyloliquefaciens* FZB45 contributes to its plant-growth-promoting effect. Microbiology 148:2097–2109

Iijima M, Griffiths B, Bengough AG (2000) Sloughing of cap cells and carbon exudation from maize seedlin roots in compacted sand. New Phytol 145:477–482

Jaeger CH III, Lindow SE, Miller W, Clark E, Firestone MK (1999) Mapping of sugar and amino acid availability in soil around roots with bacterial sensors of sucrose and tryptophan. Appl Environ Microbiol 65:2685–2690

Jasper MCM, Meier C, Zehnder AJB, Harms H, van der Meer JR (2001) Measuring mass transfer processes of octane with the help of an alkS-alkB gfp-tagged *Escherichia coli*. Environ Microbiol 3:512–524

Jenkinson DS, Fox RH, Rayner JH (1985) Interactions between fertilizer nitrogen and soil nitrogen – the so-called "priming" effect. J Soil Sci 36:425–444

Johnsen K, Jacobsen CS, Torsvik V, Sørensen J (2001) Pesticide effects on bacterial diversity in agricultural soils-a review Biol Fertil Soils 33:443–453

Johnson JF, Allan DL, Vance CP, Weiblen G (1996) Root carbon dioxide fixation by phosphorus deficient *Lupinus albus*. Contribution to organic acid exudation by proteoid roots. Plant Physiol 112:19–30

Jones DL, Darrah PR (1993) Re-sorption of organic compounds by roots of *Zea mays* L. and its consequences in the rhizosphere: I. Re-sorption of ^{14}C labelled glucose, mannose and citric acid. Plant Soil 153:47–59

Jones DL, Healey JR, Willett VB, Farrar JF, Hodge A (2005) Dissolved organic nitrogen uptake by plants-an important N uptake pathway? Soil Biol Biochem 37:413–423

Kamh M, Horst WJ, Amer F, Mostafa H, Maier P (1999) Mobilization of soil and fertilizer phosphate by cover crops. Plant Soil 211:19–27

Keister DL, Creagan PB (1991) The rhizosphere and plant growth. Kluwer Academic Publishers, Dordrecht.

Kim J, Verma SB (1992) Soil surface CO_2 flux in a flux in a Minnesota peatland. Biogeochem 18:37–51

Kourtev PS, Ehrenfed JG, Häggblom M (2003) Experimental analysis of the effect of exotic and native plant species on the structure and function of soil microbial communities. Soil Biol Biochem 35:895–905

Kowalchuk GA (1999) New perspectives towards analysing fungal communities in terrestrial environments. Curr Opin Biotechnol 10:247–251

Kowalchuk GA, Gerards S, Woldendorp JW (1997) Detection and characterization of fungal infections of *Ammophila arenaria* (marram grass) roots by denaturing gradient gel electrophoresis of specifically amplified 18 S rDNA. Appl Environ Microbiol 63:3858–3865

Kowalchuk GA, Bruinsma M, van Veen JA (2003) Assessing responses of soil microorganisms to GM plants. Trends Ecol Evol 18:403–410

Kozdroj J, van Elsas JD (2000) Response of the bacterial community to root exudates in soil polluted with heavy metals assessed by molecular and cultural approaches. Soil Biol Biochem 32:1405–1417

Kuchenbuch R, Jungk A (1982) Method for determining concentration profiles at the soil root interface by thin slicing rhizospheric soil. Plant Soil 68:391–394

Kuikman PJ, Jansen AG, vanVeen JA, Zehnder JB (1990) Protozoan predation and the turnover of soil organic carbon and nitrogen in the presence of plants. Biol Fertil Soils 10:22–28

Kuske CR, Lawrence OT, Mark EM, John MD, Jody AD, Susan MB, Jayne B (2002) Comparison of soil bacterial communities in rhizosphere of three plant species and the interspaces in an arid grassland. Appl Environ Microbiol 68:1854–1863

Kuzyakov Y (2002a) Review: factors affecting rhizosphere priming effects. J Plant Nutr Soil Sci 165:382–396

Kuzyakov Y (2002b) Separating microbial respiration of exudates from root respiration in a nonsterile soil: a comparison of four methods. Soil Biol Biochem 34:1621–1651

Kuzyakov Y, Cheng W (2001) Photosynthesis controls of rhizosphere respiration and organic matter decomposition. Soil Biol Biochem 33:1915–1925

Kuzyakov Y, Friedel JK, Stahr K (2000) Review of mechanisms and quantification of priming effects. Soil Biol Biochem 32:1485–1498

Ladd JN, Foster RC, Nannipieri P, Oades JM (1996) Soil structure and biological activity. In: Stotzky G, Bollag J-M (eds) Soil biochemistry, vol 9. Marcel Dekker, New York, pp 23–78

Landi L, Valori F, Ascher J, Renella G, Falchini L, Nannipieri P (2005) Root exudates effects on the bacterial communities, CO_2 evolution, nitrogen transformations and ATP content of rhizosphere and bulk soils. Soil Biol Biochem 38:509–516

Li X, George E, Marschner H (1991) Phosphorus depletion and pH decrease at the root–soil and hyphae soil interfaces of VA mycorhizal white clover fertilized with ammonium. New Phytol 119:397–404

Li Z, Yagi K (2004) Rice root-derived carbon input and its effect on decomposition of old soil carbon pool under elevated CO_2. Soil Biol Biochem 36:1967–1973

Liljeroth E, van Veen JA, Miller HJ (1990) Assimilate translocation to the rhizosphere of two wheat lines and subsequent utilization by rhizosphere microorganisms at two soil nitrogen concentrations. Soil Biol Biochem 2:1015–1021

Liljeroth E, Kuikamn PJ, van Veen JA (1994) Carbon translocation to the rhizosphere of maize and wheat and influence of the turnover of native soil organic matter at different soil nitrogen levels Plant Soil 161:233–240

Lung S-C, Chan W-L, Yip W, Wang L, Yeung EC, Lim BL (2005) Secretion of beta-propeller phytase from tobacco and *Arabidopsis* roots enhances phosphorus utilization. Plant Sci 169:341–349

Lynch JM (1990a) The rhizosphere. Wiley, New York

Lynch JM (1990b) Microbial metabolites. In: Lynch JM (ed) The rhizosphere. Academic Press, London, pp 177–206

Lynch JM, Whipps JM (1991) Substrate flow in the rhizosphere. In: Keister DL, Creagan PB (eds) The rhizosphere and plant growth. Kluwer Academic Publishers, Dordrecht, pp 15–45

Lynch JM, Benedetti A, Insam H, Nuti PM, Smalla K, Torsvik V, Nannipieri P (2004) Microbial diversity in soil: ecological theories, the contribution of molecular techniques and the impact of transgenic plants and transgenic microorganisms. Biol Fertil Soils 40:363–385

Marschner H (1995) Mineral nutrition of higher plants, 2nd edn. Academic Press, London

Marschner P, Yang C-H, Lieberei R, Crowley DE (2001) Soil plant specific effects on bacterial community composition in the rhizosphere. Soil Biol Biochem 33:1437–1445

Marschner P, Günter N, Angelika K, Laure W, Reinhard L (2002) Spatial and temporal dynamics of the microbial community structure in the rhizosphere of cluster roots of white lupin (*Lupinus albus* L.). Plant Soil 246:167–174

Marschner P, Grierson P, Rengel Z (2005) Microbial community composition and functioning in the rhizosphere of three species of Banksia species in native woodland in western Australia. Appl Soil Ecol 28:191–201

Mary B, Fresneau C, Morel L, Mariotti A (1993) C and N cycling during decomposition of root mucilage, roots and glucose in soil. Soil Biol Biochem 25:1005–1014

Maurhofer M, Baehler E, Notz R, Martinez V, Keel C (2004) Cross talk between 2,4-diacetylphloroglucinol- producing biocontrol pseudomonads on wheat roots. Appl Environ Microbiol 70:1990–1998

McCully M (1989) Cell separation: a developmental feature of root caps which may be of fundamental functional significance. In: Osborne DJ, Jackson MB (eds) Cell separation in plants. Springer, Berlin Heidelberg New York, pp 241–280

McCully M (1995) How do real roots work? Some new views of root structure. Plant Physiol 109:1–6

Merckx R, Dijkstra A, den Hartog A, van Veen JAA (1987) Production of root-derived material and associated microbial growth in soil at different nutrient levels. Biol Fertil Soils 5:126–132

Miethling R, Ahrends K, Tebbe CC (2003) Structural differences in the rhizosphere communities of legumes are not equally reflected in community-level physiological profiles Soil Biol Biochem 35:1405–1410

Morgan JAW, Whipps JM (2001) Methodological approaches to the study of rhizosphere carbon flow and microbial population dynamics. In: Pinton R, Varanini Z, Nannipieri P (eds) The rhizosphere. Biochemistry and organic substances at the soil-plant interface. Marcel Dekker, New York, pp 373–409

Muyzer G, Smalla K (1998) Application of denaturing gradient gel electrophoresis (DGGE) and temperature gradient gel electrophoresis (TGGE) in microbial ecology. Antonie Van Leeuwenhock J Microbiol Serol 73:127–141

Nannipieri P (1994) The potential use of enzymes as indicators of productivity, sustainability and pollution. In: Pankhurst CE, Doube BM, Gupta VVSR, Grace PR (eds) Soil biota – management in sustainable farming systems. CSIRO, East Melbourne Australia, pp 238–244

Nannipieri P, Grego S, Ceccanti B (1990) Ecological significance of the biological activity in soil. In: Bollag J-M, Stotzky G (eds) Soil biochemistry, vol 6. Marcel Dekker, New York, pp 293–355

Nannipieri P, Kandeler E, Ruggiero P (2002) Enzyme activities and microbiological and biochemical processes in soil. In: Burns RG, Dick RP (eds) Enzymes in the environment. Marcel Dekker, New York, pp 1–33

Nannipieri P, Ascher J, Ceccherini MT, Landi L, Pietramellara G, Renella G (2003) Microbial diversity and soil functions. Eur J Soil Sci 54:655–670

Naseby DC, Lynch JM (1997) Rhizosphere soil enzymes as indicators of perturbations caused by enzyme substrate addition and inoculation of a genetically modified strain of *Pseudomonas fluorescens* on wheat seed. Soil Biol Biochem 29:1353–1362

Naseby DC, Lynch JM (1998) Impact of wild-type and genetically modified *Pseudomonas fluorescens* on soil enzyme activities and microbial population structure in the rhizosphere of pea. Mol Ecol 7:617–625

Naseby DC, Lynch JM (2002) Enzymes and microorganisms in the rhizosphere. In: Burns RG, Dick RP (eds) Enzymes in the environment. Activity, ecology and applications. Marcel Dekker, New York, pp 109–123

Naseby DC, Moënne-Loccoz JP, O'Gara F, Lynch JM (1998) Soil enzyme activities in the rhizosphere of field-grown sugar beet inoculated with the biocontrol agent *Pseudomonas fluorescens* F113. Biol Fertil Soils 27:39–43

Neumann G, Römheld V (2001) The release of root exudates as affected by plant's physiological status. In: Pinton R, Varanini Z, Nannipieri P (eds) Marcel Dekker, New York, pp 41–93

Neumann G, Massonneau A, Martinoia E, Römheld V (1999) Physiological adaptations to phosphorus deficiency during proteoid root development in white lupin. Planta 208:373–382

Nicol GW, Glover LA, Prosser JI (2003) Spatial analysis of archaeal community structure in grassland soil. Appl Environ Microbiol 69:7420–7429

Niemi RM, Heiskanen I, Wallenus K, Lindström K (2001) Extraction and purification of DNA in rhizosphere soil samples for PCR-DGGE analysis of bacterial consortia. J Microbial Methods 45:155–165

Norton JM, Firestone MK (1996) N dynamics in the rhizosphere of *Pinus ponderosa* seedling. Soil Biol Biochem 28:351–362

Norvel WA, Cary EE (1992) Potential errors caused by roots in analyses of rhizosphere soil. Plant Soil 143:223–231

Nunan N, Daniell TJ, Singh BK, Papert A, McNicol JW, Prosser JI (2005) Links between plant and rhizoplane bacterial communities in grassland soils, characterized using molecular techniques. Appl Environ Microbiol 71:6784–6792

Phillips DA, Fox TC, King MD, Bhuvaneswari TV, Teuber LR (2004) Microbial products trigger amino acid exudation from plant roots. Plant Physiol 136:2887–2894

Pinay G, Barbera P, Carreras-Palou A, Fromin N, Sonié L, Couteaux MM, Roy J, Philippot L, Lensi R (2007) Impact of atmospheric CO_2 and plant life forms on soil microbial activities Soil Biol Biochem 39:33–42

Pinton R, Varanini Z, Nannipieri P (2001) The rhizosphere. Biochemistry and organic substances at the soil-plant interface. Marcel Dekker, New York

Pinton R, Varanini Z, Nannipieri P (2002) The rhizosphere. Biochemistry and organic substances at the soil-plant interface. CRC press, Boca Raton, Fl. Second edition

Prikryl Z, Vancura V (1980) Root exudates in plants. VI Wheat exudation as dependent on growth, concentration gradient of exudates and the presence of bacteria. Plant Soil 57:69–83

Qian JH, Doran JW, Walters DT (1997) Maize plant contributions to root zone available carbon and microbial transformations of nitrogen. Soil Biol Biochem 29:1451–1462

Raich JW, Mora G (2005) Estimating root plus rhizosphere contributions to soil respiration in annual cropland. Soil Sci Soc Am J 69:634–639

Renella G, Michel M, Landi L, Nannipieri P (2005) Microbial activity and hydrolase activities during decomposition of model root exudates released by a model root surface in Cd-contaminated soils. Soil Biol Biochem 37:133–139

Richardson AE, Hadobas PA, Hayes JE (2001) Extracellular secretion of *Aspergillus* phytase from *Arabidopsis* roots enables plants to obtain phosphorus from phytate. Plant J 25:641–649

Rochette P, Flanagan LB (1997) Quantifying rhizosphere respiration in a corn crop under field conditions. Soil Sci Soc Am J 61:466–474

Seeling B, Jungk A (1996) Utilization of organic phosphorus in calcium chloride extracts of soil by barley plants and hydrolysis and alkaline phosphatases Plant Soil 178:179–184

Sharma S, Radl V, Hai B, Kloos K, Fuka MM, Engel M, Schauss K, Schloter M (2007) Quantification of functional genes from procaryotes in soil by PCR. J Microbiol Methods 68:445–452

Sliwinski MK, Goodman RM (2004) Comparison of Crenarchael consortia inhabiting the rhizosphere of diverse terrestrial plants with those in bulk soil in native environments. Appl Environ Microbiol 70:1821–1826

Smalla K, Wieland G, Buchner A, Zock A, Parzy J, Kaiser S, Roskot N, Heuer H, Berg G (2001) Bulk and rhizosphere soil bacterial communities studied by denaturing gradient gel electrophoresis: plant-dependent enrichment and seasonal shifts revealed. Appl Environ Microbiol 67:4742–4751

Smit E, Leeflang P, Glandorf B, van Elsas JD, Wernars K (1999) Analysis of fungal diversity in the wheat rhizosphere by sequencing of cloned PCR-amplified genes encoding 18 S rRNA and temperature gradient gel electrophoresis. Appl Environ Microbiol 65:2614–2621

Söderberg KH, Bååth E (2004) The influence of nitrogen fertilisation on bacterial activity in the rhizosphere of barley. Soil Biol Biochem 36:195–198

Sorensen J (1997) The rhizosphere as a habitat for soil microorganisms. In: van Elsas JD, Trevors JT, Wellington EMH (eds) Modern soil microbiology. Marcel Dekker, New York, pp 21–45

Stark C, Condron LM, Stewart A, Di HJ, O'Callaghan M (2007) Influence of organic and mineral amendments on microbial soil properties and processes. Appl Soil Ecol 35:79–93

Talbot NJ (2003) Functional genomics of plant–pathogen interactions. New Phytol 159:1–10

Tarafdar JC, Jungk A (1987) Phosphatase activity in the rhizosphere and its relation to the depletion of soil organic phosphorus. Biol Fertil Soils 3:199–204

Tarafdar JC, Marschner H (1994) Phosphatase activity in the rhizosphere and hyposphere of VA mycorrhizal wheat supplied with inorganic and organic phosphorus Soil Biol Biochem 26:387–395

Tinker PB, Nye PH (2000) Solute movement in the rhizosphere. Oxford University Press, New York

Torsvik V, Øvreas L (2002) Microbial diversity and function in soil: from genes to ecosystems. Curr Opin Microbiol 5:240–245

Torsvik V, Sørheim R, Gorksøyr J (1996) Total bacterial diversity in soil and sediment communities-a review. J Ind Microbiol 17:170–178

Trofymow JA, Coleman DC, Cambardella C (1987) Rates of rhizodeposition and ammonium depletion in the rhizosphere of axenic oat roots. Plant Soil 97:333–344

Uren NC (2007) Types, amounts, and possible function of compounds released into the rhizosphere by soil-grown plants. In: Pinton R, Varanini Z, Nannipieri P (eds) The rhizosphere. Biochemistry and organic substances at the soil-plant interface. Marcel Dekker, New York (pp 1–21)

Uren NC, Reisenauer HM (1988) The role of root exudates in nutrient acquisition. Adv Plant Nutr 3:79–144

Valè M, Nguyen C, Dambrine E, Dupouey JL (2005) Microbial activity in the rhizosphere soil of six herbaceous species cultivated in a greenhouse is correlated with shoot biomass and root C concentrations. Soil Biol Biochem 37:2329–2333

Van der Krift TAJ, Kuikman PJ, Moller F, Berendse F (2001) Plant species and nutritional-mediated control over rhizodeposition and root decomposition. Plant Soil 228:191–200

Vandenkoornhuyse P, Baldauf SL, Leyval C, Straczek J, Young JP (2003) Extensive fungal diversity in plant roots. Science 295:2051

Waisel Y, Eshel A, Kafkafi U (1991) Plant roots. The hidden half. Marcel Dekker, New York

Warembourg FR, Roumet C, Lafont F (2003) Differences in rhizosphere carbon-partitioning among plant species of different families. Plant Soil 256:347–357

Wasaki J, Ando M, Ozawa K, Omura M, Osaki M, Ito H, Matsui H, Tadano T (1997) Properties of secretory acid phosphatase from lupin roots under phosphorus-deficient conditions. In: Ando T, Fujita K, Mae T, Matsumoto H, Mori S, Sekiya J (eds) Plant nutrition for sustainable food production and environment. Kluwer Academic Publishers, Dordrecht, The Netherlands, pp 295–300

Wenzel WW, Wieshammer G, Fitz WJ, Puschenreiter M (2001) Novel rhizobox design to assess rhizosphere characteristics at high spatial resolution. Plant Soil 237:37–45

Whipps JM (1990) Carbon economy. In: Lynch JM (ed) The rhizosphere. Wiley, Chichester, pp 59–97

Zagal E, Bjarnason S, Olsson UFL (1993) Carbon and nitrogen in the root-zone of barley (*Hordeum vulgare* L) supplied with nitrogen fertilizer at two rates. Plant Soil 157:51–63

Zelles L (1999) Fatty acid patterns of phospholipids and lipopolysaccharides in the characterization of microbial communities in soil: a review. Biol Fertil Soils 29:111–112

Part IV
Methods to Study Plant and Microbe Coexistence

Chapter 15
Siderotyping, a Straightforward Tool to Identify Soil and Plant-Related Pseudomonads

Jean-Marie Meyer(✉), Christelle Gruffaz, and Marion Fischer-LeSaux

15.1 Introduction

Siderotyping is a method recently developed to characterize bacterial strains by the siderophore(s) they produce when grown under iron deficiency. First applied to fluorescent pseudomonads and their main siderophores, the pyoverdines, the method was primarily used for the recognition of new molecules among pyoverdines. Because of the huge diversity of molecules encountered among this siderophore family, the method became rapidly a useful prerequisite for starting novel structure investigations. Close to 50 structures have been already established and a total of more than 110 structurally different compounds are presently recognized by siderotyping.

Interest for siderotyping considerably increased when it became evident that all strains belonging to a well defined *Pseudomonas* species produce an identical pyoverdine and, furthermore, that most species are characterized by specific pyoverdines. Therefore, beside their interest as powerful siderophores, pyoverdines are also potent taxonomic markers, opening a new and valuable way for bacterial identification and taxonomy within this major genus.

In the present chapter, the chemical as well as the physiological basis of the siderotyping methodology and details on the different methods used to validly differentiate pyoverdines are presented. Our present knowledge on siderophore diversity among *Pseudomonas*, with a particular focus on pyoverdine diversity among fluorescent *Pseudomonas*, as well as a brief overview on what is known on siderophores of non-fluorescent *Pseudomonas*, is summarized. Moreover, a brief analysis of the taxonomic methods presently in use for an efficient *Pseudomonas* identification and classification are developed for comparison purposes with siderotyping methods. As an example, strain clustering obtained by numerical taxonomy

J.-M. Meyer

Département Génétique Moléculaire, Génomique et Microbiologie, UMR 7156 Université Louis-Pasteur/CNRS, Strasbourg, France

e-mail: meyer@gem.u-strasbg.fr

C.S. Nautiyal, P. Dion (eds.) *Molecular Mechanisms of Plant and Microbe Coexistence.* Soil Biology 15, DOI: 10.1007/978-3-540-75575-3

and illustrated by a dendrogram of phenotypic distances of 85 type-strains and phytopathogen *Pseudomonas* is compared with clusters reached by siderotyping of the same collection. The numerous advantages of the siderotyping method, but also its limits, will be discussed.

15.2   Soil- and Plant-related Pseudomonads: A World within the Microbial World

The genus *Pseudomonas* is widely distributed in nature. These bacteria rank among the major bacterial population in soil and natural water samples, representing very often 2–10% or more of soil isolates as obtained by colony counting (Janssen 2006), whereas close to half of the natural isolates present in mineral waters are pseudomonads (Guillot and Leclerc 1993). Thanks to not yet fully understood attraction mechanisms (Espinosa-Urgel et al. 2002), *Pseudomonas* are also well distributed in plant root environments, thus contributing to a large proportion of the plant-related microbial population. Moreover, many of them demonstrate properties of biotechnological values: *Pseudomonas* isolates are used as biocontrol agents, able to lower or suppress plant diseases of fungal origin (Lemanceau and Alabouvette 1993) thanks to various mechanisms, among them the production of siderophores (Kloepper et al. 1980a,b) or antibiotics (Haas and Defago 2005). Others are successfully competing with saprophytic fungi at the plant rhizosphere level and, therefore, can be used as stimulating agents resulting in crop yield increase. Indeed, thanks to their high metabolic versatility, many pseudomonads have been successfully used in bioremediation of chemicals like nitrates or pesticides, including the degradation of toxic organic compounds such as carbon tetrachloride (Lee et al. 1999).

Pathogenicity to plants and mushrooms is also a trait of interest of many *Pseudomonas* species: 23 species are presently listed by the Taxonomy Committee of the International Society of Plant Pathology. Moreover, some of these species encompass many pathovars: *P. syringae* for instance includes more than 55 pathovars. A huge host range of plant species is attacked by *Pseudomonas* strains with a great variety of symptoms (necrosis, cankers, tumors, maceration). Thus this genus is considered under temperate climates as the major group of phytopathogenic bacteria. Species of the *Pseudomonas syringae* group are known to be epiphytic bacteria whereas most other phytopathogenic *Pseudomonas*, including *P. corrugata*, *P. marginalis,* and *P. tolaasii,* are soil inhabitants.

The multitude of valuable characteristics of *Pseudomonas* has inspired many studies resulting in the isolation of collections of natural isolates from various environments and including in most cases the characterization and identification of the bacterial isolates of interest. However, although the concomitant use of many different phenotypic and genotypic methods (see below for details), the identification at the species level of the numerous isolates worked out in such studies usually fails dramatically. The general conclusion reached after much effort and investment is at the best that a high genetic polymorphism exists within such collections, but without being able to specify in detail the different bacterial species causing that diversity. This is particularly frustrating when studies reveal specific sub-populations presenting valuable particular features and which would indeed be of interest to characterize precisely.

Such difficulty in determining species affiliation is due in part to the lack of precision of taxonomical methods presently in use and also to the great diversity encountered among pseudomonads which, moreover, often has no standing in a nomenclatural frame. Although 16S rDNA sequencing has clarified the taxonomical position of pseudomonads and limited the number of *Pseudomonas* species to those belonging to the DNA-RNA hybridization group I of Palleroni (1984) (Kersters et al. 1996; Anzai et al. 2000), taking out from the *Pseudomonas* sensu lato listing more than 60 species, the number of sensu stricto species presently recognized is still high with 55 species identified as fluorescent *Pseudomonas* and 53 belonging to the non-fluorescent species (personal compilation of the authors). The total number of 108 species should, moreover, increase considerably in the future. Many species delineated at the early times of phenotypic taxonomy, e.g., *P. fluorescens*, *P. putida*, *P. syringae*, and *P. stutzeri*, have since proved to be very heterogeneous at the genomic level: *P. stutzeri* has recently been split into 18 genomospecies based on DNA-DNA hybridization (Sikorski et al. 2005), while the numerous pathovars of *P. syringae* were separated based on the same criteria into 9 genomovars (Gardan et al. 1999). Moreover, it is well established that *P. fluorescens* and *P. putida* are very heterogeneous species at the phenotypic level as suggested by the recognition of 5 biovars (I–V) within the *P. fluorescens* species and three biovars (A–C) within the *P. putida* species (Palleroni 1984), and also at the genomic level (Hilario et al. 2004). According to the siderotyping method described below, 28 strains belonging to the biovar I of the *P. fluorescens* species are dispatched among 10 siderovars (Meyer et al. 2002), while 144 *P. putida* isolates can be divided into 35 siderovars (Meyer et al., in preparation). The general rule being that one siderovar corresponds usually to one species (Meyer et al. 2002), and even if some already published species are now recognized as junior synonyms (Cladera et al. 2006; Lang et al. 2007), we could easily expect the recognition of close to 200 *Pseudomonas* species in the near future. In such a perspective, siderotyping as a simple and powerful method for strain differentiation, identification and grouping, will be of great interest compared to the taxonomic methods presently in use.

15.3 Conventional Tools for Pseudomonad Characterization and Identification

15.3.1 Phenotypic Tools

Conventional bacteriological tests used to characterize *Pseudomonas* strains include cellular morphology and flagella typing, Gram staining, glucose metabolism, presence of cytochrome C oxidase. Other phenotypic characters of particular interest in this genus are:

- *Accumulation of Endocellular Granules of Poly-Beta-Hydroxybutyrate (PHB).* Strains of *Pseudomonas* sensu stricto do not accumulate PHB, at the opposite of former pseudomonads which were afterward reclassified in other genera. *P. corrugata* was thought to be an exception, but it was later on demonstrated that this species does not accumulate PHB but medium-chain-length poly-hydroxyalkanoates (Kessler and Palleroni 2000).
- *Production of Specific Pigments.* The best known and first studied is pyocyanin, a phenazine blue pigment that gives its typical blue color to the pus produced in some *P. aeruginosa* infections. Pyocyanin production on King's A medium (King et al. 1954) is a key character in identification of *P. aeruginosa*. Since then, other pigments were characterized and are used in species or biovar identification, among them lemonnierin produced by strains of biovar IV of *P. fluorescens* (Starr et al. 1967) and chlororaphine produced by *P. chlororaphis* (Breed et al. 1957). The most common studied pigment remains pyoverdine, the green fluorescent siderophore produced by fluorescent *Pseudomonas* grown under iron-deficiency, usually detected thanks to the King's B medium (King et al. 1954; Meyer 2000).

To identify plant pathogenic *Pseudomonas*, bacteriologists rely on a combination of five phenotypic tests proposed by Lelliott et al. (1966). This identification key is called LOPAT for *L*evane production from sucrose, presence of cytochrome C *O*xidase, *P*ectinase, *A*rginine dihydrolase and hypersensitive reaction on *T*obacco leaves. It allows one to define five groups: groups I and II correspond to oxidase negative phytopathogenic species (*P. syringae* and related species, and *P. viridiflava*, respectively), groups III and IV to oxidase positive phytopathogenic species (*P. cichorii* and *P. marginalis*, respectively) and group V corresponds to *P. fluorescens* and other saprophytic strains. This determinative key is still very useful but is insufficient and suffers of the failings that characterize identification schemes based on few characters. To identify strains at the pathovar level, phenotypic tests are completed by pathogenicity tests in order to determine host range and symptoms.

Extensive phenotypic studies including 146 nutritional tests (Stanier et al. 1966) demonstrated the extreme nutritional versatility of pseudomonads and allowed their differentiation at the species level with the recognition of biovars for the most heterogeneous *P. fluorescens* and *P. putida* species. Numerical analysis of these data by Sneath et al. (1981) confirmed the discriminative capacity of this approach. Since then, auxanograms were miniaturized and different commercial kits are now available. Assimilation of carbon compounds of three different chemical families could be studied with strips commercialized by BioMérieux: API 50CH (carbohydrates), API 50AA (amino acids) and API 50AO (organic acids). These API systems were replaced by Biotype 100 strips (BioMérieux) which is designed to test carbon assimilation from 99 different sources. The Biolog GN MicroPlate System (Biolog Inc.) allows one to test oxidation of 95 substrates. These kits are very useful for numerical taxonomic analysis and have been used in polyphasic approach to identify discriminative characters for the description of new species (Grimont et al. 1996; Gardan et al. 2002). Analysis of fatty acid methyl esters of whole cells by high resolution gas chromatography is used in the Microbial Identification System (MIDI, Microbial ID Inc.). The main disadvantage of these kits, beside their cost, is the maintenance of up-to-date databases. Therefore, while descriptions of new species increase, identification scores may decrease.

Methods used for epidemiological purposes involve serotyping, production and sensibility to phages, antibiograms, whole cell protein fingerprints. Most of the time, such methods have been used to characterize species of clinical interest like *P. aeruginosa* (Palleroni 2005).

15.3.2 Genotypic Tools

15.3.2.1 DNA-DNA Hybridization

Since 1987, DNA-DNA hybridization has been the reference method for species delineation (Wayne et al. 1987). First results evidenced very low genomic relatedness within pseudomonads, a result confirmed by rRNA-DNA hybridization (Palleroni et al. 1973) and subsequent affiliation of several species to different classes of *Proteobacteria* (Anzai et al. 2000). DNA-DNA hybridization also showed that some historic species were constituted of several genomospecies (see above). Nowadays, this method – which consists of pair wise comparison of whole genome to determine DNA relatedness – is systematically used in polyphasic taxonomic studies to define new species in the genus. Such defined species represent groups of strains sharing more than 70% of DNA-DNA homology, clearly separated from neighbouring species by lower values. DNA-DNA hybridization, which is cumbersome and time consuming, is not per se an identification tool, but it allows one to delineate species as genetically homogeneous groups for which molecular identification tools can easily be defined.

15.3.2.2 Sequencing of Conserved Genes

Based on *rrs* gene sequences, phylogenetic relationships between *Pseudomonas* species were elucidated (Moore et al. 1996; Anzai et al. 2000). Sequencing of *rrs* gene and comparison with international databases is a convenient way to achieve isolate identification. However, effective identity of *rrs* sequence is not necessarily a sufficient criterion to guarantee species identity because of the highly conservative nature of the *rrs* gene. To overcome this problem and to refine phylogenies, less conserved housekeeping genes were studied. Yamamoto et al. (2000) sequenced *gyrB* and *rpoD* genes. Discrepancies with *rrs* phylogeny were evidenced but the resolution level was correlated with DNA-DNA hybridization data. Recently, Ait Tayeb et al. (2005) showed that the resolution power of *rpoB* tree was three times higher than those of *rrs* and that partial sequence of *rpoB* was a good identification marker. There is no doubt that multilocus sequence analysis of housekeeping genes will play a major role in future taxonomic studies and identification schemes, as already shown in a recent study involving 10 housekeeping genes (Frapolli et al. 2007).

15.3.2.3 DNA Fingerprinting

Taxonomic studies within the *Pseudomonas* genus took advantages of the numerous molecular methods developed to investigate genomic polymorphism. Methods based on PCR amplification of whole genome used random priming (RAPD) or primers in repetitive elements (rep-PCR). Others used restriction of the genome (AFLP, PFGE, ribotyping). Among these, ribotyping was developed as an identification tool, with commercialization of an automate (Riboprinter, Qualicon) and development of a database called Taxotron by the Institut Pasteur (Paris, France). Other DNA finger-printing methods were mainly used to investigate genetic diversity at sub-specific level (Louws et al. 1994; Clerc et al. 1998). In conclusion, characterization and identification of *Pseudomonas* species followed the general evolution of methods used for bacterial identification. Despite the splitting of *Pseudomonas* sensu lato and restriction of *Pseudomonas* genus to species belonging to the DNA-RNA hybridiza-tion group I, this genus remains extremely heterogeneous. Identification of species – the majority of which are saprophytic – has for a long time been an awkward task and limits of classical tests for diagnosis were reached. Extensive improvement of characterization and identification of pseudomonads was achieved with the develop-ment of molecular methods based on genotypic data like *rrs* gene sequencing. However, at present, there is still a need for routine techniques that allow one to resolve easily the *Pseudomonas* diversity at the species level.

15.4 Siderotyping, or How to Identify Pseudomonads Through a Unique Phenotypic Character

One phenotypic character easy to observe, and quite important since it concerns about half of the *Pseudomonas* sensu stricto species, is the production of pyoverdine, the water-soluble yellow-green and fluorescent pigment previously called fluorescein, produced by the so-called fluorescent *Pseudomonas* (Elliot 1958). This pigment is observable when growing the bacteria on King's B medium where its production results in the appearance of a bright fluorescence in the medium (King et al. 1954). Only a few other genera, e.g., *Azotobacter* spp. or *Azospirillum* spp., are able to produce similar compounds. However, very specific features like nitrogen fixation allow an easy discrimination of these bacteria from the fluorescent pseudomonads.

Pyoverdines are chromopeptides made of a quinolein-based chromophore con-ferring to the molecule its color and bright fluorescence and, branched to it, a small peptide as well as a dicarboxylic acid side chain (see Fig. 15.1). So far, 50 different

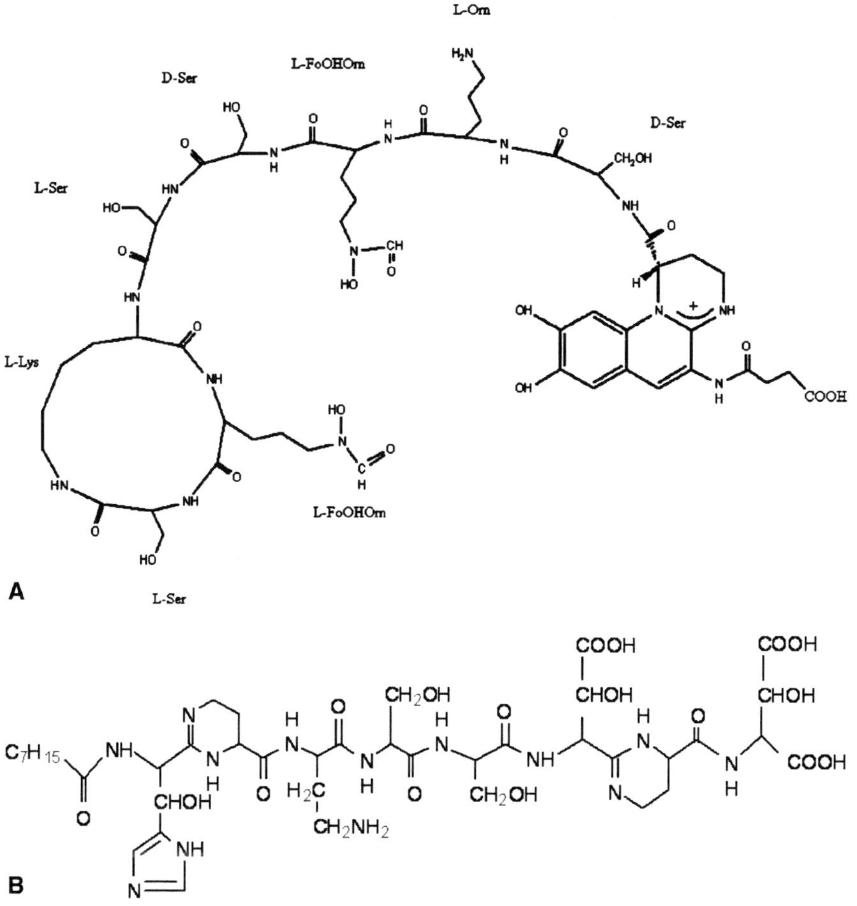

Fig. 15.1 Structures of *Pseudomonas* siderophores. **A** Structure of a pyoverdine. **B** Structure of corrugatin

pyoverdines have been structurally determined. That huge diversity results mainly from variations at the peptide chain level which vary by the number and composition of the aminoacyl residues (for details the reader is referred to the recent exhaustive compilation of Budzikiewicz 2004). It is important to note that pyoverdine structural features are strain specific, which means that one iron-starved strain, whatever the other growth conditions, always produces pyoverdine molecules with the same peptide chain (a very few exceptions have been claimed in the literature; see Budzikiewicz 2004). However, different pyoverdine molecules can be found in a culture supernatant, differing by the nature of the carboxylic acid side chain (succinic or malic acid or their amides, α-ketoglutaric or fumaric acid, for the most frequent ones). Some of these so-called pyoverdine isoforms are produced by iron-starved cells, some others, however, are resulting from chemical modifications occurring during growth, e.g., the formation of the succinyl isoform resulting from the hydrolysis of the biosynthetised succinamide isoform.

Thus, applying a method which could separate the different isoforms like isoelectrophoresis (electrophoresis in presence of ampholins which determine a pH gradient in the gel), each fluorescent *Pseudomonas* strain producing a particular pyoverdine could be identified through its pyoverdine-isoelectrofocusing (PVD-IEF) pattern, depending on the amino acid content of its pyoverdine and also depending on the number of pyoverdine isoforms present in its culture supernatant. An example is given in Fig. 15.2 which illustrates the PVD-IEF patterns obtained with *P. salomonii, P. palleroniana, P. tolaasii, P. costantinii, P. fuscovaginae* and *P. syringae.*

The experimental procedure to reach such patterns is very simple: 1 mL of a 24-h culture at 25 °C in CAA medium (a Casamino acid-based medium with low iron content, (see Meyer et al. 2002 for detailed formula) is centrifuged and 400 μL of the clear supernatant are lyophilized. The dry residue is dissolved in 20 μL of water and 1 μL solution is deposited on the ampholin-containing polyacrylamide gel for the mini-IEF isoelectrophoresis procedure developed according to the manufacturer recommendations (Biorad). The natural fluorescence of pyoverdines under UV light (350 nm) is used for the revelation of the bands. As experimented in our laboratory, one person using two Mini-IEF gel apparatus can run up to 10 gels, thus determining the PVD-IEF patterns of 140 strains within a day.

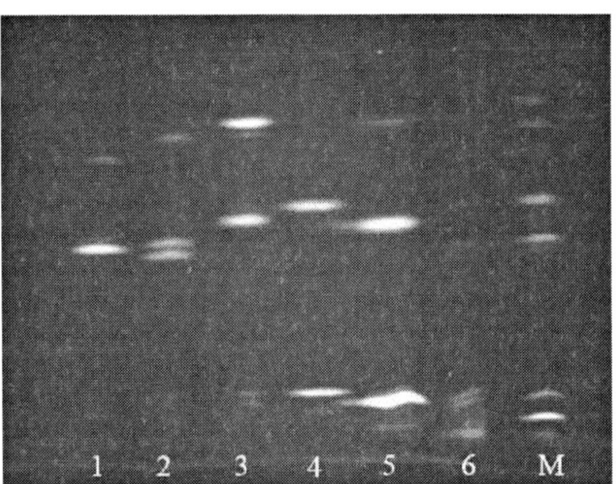

Fig. 15.2 Isoelectrophoretic patterns of some fluorescent *Pseudomonas* pyoverdines. Lane 1, *Pseudomonas salomonii* CFBP 2022[T]; lane 2, *Pseudomonas palleroniana* CFBP 4389[T]; lane 3, *Pseudomonas tolaasii* CFBP 2068[T]; lane 4, *Pseudomonas costantinii* CFBP 5705[T]; lane 5, *Pseudomonas fuscovaginae* CFBP 2065[T]; lane 6, *Pseudomonas syringae* CFBP 1392[T]; lane M corresponds to standard pyoverdine markers for pHi measurements (see Meyer et al. 2002). Abbreviation: CFBP, Collection Française de Bactéries Phytopathogènes (INRA-Angers, France)

The next step of the IEF-siderotyping procedure is then to group strains presenting the same PVD-IEF pattern as seen by visual comparison. A control is done by co-migrating on a same gel pyoverdines of a same IEF group. A second control is then reached by using another siderotyping procedure, based on the usually high specificity of recognition between ferri-pyoverdines and their respective outer membrane receptors (Hohnadel and Meyer 1988). It is thus controlled that each strain of the IEF group is able to use as iron transporter any of the pyoverdines produced by strains belonging to the same group. Although some pyoverdines have been seen to share very closely related, if not identical, IEF patterns, this second control is usually very discriminative and has allowed so far the discrimination of more than 110 different pyoverdines characterizing as many siderovars, i.e., groups of strains sharing an identical pyoverdine.

One major conclusion reached by siderotyping is that pyoverdine molecules could be used as powerful taxonomic markers. It effectively became evident, once well polyphasic defined species were available to siderotyping analysis, that, as a general rule, strains belonging to one species produce an identical pyoverdine while strains belonging to different species produce structurally different pyoverdines. This has been well established for several well-circumscribed species of fluorescent *Pseudomonas*, i.e., *Pseudomonas monteilii*, *Pseudomonas rhodesiae*, *Pseudomonas mandelii*, *Pseudomonas veronii*, *Pseudomonas tolaasii*, and *Pseudomonas syringae* (Meyer et al. 2002). Moreover, the method successfully contributed to the definition of recently described new species, namely *Pseudomonas brassicacearum* and *Pseudomonas thivervalensis* (Achouak et al. 2000), *Pseudomonas lini* (Delorme et al. 2002), *Pseudomonas mosselii* (Dabboussi et al. 2002), *Pseudomonas salomonii* and *Pseudomonas palleroniana* (Gardan et al. 2002), *Pseudomonas costantinii* (Munsch et al. 2002), and *Pseudomonas lurida* (Behrendt et al. 2007). Furthermore, the assignation, as postulated by siderotyping, of a phenotypic cluster to a given species, was positively verified by DNA-DNA-hybridization (Meyer et al. 2002), proving that the method was particularly efficient for the detection of potential new species. Interestingly, the method was successfully extended to the non-fluorescent species *Pseudomonas corrugata*, *Pseudomonas fredericksbergensis*, *Pseudomonas graminis* and *Pseudomonas plecoglossicida*. Such bacteria do not synthesize pyoverdines as siderophores but other structurally different compounds sharing in common the ability to tightly bind iron(III) and to transport it into the cells thanks to specific outer membrane receptors. These siderophores and their respective iron transport systems are still unknown for a majority of them. Of the four species cited above, only corrugatin, the siderophore of *P. corrugata*, has been identified at the structure level (Risse et al. 1998; Fig. 15.1). The three others, each defined by a specific pHi value (Meyer et al. 2002), remain to be characterized, as well as the ferri-siderophore receptors of the four species. The IEF-analysis procedure for siderophores produced by such non-fluorescent pseudomonads is identical to the one described above for pyoverdines, except that the revelation of siderophores is done by the CAS-overlay method as described by Koedam et al. (1994).

15.5 An Application Within Plant-Pathogen Pseudomonads: Correlation Between Siderotyping and Numerical Taxonomy

A collection of 85 phytopathogenic *Pseudomonas* strains were analyzed through siderotyping and numerical taxonomy. The resulting groups reached by the two methods are compared in Fig. 15.3 as illustrated by PVD-IEF patterns and by a

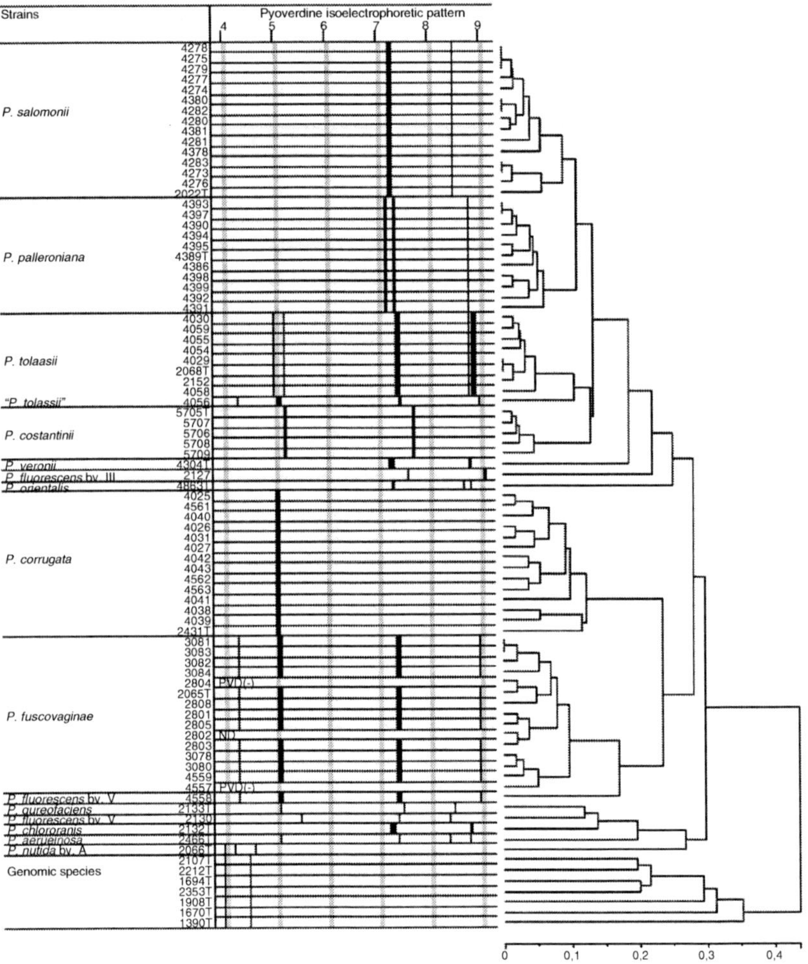

Fig. 15.3 Correlation between siderotyping and numerical taxonomy. The pyoverdine-isoelectrophoretic patterns of 85 *Pseudomonas* strains are shown in the *middle panel* by *bars* representing the different pyoverdine isoforms found in the respective CAA culture supernatants. The *thickness of the bars* reflects the intensity of fluorescence of the pyoverdine-isoform bands, as visualized under UV light at 350 nm after electrophoresis. Phenotypic dendrogram is depicted in the *right panel*. For details on pyoverdine isoelectrophoresis and on the construction of the dendrogram, see Meyer et al. (2002) and Gardan et al. (2002), respectively

dendrogram representing phenotypic distances between strains. Six phenotypic clusters can easily be distinguished at a phenotypic distance of 0.11 in the dendrogram, corresponding to the species *P. salomonii, P. palleroniana, P. tolaasii, P. costantinii* and *P. fuscovaginae*. It is evident that these groups match perfectly with the grouping reached by siderotyping. This conclusion is also valid for the non-fluorescent *P. corrugata* species. Some discrepancies, however, concerns *P. syringae* strains which were characterized by a unique PVD-IEF pattern while demonstrating some phenotypic heterogeneity (phenotypic distance of 0.35) which is in agreement with the genomovar multiplicity found in this group (Gardan et al. 1999). It is also evident that unclustered strains with high phenotypic distances, e.g., strains identified in Fig. 15.3 as *P. veronii, P. fluorescens* bv. III, *P. orientalis, P. fluorescens* bv. V, *P. aureofaciens, P. chlororaphis, P. putida* bv. A and *P. aeruginosa*, are also characterized by as many specific PVD-IEF patterns, thus confirming the strong correlation between siderovars and phenotypic clusters. Interestingly, among the *P. tolaasii* isolates, strain CFBP 4056 demonstrated a surprisingly high phenotypic distance when compared to the other strains of the group. According to its PVD-IEF pattern, it is evident that the strain is related to the *P. fuscovaginae* siderotype, suggesting that its present taxonomic position as a *P. tolaasii* isolate needs to be revised.

A difficulty encountered in siderotyping concerns isolates that are deficient in siderophore production and which cannot be characterized, indeed, by a siderophore-IEF profile. This problem can be overcome by developing siderophore-mediated (^{59}Fe) iron uptake experiments, as was done for the pyoverdine-deficient *P. fuscovaginae* strains. These strains were grouped within the *P. fuscovaginae* siderovar based on their capacity to specifically use the *P. fuscovaginae* pyoverdine as iron transporter. This grouping was consistent with the dendrogram which shows that these strains are closely related within the species.

15.6 Conclusions

By its simplicity and rapidity of execution, siderotyping is a particularly promising method for the characterization and identification at the species level of fluorescent and non-fluorescent *Pseudomonas*. As illustrated here, the method could advantageously replace a numerical analysis of phenotypic data, thus saving time and money while results obtained by the two methods are in most cases in full agreement.

We are presently developing the use of the method in studies relevant to different ecological topics which are usually investigated through conventional methods such as the study of survival capacity of biocontrol strains (Molina et al. 1998; Nautiyal et al. 2002), the search and characterization of strains with specific biological features (Landa et al. 2002), the influence of soil factors and agricultural practices on the diversity and distribution of pseudomonads in soils (Lemanceau et al. 1995; Frey et al. 1997; Aagot et al. 2001; Kwon et al. 2005), or the selection of specific pseudomonad populations (Ross et al. 2000; Founoune et al. 2002).

References

Aagot N, Nybroe O, Nielsen P, Johnsen K (2001) An altered *Pseudomonas* diversity is recovered from soil by using nutrient-poor *Pseudomonas*-selective soil extract media. Appl Environ Microbiol 67:5233–5239

Achouak W, Sutra L , Heulin T, Meyer JM, Fromin N, Degreave S, Christen R, Gardan L (2000) Description of *Pseudomonas brassicacearum* sp. nov. and *Pseudomonas thivervalensis* sp. nov., root-associated bacteria isolated from *Arabidopsis thaliana* and *Brassica napus*. Int J Syst Evol Microbiol 50:9–18

Ait Tayeb L, Ageron E, Grimont F, Grimont PAD (2005) Molecular phylogeny of the genus *Pseudomonas* based on *rpoB* sequences and application for the identification of isolates. Res Microbiol 156:763–773

Anzai Y, Kim H, Park JY, Wakabayashi H, Oyaizu H (2000) Phylogenetic affiliation of the pseudomonads based on 16 S rRNA sequence. Int J Syst Evol Microbiol 50:1563–1589

Behrendt U, Ulrich A, Schumann P, Meyer JM, Spröer C (2007) *Pseudomonas lurida* sp. nov., a fluorescent species associated with the phyllosphere of grasses. Int J Syst Evol Microbiol 57:979–985

Breed RS, Murray EGD, Smith NR (1957) Bergey's manual of determinative bacteriology, 7th edn. The Williams and Wilkins Co. Baltimore, p 103

Budzikiewicz H (2004) Siderophores of the Pseudomonadaceae sensu stricto (fluorescent and non-fluorescent *Pseudomonas* spp.) Prog Chem Org Nat Prod 87:81–235

Cladera AM, Garcia-Valdes E, Lalucat J (2006) Genotype versus phenotype in the circumscription of bacterial species: the case of *Pseudomonas stutzeri* and *Pseudomonas chloritidismutans*. Arch Microbiol 184:353–361

Clerc A, Manceau C, Nesme X (1998) Comparison of randomly amplified polymorphic DNA with amplified fragment length polymorphism to assess genetic diversity and genetic relatedness within genospecies III of *Pseudomonas syringae*. Appl Environ Microbiol 64:1180–1187

Dabboussi F, Hamzé M, Singer E, Geoffroy V, Meyer JM, Izard D (2002) *Pseudomonas mosselii* sp. nov., a new species isolated from clinical specimens. Int J Syst Evol Microbiol 52:363–376

Delorme S, Lemanceau P, Christen R, Corberand T, Meyer JM, Gardan L (2002) *Pseudomonas lini* sp. nov., a novel species from bulk and rhizospheric soils. Int J Syst Evol Microbiol 52:513–523

Elliot RP (1958) Some properties of pyoverdine, the water-soluble pigment of the *Pseudomonas*. Appl Microbiol 6:241–246

Espinosa-Urgel M, Salido A, Ramos JL (2002) Genetic analysis of functions involved in adhesion of *Pseudomonas putida* to seeds. J Bacteriol 182:2363–2369

Founoune H, Duponnois R, Meyer JM, Thioulouse J, Masse D, Chotte JL, Neyra M (2002) Interactions between ectomycorrhizal symbiosis and fluorescent pseudomonads on *Acacia holosericea*: isolation of Mycorrhiza Helper Bacteria (MHB) from a soudano-sahelian soil. FEMS Microbiol Ecol 41:37–46

Frapolli M, Défago G, Moënne-Loccoz Y (2007) Multilocus sequence analysis of biocontrol fluorescent *Pseudomonas* spp. producing the antifungal compound 2,4-diacetylphloroglucinol. Env Microbiol (in press)

Frey P, Frey-Klett P, Garbaye J, Berge O, Heulin T (1997) Metabolic and genotypic fingerprinting of fluorescent pseudomonads associated with the Douglas Fir-*Laccaria bicolor* myccorrhizosphere. Appl Environ Microbiol 63:1852–1860

Gardan L, Shafik H, Belouin S, Grimont F, Grimont PAD (1999) DNA relatedness among the pathovars of *Pseudomonas syringae* and description of *Pseudomonas tremae* sp. nov. and *Pseudomonas cannabina* sp. nov. (ex Sutic and Dowson 1959). Int J Syst Bacteriol 49:469–478

Gardan L, Bella P, Meyer JM, Christen R, Rott P, Achouak W, Samson R (2002) *Pseudomonas salomonii* sp. nov. pathogenic on garlic, and *Pseudomonas palleroniana* sp. nov., isolated from rice. Int J Syst Evol Microbiol 52:2065–2074

Grimont PAD, Vancanneyt M, Lefevre M, Vandemeulebroecke K, Vauterin L, Brosch R, Kersters K, Grimont F (1996) Ability of Biolog and Biotype-100 systems to reveal the taxonomic diversity of the pseudomonads. Syst Appl Microbiol 19:510–527

Guillot E, Leclerc H (1993) Bacterial flora in natural mineral waters: characterization by ribosomal ribonucleic acid gene restriction patterns. Syst Appl Microbiol 16:483–493

Haas D, Défago G (2005) Biological control of soil-borne pathogens by fluorescent pseudomonads. Nat Rev Microbiol 3:307–319

Hilario E, Buckley TR, Young JM (2004) Improved resolution on the phylogenic relationship by the combined analysis of *atp*D, *car*A, *rec*A and 16SrDNA. Antonie van Leeuwenhoek 86:51–64

Hohnadel D, Meyer JM (1988) Specificity of pyoverdine-mediated iron uptake among fluorescent *Pseudomonas* strains. J Bacteriol 170:4865–4873

Janssen PH (2006) Identifying the dominant soil bacterial taxa in libraries of 16s rRNA and 16S rRNA genes. Appl Environ Microbiol 72:1719–1728

Kersters K, Ludwig W, Vancanneyt M, De Vos P, Gillis M, Schleifer KH (1996) Recent changes in the classification of the pseudomonads: an overview. Syst Appl Microbiol 19:465–477

Kessler B, Palleroni NJ (2000) Taxonomic implications of synthesis of poly-beta-hydroxybutyrate and other poly-beta-hydroxyalkanoates by aerobic pseudomonads Int J Syst Evol Microbiol 50:711–713

King EO, Ward MK, Raney DF (1954) Two simple media for the demonstration of pyocyanin and fluorescein. J Lab Clin Med 44:301–307

Kloepper JW, Leong J, Teintze M, Schroth MN (1980a) Enhanced plant growth by siderophores produced by plant growth promoting rhizobacteria. Nature 286:885–886

Kloepper JW, Leong J, Teintze M, Schroth MN (1980b) *Pseudomonas* siderophores: a mechanism explaining disease suppressive soils. Curr Microbiol 4:317–320

Koedam N, Wittouck E, Gaballa A, Gillis A, Höfte M, Cornelis P (1994) Detection and differenciation of microbial siderophores by isoelectric focusing and chrome azurol S overlay. BioMetals 7:287–291

Kwon SW, Kim JS, Crowley DE, Lim CK (2005) Phylogenetic diversity of fluorescent pseudomonads in agricultural soils from Korea. Lett Appl Microbiol 41:417–423

Landa BB, Mavrodi OV, Raaijmakers JM, NcSpadden Gardener BB, Thomashow LS, Weller DM (2002) Differential ability of genotypes of 2,4-diacetylphloroglucinol-producing *Pseudomonas fluorescens* strains to colonize the roots of pea plants. Appl Environ Microbiol 68:3226–3237

Lang E, Griese B, Spröer C, Schumann P, Steffen M, Verbarg S (2007) Characterization of 'Pseudomonas azelaica' DSM 9128, leading to emended descriptions of *Pseudomonas citronellolis* Seubert 1960 (Approved Lists 1980) and *Pseudomonas nitroreducens* Iizuka and Komagata 1964 (Approved Lists 1980), including *Pseudomonas multiresinivorans* as its later heterotypic synonym. Int J Syst Evol Microbiol 57:878–882

Lee CH, Lewis TA, Paszczynski A, Crawford RL (1999) Identification of an extracellular catalyst of carbon tetrachloride dehalogenation from *Pseudomonas stutzeri* strain KC as pyridine-2, 6-bis(thiocarboxylate). Biochem Biophys Res Commun 261:562–566; erratum 265:770

Lelliott RA, Billing E, Hayward AC (1966) A determinative scheme for the fluorescent plant pathogenic pseudomonads. J Appl Bacteriol 29:470–489

Lemanceau P, Alabouvette C (1993) Biological control of *Fusarium* diseases by fluorescent *Pseudomonas* and non-pathogenic *Fusarium*. Crop Prot 10:279–286

Lemanceau P, Corberand T, Gardan L, Latour X, Laguerre G, Boeufgras JM, Alabouvette C (1995) Effect of two plant species, Flax (*Linum usitatissinum* L.) and tomato (*Lycopersicon esculentum* Mill.), on the diversity of soilborne populations of fluorescent pseudomonads. Appl Environ Microbiol 61:1004–1012

Louws FJ, Fulbright DW, Stephens CT, Debruijn FJ (1994) Specific genomic fingerprints of phytopathogenic *Xantomonas* and *Pseudomonas* pathovars and strains generated with repetitive sequences and PCR. Appl Environ Microbiol 60:2286–2295

Meyer J-M (2000) Pyoverdines: pigments, siderophores and potential taxonomic markers of fluorescent *Pseudomonas* species. Arch Microbiol 174:135–142

Meyer J-M, Geoffroy VA, Baida N, Gardan L, Izard D, Lemanceau P et al. (2002) Siderophore typing, a powerful tool for the identification of fluorescent and non-fluorescent *Pseudomonas*. Appl Environ Microbiol 68:2745–2753

Molina L, Ramos C, Ronchel MC, Molin S, Ramos JL (1998) Construction of an efficient biologically contained *Pseudomonas putida* strain and its survival in outdoor assays. Appl Environ Microbiol 64:2072–2078

Moore ERB, Mau M, Arnscheidt A, Böttger EC, Hutson RA, Collins MD, Van De Peer Y, De Watcher R, Timmis KN (1996) The determination and comparison of the 16 S rRNA gene sequences of species of the genus *Pseudomonas* (sensu stricto) and estimation of the natural intrageneric relationships. Syst Appl Microbiol 19:478–492

Munsch P, Alatossava T, Meyer JM, Marttinen N, Christen R, Gardan L (2002) *Pseudomonas costantinii* sp. nov., another causal agent of brown blotch disease, isolated from cultivated mushroom sporophores in Finland. Int J Syst Evol Microbiol 52:1973–1983

Nautiyal CS, Johri JK, Singh HB (2002) Survival of the rhizosphere-competent biocontrol strain *Pseudomonas fluorescens* NBRI2650 in the soil and phytosphere. Can J Microbiol 48:588–601

Palleroni NJ (1984) *Pseudomonas*. In: Krieg NR (ed.) Bergey's manual of systematic bacteriology, vol 1. Williams and Wilkins, Baltimore, pp 141–199

Palleroni NJ (2005) Genus I. *Pseudomonas* Migula 1894. In: Brenner DJ, Krieg NR, Staley JT, Garrity GM (eds) Bergey's manual of systematic bacteriology, vol 2, pt B, 2nd edn. Springer, Berlin Heidelberg New York, pp 323–379

Palleroni NJ, Kunisawa R, Contopoulos R, Doudoroff M (1973) Nucleic acid homologies in the genus *Pseudomonas*. Int J Syst Bacteriol 23:333–339

Risse D, Beiderbeck H, Taraz K, Budzikiewicz H, Gustine D (1998) Corrugatin, a lipopeptide siderophore from *Pseudomonas corrugata*. Z Naturforsch 53c:295–304

Ross IL, Alami Y, Harvey PR, Achouak W, Ryder MH (2000) Genetic diversity and biological control activity of novel species of closely related pseudomonads isolated from wheat field soils in South Australia. Appl Environ Microbiol 66:1609–1616

Sikorski J, Lalucat J, Wackernagel W (2005) Genomovars 11 to 18 of *Pseudomonas stutzeri*, identified among isolates from soil and marine sediment. Int J Syst Evol Microbiol 55:1767–1770

Sneath PHA, Stevens M, Sackin MJ (1981) Numerical taxonomy of *Pseudomonas* based on published records of substrate utilization. Antonie van Leeuwenhoek 47:423–448

Stanier RY, Palleroni NJ, Doudoroff M (1966) The aerobic pseudomonads: a taxonomic study. J Gen Microbiol 43:159–271

Starr MP, Knackmuss HJ, Cosens G (1967) The intracellular blue pigment of *Pseudomonas lemonnieri*. Arch Mikrobiol 59:287–294

Wayne LG, Brenner DJ, Colwell RR, Grimont PAD, Kandler O, Krichevsky MI, Moore LH, Moore WEC, Murray RGE, Stackebrandt E, Starr MP, Truper HG (1987) Report of the Ad Hoc committee on reconciliation of approaches to bacterial systematics. Int J Syst Bacteriol 37:463–464

Yamamoto S, Kasai H, Arnold DL, Jackson RW, Vivian A, Harayama S (2000) Phylogeny of the genus *Pseudomonas*: intrageneric structure reconstructed from the nucleotide sequences of *gyrB* and *rpoD* genes. Microbiology 146:2385–2394

Chapter 16
Molecular Strategies for Identifying Determinants of Oomycete Pathogenicity

Howard S. Judelson(⊠) and Audrey M.V. Ah-Fong

16.1 Introduction

Oomycetes are a diverse group of fungus-like eukaryotes encompassing both saprophytes and pathogens of plants and animals. Of the approximately 500 known species, those with the greatest human impact are the plant pathogens. These infect a wide range of crops, ornamentals, and native species, resulting in tens of billions of dollars of losses annually. Understanding factors required to be successful pathogens is a priority in oomycete research, since these may be targets for crop protection chemicals or plant-based resistance strategies. As illustrated in Fig. 16.1, such features may include the processes used to form the major infective propagules (spores), to breach physical barriers of the host (such as appressoria and hydrolytic enzymes), to acquire nutrients (transporters), and to alter the physiology of the plant (cytoplasmic or apoplastic effectors).

The aim of this chapter is to describe the approaches used to study oomycetes and progress in understanding the molecular bases of their pathogenicity. In particular, genomics and proteomics-based strategies have recently become feasible for several oomycetes, adding to traditional methods of gene cloning and classical genetics. Some of these resources have existed for years in other pathogens such as the ascomycete and basidiomycete "true" fungi, but their impact on the oomycete field is relatively new. These advances will be addressed after introducing the reader to the taxonomy, biology, and pathology of oomycetes. Thorough coverage of these introductory topics is impractical due to the diversity of oomycetes, however the reader is referred to several comprehensive descriptions of their biology (Spencer 1981; van der Plaats-Niterink 1981; Erwin and Ribeiro 1996; Hardham and Hyde 1997; Dick 2001).

H.S. Judelson
Department of Plant Pathology, University of California, Riverside, California 92521, USA
e-mail: howard.judelson@ucr.edu

C.S. Nautiyal, P. Dion (eds.) *Molecular Mechanisms of Plant and Microbe Coexistence.* Soil Biology 15, DOI: 10.1007/978-3-540-75575-3
© Springer-Verlag Berlin Heidelberg 2008

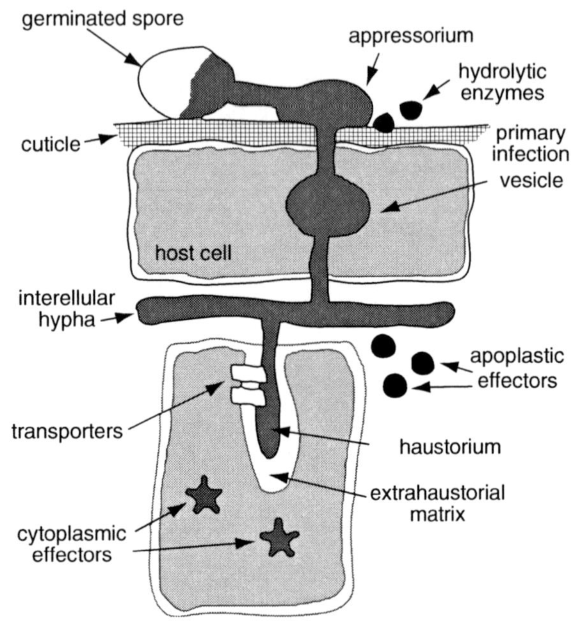

Fig. 16.1 Selected developmental stages and proteins relevant to disease. Illustrated is a spore that has penetrated a host using an appressorium. Extracellular and membrane-bound proteins of interest include hydrolytic enzymes used to degrade the plant cuticle and cell wall, effectors released to the plant apoplast or transported into the host cytoplasm, and transporters for moving nutrients from plant into hyphae and haustoria. Image inspired by a slide by S. Kamoun

16.1.1 Diseases Caused by Oomycetes

Oomycetes infect a wide range of monocots and dicots. The most important plant pathogens reside in the orders Peronosporales and Saprolegniales. Belonging to the Peronosporales is the most notorious and best-studied oomycete, *Phytophthora infestans*, which as the cause of potato late blight was responsible for the Irish Famine. *Phytophthora* includes more than 60 other species that infect many crop, forest, and ornamental plants, such as *P. sojae* which causes soybean root rot, *P. palmivora* and *P. megakarya* which cause black pod of cacao, and *P. parasitica* which affects both herbaceous and deciduous hosts (Erwin and Ribeiro 1996). Other important Peronosporales include *Bremia, Hyaloperonospora, Peronospora, Plasmopora, Pseudoperonospora, and Sclerospora* which cause downy mildew on crops and ornamentals; *Albugo,* which causes white rust on crucifers; and more than 100 species of *Pythium* which cause root and seed rots plus foliage diseases. Interestingly, *Pythium* also includes one vertebrate pathogen (*Py. ultimum*) and a parasite of fungi (*Py. oligandrum*). Within the Saprolegniales, significant damping-off pathogens belong to the genus *Aphanomyces*, which also includes animal pathogens. Phylogenies of these species suggest that pathogenic specialization on plant and animal hosts has evolved more than once (Fig. 16.2).

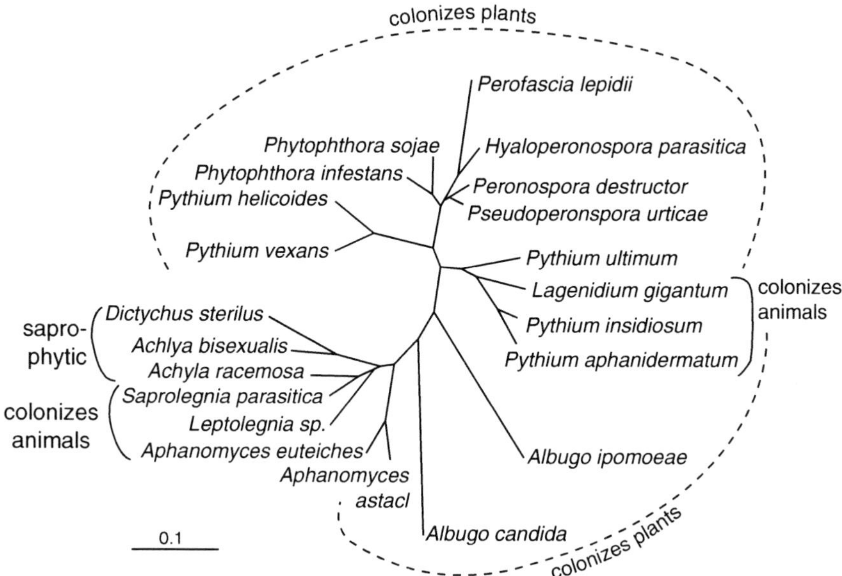

Fig. 16.2 Neighbor-joining tree of selected oomycete species, based on internal transcribed spacer and 5.8 S rRNA sequences. Data is based on 1000 bootstrap replicates; each branch point is supported by >70% of trees from 1000 bootstrap replicates

16.1.2 Taxonomy

Oomycetes were once grouped with the fungi, due to their typical filamentous growth habits. However, oomycetes are more accurately placed in another branch of the eukaryotic tree along with diatoms and brown algae (kelps), forming the kingdom Stramenopilia (Baldauf et al. 2000). Characteristics distinguishing oomycetes from true fungi include the use of the β-1,3 glucan mycolaminarin as the major storage carbohydrate, a feature also observed in brown algae, diploidy in the vegetative stage, and the predominance in the cell wall of β-1,3-glucan polymers, not chitin.

Understanding the taxonomic placement of oomycetes is important since this impacts approaches for studying genes relevant to disease. For example, mutagenesis strategies used to identify genes in true fungi are difficult due to the diploidy of oomycetes. Moreover, although the disease cycles of many oomycetes resemble those of true fungi, the underlying mechanisms of pathogenesis may be genetically and biochemically distinct.

16.2 Life Cycles

Understanding the life and disease cycles of oomycetes provides a foundation for learning what factors may be important in their pathology. Most oomycetes display sexual and asexual cycles relevant to disease (Fig. 16.3). However, only some

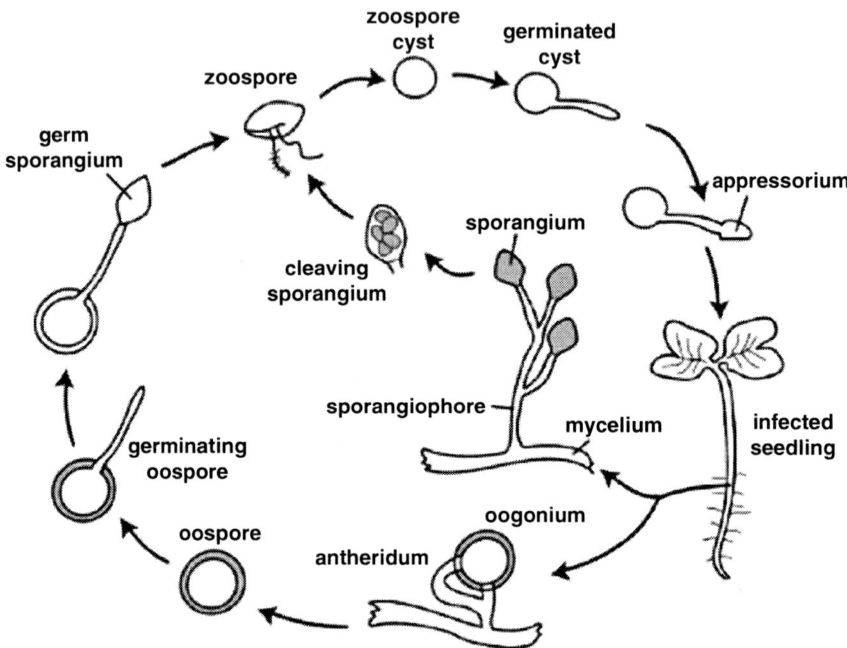

Fig. 16.3 Idealized life cycle of a homothallic root-infecting oomycete. Other species are heterothallic, requiring the interaction of two mating types to form oospores

species can be cultured apart from their hosts. Downy mildews and white rusts are obligate pathogens, while *Aphanomyces*, *Phytophthora*, *Pythium* and *Saprolegnia* can be propagated on artificial media. Attempts to culture axenically the obligate pathogens, for example by adding plant hormones or extracts to media, have failed except for *Sclerophthora macrospora*, a species intermediate between the downy mildews and *Phytophthora*. However, attempts to infect plants using the cultured *S. macrospora* strains were unsuccessful (references in Michelmore et al. 1988).

16.2.1 Pathogenic Lifestyles

Oomycetes exhibit biotrophic, hemibiotrophic, or necrotrophic interactions with plants. At one end of the spectrum are downy mildews and white rusts, which maintain biotrophy throughout infection. In common with biotrophs in other taxa, these do not appear to produce toxins or large amounts of cell wall degrading enzymes, and sporulate from green tissue. Some penetrated plant cells may die, but usually as the result of secondary infections or host defense reactions. As these species are not only biotrophs but also obligate pathogens, perhaps defects accumulated in their metabolic pathways during coevolution with their hosts. The other extreme of pathogenic behavior is illustrated by *Aphanomyces* and most *Pythium* spp., which are usually strong necrotrophs. These kill their hosts rapidly and then feed like saprophytes. Hemibiotrophs show intermediate behavior, first infecting living tissue, but then shifting towards necrotrophy. This is the case for *Phytophthora* and some *Pythium* spp.. As will be discussed in Sect. 16.4.4.5, proteinaceous toxins that may explain the transition to necrotrophy have been identified in *Phytophthora*.

Some oomycetes colonize only a few plant species, while others are more cosmopolitan. Most downy mildews and some *Phytophthora* species infect a small number of hosts. *Bremia lactucae,* for example, causes downy mildew only on lettuce (*Lactuca*) species. In contrast, *P. cinnamomi* and *Py. aphanidermatum* each have hundreds of hosts. Specialization is also observed in the portion of plants colonized. For example, of the two species causing downy mildew on lettuce, *Plasmopara lactucae-radicis* is exclusively a root pathogen while *B. lactucae* only colonizes foliage (Stanghellini and Gilbertson 1988). *P. infestans* colonizes foliage and tubers (which are modified stems) but not true roots, while *P. cinnamomi* is known mostly for root or crown rots. The determinants of host and tissue-specificity are largely unknown, but presumably reflect different complements of enzymes used to break down physical barriers, variation in effectors of plant metabolism, and developmental stages suited for recognizing only certain plant structures.

16.2.2 Importance of Spores

Both sexual spores (oospores) and asexual spores (mostly sporangia or conidia) are central to the disease cycle, playing roles in dissemination, survival, and infection. Many studies of oomycete pathogenicity therefore analyze these stages, as described in Sect. 16.4. Generalizations spanning the breadth of oomycetes are challenging, but asexual spores are typically short-lived and responsible for starting most infections in a growing season. In contrast, oospores can usually survive between growing seasons to initiate disease each year. However, the relative importance of the sexual and asexual spores varies for each pathosystem. For example, the sexual cycle is assumed to be relatively unimportant in most temperate downy mildews, while in most graminaceous downy mildews the sexual cycle predominates with asexual sporulation undescribed for some species (Michelmore et al. 1988). In other downy mildews and homothallic *Phytophthora,* asexual and sexual sporulation can occur in parallel (Koch and Slusarenko 1990; Erwin and Ribeiro 1996).

16.2.2.1 Spore Formation

Asexual spores typically develop at the termini or side branches of sporangiophores (or conidiophores) that emerge from stomata or other plant openings. Sporulation of root pathogens generally occurs within air spaces between soil particles, which enables dissemination through rain or irrigation water. Spores from foliar lesions are spread by both wind and water.

Oospores form within tissue colonized by homothallics, or in the case of heterothallics when lesions of opposite mating types merge; homothallic and heterothallic species are found within most oomycete orders. Mating hormones induce the differentiation of oogonia and antheridia, in which meiosis occurs. Male nuclei migrate into the oogonia, which matures into a oospore containing the zygote. In *Phytophthora, Pythium,* and most downy mildews these oospores remain dormant and viable in plant debris or soil for a decade or more, but in some species such as *Plasmopara viticola* survival through only one winter is more typical (Michelmore et al. 1988).

Several *Phytophthora* and *Pythium* spp. produce chlamydospores, which are globose and often thick-walled structures delimited from hyphae by septa (van der Plaats-Niterink 1981; Erwin and Ribeiro 1996). Some species produce asexual sporangia, chlamydospores, and oospores, and in such cases the relative survivability of each can be compared. In *P. cactorum,* survival times of 14 days were reported for hyphae, 35 days for sporangia, >105 days for chlamydospores, and >1 year for oospores (Malajczuk 1983).

16.2.2.2 Pathways of Spore Germination and Host Penetration

Asexual spores germinate through two distinct pathways (Hardham and Hyde 1997). In *Phytophthora, Albugo,* some *Pythium* spp., and downy mildews this sometimes occurs through the extension from the sporangium or conidium of a germ tube (direct germination), which can then contact and penetrate a host. The second mode of germination (indirect germination) involves zoosporogenesis. This entails partitioning the multinucleate sporangial cytoplasm into several uninuclear and biflagellated zoospores, which can swim toward plants and then generate cysts from which infective germ tubes form. While many oomycetes are capable of both direct and indirect germination, many downy mildews such as most *Bremia, Peronospora,* and *Peronosclerospora* spp. only exhibit direct germination.

The details of zoosporogenesis vary somewhat in different taxa. For example, zoospores of *Phytophthora* and downy mildews form in the sporangium prior to their release. In contrast, in *Pythium* the undivided cytoplasm moves out of the sporangium into a membrane vesicle, in which zoospores form and from which zoospores then escape. In all species the process is favored by cool conditions, while direct germination predominates at higher temperatures. Zoospores are potent infectious propagules since they can swim in search of a suitable host through surface or subterranean water. Chemotaxis helps guide zoospores to infection sites, as illustrated by the abilities of daidzein and genistein, isoflavones exuded from soybean roots, to attract *P. sojae* (Tyler 2002). The weak electric fields of plant roots may provide alternative homing information. For example, *P. palmivora* is attracted to anodic zones of roots, while *Py. aphanidermatum* moves towards cathodic regions (van West et al. 2002).

Physical and chemical signals trigger zoospores to encyst. This entails the detachment of flagella, cell wall formation, and adhesion to the plant by means of extruded mucilage. Intracellular phospholipid and Ca^{2+} signals regulate encystment, when triggered by certain amino acids, isoflavones, and pectin (Warburton and Deacon 1998; Latijnhouwers et al. 2002). The distribution of these compounds may explain why several species encyst preferentially at certain sites, as in the case of *Pl. viticola* which usually targets stomata (Kiefer et al. 2002). Cysts germinate through a single germ tube, which can produce an appressorium which uses mechanical pressure and cell wall degrading enzymes to penetrate epidermal cells of the plant (Hardham 2001). Alternatively, germ tubes may enter between the anticlinal walls of root epidermal cells or other openings. For example, *P. infestans* commonly penetrates epidermal cells using appressoria, but its germ tubes can also enter through stomata, wounds, and lenticels. In contrast, *Pl. viticola* germ tubes do not penetrate host cells, instead entering the plant through stomata (Michelmore et al. 1988).

After penetration, most oomycetes proliferate by producing primarily intercellular hyphae. If infection occurred through an appressorium this is typically preceded by the formation of vesicles within cells adjacent to the entry site. In biotrophic and hemibiotrophic interactions, varying numbers of haustoria extend into host cells, but these are not described for necrotrophs. Downy mildews such as *Pl. viticola* and the white rust *A. candida* typically produce many haustoria, while *P. infestans* produces a lesser amount (Hohl and Suter 1976; Woods and Gay 1983; Kiefer et al. 2002).

Chlamydospores and oospores germinate and infect plants through processes similar to those described above. Chlamydospores typically germinate in the presence of plant exudates to form hyphal-like germ tubes. These may directly enter plants or terminate in a sporangium capable of releasing zoospores (Hohl and Suter 1976). Oospores may also germinate to form an infective hyphae or a sporangium.

Most oomycete diseases are initiated by asexual sporangia, oospores, or chlamydospores, but there are exceptions. For example, some necrotrophs such as *Pythium* persist as saprophytic hyphae in soil, and some biotrophic downy mildews survive as mycelium in plant debris. When these contact seedlings or damaged plants, new infections can occur.

16.3 Tools for Molecular Analyses

This section focuses on several of the technologies and resources used to study oomycetes, especially related to identifying factors involved in spore biology and plant infection. Some classical methods are mentioned, but the emphasis will be on structural and functional genomic tools developed over the past decade. Most have been applied only to limited species, particularly *Phytophthora*, but have relevance to all oomycetes. Examples of their use to analyze genes and proteins relevant to oomycete pathogenesis will be described in Sect. 16.4.

16.3.1 DNA-Mediated Transformation

Efficient transformation methods are critical for assigning function to cloned genes, useful for discovering new sequences through insertional mutagenesis or promoter trapping, and can aid microscopic studies of plant colonization by expressing reporter genes. Only in the 1990s was a reliable transformation method developed for an oomycete. Transformation came slowly to the oomycete field due to the small size of the research community and an absence of promoters suitable for expressing selectable markers.

Oomycete promoters were first isolated from the downy mildew *B. lactucae* as part of attempts to transform that species. Due to the evolutionary distance between oomycetes and other taxa, constructing vectors using transcriptional regulators from oomycetes was assumed important, although this was not confirmed until later studies in *P. infestans* (Judelson et al. 1992). Promoters from the *ham34* and *hsp70* genes were used successfully to transiently express the β-glucuronidase (GUS) reporter in *B. lactucae*. This was accomplished by bombarding conidia with DNA-coated microprojectiles, which were inoculated onto lettuce cotyledons (Judelson and Michelmore, unpublished). Stably transformed lines were not obtained, likely due to complications associated with establishing an in planta selection for this obligate pathogen. However, the *ham34* and *hsp70*-based expression cassettes have remained integral to all vectors used to transform other oomycetes.

The first reliable method for stable transformation involved *P. infestans* (Judelson et al. 1991). Plasmids expressing selectable markers driven by *ham34* or *hsp70* promoters were introduced into protoplasts using polyethylene glycol-CaCl$_2$ treatment. Selection was achieved using genes for resistance to geneticin, hygromycin, or streptomycin. Gene transfer is also possible using microprojectile bombardment, electroporation, or *Agrobacterium*, although the protoplast method is employed most often (Cvitanich and Judelson 2003; Vijn and Govers 2003; Latijnhouwers et al. 2004).

With *P. infestans*, a talented worker can generate hundreds of transformants per week. Lower rates are more typical but adequate for testing the function of genes. It is poorly understood which method for transformation is best for a given application, such as overexpression or gene silencing, but the characteristics of DNA integration are known to vary. *Agrobacterium* transformation results in single-copy integrations, while the others usually insert multiple copies of transgenes into chromosomes, typically in tandem arrays (Judelson 1993; Vijn and Govers 2003). Transgenes appear to be structurally stable, but many lose activity due to position effects (Judelson et al. 1993b).

Although most experiments have involved *P. infestans*, transformation has also been reported for several other pathogenic oomycetes. These include *P. palmivora, P. nicotianae, P. sojae,* and *Py. aphanidermatum* (Judelson et al. 1993a; van West et al. 1999b; Lin et al. 2002; Weiland 2003). Selection for geneticin or hygromycin-resistance is usually preferred.

A gene disruption method is not yet demonstrated, due to infrequent homologous integration and the complication of dealing with a diploid where two genes need to be mutated. Stable silencing based on homology-dependent methods has been demonstrated by several laboratories, using sense, antisense, or hairpin constructs. This was accomplished first for GUS (Judelson et al. 1993b), and then endogenous *Phytophthora* genes. Silencing has also been reported for *P. infestans* genes encoding G-protein subunits, a *Cdc14* mitotic regulator, a family of transcriptional regulators in the NIF family, and a bZIP transcription factor, and in *P. nicotianae* for the CBEL elicitor (Lin et al. 2002; Ah Fong and Judelson 2003; Latijnhouwers and Govers 2003; Latijnhouwers et al. 2004; Blanco and Judelson 2005; Judelson and Tani 2007). Silencing in these studies occurred in 3–75% of transformants, but failures to silence certain genes are also reported. Although the molecular basis of silencing is poorly understood, it appears to occur at the level of transcription based on studies of the *infl* elicitin (van West et al. 1999a) and the NIF transcriptional regulator family (Judelson and Tani 2007). In the latter case, silencing was associated with the establishment of a tighter configuration of chromatin, which spread slightly outwards from the targeted gene.

A transient silencing system has been reported by Whisson et al. (2005). They treated protoplasts of *P. infestans* with double-stranded RNA matching a target gene, and found that its mRNA was reduced in some regenerants. The onset of silencing was about 12 days after dsRNA treatment, slower than in analogous studies of plants and animals, and weakened after 17 days. However, this allowed time for colonies to form in which phenotypes could be assessed. Little is known of the molecular basis of transient silencing. For example, whether silencing involves small interfering RNA (siRNA) is unknown as is the extent of off-target effects.

Uses of transformation for gain-of-function assays are not widely described. In one of the few examples, a constitutively active G-protein α subunit was expressed in *P. infestans*, although this resulted in no obvious phenotypic change (Latijnhouwers et al. 2004). Another case involved expressing a gene encoding an elicitor from *P. cryptogea* in *P. infestans*, which altered its interaction with tobacco (Panabieres et al. 1998).

16.3.2 *Heterologous Systems for Functional Studies*

In some cases it is not necessary, or preferable, to use an oomycete transformation system for functional analyses. This may be the case when another species offers higher rates of transformation or useful genetically marked strains. For example, the ras-like gene *Piypt* of *P. infestans* was confirmed to participate in vesicle transport based on its ability to complement a mutant of *S. cerevisiae* (Chen and Roxby 1997). Similarly, the function of an ABC transporter from *P. sojae* in toxicant defense was tested using transporter mutants of *S. cerevisisae* (Connolly et al. 2005). Also using complementation in *S. cerevisiae*, the *PiCdc14* phosphatase from *P. infestans* was confirmed to be a regulator of mitosis (Ah Fong and Judelson 2003).

Transient in planta expression systems have proved helpful in testing genes believed to influence host defenses. As described in Sect. 16.4, these assays were instrumental in identifying several proteins that elicit defense responses in plants. These are currently only applicable to leaves, however. In agroinfiltration, a gene is placed in a T-DNA vector, and transformed into *A. tumefaciens* which is infiltrated into plant tissue (Huitema et al. 2004). This allows expression of the gene in plant cells, which are checked for responses. Agroinfection involves inserting a gene within the Potato Virus X (PVX) genome, which is also expressed in plants using *A. tumefaciens*. Initially only a few cells are infected, but macroscopic zones of phenotypic response result from spread of the virus. Large numbers of virus-encoding *A. tumefaciens* can be stab-inoculated in parallel, making the assay useful for high-throughput testing of genes identified from oomycete genome projects.

16.3.3 Genomics Data

Sequencing projects have been recently completed, or are in progress, for several oomycete plant pathogens. These provide new opportunities for discovering genes relevant to disease, as will be shown in Sect. 16.4. Genomic resources are particularly useful in analyses of the obligate pathogens, where the inability to manipulate the organism apart from the plant poses challenges to traditional approaches for identifying genes. Prior to the genome projects, only a handful of oomycete genes had been identified using conventional approaches such as heterologous hybridization, immunoscreening, subtraction cloning, or differential display (for examples see Pieterse et al. 1994; Goernhardt et al. 2000; Fabritius et al. 2002).

16.3.3.1 Expressed Sequence Tags (ESTs)

The first oomycete genomics project generated 1000 ESTs from *P. infestans* (Kamoun et al. 1999). EST sequencing was a good first option for oomycetes since their genomes are rich in repetitive sequences and relatively large, between 60 and 240 Mb (Judelson and Randall 1998; Voglmayr and Greilhuber 1998). The *P. infestans* EST data has been expanded to 94,121 ESTs, representing about 18,256 unigenes (Randall et al. 2005). Libraries from 20 distinct tissues were used to maximize sampling of the transcriptome, including mating cultures, infected plants, asexual sporangia, zoospores, germinated cysts, and hyphae from defined, rich, and starvation media, and hyphae exposed to plant exudates. Bioinformatics distinguished plant from pathogen sequences in the case of the infection ESTs, which was aided by the fact that host transcripts are about 46% G+C while *Phytophthora* transcripts average 57%.

ESTs have also been obtained from other oomycetes. These include 28,913 *P. sojae* ESTs representing 13,234 genes based on clones from zoospores, infected soybean, and hyphae grown on rich and nutrient-limited media, 755 from germinated cysts or zoospores from *P. nicotianae*, 3568 from *P. parasitica* hyphae grown in defined media, and 1500 from hyphae and plant tissue infected with the sugarbeet pathogen, *Aphanomyces cochlioides* (Qutob et al. 2000; Shan et al. 2004b; Skalamera et al. 2004; Panabieres et al. 2005; Weiland and McGrath, unpublished).

16.3.3.2 Genome Sequence Data

Draft genome sequences based on seven- to ninefold coverage were generated for *P. ramorum* and *P. sojae* by the U. S. Department of Energy (http://genome. jgi-psf.org), which is also sequencing *P. capsici*. For *P. infestans*, pilot projects achieved onefold coverage (Randall et al. 2005) and an eightfold draft has been completed by the Broad Institute of MIT and Harvard. A draft has also been completed for the H. parasitica genome by Washington University (St. Louis, USA) and the Sanger Institute (UK). *P. capsici* sequences have also been generated by the JGI and initiatives to analyze other oomycetes are progressing. Future integration of the draft sequences of these species with genetic maps, finger-printed BAC libraries, and analyses of synteny should prove useful in finishing and annotating the genomes.

For *P. ramorum*, the assembly data indicate a genome size of 65 Mb and gene prediction programs identified 15,743 genes. For *P. sojae*, a 95-Mb genome containing 19,027 genes is predicted. The expressed content of the 237-Mb *P. infestans* genome appears to be about 18,500 genes, which is similar to an prediction made earlier based on EST data (Randall et al. 2005). These values should be taken as estimates, as prediction methods are being refined, but the gene content of *Phytophthora* is clearly much higher than that of phytopathogenic true fungi. For example, the ascomyceteous rice blast agent *Magnaporthe grisea* is predicted to encode 11,109 genes (Dean et al. 2005). The larger number of genes in *Phytophthora* can be attributed to the expansion of gene families and the presence of oomycete-specific sequences, many of which may play roles in disease. For example, the genome contains a large superfamily of ABC transporters, which may participate in the efflux of phytoalexins. Large families also exist of factors that potentially induce plant necrosis, such as the elicitin and crinkler (CRN) proteins (Torto et al. 2003) which are discussed in more detail in Sect. 16.4. Approximately 1500 *Phytophthora* genes lack significant similarity with non-oomycete sequences, of which many encode secreted proteins. Overall, the secretomes of *P. ramorum* and *P. sojae* are predicted at 1256 and 1570 proteins, respectively (Jiang et al. 2005b). This is close in size to the 1258-protein secretome of *M. grisea* (Dean et al. 2005).

As more oomycete genomes are sequenced, comparative genomics should lead to a better understanding of their pathogenic lifestyles. Even between *P. infestans*, *P. ramorum*, and *P. sojae*, substantial differences are evident. When *P. ramorum* and *P. sojae* were compared, for example, the former appeared to have 624 unique genes and the latter 1755 (Tyler et al. 2006), and their analysis may reveal the basis of host-species specificity.

16.3.4 Classical Genetics

In an era focused on genomics some might downplay traditional genetics, but this is still an important tool. For example, crosses are needed to determine the number of avirulence (AVR) loci, which interact with plant resistance loci as part of gene-for-gene interactions. Classical genetics identified 12 dominant AVR loci in the interaction between *P. infestans* and potato, seven dominant loci influencing the association of *P. sojae* with soybean, and over 14 dominant loci plus modifier genes involved in the lettuce-*Bremia lactucae* interaction (Al-Kherb et al. 1995; Whisson et al. 1995; MacGregor et al. 2002; Sicard et al. 2003). Genetic maps are also a prerequisite for positional cloning which, as detailed in Sect. 16.4, has been used to clone AVR genes from several oomycetes. Many traits relevant to disease may be determined by multiple genes (quantitative trait loci, QTLs), which in the future may be mapped and cloned. In *P. infestans,* for example, QTLs determine pathogenic specialization on potato or tomato (Legard et al. 1995).

16.3.4.1 Crossing Techniques

Oomycetes are not the easiest systems for genetic analysis, but methods are developed for several *Phytophthora* species, downy mildews, and *Pythium* (Shattock et al. 1986; Michelmore et al. 1988; Martin 1989). Crosses can often be performed in a matter of weeks by pairing isolates in culture media or in planta in the case of obligate pathogens, and then extracting and germinating oospores. Some species (or isolates within a species) may be recalcitrant to genetics, however. Germinating oospores is often a challenge, not all progeny may be viable, a fraction of offspring may be reduced in pathogenic aggressiveness, and linkage studies may be confused by distorted segregation. Such limitations do not mean that oomycetes are permanently unsuited for genetics. Field isolates of *M. grisea* also showed low fertility, but a backcrossing program generated strains highly amenable to genetic analysis and helped the species develop into a model system (Valent and Chumley 1991).

Markers such as RAPDs, AFLPs, and isozymes play important roles in analyzing crosses. For example, markers can distinguish hybrid from selfed offspring when mating heterothallics (Judelson et al. 1995); selfs develop since mating hormones diffusing from the opposite mating type can stimulate an isolate to self (Ko 1988). Markers are also useful when outcrossing homothallics. As shown in *P. sojae* and *Py. ultimum,* growing mixed isolates results in oospores that are mostly selfs but occasionally hybrids. Once the latter are identified using markers, F_2 populations can be established (Francis and St Clair 1993; Whisson et al. 1994).

The ability to obtain selfed or out-crossed progeny is also a prerequisite for a potentially promising method for functional analysis in oomycetes named TILLING. This combines chemical mutagenesis with screens of pooled PCR products for mutations, resulting in the isolation of strains heterozygous for missense or nonsense alleles of the targeted genes. Crossing schemes can then generate strains homozygous for the mutation, which can be tested for function. First developed for plants, this method has been recently adopted to *Phytophthora* (Lamour et al. 2006).

16.3.4.2   Genetic Maps

Loci used for mapping have been limited to AVR and mating type genes, and DNA markers such as RAPDs, AFLPs, and RFLPs. Since oomycetes are diploid and do not produce pigments, auxotrophic or visible makers are rare. Linkage maps are developed for *P. infestans, P. sojae,* and *B. lactucae,* which contain 508, 386, and 430 DNA markers, respectively, plus avirulence genes (May et al. 2002; Sicard et al. 2003; van der Lee et al. 2004). Genetic sizes are estimated at 1200 cM, 2590 cM, and 835 cM, respectively, averaging 197, 35, and 70 kb per cM.

16.4   Molecular Insights into Pathogenicity

Two main strategies have been used to identify genes and proteins important for disease. One focuses on understanding developmental stages such as spores, which are needed for dispersal and which germinate to enable host penetration. The second concentrates on factors influencing interactions with plants, such as cell wall degrading enzymes and proteins which activate or disrupt host defenses. These two approaches may appear philosophically distinct, but are in fact complementary and overlapping. For example, a study of gene expression during spore germination should reveal proteins involved in making infection structures as well as effectors of plant physiology. Examples of both approaches are featured below.

16.4.1   Developmental Biology of Spores

Early studies described the physiology and cytology of these stages, with asexual sporulation and zoosporogenesis being examined in the most detail (Hardham 2001; Hardham and Hyde 1997). Molecular approaches are now helping to understand spores, the main infectious propagules of oomycetes, in more detail.

16.4.1.1   Differential Expression during the Spore Cycle

Many genes activated during the asexual spore cycle have been discovered, mostly through array studies in *Phytophthora.* The first large-scale study was performed in *P. infestans,* where 4 100-gene arrays were used to identify 60 and 71 genes induced >5-fold during sporulation and zoosporogenesis, respectively (Kim and Judelson 2003; Tani et al. 2004). One-third were expressed only during those stages. More recently, Affymetrix GeneChip studies of 15 645 *P. infestans* genes (about 75% of the transcriptome) compared expression in hyphae, sporulating hyphae, sporangia, zoospores, germinating zoospore cysts and appressoria. This expanded the number of genes known to be activated during the spore cycle to over 1600 (Judelson et al., 2008). Since such a high fraction are induced, even small projects have discovered stage-specific genes, such as a study of 386 genes from germinated cysts and zoospores of *P. nicotianae* (Shan et al. 2004b; Skalamera et al. 2004).

The stage-specific genes appear to serve a range of metabolic, regulatory, and structural functions needed by oomycetes to colonize their hosts. For example, many genes induced during sporulation appear to play roles in zoospores. These include flagella-associated proteins such as centrin and dynein, phosphagen and adenylate kinases which may buffer or channel ATP during the energy-intensive swimming stage, a proline biosynthesis enzyme likely used to osmotically stabilize zoospores, and proteins stored in zoospore vacuoles as a carbon source (Judelson et al., unpublished; Marshall et al. 2001b; Ambikapathy et al. 2002; Kim and Judelson 2003; Randall et al. 2005). Others with potential roles in disease encode mucin-like proteins, which may confer protection against desiccation or enable adherence to plant surfaces. One mucin-like gene cloned using a subtraction method is specific to germinated cysts (Goernhardt et al. 2000), and others from the EST projects were found to be specific to zoospore, cyst, and appressorium stages. Many genes encoding potential effectors of plant structure or physiology were also within the datasets of induced genes. However, in most cases clues to their function remained unknown until subsequent bioinformatics studies, as described in Sects. 16.4.3. and 16.4.4.

Changes during spore development and germination have been documented by several two-dimensional protein gel studies, including one that concluded that 1% of approximately 700 resolvable proteins were specific for the sporangial, zoospore, cycle and germinated cyst stages of *P. palmivora* (Kraemer et al. 1997; Shepherd et al. 2003). Another study compared the cyst, germinating cyst and appressorium stages of *P. infestans* and sequenced several of the detected stage-specific proteins. Matches were found in the *P. infestans* databases and GenBank for 13 proteins, which were assigned roles in protein synthesis, amino acid metabolism, energy metabolism, and reactive oxygen scavenging (Ebstrup et al. 2005).

Genes induced during oosporogenesis in *P. infestans* are also identified, using subtraction cloning and microarray studies of ESTs. A total of 97 genes induced over 10-fold during mating were identified, of which 46 were totally mating-specific (Fabritius et al. 2002; Fabritius and Judelson 2003; Prakob and Judelson, 2007). A disproportionate number encode RNA-binding or metabolizing proteins, suggesting the importance of post-transcriptional regulation. Comparisons of gene expression during the germination of sexual and asexual spores have not yet been possible, due to difficulties in obtaining large amounts of synchronously germinating oospores.

One study examined several genes involved in amino acid biosynthesis, many of which appeared to be regulated during spore development, and tested whether RNA and protein levels were correlated (Grenville-Briggs et al. 2005). Good correspondences were observed. However, it is important to remember that this is not necessarily true for all stage-specific genes, as exceptions exist (Marshall et al. 2001a). Also, only in a few cases have the cellular roles of the genes been proved by gene silencing. These are described below.

16.4.1.2 Cdc14 Protein Phosphatase

Silencing of this sporulation-specific gene in *P. infestans* blocked sporulation, thereby defining a pathway required for oomycetes to be successful pathogens (Ah Fong and Judelson 2003). Cdc14 in other species such as budding yeast are regulators of mitosis that are transcribed constitutively, in contrast to the situation in *P. infestans* (Stegmeier and Amon 2004). Complementation in *cdc14^{ts}* budding yeast demonstrated that the *P. infestans* protein can regulate mitosis. In *P. infestans*, the protein might regulate a specific aspect of nuclear behavior during early sporulation or use its phosphatase activity to control other pathways.

16.4.1.3 G-Protein Signaling

Genes encoding α and β G-protein subunits, *Pigpa1* and *Pigpb1*, are up-regulated during asexual sporulation in *P. infestans* (Laxalt et al. 2002; Kim and Judelson 2003). Silencing studies confirmed that they play important roles in disease. Strains no longer expressing *Pigpb1* fail to sporulate (Latijnhouwers and Govers 2003), while strains silenced for *Pigpa1* exhibit impaired zoospore swimming and chemotaxis, and are reduced in virulence (Latijnhouwers et al. 2004). Potential downstream targets of G-protein signaling were identified by cDNA-AFLP (Dong et al. 2005).

16.4.1.4 bZIP-Regulated Transcriptional Network

A network important for pathogenesis was identified by studying developmentally-regulated protein kinases, which were of interest since kinase assays revealed significant changes in the spore cycle (Fig. 16.4). Stage-specific kinases with calcium, phospholipid, or cyclic-nucleotide regulatory domains, or no obvious regulatory domain, were identified (Kim and Judelson 2003; Tani et al. 2004) and substrates discovered by yeast two-hybrid. One zoosporogenesis-specific kinase, *PiPkz1*, bound a bZip transcription factor *(PiBzp1)* which when silenced resulted in a swimming defect (zoospores that perpetually turn) and an appressorium-minus phenotype (Blanco and Judelson 2005). Virulence was therefore essentially eliminated, as plant infection could only occur through wounds. Candidates for genes regulated by *PiBzp1* were identified using microarrays (Blanco and Judelson, unpublished).

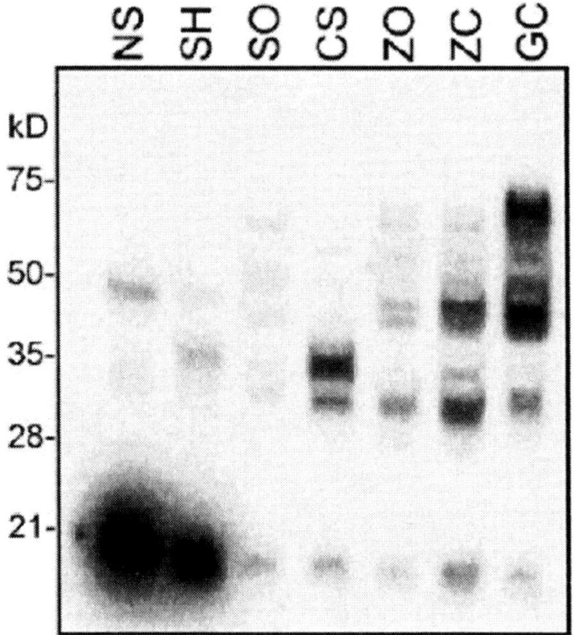

Fig. 16.4 In-gel kinase assay of proteins from developmental stages of *P. infestans* using myelin basic protein as non-specific substrate (Blanco and Judelson, unpublished). Proteins are from nonsporulating hyphae (NS), sporulating hyphae (SH), purified sporangia (SO), sporangia placed at 4 °C to induce cytoplasmic cleavage (CS), motile zoospores (ZO), cysts (ZC), and germinating cysts (GC)

16.4.2 Genes Expressed in Colonized Plants

Genes expressed in planta have also been characterized for *P. infestans* on tomato, *P. sojae* on soybean, and *H. parasitica* on *Arabidopsis* using subtraction cloning, cDNA-AFLP, or array methods (Pieterse et al. 1994; Goernhardt et al. 2000; Beyer et al. 2001; Bittner-Eddy et al. 2003; Avrova et al. 2004; Moy et al. 2004; Wang et al. 2006; Chen X et al. 2007). Despite the challenge of accurately detecting oomycete transcripts in infected tissue, multiple genes of interest were identified. These included many potential effectors including the CRN, elicitin, CBEL, and cell-wall degrading proteins that are described in Sects. 16.4.3 and 16.4.4. Another common class of genes identified were transporters. These included transporters of amino acids, phosphate, and sucrose that may potentially act at the plant-oomycete interface, and ABC transporters possibly involved in the efflux of plant defense compounds.

16.4.3 Cell Wall Degrading Enzymes (CWDEs)

Genes for enzymes involved in degrading plant cell walls were discovered through studies focused directly on cloning the CWDE genes, analyses of genes expressed during spore germination or infection, and mining sequence databases. This identified extracellular cellulases, cutinases, pectate lyases, and polygalacturonases (Gotesson et al. 2002; Torto et al. 2002; Randall et al. 2005; Yan and Liou 2005). Many belong to large families, for example over 20 predicted pectate lyases were found in the *P. ramorum* database (Judelson 2007) and 19 polygalacturonases were cloned by traditional methods from *P. cinnamomi* (Gotesson et al. 2002; Torto et al. 2002). Polygalacturonases from *P. infestans* and *P. parasitica* were shown to be induced during cyst germination and exposure to plant cell walls, respectively (Yan and Liou 2005). This implies a role in infection; however gene silencing tests are not yet performed. This may be a challenge, however, since disrupting such genes in true fungi often have little effect on pathogenicity due to functional redundancy.

16.4.4   Effectors: Avirulence Factors, Elicitors, and Others

Many early studies of oomycete-plant interactions focused on molecules eliciting host defenses in non-host and host plants. Several early studies emphasized nonprotein elicitors, such as fragments of cell walls released by host glucanases (Tyler 2002). Genomics data has now shifted the emphasis to proteins. Elicitors and avirulence proteins are often considered within a broader class termed "effectors," which may have a range of effects (van Dijk et al. 1999). Once transported from pathogen to plant, effectors trigger or suppress plant defenses or modulate other aspects of plant physiology. Hundreds of such proteins have been identified from oomycetes through biochemical studies, positional cloning, and bioinformatic approaches. Most appear to target distinct sites in the plant apoplast or cytoplasm, although a few may imported into plant nuclei (Kanneganti et al. 2007).

16.4.4.1   PEP13 Transglutaminase

This 42-kDa protein was the first known proteinaceous elicitor from an oomycete, which was discovered and purified based on the ability of *P. sojae* to cause necrosis in a non-host, parsley, though binding to a 91-kDa receptor in the plant apoplast (Nurnberger et al 1995). Work in *P. infestans* showed that it belongs to a multigene family in which different members are expressed during hyphal growth, zoosporogenesis, or oosporogenesis (Fabritius and Judelson 2003). The proteins are transglutaminases, which probably strengthen walls in each developmental stage (Brunner et al. 2002). Its sequence is distinct from transglutaminases in other kingdoms, and is oomycete-specific. A 13-aa fragment, Pep-13, is needed for both elicitor and transglutaminase activity. Pep-13 was one of the early examples from a phytopathogen of a "pathogen-associated molecular pattern" or PAMP (Nürnberger and Brunner 2002); however its role in *Phytophthora* fitness or pathogenicity is not yet proven.

16.4.4.2 Elicitins

Conventional purification strategies also identified these proteins, which induce programmed cell death in a narrow range of plants such as *Nicotiana* and radish (Ponchet et al. 1999). They were first identified in *Phytophthora* and *Pythium* as small (100 aa) cysteine-rich secreted proteins. They may help define host range since *P. infestans* silenced for *inf1* acquired the ability to partially colonize *N. benthamiana* (Kamoun et al. 1998). Molecular cloning and database mining showed that a diverse family of elicitin (ELI) and elicitin-like (ELL) proteins exist, with 48 and 57 members in *P. ramorum* and *P. sojae*, respectively (Jiang et al. 2005a). Some are secreted and others membrane-bound. Their intrinsic biological targets are lipid-like molecules, since the major elicitin from *P. cryptogea* has sterol-binding activity and ELI-4 of *P. capsici* is a phospholipase (Nespoulous et al. 1999; Osman et al. 2001). A role in assimilating sterols has been suggested, since oomycetes can not synthesize such compounds. However, *P. infestans* strains silenced for the major *inf1* elicitin grow normally. Like the PEP-13 protein, elicitins are unique to oomycetes.

16.4.4.3 PcF-Like Proteins

This family of oomycete-specific apoplastic effectors was first identified by purifying from *P. cactorum* a secreted protein that induces necrosis in strawberry and tomato, producing symptoms similar to that of the authentic disease (Orsomando et al. 2001). Relatives can also be found in *P. infestans, P. ramorum, and P. sojae* (Fig. 16.5). All are cysteine-rich, a feature also seen in elicitins and avirulence genes from true fungi. In fact, searching sequence databases for small cysteine-rich proteins is a common way of identifying effectors. For example, this independently identified the *P. infestans* PcF-like proteins, SCR74 and SCR91 (Torto et al. 2003). While the functions of the *P. infestans* proteins are unknown, SCR74 is a member of a 21-gene family, is upregulated during plant colonization, and has been subject to strong diversifying selection; these are all typical features of proteins involved in host-pathogen interactions (Liu et al. 2005).

16.4.4.4 CBEL Elicitor

This 34-kDa cell-wall protein was purified from *P. nicotianae* based on its elicitation of necrosis in its normal host, tobacco (Sejalon-Delmas et al. 1997). Gene silencing indicated that it is required for attachment to cellulosic surfaces, probably through a PAN module which is associated with protein-protein or protein-carbohydrate interactions. However, the silenced strains retained their ability to infect tobacco (Gaulin et al. 2002).

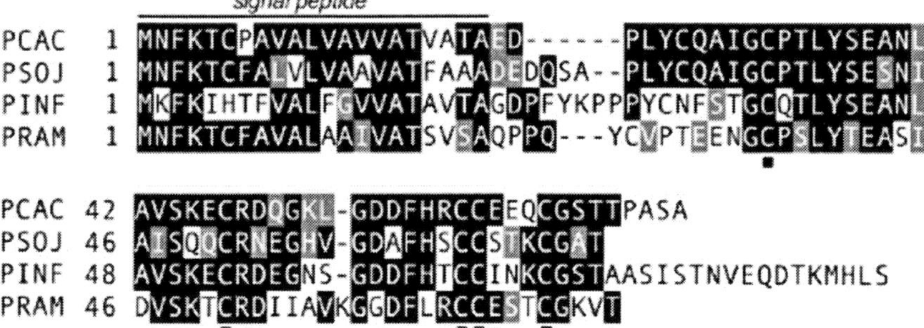

Fig. 16.5 PcF proteins from *P. cactorum*, *P. sojae*, *P. infestans*, and *P. ramorum* (*top to bottom*). Shown are the closest relatives in each species, with *shading* representing regions of >50% conservation. *Black boxes* underneath the alignment indicate positions of conserved cysteines. These are presumably important in establishing a compact structure resistant to extracellular proteases, and are a feature of many other oomycete effectors including elicitins (Sect. 16.4.4.2) and the EPI proteins (Sect. 16.4.4.6)

16.4.4.5 Nep1-Like Proteins (NLPs)

This is one of the few necrosis-inducing effectors with homologues in other kingdoms. It was identified nearly simultaneously from three oomycetes and found to be related to the Nep1 necrosis-inducing protein of a *Fusarium* pathogen of cacao. The oomycete relatives were discovered by database mining in the case of *P. infestans* and *P. sojae* (Qutob et al. 2002; Randall et al. 2005) and by purifying toxins secreted by *P. parasitica* and *Py. aphanidermatum* (Veit et al. 2001; Fellbrich et al. 2002). Relatives are now known in many fungi and bacteria, causing cell death in at least twenty diverse species of dicots (Pemberton and Salmond 2004). Assays using the PVX system in *N. benthamiana* and particle bombardment in soybean showed that a NLP from *P. sojae*, a 237-aa cysteine-rich secreted protein called PsojNIP, induces necrosis (Qutob et al. 2002). PsojNIP expression starts late in infection, potentially explaining how the transition to necrotrophy occurs in this hemibiotroph. Analyses of the *P. ramorum* and *P. sojae* genomes detect 50 to 60 relatives, although some may be pseudogenes.

16.4.4.6 Apoplastic Enzyme Inhibitors

Oomycetes secrete inhibitors that may hinder plant defense enzymes in the apoplast. These include two Kazal-like protease inhibitors (EPI1 and EPI10) and several cystatin-like inhibitors (EPIC1, EPIC2, EPIC3, and EPIC4) that were discovered using a bioinformatic search for secreted proteins from *P. infestans* (Song et al. 2005; Tian et al. 2004, 2005). One EPIC2 protein appears to interact with an extracellular protease from tomato that is closely related to Rcr3, a cysteine protease shown to function in resistance to fungi. EPI1 and EPI10 inhibit a subtilisin-like serine protease that tomato secretes as part of its defense response, and silencing of EPI1 expression in *P. infestans* transformants is reported to affect virulence on tomato (Tian et al. 2007).

Phytophthora spp. also appear to have counter-defenses against plant endoglucanases. These plant enzymes are thought to impair the pathogen directly by weakening its cell wall and indirectly by releasing 1,6-α-linked and 1,3-β-branched heptaglucosides which activate phenylpropanoid defenses (Rose et al. 2002). First purified biochemically from *P. sojae*, the inhibitors are encoded by gene families in each *Phytophthora* spp. that appear to have coevolved with the matching host glucanases (Bishop et al. 2005).

16.4.4.7 Crinkler (CRN) Proteins

While the proteins listed above are thought to act in the plant apoplast, members of the CRN family of proteins appear to act in the cytoplasm (Torto et al. 2003). They were identified by searching *P. infestans* ESTs for sequences encoding signal peptides, followed by functional testing *in planta* using agroinfection. Two related proteins, CRN1 and CRN2, were found to induce leaf crinkling and necrosis on *N. benthamiana*, tobacco, and tomato. At least 50 CRN-like proteins are found in each *Phytophthora* genome, ranging in size from 450 to 850 aa, and these show signs of strong diversifying selection. The proteins may be regulated by phosphorylation (Ebstrup et al. 2005).

16.4.4.8 RXLR Effectors

Several oomycete proteins thought to act within the plant cytoplasm, including all known AVR gene products, share an N-terminal targeting domain comprised of a signal peptide followed by a motif with the consensus of RXLR-X_{5-21}-ddEER (Fig. 16.6). The latter resembles a host-targeting signal from the malaria parasite *Plasmodium*, where it mediates transport across the outer membrane of host blood cells (Lingelbach and Przyborski 2006). Presumably the RXLR motif enables the trafficking of secreted oomycete proteins into the plant. Deletion analyses indicate that once introduced into a plant cell, the RXLR region itself is not required for effector activity (Bos et al. 2006).

Two AVR genes are identified from *H. parasitica*. *ATR1*[NdWsB] determines avirulence against *A. thaliana* ecotypes carrying the *RPP1-Nd* and *RPP1-Wd* resistance genes (Rehmany et al. 2005). The locus, positionally cloned using AFLP markers obtained by bulked segregant analysis, encodes a 311-aa inducer of cell death in resistant plants. Alleles in *H. parasitica* isolates are highly polymorphic, suggestive of pressure to evolve to virulence. *ATR13* was discovered as a pathogen gene in a study of gene expression in infected *A. thaliana* (Bittner-Eddy et al. 2003), and its AVR function was suggested by cosegregation analysis (Allen et al. 2004). That its 187-aa

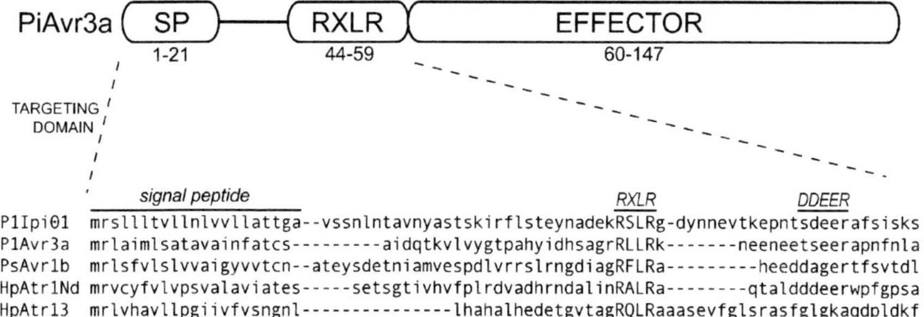

Fig. 16.6 Typical RXLR effectors from *Phytophthora*. Indicated are the N-terminal portions of Avr3a and IpiO1 of *P. infestans*, ATR1[NdWsB] and Atr13 of *H. parasitica*, and Avr1b-1 from *P. sojae*, with their signal peptide RXLR and ddEER blocks marked as defined by Rehmany et al. (2005)

protein triggers *RPP13*-dependent defense reactions was confirmed by in planta expression. Like *ATR1[NdWsB]* and other effectors such as CRN and PcF, *ATR13* exhibits high polymorphism with evidence of diversifying selection.

Two *Phytophthora* AVR genes are also identified. *Avr1-b* of *P. sojae* was isolated by positional cloning, using the *P. sojae* genetic map and a large F_2 mapping population (Shan et al. 2004a). The gene encodes a 138-aa protein that causes cell death when infiltrated into *Rps1b* soybean; it is assumed but not proved to enter plant cells, since Rps1b is cytoplasmic. A closely linked gene, *Avr1b-2*, regulates *Avr1b-1* expression through an unknown mechanism. Also cloned is *Avr3a* from *P. infestans*, which was discovered using association genetics (Armstrong et al. 2005). This entailed checking ESTs encoding small secreted proteins for polymorphisms associated with avirulence phenotypes of *P. infestans* isolates. The 147-aa Avr3a protein was shown to determine R3a-specific cell death using *in planta* expression assays.

At least 100 other RXLR genes have been found in the genomes of *P. infestans*, *P. ramorum*, and *P. sojae*. Some, such as *ipiO* of *P. infestans*, had also been identified as *in planta* expressed genes using differential cloning methods (Pieterse et al. 1994). While the avirulence functions of several of the oomycete proteins are demonstrated, none are yet shown to confer an advantage to virulent isolates.

16.5   Conclusion

Genomics data is revealing a wealth of information about factors needed by oomycetes to succeed as pathogens, including genes involved in forming the infectious propagules and influencing the physiology of their plant hosts. The application of transformation for functional analyses of such genes is now firmly established in *Phytophthora*, although improvements in transformation rates and gene silencing efficiency would be useful. The extension of such methods to obligate pathogens such as downy mildews remains an elusive goal. Doing so is important since comparative studies of different oomycetes should be informative. Ultimately, this should lead to breakthroughs in understanding and controlling these important pathogens.

Acknowledgements The authors' laboratory has been supported by the Department of Agriculture-CSREES and National Science Foundation of the United States, the University of California Discovery Grant program, and Syngenta Corporation.

References

Ah Fong A, Judelson HS (2003) Cell cycle regulator *Cdc14* is expressed during sporulation but not hyphal growth in the fungus-like oomycete *Phytophthora infestans*. Mol Microbiol 50:487–494

Al-Kherb SM, Fininsa C, Shattock RC, Shaw DS (1995) The inheritance of virulence of *Phytophthora infestans* to potato. Plant Pathol 44:552–562

Allen RL, Bittner-Eddy PD, Grenville-Briggs LJ, Meitz JC, Rehmany AP, Rose LE, Beynon JL (2004) Host-parasite coevolutionary conflict between *Arabidopsis* and downy mildew. Science 306:1957–1960

Ambikapathy J, Marshall JS, Hocart CH, Hardham AR (2002) The role of proline in osmoregulation in *Phytophthora nicotianae*. Fungal Genet Biol 35:287–299

Armstrong MR, Whisson SC, Pritchard L, Bos JI, Venter E, Avrova AO, Rehmany AP, Bohme U, Brooks K, Cherevach I, Hamlin N, White B, Fraser A, Lord A, Quail MA, Churcher C, Hall N, Berriman M, Huang S, Kamoun S, Beynon JL, Birch PR (2005) An ancestral oomycete locus contains late blight avirulence gene *Avr3a*, encoding a protein that is recognized in the host cytoplasm. Proc Natl Acad Sci USA 102:7766–7771

Avrova AO, Venter E, Birch PRJ, Whisson SC (2004) Profiling and quantifying differential gene transcription in *Phytophthora infestans* prior to and during the early stages of potato infection. Fungal Genet Biol 40:4–14

Baldauf SL, Roger AJ, Wenk-Siefert I, Doolittle WF (2000) A kingdom-level phylogeny of eukaryotes based on combined protein data. Science 290:972–977

Beyer K, Binder A, Boller T, Collinge M (2001) Identification of potato genes induced during colonization by *Phytophthora infestans*. Mol Plant Pathol 2:125–134

Bishop JG, Ripoll DR, Bashir S, Damasceno CM, Seeds JD, Rose JK (2005) Selection on *Glycine* ß-1,3-endoglucanase genes differentially inhibited by a *Phytophthora* glucanase inhibitor protein. Genetics 169:1009–1019

Bittner-Eddy PD, Allen RL, Rehmany AP, Birch P, Beynon JL (2003) Use of suppression subtractive hybridization to identify downy mildew genes expressed during infection of *Arabidopsis thaliana*. Mol Plant Pathol 4:501–507

Blanco FA, Judelson HS (2005) A bZIP transcription factor from *Phytophthora* interacts with a protein kinase and is required for zoospore motility and plant infection. Mol Microbiol 56:638–648

Bos JI, Kanneganti TD, Young C, Cakir C, Huitema E, Win J, Armstrong MR, Birch PR, Kamoun S (2006) The C-terminal half of *Phytophthora infestans* RXLR effector AVR3a is sufficient to trigger R3a-mediated hypersensitivity and suppress INF1-induced cell death in *Nicotiana benthamiana*. Plant J 48:165–176

Brunner F, Rosahl S, Lee J, Rudd JJ, Geiler C, Kauppinen S, Rasmussen G, Scheel D, Nuernberger T (2002) Pep-13, a plant defense-inducing pathogen-associated pattern from *Phytophthora* transglutaminases. EMBO J 21:6681–6688

Chen X, Shen G, Wang Y, Zheng X, Wang Y (2007) Identification of *Phytophthora sojae* genes upregulated during the early stage of soybean infection. FEMS Microbiol Lett 269:280–288

Chen Y, Roxby R (1997) Identification of a functional CT-element in the *Phytophthora infestans piypt1* gene promoter. Gene 198:159–164

Connolly MS, Sakihama Y, Phuntumart V, Jiang Y, Warren F, Mourant L, Morris PF (2005) Heterologous expression of a pleiotropic drug resistance transporter from *Phytophthora sojae* in yeast transporter mutants. Curr Genet 48:356–365

Cvitanich C, Judelson H (2003) Stable transformation of the oomycete, *Phytophthora infestans*, using microprojectile bombardment. Curr Genet 42:228–235

Dean RA, Talbot NJ, Ebbole DJ, Farman ML, Mitchell TK, Orbach MJ, Thon M, Kulkarni R, Xu JR, Pan H, Read ND, Lee YH, Carbone I, Brown D, Oh YY, Donofrio N, Jeong JS, Soanes DM, Djonovic S, Kolomiets E, Rehmeyer C, Li W, Harding M, Kim S, Lebrun MH, Bohnert H, Coughlan S, Butler J, Calvo S, Ma LJ, Nicol R, Purcell S, Nusbaum C, Galagan JE, Birren BW (2005) The genome sequence of the rice blast fungus *Magnaporthe grisea*. Nature 434:980–986

Dick MW (2001), Straminipilous fungi. Kluwer Academic Publishers, Dordrecht

Dong W, Latijnhouwers M, Jiang RHY, Meijer HJG, Govers F (2005) Downstream targets of the *Phytophthora infestans* G-alpha subunit PiGPA1 revealed by cDNA-AFLP. Mol Plant Pathol 5:483–494

Ebstrup T, Saalbach G, Egsgaard H (2005) A proteomics study of in vitro cyst germination and appressoria formation in *Phytophthora infestans*. Proteomics 5:2839–2848

Erwin DC, Ribeiro OK (1996), *Phytophthora* diseases worldwide. APS Press, St Paul, MN

Fabritius AL, Judelson HS (2003) A mating-induced protein of *Phytophthora infestans* is a member of a family of elicitors with divergent structures and stage-specific patterns of expression. Mol Plant-Microbe Intl 16:926–935

Fabritius AL, Cvitanich C, Judelson HS (2002) Stage-specific gene expression during sexual development in *Phytophthora infestans*. Mol Microbiol 45:1057–1066

Fellbrich G, Romanski A, Varet A, Blume B, Brunner F, Engelhardt S, Felix G, Kemmerling B, Krzymowska M, Nuernberger T (2002) NPP1, a *Phytophthora*-associated trigger of plant defense in parsley and *Arabidopsis*. Plant J 32:375–390

Francis DM, St Clair DA (1993) Outcrossing in the homothallic oomycete, *Pythium ultimum*, detected with molecular markers. Curr Genet 24:100–106

Gaulin E, Jauneau A, Villalba F, Rickauer M, Esquerre-Tugaye MT, Bottin A (2002) The CBEL glycoprotein of *Phytophthora parasitica* var. *nicotianae* is involved in cell wall deposition and adhesion to cellulosic substrates. J Cell Sci 115:4565–4575

Goernhardt B, Rouhara I, Schmelzer E (2000) Cyst germination proteins of the potato pathogen *Phytophthora infestans* share homology with human mucins share homology with human mucins. Mol Plant-Microbe Interact 13:32–42

Gotesson A, Marshall JS, Jones DA, Hardham AR (2002) Characterization and evolutionary analysis of a large polygalacturonase gene family in the oomycete plant pathogen *Phytophthora cinnamomi*. Mol Plant-Microbe Interact 15:907–921

Grenville-Briggs LJ, Avrova AO, Bruce CR, Williams A, Whisson SC, Birch PR, van West P (2005) Elevated amino acid biosynthesis in Phytophthora infestans during appressorium formation and potato infection. Fungal Genet Biol 42:244–256

Hardham AR (2001) The cell biology behind *Phytophthora* pathogenicity. Austral Plant Pathol 30:91–98

Hardham AR, Hyde GJ (1997) Asexual sporulation in the oomycetes. Adv Bot Res 24:353–398

Hohl HR, Suter E (1976) Host-parasite interfaces in a resistant and a susceptible cultivar of *Solanum tuberosum* inoculated with *Phytophthora infestans*: leaf tissue. Can J Bot 54:1956–1970

Huitema E, Bos JI, Tian M, Win J, Waugh ME, Kamoun S (2004) Linking sequence to phenotype in *Phytophthora*-plant interactions. Trends Microbiol 12:193–200

Jiang RH, Tyler BM, Whisson SC, Hardham AR, Govers F (2005a) Ancient origin of elicitin gene clusters in *Phytophthora* genomes. Mol Biol Evol 23:338–351

Jiang RHY, Tyler B, Govers F (2005b) Comparative genomics and synteny studies revealing the reservoir of secreted proteins in *Phytophthora*. *Phytophthora* Molecular Genetics Workshop, unnumbered abstract

Judelson HS (1993) Intermolecular ligation mediates efficient cotransformation in *Phytophthora infestans*. Mol Gen Genet 239:241–250

Judelson HS (2007) Genomics of the plant pathogenic oomycete *Phytophthora*: insights into biology and evolution. Adv Genet 57:98–142

Judelson HS, Randall TA (1998) Families of repeated DNA in the oomycete *Phytophthora infestans* and their distribution within the genus. Genome 41:605–615

Judelson HS, Ah-Fong AMV, Aux G, Avrova AO, Bruce C, Cakir C, da Cunha L, Grenville-Briggs L, Latijnhouwers M, Ligterink W, Meijer HJG, Roberts S, Thurber CS, Whisson SC, Birch PRJ, Govers F, Kamoun S, van West P, Windass J (2008) Gene expression profiling during asexual development of the late blight pathogen Phytophthora infestans reveals a highly dynamic transcriptome. Mol Plant-Microbe Interact 21:433–447

Judelson HS, Tani S (2007) Transgene-induced silencing of the zoosporogenesis-specific PiNIFC gene cluster of *Phytophthora infestans* involves chromatin alterations. Eukaryot Cell 6:1200–1209

Judelson HS, Tyler BM, Michelmore RW (1991) Transformation of the oomycete pathogen, *Phytophthora infestans*. Mol Plant-Microbe Inter 4:602–607

Judelson HS, Tyler BM, Michelmore RW (1992) Regulatory sequences for expressing genes in oomycete fungi. Mol Gen Genet 234:138–146

Judelson HS, Coffey MD, Arredondo FR, Tyler BM (1993a) Transformation of the oomycete pathogen *Phytophthora megasperma* f. sp. *glycinea* occurs by DNA integration into single or multiple chromosomes. Curr Genet 23:211–218

Judelson HS, Dudler R, Pieterse CMJ, Unkles SE, Michelmore RW (1993b) Expression and antisense inhibition of transgenes in *Phytophthora infestans* is modulated by choice of promoter and position effects. Gene 133:63–69

Judelson HS, Spielman LJ, Shattock RC (1995) Genetic mapping and non-Mendelian segregation of mating type loci in the oomycete, *Phytophthora infestans*. Genetics 141:503–512

Kamoun S, van West P, Vleshouwers VGAA, De Groot KE, Govers F (1998) Resistance of *Nicotiana benthamiana* to *Phytophthora infestans* is mediated by the recognition of the elicitor protein INF1. Plant Cell 10:1413–1425

Kamoun S, Hraber P, Sobral B, Nuss D, Govers F (1999) Initial assessment of gene diversity for the oomycete pathogen *Phytophthora infestans* based on expressed sequences. Fungal Genet Biol 28:94–106

Kanneganti TD, Bai X, Tsai CW, Win J, Meulia T, Goodin M, Kamoun S, Hogenhout SA (2007) A functional genetic assay for nuclear trafficking in plants. Plant J 50:149–158

Kiefer B, Riemann M, Bueche C, Kassemeyer H-H, Nick P (2002) The host guides morphogenesis and stomatal targeting in the grapevine pathogen *Plasmopara viticola*. Planta 215:387–393

Kim KS, Judelson HS (2003) Sporangia-specific gene expression in the oomyceteous phytopathogen *Phytophthora infestans*. Eukaryot Cell 2:1376–1385

Ko WH (1988) Hormonal heterothallism and homothallism in *Phytophthora*. Ann Rev Phytopathol 26:57–73

Koch E, Slusarenko A (1990) *Arabidopsis* is susceptible to infection by a downy mildew fungus. Plant Cell 2:437–445

Kraemer R, Freytag S, Schmelzer E (1997) In vitro formation of infection structures of *Phytophthora infestans* in associated with synthesis of stage specific polypeptides. Eur J Plant Pathol 103:43–53

Lamour KH, Finley L, Hurtado-Gonzales O, Gobena D, Tierney M, Meijer HJ (2006) Targeted gene mutation in *Phytophthora* spp. Mol Plant Microbe Interact 19:1359–1367

Latijnhouwers M, Govers F (2003) A *Phytophthora infestans* G-Protein ß subunit is involved in sporangium formation. Euk Cell 2:971–977

Latijnhouwers M, Munnik T, Govers F (2002) Phospholipase D in *Phytophthora infestans* and its role in zoospore encystment. Mol Plant-Microbe Interact 15:939–946

Latijnhouwers M, Ligterink W, Vleeshouwers VGAA, van West P, Govers F (2004) A G-alpha subunit controls zoospore motility and virulence in the potato late blight pathogen *Phytophthora infestans*. Mol Microbiol 51:925–936

Laxalt AM, Latijnhouwers M, van Hulten M, Govers F (2002) Differential expression of G protein alpha and beta subunit genes during development of *Phytophthora infestans*. Fungal Genet Biol 36:137–146

Legard DE, Lee TY, Fry WE (1995) Pathogenic specialization in *Phytophthora infestans:* Aggressiveness on tomato. Phytopathology 85:1355–1361

Lin E, Jauneau A, Villalba F, Rickauer M, Esquerre-Tugaye MT, Bottin A (2002) The CBEL glycoprotein of *Phytophthora parasitica* var. *nicotianae* is involved in cell wall deposition and adhesion to cellulosic substrates. J Cell Sci 115:4565–4575

Lingelbach K, Przyborski JM (2006) The long and winding road: protein trafficking mechanisms in the *Plasmodium falciparum* infected erythrocyte. Mol Biochem Parasitol 147:1–8

Liu Z, Bos JI, Armstrong M, Whisson SC, da Cunha L, Torto-Alalibo T, Win J, Avrova AO, Wright F, Birch PR, Kamoun S (2005) Patterns of diversifying selection in the phytotoxin-like scr74 gene family of *Phytophthora infestans*. Mol Biol Evol 22:659–672

MacGregor T, Bhattacharyya M, Tyler B, Bhat R, Schmitthenner AF, Gijzen M (2002) Genetic and physical mapping of *Avr1a* in *Phytophthora sojae*. Genetics 160:949–959

Malajczuk N (1983) Microbial antagonism to *Phytophthora*. In: Erwin DC, Bartnicki-Garcia S, Tsao PH (eds) *Phytophthora,* its biology, taxonomy, ecology, and pathology. APS Press, pp 197–218

Marshall JS, Ashton AR, Govers F, Hardham AR (2001a) Isolation and characterization of four genes encoding pyruvate, phosphate dikinase in the oomycete plant pathogen *Phytophthora cinnamomi*. Curr Genet 40:73–81

Marshall JS, Wilkinson JM, Moore T, Hardham AR (2001b) Structure and expression of the genes encoding proteins resident in large peripheral vesicles of *Phytophthora cinnamomi* zoospores. Protoplasma 215:226–239

Martin FN (1989) Maternal inheritance of mitochondrial DNA in sexual crosses of *Pythium sylvaticum*. Curr Genet 16:373–374

May KJ, Whisson SC, Zwart RS, Searle IR, Irwin JA, Maclean DJ, Carroll BJ, Drenth A (2002) Inheritance and mapping of 11 avirulence genes in *Phytophthora sojae*. Fungal Genet Biol 37:1–12

Michelmore RW, Ilott T, Hulbert SH, Farrara B (1988) The downy mildews. Adv Plant Pathol 6:53–79

Moy P, Qutob D, Chapman BP, Atkinson I, Gijzen M (2004) Patterns of gene expression upon infection of soybean plants by *Phytophthora sojae*. Mol Plant-Microbe Interact 17:1051–1062

Nespoulous C, Gaudemer O, Huet JC, Pernollet JC (1999) Characterization of elicitin-like phospholipases isolated from *Phytophthora capsici* culture filtrate. FEBS Lett 452: 400–406

Nürnberger T, Brunner F (2002) Innate immunity in plants and animals: emerging parallels between the recognition of general elicitors and pathogen-associated molecular patterns. Curr Opin Plant Biol 5:318–324

Nürnberger T, Nennstiel D, Hahlbrock K, Scheel D (1995) Covalent cross-linking of the *Phytophthora megasperma* oligopeptide elicitor to its receptor in parsley membranes. Proc Natl Acad Sci USA 92:2338–2342

Orsomando G, Lorenzi M, Raffaelli N, Dalla Rizza M, Mezzetti B, Ruggieri S (2001) Phytotoxic protein PcF, purification, characterization, and cDNA sequencing of a novel hydroxyproline-containing factor secreted by the strawberry pathogen *Phytophthora cactorum*. J Biol Chem 276:21578–21584

Osman H, Mikes V, Milat ML, Ponchet M, Marion D, Prange T, Maume BF, Vauthrin S, Blein JP (2001) Fatty acids bind to the fungal elicitor cryptogein and compete with sterols. FEBS Lett 489:55–58

Panabieres F, Birch PR, Unkles SE, Ponchet M, Lacourt I, Venard P, Keller H, Allasia V, Ricci P, Duncan JM (1998) Heterologous expression of a basic elicitin from *Phytophthora cryptogea* in *Phytophthora infestans* increases its ability to cause leaf necrosis in tobacco. Microbiology 144:3343–3349

Panabieres F, Amselem J, Galiana E, Le Berre J-Y (2005) Gene identification in the oomycete pathogen *Phytophthora parasitica* during in vitro vegetative growth through expressed sequence tags. Fung Genet Biol 42:611–623

Pemberton CL, Salmond GPC (2004) The Nep1-like proteins—a growing family of microbial elicitors of plant necrosis. Mol Plant Pathol 5:353–359

Pieterse CMJ, Derksen A-MCE, Folders J, Govers F (1994) Expression of the *Phytophthora infestans ipiB* and *ipi0* genes in planta and in vitro. Mol Gen Genet 244:269–277

Ponchet M, Panabieres F, Milat ML, Mikes V, Montillet JL, Suty L, Triantaphylides C, Tirilly Y, Blein JP (1999) Are elicitins cryptograms in plant-oomycete communications? CMLS Cell Mol Life Sci 56:1020–1047

Prakob W, Judelson HS (2007) Gene expression during oosporogenesis in heterothallic and homothallic Phytophthora. Fungal Genet Biol 44:726–739

Qutob D, Hraber PT, Sobral BWS, Gijzen M (2000) Comparative analysis of expressed sequences in *Phytophthora sojae*. Plant Physiol 123:243–253

Qutob D, Kamoun S, Gijzen M (2002) Expression of a *Phytophthora sojae* necrosis-inducing protein occurs during transition from biotrophy to necrotrophy. Plant J 32:361–373

Randall TA, Dwyer RA, Huitema E, Beyer K, Cvitanich C, Kelkar H, Ah Fong AMV, Gates K, Roberts S, Yatzkan E, Gaffney T, Law M, Testa A, Torto T, Zhang M, Zheng L, Mueller E, Windass J, Binder A, Birch PRJ, Gisi U, Govers F, Gow NAR, Mauch F, van West P, Waugh ME, Yu J, Boller T, Kamoun S, Lam ST, Judelson HS (2005) Large-scale gene discovery in the oomycete *Phytophthora infestans* reveals likely components of phytopathogenicity shared with true fungi. Mol Plant-Microbe Interact 18:229–243

Rehmany AP, Gordon A, Rose LE, Allen RL, Armstrong MR, Whisson SC, Kamoun S, Tyler BM, Birch PR, Beynon JL (2005) Differential recognition of highly divergent downy mildew avirulence gene alleles by RPP1 resistance genes from two *Arabidopsis* lines. Plant Cell 17:1839–1850

Rose JKC, Ham K-S, Darvill AG, Albersheim P (2002) Molecular cloning and characterization of glucanase inhibitor proteins: coevolution of a counterdefense mechanism by plant pathogens. Plant Cell 14:1329–1345

Sejalon-Delmas N, Villalba Mateos F, Bottin A, Rickauer M, Dargent R, Esquerre-Tugaye MT (1997) Purification, elicitor activity, and cell wall localization of a glycoprotein from *Phytophthora parasitica* var. *nicotianae*, a fungal pathogen of tobacco. Phytopathology 87:899–909

Shan W, Cao M, Leung D, Tyler BM (2004a) The *Avr1b* locus of *Phytophthora sojae* encodes an elicitor and a regulator required for avirulence on soybean plants carrying resistance gene *Rps1b*. Mol Plant Microbe Interact 17:394–403

Shan W, Marshall JS, Hardham Adrienne R (2004b) Gene expression in germinated cysts of *Phytophthora nicotianae*. Mol Plant Pathol 5:317–330

Shattock RC, Tooley PW, Fry WE (1986) Genetics of *Phytophthora infestans*: determination of recombination, segregation, and selling by isozyme analysis. Phytopathology 76:410–413

Shepherd SJ, van West P, Gow NAR (2003) Proteomic analysis of asexual development of *Phytophthora palmivora*. Mycol Res 107:395–400

Sicard D, Legg E, Brown S, Babu NK, Ochoa O, Sudarshana P, Michelmore RW (2003) A genetic map of the lettuce downy mildew pathogen, *Bremia lactucae*, constructed from molecular markers and avirulence genes. Fungal Genet Biol 39:16–30

Skalamera D, Wasson AP, Hardham AR (2004) Genes expressed in zoospores of *Phytophthora nicotianae*. Mol Genet Genom 270:549–557

Song J, Champouret N, Win J, Tian M, Kamoun S (2005) Genetic evidence for a role of *Phytophthora infestans* protease inhibitors in disease. The 23rd Fungal Genetics Conference, Abstract 348

Spencer DM (1981), The downy mildews. Academic Press, New York

Stanghellini ME, Gilbertson RL (1988) *Plasmopara lactucae-radicis* new species on roots of hydroponically grown lettuce. Mycotaxon 31:395–400

Stegmeier F, Amon A (2004) Closing mitosis: the functions of the Cdc14 phosphatase and its regulation. Annu Rev Genet 38:203–232

Tani S, Yatzkan E, Judelson HS (2004) Multiple pathways regulate the induction of genes during zoosporogenesis in *Phytophthora infestans*. Mol Plant-Microbe Inter 17:330–337

Tian M, Huitema E, da Cunha L, Torto-Alalibo T, Kamoun S (2004) A Kazal-like extracellular serine protease inhibitor from *Phytophthora infestans* targets the tomato pathogenesis-related protease P69B. J Biol Chem 279:26370–26377

Tian M, Benedetti B, Kamoun S (2005) A second Kazal-like protease inhibitor from *Phytophthora infestans* inhibits and interacts with the apoplastic pathogenesis-related protease P69B of tomato. Plant Physiol 138:1785–1793

Tian M, Win J, Song J, van der Hoorn R, van der Knaap E, Kamoun S (2007) A Phytophthora infestans cystatin-like protein targets a novel tomato papain-like apoplastic protease. Plant Physiol 143:364–377

Torto TA, Rauser L, Kamoun S (2002) The *pipg1* gene of the oomycete *Phytophthora infestans* encodes a fungal-like endopolygalacturonase. Curr Genet 40:385–390

Torto TA, Li S, Styer A, Huitema E, Testa A, Gow NA, van West P, Kamoun S (2003) EST mining and functional expression assays identify extracellular effector proteins from the plant pathogen *Phytophthora*. Genome Res 13:1675–1685

Tyler BM (2002) Molecular basis of recognition between *Phytophthora* pathogens and their hosts. Ann Rev Phytopathol 40:137–167

Tyler BM, Tripathy S, Zhang X, Dehal P, Jiang RH, Aerts A, Arredondo FD, Baxter L, Bensasson D, Beynon JL, Chapman J, Damasceno CM, Dorrance AE, Dou D, Dickerman AW, Dubchak IL, Garbelotto M, Gijzen M, Gordon SG, Govers F, Grunwald NJ, Huang W, Ivors KL, Jones RW, Kamoun S, Krampis K, Lamour KH, Lee MK, McDonald WH, Medina M, Meijer HJ, Nordberg EK, Maclean DJ, Ospina-Giraldo MD, Morris PF, Phuntumart V, Putnam NH, Rash S, Rose JK, Sakihama Y, Salamov AA, Savidor A, Scheuring CF, Smith BM, Sobral BW, Terry A, Torto-Alalibo TA, Win J, Xu Z, Zhang H, Grigoriev IV, Rokhsar DS, Boore JL (2006) Phytophthora genome sequences uncover evolutionary origins and mechanisms of pathogenesis. Science 313:1261–1266

Valent B, Chumley FG (1991) Molecular genetic analysis of the rice blast fungus *Magnaporthe grisea*. Ann Rev Phytopathol 29:443–468

van der Lee T, Testa A, Robold A, van 't Klooster JW, Govers F (2004) High density genetic linkage maps of *Phytophthora infestans* reveal trisomic progeny and chromosomal rearrangements. Genetics 157:949–956

van der Plaats-Niterink AJ (1981) Monograph of the genus *Pythium*. Centraalbureau Voor Schimmelcultures, Bamm, The Netherlands

van Dijk K, Fouts DE, Rehm AH, Hill AR, Collmer A, Alfano JR (1999) The Avr (effector) proteins HrmA (HopPsyA) and AvrPto are secreted in culture from *Pseudomonas syringae* pathovars via the Hrp (type III) protein secretion system in a temperature- and pH-sensitive manner. J Bacteriol 181:4790–4797

van West P, Kamoun S, van 't Klooster JW, Govers F (1999a) Internuclear gene silencing in *Phytophthora infestans*. Mol Cell 3:339–348

van West P, Reid B, Campbell TA, Sandrock RW, Fry WE, Kamoun S, Gow NAR (1999b) Green fluorescent protein (GFP) as a reporter gene for the plant pathogenic oomycete *Phytophthora palmivora*. FEMS Microbiol Lett 178:71–80

van West P, Morris BM, Reid B, Appiah AA, Osborne MC, Campbell TA, Shepherd SJ, Gow NAR (2002) Oomycete plant pathogens use electric fields to target roots. Mol Plant-Microbe Interact 15:790–798

Veit S, Worle JM, Nurnberger T, Koch W, Seitz HU (2001) A novel protein elicitor (PaNie) from *Pythium aphanidermatum* induces multiple defense responses in carrot, *Arabidopsis*, and tobacco. Plant Physiol 127:832–841

Vijn I, Govers F (2003) *Agrobacterium tumefaciens* mediated transformation of the oomycete plant pathogen *Phytophthora infestans*. Mol Plant Pathol 4:456–467

Voglmayr H, Greilhuber J (1998) Genome size determination in peronosporales (Oomycota) by Feulgen image analysis. Fung Genet Biol 25:181–195

Wang Z, Wang Y, Chen X, Shen G, Zhang Z, Zheng X (2006) Differential screening reveals genes differentially expressed in low- and high-virulence near-isogenic Phytophthora sojae lines. Fungal Genet Biol 43:826–839

Warburton AJ, Deacon JW (1998) Transmembrane Ca^{2+} fluxes associated with zoospore encystment and cyst germination by the phytopathogen *Phytophthora parasitica*. Fungal Genet Biol 25:54–62

Weiland JJ (2003) Transformation of *Pythium aphanidermatum* to geneticin resistance. Curr Genet 42:344–352

Whisson SC, Drenth A, Maclean DJ, Irwin JA (1994) Evidence for outcrossing in *Phytophthora sojae* and linkage of a DNA marker to two avirulence genes. Curr Genet 27:77–82

Whisson SC, Drenth A, MacLean DJ, Irwin JAG (1995) *Phytophthora sojae* avirulence genes, RAPD, and RFLP markers used to construct a detailed genetic linkage map. Mol Plant-Microbe Interact 8:988–995

Whisson SC, Avrova AO, van West P, Jones JT (2005) A method for double-stranded RNA-mediated transient gene silencing in *Phytophthora infestans*. Mol Plant Pathol 6:153–163

Woods AM, Gay JL (1983) Evidence for a neckband delimiting structural and physiological regions of the host plasma membrane associated with haustoria of *Albugo candida*. Physiol Plant Pathol 23:73–88

Yan HZ, Liou RF (2005) Cloning and analysis of *pppg1*, an inducible endopolygalacturonase gene from the oomycete plant pathogen *Phytophthora parasitica*. Fungal Genet Biol 42:339–350

Chapter 17
Molecular Methods for Studying Microbial Ecology in the Soil and Rhizosphere

Janice E. Thies

This work is dedicated to my brother, Eric, whose incredible talents are now lost to this world.

17.1 Introduction

As described throughout this book, soil and rhizosphere microorganisms are responsible for a wide range of ecosystem services, including decomposing organic matter, cycling and immobilizing nutrients, aggregating soil, filtering and bioremediating pollutants, suppressing and causing plant disease, and producing and releasing greenhouse gasses. A long-standing challenge for studies in soil and rhizosphere ecology has been developing effective methods that can be used to describe the diversity, function and abundance of soil and plant-associated microbial populations. Enormous advances have been made since the first report by Torsvik (1980) that deoxyribonucleic acids (DNA) could be extracted from soil and subsequently characterized and that there may be as many as 6000–10,000 different genomes in 1 g of soil (Torsvik et al. 1990). A recent analysis based on reassociation kinetics done by Gans et al. (2005) suggests that this number is conservative and that the number of individual genomes per 1 g of soil may approach 277,000. This number far exceeds diversity estimates from any other matrix, making soil the most complex and diverse environment on earth.

Characterizing this diversity is an intricate and difficult task, in which all current methods fall short. Each month better approaches are being developed and published, allowing us to continue to explore this biologically rich environment. These new approaches have enabled us to not only ask who is living in soil, but to also determine how populations respond to management and consider how we might develop better soil management practices to encourage beneficial associations between plants and the soil biota and discourage detrimental ones.

J.E. Thies
Department of Crop and Soil Sciences, Cornell University, Ithaca, NY
e-mail: jet25@cornell.edu

C.S. Nautiyal, P. Dion (eds.) *Molecular Mechanisms of Plant and Microbe Coexistence.* Soil Biology 15, DOI: 10.1007/978-3-540-75575-3

Measures that describe population diversity attempt to capture (i) the genetic variability within a species, (ii) the number (richness) and relative abundance (evenness) of species, and/or (iii) the number of different functional groups within studied communities (Torsvik and Ovreas 2002). Describing diversity at the ecosystem scale often involves (i) identifying the variety of processes occurring, (ii) characterizing the interactions taking place between different organisms and/or (iii) assessing the number of trophic levels represented within the community. A major difficulty with describing diversity for microorganisms is that the species concept, derived from plant and animal community ecology, does not translate well to microbial populations. As yet, there is no satisfactory species concept for bacteria or fungi (Ward 1998; Liu and Stahl 2002), making it somewhat difficult to characterize the diversity of these populations in ecologically meaningful ways. The advent of molecular ecology has not resolved, but rather complicated the picture as more has become known about the lateral transfer of genetic elements between bacteria in the environment (Smalla and Sobecky 2002).

Assessing microbial population function frequently involves measuring rates of different processes, such as organic matter decomposition, respiratory activity or denitrification; or detecting the presence of genes needed to carry out biochemical reactions of ecological relevance, such as nitrogen fixation (e.g., *nif*H), ammonia or methane oxidation (*amo*A or *pmo*A, respectively), or denitrification (*nar*G, *nap*A, *nir*S, *nir*K, *nor*B, *nor*Z, *nos*Z). Molecular biological approaches have contributed substantially to our understanding of how microbial functions vary in relation to space, time and soil management practices (Handelsman and Smalla 2003). Processes that are unique to particular groups of organisms and are catalyzed by well-described enzyme systems and for which sequence information is known, such as those noted immediately above, have been particularly tractable to study with molecular methods. However, key ecosystem processes, such as depolymerization of organic matter, carbon (C) metabolism or sulfur (S) oxidation are so common across diverse lineages of bacteria and fungi (and other soil eukaryotes), and are carried out by such a diversity of enzyme systems, that molecular approaches may cloud, rather than clarify who or what the key system drivers may be.

Measuring abundance or population density normally involves (i) counting individuals within target groups, such as total bacteria, protozoa or nematodes using microscopy or culturing techniques; (ii) using molecular probes combined with microscopy to enumerate target groups of interest, e.g., the alpha-Proteobacteria or Planctomycetes, and/or (iii) measuring the concentration or content of general or unique biochemical markers, such as microbial biomass, adenosine triphosphate (ATP, the total energy charge of soil) or ergosterol (fungi). It is well known that traditional culturing methods detect only a small fraction of the extant microbial abundance and diversity in soil (Torsvik et al. 1990). Molecular techniques in which DNA is extracted from soil, then cloned and sequenced invariably reveal the presence of populations that are not recovered by traditional culturing (Liu and Stahl 2002). The rapidly expanding sequence databases, such as GeneBank (http://www.ncbi.nlm.nih.gov/), the Ribosomal Database Project II (http://rdp.cme.msu.edu/), EMBL-EBI (http://www.ebi.ac.uk/) and TIGR (http://www.tigr.org/tdb/) assist users in designing probes for use in detecting and quantifying specific populations in environmental samples. However, these approaches require that definitive sequence information is available; hence, populations that may have significant ecological relevance may still be overlooked when using targeted molecular approaches for abundance estimates.

Despite increased access to soil biodiversity by use of molecular methods, community members detected by these approaches may not necessarily correspond to populations responsible for significant biogeochemical processes in situ, especially when these populations constitute only a small proportion of the total community. Likewise, knowing the taxonomic identity or phylogenetic affiliation of a cloned sequence does not necessarily mean that we will know the function of the organism in situ. Even if we can confirm the presence of an organism in a sample whose function is known, it does not necessarily mean that the organism is active. Recently developed RNA-based techniques (e.g., Aneja et al. 2004) and use of stable isotope labeling and tracing (Radajewski et al. 2003) have helped to address this latter point because they can be used to identify those members of a community that are most active under a given set of environmental conditions (see below for approaches and applications).

In studying soil and rhizosphere ecology, one must recognize that organisms residing in these environments are physiologically and phylogenetically diverse. A holistic understanding of microbial communities and their interactions with plant roots requires a polyphasic approach; one that employs culturing and activity measures combined with molecular approaches. Describing and discussing the variety of methods used for polyphasic analysis of rhizosphere communities is beyond the scope of this chapter. Here I focus on recent molecular methods that are being used to characterize soil and rhizosphere microbial community composition and, in some cases, identify the functions of select members of these communities.

The relationship between many of the techniques described and how each is used in microbial community studies is shown in Fig. 17.1. Amplifying and analyzing rRNA genes present in DNA extracted from soil samples forms the basis for many of these techniques. Figure 17.1 also includes references to traditional techniques used in soil microbiology and biochemistry and illustrates how the new molecular approaches support and augment the traditional approaches. A general introduction to soil and rhizosphere ecology is given in Thies and Grossman (2006). An introduction to soil molecular ecology, nucleic acid structure and basic molecular methods is given in Thies (2007a). In this chapter, I describe more advanced analytical methods and how they are being used to better understand soil and rhizosphere microbial ecology.

17.2 Analyzing Nucleic Acids

Nucleic acid sequences define an organism's typical morphology and what activities it can carry out (its genotype). The organism's interaction with its environment and how the genotype is expressed define the organism's phenotype. As more has become known about nucleic acid sequences, particularly the sequences of the rRNA genes, a new phylogeny of the living world has emerged (Woese 1987). This new phylogeny

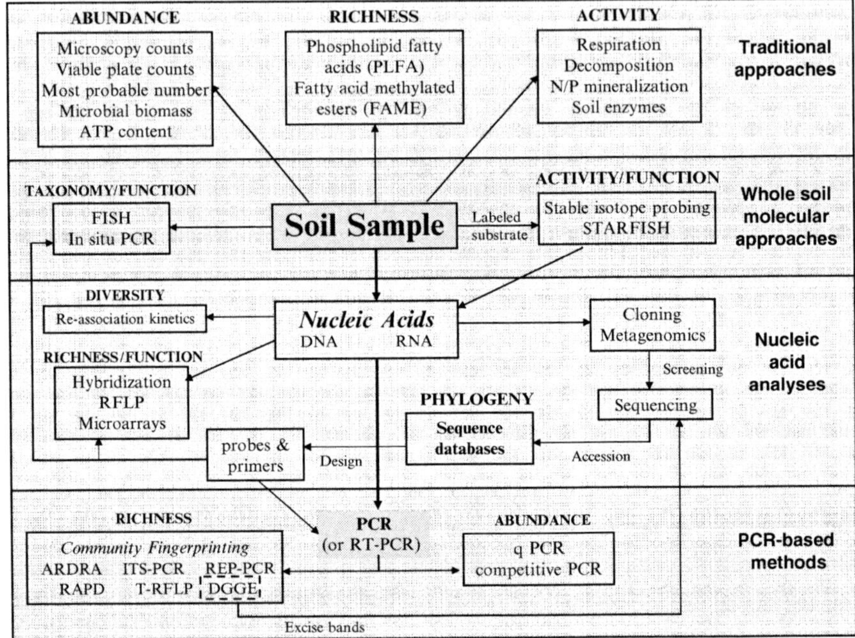

Fig. 17.1 Overview of approaches used to characterize soil microbial communities (adapted from Thies 2007a)

aids in understanding the evolutionary history and relatedness of different organisms to each other (Woese 1987). The ability to predict function from known sequences and to place organisms within a phylogenetic framework make the nucleic acid complement of cells particularly information rich targets for analysis.

A wide range of techniques are now available for analyzing nucleic acids. These techniques fall into three basic categories: (i) methods used to analyze nucleic acids in situ; (ii) those used to analyze extracted DNA/RNA directly; and (iii) those that employ polymerase chain reaction (PCR) to amplify, and subsequently analyze, target DNA sequences or RNA that has been reverse-transcribed to copy DNA (cDNA) (see Fig. 17.1). Applying these techniques to microbial community ecology studies has enabled us to overcome the limitations inherent with traditional enrichment and isolation techniques, thereby allowing us to detect organisms yet to be cultivated and, in some cases, infer their ecological functions.

17.2.1   *Extracting DNA and RNA*

Most molecular approaches require that nucleic acids be extracted from the soil matrix before analysis. A variety of methods have been developed to extract nucleic acids from soils of varying texture and these have been summarized recently by

Bruns and Buckley (2002). Two main approaches are used: (i) cell fractionation and (ii) direct lysis. In cell fractionation, intact microbial cells are released and separated from the soil matrix. After extraction, the cells are subsequently lysed and the DNA separated from the cell debris. In the direct lysis methods, microbial cells are lysed directly in the soil and then the nucleic acids are separated from the soil matrix. The main considerations when choosing a suitable protocol are extraction efficiency, obtaining a sample that is representative of the resident community, and obtaining an extract free of contaminants that could interfere with either PCR or probe hybridization.

Extraction efficiency of both cell fractionation and direct lysis procedures can be assessed by direct microscopy using vital stains, where extracted soil is examined for intact, viable microbial cells. Alternatively, soil samples may be spiked with a known quantity of DNA (or bacterial cells) and then the recovery of the added DNA assessed. In general, DNA recovery is generally much higher when using direct lysis as compared to cell fractionation protocols (Courtois et al. 2001).

DNA is normally extracted from very small quantities of soil, typically 500–1000 mg. This alone makes obtaining a representative sample difficult. In addition, cell walls of different organisms lyse with varying efficiencies. Cell walls of high G+C Gram-positive bacteria are often difficult to lyse, whereas those of Gram-negative bacteria lyse more readily. Hence, DNA or rRNA recovered may contain an artificially greater amount of DNA derived from Gram-negative bacteria. In characterizing microbes colonizing bulk or rhizosphere soil, the rhizoplane and the endorhizosphere, care must be taken to attribute extracted nucleic acids to their associated habitat. For rhizosphere communities, soil adhering to roots can be removed by soaking the roots in buffer with moderate agitation. Roots are removed and nucleic acids extracted from the soil remaining. Roots are then subjected to several rounds of sonication to remove microbes colonizing the rhizoplane. Finally, enzymatic hydrolysis in an appropriate buffer can be used to enrich extracts for nucleic acids from endophytes (Jiao et al. 2006).

Contaminants, such as humic and fulvic acids, have a similar solubility to nucleic acids and hence are often co-extracted. These contaminants interfere with PCR amplification and hybridization experiments. Co-extracted contaminants can be reduced or removed by use of a post-PCR DNA clean-up kit, such as the QIAquick® PCR purification kit (Qiagen, Chatsworth, CA) or GENECLEAN Spin kit (Qbiogene, Inc., Carlsbad, CA); or by washing the nucleic acid extract with dilute EDTA or passing it through a Sephadex G-75 column (Bruns and Buckley 2002). While improving PCR amplification, extra cleaning steps can also lead to a loss of nucleic acids and hence sparsely represented members of the community may be lost from subsequent analyses. In addition, all post-extraction clean-up procedures add cost and processing time, and thus reduce the number of samples that can be analyzed within the scope of any experiment.

Commercial soil DNA/RNA extraction kits based on direct lysis by bead-beating, such as the FastDNA® SPIN Kit for Soil and the FastRNA® Pro Soil-Direct Kit (Qbiogene, Inc., Carlsbad, CA) and the Ultraclean™ and PowerSoil™ DNA isolation kits (MoBio Laboratories, Solana Beach, CA) have recently become available. The DNA/RNA extracted is of high molecular weight and of sufficient quality to be used in PCR or nucleic acid hybridization experiments for most soils. The PowerSoil™ DNA isolation kit (MoBio Laboratories) has been specifically recommended, by the manufacturer, for use with high organic matter samples.

For any given study, the type of molecule(s) that will be extracted, e.g., DNA, RNA, both types of nucleic acids and/or PLFAs, must be determined prior to molecular microbial analysis. DNA is extracted and analyzed most commonly because it is more stable and easier and less costly to extract from soil. Post-extraction analyses are straight-forward and information obtained reflects the whole community at the time of sampling. The key issue with DNA analysis is that it does not reflect the abundance of viable organisms or their level of activity. DNA that is free in soil is readily hydrolyzed by nucleases; however, it can be protected from hydrolysis when present in dead cells or protected within soil aggregates. Protected, free DNA is extracted along with that from moribund and active cells. RNA, on the other hand, is highly labile and more difficult to extract. Methods for extracting rRNA from soil are given in Felske et al. (1999) and Sessitsch et al. (2002) and methods for the simultaneous extraction of DNA and RNA are given in Griffiths et al. (2000) and Hurt et al. (2001). Commercial kits for extracting RNA from soil are also now available (Qbiogene, Inc.). Extracting mRNA is still fraught with difficulty, but some success has been reported (Hurt et al. 2001; Sessitsch et al. 2002). Most post-extraction analyses require that RNA is first reverse-transcribed (RT) into cDNA and then the cDNA is used in downstream analyses. The advantage of extracting and analyzing RNA is that it is generally only present in high amounts in actively metabolizing cells. As substrate becomes limiting, cell processes slow down, along with rDNA transcription. Thus, rRNA analysis is more reflective of the portion of the soil microbial community that is either active at the time of sampling or has recently been active. When mRNA can be recovered, insights into genes that are being actively transcribed under a given set of environmental conditions can be obtained and, hence, is most desirable for studies of microbial community function.

17.2.2   *Re-Association Kinetics*

For assessing total diversity of microorganisms in an environmental sample, re-association kinetics is considered the 'gold standard'. Yet it is rarely performed in research laboratories because the equipment needed is very costly and sample processing times are high, thus it does not lend itself to high sample throughput. In addition, it does not provide any information on identity or function of any member of the microbial community.

For the analysis, DNA is denatured by either heating or use of a denaturant (e.g., urea). Under highly controlled conditions, the denaturant is removed or the temperature is lowered, and complementary DNA strands are allowed to re-anneal. When genome complexity is low, the time it takes for all single DNA strands to find their complement is brief. As complexity increases, the time it takes for complementary strands to re-anneal increases. Experimentally, this is referred to as a C_ot curve, where C_o is the initial molar concentration of nucleotides in single-stranded DNA and t is time. This measure reflects both the total amount of information in the system (richness or number of unique genomes) and the distribution of that information (evenness or the relative abundance of each unique genome) (Liu and Stahl 2002), thus making it among the more robust methods for estimating extant diversity in a given sample.

The genetic complexity or genome size of several soil microbial communities was assessed using re-association kinetics by Torsvik et al. (1990, 1998). They estimated that the community genome size in undisturbed organic soils was equivalent to 6,000–10,000 *Escherichia coli* genomes, while a heavy metal-polluted soil contained 350–1500 genome equivalents. Culturing yielded less than 40 genome equivalents. These data and studies employing epifluorescence microscopy to obtain direct cell counts, are what verify that culturing methods capture only the tip of the iceberg of the diversity within soil microbial communities. Gans et al. (2005) recently reported re-association kinetics data analyzed by an improved analytical approach, which yielded an estimate of the extant diversity contained in an undisturbed soil sample of 8.3 million distinct genomes in 30 g of soil, an order of magnitude greater than that reported by Torsvik et al. (1990). In contrast, a heavy metal-polluted soil was estimated to contain only 7900 genome equivalents, 99.9% fewer than in the undisturbed soil.

17.2.3 Cloning, Sequencing and Metagenomics

DNA (or RNA) sequence information can be obtained from environmental samples in two main ways: (i) cloning DNA extracted from soil directly or (ii) cloning PCR-amplified DNA (or reverse-transcribed RNA), followed in both cases by sequencing of the cloned DNA (or cDNA). In direct cloning, purified DNA extracted from soil is ligated into a vector, most frequently a self-replicating plasmid. The vector is then transformed into a competent host bacterium, such as commercially available *E. coli* competent cells, where it is maintained and multiplied (Lane 1991). Recombinant DNA clone libraries are produced in this way. Once a clone library is obtained, DNA inserts contained in the clones can be re-isolated from the host cells, purified and sequenced. The clone library can also be screened for biological activity expressed directly in *E. coli* or probed for sequences of interest using various genomics applications. This approach circumvents the need to culture microorganisms from environmental samples, although cloning itself is subject to its own inherent biases (Handelsman 2004).

Recently it became possible to clone large (100–300-kb) fragments of genomic DNA isolated directly from soil into bacterial artificial chromosome (BAC) vectors (Handelsman et al. 1998; Rondon et al. 2000). BAC vectors are low-copy number plasmids that can readily maintain large DNA inserts. When Rondon et al. (2000) analyzed their two BAC libraries, sequences homologous to the low-G+C Gram-positive *Acidobacterium*, *Cytophagales*, and *Proteobacteria* were found. They also identified clones that expressed amylase, nuclease, lipase, hemolytic and antibacterial activities. This study heralded in the field of metagenomics, which is the genomic analysis of a population of microorganisms (Handelsman 2004). Metagenomic libraries are useful for phylogenetic studies, analyses of microbial function and as a tool for natural product discovery (Handelsman et al. 1998, Handelsman 2004, 2005). When Treusch et al. (2005) probed metagenomic libraries derived from a range of environments, they discovered that uncultivated members of the Crenarchaeota contained gene sequences homologous to the ammonia monooxygenase (*amo*A) gene in nitrifying bacteria. In a follow-on study, Leininger et al. (2006) examined 12 soils from different climatic zones from both agricultural and unmanaged systems and demonstrated that the number of Crenarchaeota *amo*A sequences (AOA) was consistently higher than those from ammonia oxidizing Bacteria (AOB), with the ratio of AOA to AOB ranging from 1.5 to 232. These results suggest that the Crenarchaeota could be playing a more significant role in global N cycling than thought previously (Nicol and Schleper 2006). Metagenomic libraries are powerful tools for exploring soil microbial diversity and will form the basis for future genomic studies that link phylogenetic information with soil microbial function (Handelsman 2004).

An alternative method for creating large clone libraries from soil sequences that allows subsequent profiling of microbial communities is called serial analysis of ribosomal sequence tags (SARST). In this approach, a region of the 16S rRNA gene is amplified by PCR, such as the V1-region. Through a series of enzymatic and ligation (linking) steps, the various V1 region amplicons are joined together. The resulting concatemers are then purified, cloned, screened and sequenced. The sequences (RSTs) of the individual V1 amplicons are deduced by ignoring the linking sequences and analyzing each sequence tag individually (Neufeld et al. 2004). Neufeld and Mohn (2005) used SARST to analyze arctic tundra and boreal forest soils and found that overall diversity was higher in the arctic tundra soil. They suggested that the high carbon flux and low pH characteristic of the boreal forest soils might contribute to lower bacterial diversity or that the high diversity in the arctic soils may be influenced by allochthonous organisms coming in via air currents and being preserved by low temperature. Yet, the comparative diversity between the two systems did not change when singleton sequence tags were eliminated from the analysis, suggesting that the arctic may serve as an unrecognized reservoir microbial diversity and biochemical potential.

Several other PCR-based community analysis methods described below, such as denaturing gradient gel electrophoresis (DGGE) and two-dimensional polyacrylamide gel electrophoresis (2D-PAGE) allow DNA fragments to be retrieved in a selective manner and these can then be cloned and sequenced using the methods described above.

Pyrosequencing is a very recent innovation that is making a big splash in the large-scale analysis of bacterial genomes (Margulies et al. 2005) and potentially soil metagenomes. Margulies et al. (2005) have developed an entirely new approach to DNA sequencing that employs fragmenting genomic DNA, ligating the fragments to adapters and separating them into single strands. The single-stranded fragments are then bound to beads, with one fragment per bead, and the beads captured in droplets of a "PCR-reaction-in-oil emulsion". DNA amplification takes place inside each droplet, such that the bead contained in the droplet has attached to it 10 million copies of a unique DNA template. The emulsion is then dispersed, the DNA strands denatured and the beads carrying the single-stranded DNA copies placed into individual wells on a fiber-optic slide. Smaller beads that have the enzymes needed for pyrophosphate sequencing immobilized on them are then added to the wells. The prepared fiber-optic slide is placed into a chamber through which sequencing reagents flow. The base of the slide comes into optical contact with a second fiber-optic bundle that is fused to a charge-coupled device (CCD) sensor. Reagents are delivered cyclically to the chamber and flow into the wells of the fiber-optic slide. Simultaneous extension reactions occur on the template-carrying beads, such that each time a nucleotide is incorporated, inorganic pyrophosphate is released and photons are generated. Raw signals are captured, background subtracted, normalized and corrected. Post run analysis are used for base calling and sequence alignments. After an individual nucleotide is pulsed into the chamber, a wash containing apyrase is used to prevent nucleotides from remaining in the wells before the next nucleotide is introduced. Using this approach, Margulies et al. (2005) shotgun sequenced and de novo assembled the genome of *Mycoplasma genitalium* (580,069 bases) with 96% coverage and 99.96% accuracy in a single 4-h run. The average read length in their study was 110 bases per fragment. The method completely circumvents cloning of DNA fragments into bacterial vectors and handling individual clones in any way. The implications of using this approach for exploring soil metagenomes are fantastic. The capacity for sequencing 25 million bases in one run means that bioinformatics approaches are now under pressure to manage the quantity of information that can potentially be generated in meaningful ways that will allow data mining on a massive scale.

Edwards et al. (2006) used pyrosequencing to explore microbial community genomics in a deep mine borehole and in water seeping from it. The borehole water and water emerging from the borehole was described as an anoxic "black" environment with a pH of 6.7 and a redox potential of $-142\,mV$. The oxygenated seepage water a few cm from the borehole orifice was characterized as a "red" environment, with a pH of 4.37 and a redox potential of $-8.0\,mV$. Through use of pyrosequencing combined with comparative metagenomics, systems analysis, statistics, chemical analyses and hydrogeology they were able to characterize the differences in the genetic composition of the two communities and derive what their metabolic capacities were in the two environments. Comparing sequences generated with those in the Ribosomal Database II, indicated that the "black" water was dominated by Actinomycetales (*Brevibacterium* and *Corynebacterium*) and the "red" water was dominated by members of the Chromatiales (*Chromatiaceae*, *Thiobacillus* and *Halothiobacillus*). The two communities, just centimeters distant from each other along the same seep, as well as their respective environments were fundamentally different. Additionally, the "red" sample had a much higher species richness than the "black" sample. Sequences from the two pyrosequencing libraries were compared to the SEED database (http://theseed.uchicago.edu/FIG/index.cgi) of microbial genomes to identify groups of genes (subsystems) that were enriched in the two environments. Subsystems involved in iron uptake and use (siderophores and ABC transporters for ferrichrome) and denitrification were common in the "black" sample, whereas respiratory complexes and cytochrome-C oxidases were commonly found in the "red" sample. This study represents a large step toward linking phylogeny with function in two extreme environments.

17.2.4 Sequence Databases

DNA sequencing, annotating sequences and maintaining sequence databases are important activities for discovery of novel genetic properties, exploring phylogenetic affiliations, and in developing more specific primers and gene probes to address particular ecological questions. Gene sequences, once obtained, are submitted to and maintained within various databases such as GenBank or the Ribosomal Database Project II. GenBank and its collaborating databases, the European Molecular Biology Laboratory (EMBL) and the DNA databank of Japan (DDBJ) reached a milestone recently of containing 100 billion bases (100 gigabases) of sequence information from over 165,000 organisms, including bacteria, fungi, protozoa, nematodes and other fauna. The Ribosomal Database Project II, Release 9 (Cole et al. 2005), update 50 (release 9.50) contains 368,406 aligned and annotated Bacterial small subunit (16S) rRNA gene sequences (as of 5/2/07) with updated on-line analyses.

GenBank holds the data generated by over 400 whole genome shotgun (WGS) sequencing projects (http://www.ncbi.nlm.nih.gov/projects/WGS/WGSprojectlist. cgi). The WGS database contains genomes from individual organisms (more than 250 bacteria and 120 eukaryotes) and environmental metagenomes from over 30 projects (NCBI News 2006/2007). The environmental genomics projects include a farm soil, acid mine drainage biofilm and the symbionts of an ocean sediment-dwelling annelid that has no digestive tract and a reduced excretory system and thus relies on the symbionts to provide its nutritional and excretory needs. As more becomes known about the genetic subsystems dominant in these metagenomes, large leaps will begin to be made in our understanding of the functional significance of different community signatures in different environments – including the rhizosphere.

17.3 Phospholipid Fatty Acids (PLFA)

Phospholipid fatty acid (PLFA) analysis is an alternative technique for studying the soil microbial community without culturing. It is a non-selective method, where phospholipid fatty acid composition of the soil is analyzed by gas chromatography (GC) (Tunlid and White 1992). PLFAs are the basic components of cell membranes and are decomposed rapidly in soil when cells die. Consequently, extracting phospholipids from soil samples provides information about living members present in microbial communities (Fritze et al. 1998; Frostegard et al. 1993).

The entire PLFA profile can be used as a fingerprint of the whole soil community. Since phospholipid-linked branched fatty acids are characteristic of bacterial origin, lipids can be used to indicate specific subgroups within the community and physiological status of those populations (Roslev et al. 1998). For example, sulfate reducers, methane-oxidizing bacteria, mycorrhizal fungi and actinomycetes have unique lipid signatures. Also, environmental changes can induce changes in certain PLFA components, such as the ratio of saturated to unsaturated fatty acids, ratio of trans- to cis-monoenoic unsaturated fatty acids and the proportion of cyclopropyl fatty acids. Such changes herald changes in the microbial community. In addition PLFA profiles may contain information concerning the dynamics of larger groups of organisms such as eukaryotes. However, common fatty acids, e.g., polyenoic fatty acids found in eukaryotes, are less able to distinguish between groups when compared to the number of fatty acids found almost exclusively in bacteria (Tunlid and White 1992).

17.4   Whole Soil Molecular Approaches

17.4.1   Stable Isotope Probing

Nucleic acid methods have recently been coupled with stable isotope labeling and detection to provide a culture-independent means of linking the identity of bacteria with their function in the environment (Manefield et al. 2002a,b; McDonald et al. 2005; Dumont and Murrell 2005). Soil is either incubated after adding a ^{13}C-labeled substrate or a plant is labeled with ^{13}C-CO_2 and rhizosphere soil sampled after labeling. Soil DNA or RNA is then extracted and centrifuged in a density gradient to separate ^{13}C-labeled nucleic acids from those containing ^{12}C. Once separated, labeled DNA can be amplified using PCR and universal primers to Bacteria, Archaea or Eucarya. Analysis of the PCR products, through cloning and sequencing for example, allows the microbes that have assimilated the labeled substrate to be identified (Manefield et al. 2002a; Wellington et al. 2003; Griffiths et al. 2004; Leake et al. 2006).

PFLA-SIP has also been used successfully to analyze active soil communities (Treonis et al. 2004; Lu et al. 2007). Lu et al. (2007) labeled rice plants in mesocosms by incubating them with ^{13}C-CO_2. After 49 pulses of $^{13}CO_2$ over 7 days, PLFAs were extracted from soils taken from different regions of the rhizosphere. By this approach they were able to establish that Gram-negative bacteria and eukaryotes were most active in incorporating ^{13}C-labeled root exudates, whereas Gram-positive bacteria dominated in the bulk soil. Microbial community changes in relation to root depth were also readily observed.

Rangel-Castro et al. (2005) used ^{13}C-CO_2 pulse-labeling, followed by RNA-SIP, to study the effect of liming on the structure of the rhizosphere microbial community metabolizing root exudates in a grassland. Their results indicated that limed soils contained a microbial community that was more complex and more active in using ^{13}C-labeled compounds in root exudates than were those in unlimed soils.

SIP-based approaches do hold great potential for linking microbial identity with function, but at present a high degree of labelling is necessary to be able to separate labeled from unlabeled marker molecules. This need for high substrate concentrations may bias community responses. Alternatively, use of long incubation times to ensure that sufficient label is incorporated increases the risk of having cross-feeding of ^{13}C from the primary consumers to the rest of the community, complicating data interpretation. Another complicating factor is identifying enriched nucleic acids within the density gradient. The point at which a given nucleic acid molecule is retrieved from the caesium chloride gradient is a function of both the incorporation of the heavy isotope and the overall G+C content of the nucleic acids. Thus, a means to attribute band position in the gradient to either incorporated label or high G+C content must be devised.

17.4.2 *Fluorescence In Situ Hybridization (FISH)*

Nucleic acid hybridization involves binding a discrete fragment (a probe) of DNA or RNA to a target sequence. The probe is generally labeled with a radioisotope or fluorescent molecule and the target sequence is bound to a nylon membrane. A positive hybridization signal is obtained when complementary base pairing occurs between the probe and the target sequence. This signal is visualized by exposing the membrane to auto-radiographic film after removing any unbound probe or viewing by fluorescence microscopy with an appropriate filter. The type of probe used and how the probe is labeled determine the range of applications. For example, oligonucleotide probes (up to 30 nucleotides long) may be used under very stringent conditions that resolve single base-pair mismatches but these will have limited sensitivity due to the constraint on the number of labels that may be attached to the probe. In contrast, larger DNA fragments may be labeled to high specific activity but it is difficult to control hybridization conditions sufficiently to guarantee 100% stringency. Both the ARB (http://www.arb-home.de/) and RDPII (http://rdp.cme.msu.edu/) websites have probe design features. Information about probes that have already been designed for specific purposes can be found at (http://www.microbial-ecology.net/probebase/).

Techniques based on nucleic acid colony hybridization (colony blotting) are useful for rapidly screening bacterial isolates to establish identity or uniqueness, for example, identifying specific rhizobia strains occupying root nodules or screening libraries containing DNA clones obtained from a soil community. Nucleic acid probes can also be used to detect specific phylogenetic groups of bacteria in appropriately prepared soil samples. In the latter application, a specific probe is fluorescently labeled and hybridized to target sequences contained within microbial cells in situ using the fluorescence in situ hybridization (FISH) technique. These protocols have been described extensively in reviews by Amann et al. (1995), Amann and Ludwig (2000) and more recently by Zwirglmaier (2005).

In the FISH technique, an oligonucleotide probe is conjugated with a fluorescent molecule (or fluorochrome). The probe is designed to bind to complementary sequences in the 16 S rRNA subunit of the ribosomes within bacterial cells. Because metabolically active cells contain a large number of ribosomes, the concentration of fluorescently labeled probe is relatively high inside these cells causing them to fluoresce under UV light. The final result is high binding specificity and typically low background fluorescence. Early techniques suffered from low signal intensity, however, two new methods, tyramide signal amplification and multiply-labeled polynucleotide probes, increase the signal intensity and allow FISH approaches to be used for a wider range of ecological settings and questions (Zwirglmaier 2005). For simultaneous counting of sub-populations in a given sample, probes can be designed that bind to specific sequences of rRNA that are found only in a particular group of organisms (i.e., Archaea, Bacteria, or sub-divisions these domains) and used in conjunction with each other. FISH can also be combined with microautoradiography to determine specific substrate uptake profiles for individual cells within complex microbial communities in a method called STARFISH, substrate-tracking autoradiographic fluorescence in situ hybridization (Ouverney and Fuhrman 1999; Lee et al. 1999). Because these methods label mainly metabolically active cells, the samples can be labeled simultaneously with dyes that bind to nucleic acids, such as 4,6-diamidine-2-phenylindole, dihydrochloride (DAPI), to facilitate a total cell count using fluorescence microscopy (Li et al. 2004). FISH is particularly useful when used in conjunction with confocal laser scanning microscopy (CLSM, see below) as it allows the relative position of diverse populations to be visualized in three dimensions, even within complex communities such as biofilms and the surfaces of soil aggregates (Binnerup et al. 2001). The ability to visualize and identify organisms on a microscale in their natural environment is the key advantage of FISH. Such techniques have great potential for studying microbial interactions with plants and the ecology of target microbial populations in soil, however, the binding of fluorescent dyes to organic matter resulting in non-specific fluorescence is a common problem in soils with high organic matter contents, such as peats, or other particles with high surface charge, such as black carbon. Image analysis software is readily available and may be 'trained' to detect only those aspects of an image that meet specified criteria.

Confocal laser scanning microscopy (CLSM), combined with in situ hybridization techniques, has been applied with considerable success to visualize the structure of soil microbial communities (Bloemberg et al. 2000). CLSM works by first capturing an image that is composed only of emitted fluorescence signals from a single plane of focus. This is done using a pinhole aperture, which eliminates any signal that may be coming from portions of the field that are out of focus. A series of these optical sections is scanned at specific depths and then each section is 'stacked' using imaging software, giving rise to either a two-dimensional image that includes all planes of focus in the specimen, or a computer generated three-dimensional image. This approach gives us unprecedented resolution in viewing environmental specimens, allowing us to better differentiate organisms from particulate matter as well as providing insight into the three dimensional spatial relationships of microbial communities within their environment.

17.4.3 Green Fluorescent Protein (GFP) and Other Marker Gene Technologies

Introduced marker genes, such as *lux*AB (luminescence), *lac*Z (β galactosidase) and *xyl*E (catechol 2,3-dioxygenase) are used frequently in soil microbial ecology studies. One marker gene that has attracted a lot of attention in rhizosphere studies is *gfp*, which encodes the green fluorescent protein (GFP). Green fluorescent protein is a unique bioluminescent genetic marker that can be used to identify, track, and count specific organisms into which the gene has been cloned that have been reintroduced into the environment (Chalfie et al. 1994). The *gfp* gene was discovered in and is derived from the bioluminescent jelly fish, *Aequorea victoria* (Prasher et al. 1992). Once cloned into the organism of interest, GFP methods require no exogenous substrates, complex media or expensive equipment to monitor and, hence, are favored over many fluorescence methods for environmental applications (Errampalli et al. 1999). GFP-marked cells can be identified using a standard fluorescence microscope fitted with excitation and emission filters of the appropriate wavelengths. One reason for such keen interest in GFP is that there is no background GFP activity in plants or the bacteria and fungi that interact with them, thereby making *gfp* an excellent target gene that can be introduced into selected bacterial or fungal strains and used to study plant-microbe interactions (Errampalli et al. 1999). Basically, *gfp* is transformed into either the chromosome or a plasmid in a bacterial strain, where it is subsequently replicated. Various gene constructs have been made that differ in the type of promoter or terminator used and some contain repressor genes such as *lac*I for control of *gfp* expression. Once key populations in a sample are known and isolates obtained, they can be subsequently marked with *gfp* or other genes producing detectable products in order to track them and assess their functions and interactions in soil and the rhizosphere. In addition to GFP, red-shifted and yellow-shifted variants have been described. Development of *gfp* mutants with a series of different excitation and emission wavelengths makes it possible to identify multiple bacterial populations simultaneously (e.g., Bloemberg et al. 2000). The *gfp* gene has been introduced into *Sinorhizobium meliloti*, *Pseudomonas putida* and *Pseudomonas* sp., among other common soil bacteria and used widely in soil ecology studies. Marked strains can be visualized in infection threads, root nodules, colonized roots and even inside digestive vacuoles of protozoa. If the *gfp* gene is cloned along with specific promoters, such as *mel*A (α galactosidase) (Bringhurst et al. 2001) or *gus*A (β glucuronidase) (Xi et al. 1999), then the transformed bacteria can be used as biosensors to report back to the observer if the inducers, in these cases galactosides or glucuronides, respectively, are present and at what relative concentration in the surrounding environment.

Marker gene approaches are restricted to use in organisms that can be cultured. While considerable information can be gained about how marked microbes interact with soil colloids and other soil organisms; and can be used as biosensors for detecting environmental concentrations of various compounds, they do not yield information about the vast, unknown majority of soil microbes for which cultured representatives have yet to be obtained.

17.4.4 Microarrays

Microarrays represents an exciting new development in microbial community analysis. Nucleic acid hybridization is the principle on which the technique is based. The main difference between past protocols and microarrays is that the oligonucleotide probes, rather than the extracted DNA or RNA targets, are immobilized on a solid surface in a miniaturized matrix. Thus, thousands of probes can be tested for hybridization with sample DNA or RNA simultaneously. In contrast to other hybridization techniques, the sample nucleic acids to be probed are fluorescently-labeled, rather than the probes themselves. After the labeled sample nucleic acids are hybridized to the probes contained on the microarray, positive signals are detected by use of CSLM or other laser microarray scanning device. A fully-developed DNA microarray could include a set of probes encompassing virtually all known natural microbial groupings and thereby serve to simultaneously monitor the population structure at multiple levels of resolution (see Guschin et al. 1997; Ekins and Chu 1999; Wu et al. 2001; Zhou and Thompson 2002; Zhou 2003). Such an array would potentially allow for an enormous increase in sample throughput. A major drawback of microarrays for use in soil ecology studies currently is their need for a high copy number of target DNA/RNA to obtain a signal that is detectable with current technologies. Targets in concentrations less than 10^3–10^4 are difficult to detect using this approach. Non-specific binding of target nucleic acids to the probes is also a serious issue that needs to be overcome (Zhou and Thompson 2002).

There are three basic types of arrays used in soil ecology: (i) community genome arrays (CGA), used to compare the genomes of specific groups of organisms; (ii) functional gene arrays (FGA), used to detect the presence of genes of known function in microbial populations in prepared soil samples and more recently used to detect gene expression; and (iii) phylogenetic oligonucleotide arrays (POA), used to characterize the relative diversity of organisms in a sample through use of rRNA sequence-based probes. The details of microarray construction and types of arrays can be found in Ekins and Chu (1999) and ecological applications are reviewed in Zhou (2003).

17.5 PCR-based Methods

PCR involves separating a double-stranded DNA template into two strands (denaturation), hybridizing (annealing) oligonucleotide primers (short strands of nucleotides of a known sequence) to the template DNA and then elongating the primer-template hybrid by a DNA polymerase enzyme. The potential target genes for PCR are many and varied, limited only by available sequence information. The primers most frequently used for soil ecological studies are designed to target specific DNA fragments, such as 16 S or 18 S rRNA genes, functional genes, repetitive sequences, e.g., REP (Repetitive Extragenic Palindromic) sequences, or arbitrary primers, e.g., randomly amplified polymorphic DNA (RAPD). The discovery of thermal-stable DNA polymerases from organisms such as *Thermus aquaticus*

(*Taq* polymerase) has made PCR a standard protocol in laboratories around the world (Mullis and Faloona 1987; Saiki et al. 1998).

By far the more common targets for characterizing microbial communities are the rRNA genes because of their importance in establishing phylogenetic and taxonomic relationships (Woese et al. 1990). These are the small subunit (SSU) rRNA genes, 16S in Bacteria and Archaea or 18S in Eucarya; the large subunit (LSU) rRNA genes, 23S in Bacteria and Archaea or 28S in Eucarya; or the internal transcribed spacer (ITS) regions, sequences that lie between the SSU and LSU genes. Other defined targets are genes that code for ecologically significant functions, such as genes involved in nitrogen fixation, e.g., *nif*H (Chelius and Lepo 1999; Rösch et al. 2002); *amo*A which codes for ammonium monooxygenase, a key enzyme in nitrification reactions (Rotthauwe et al. 1997); and *nir*S and *nir*K which codes for nitrite reductase, a key enzyme in denitrification reactions (Rösch et al. 2002; Henry et al. 2004).

In any study where PCR is used, sources of bias must be considered (Wintzingerode et al. 1997). The main sources of bias in amplifying soil community DNA are: (i) the use of very small sample sizes (typically only 500 mg of soil); (ii) preferential amplification of some DNA templates over others; and, (iii) for amplification of the rRNA genes, the fact that many bacteria contain multiple copies of these operons (e.g., *Bacillus* and *Clostridium* species contain 15 copies), hence sequences from such species will be over-represented among the amplification products. In addition, chimeras, composed of double-stranded DNA where each strand was derived from a different organism rather than a single organism, may be generated. Acknowledged biases associated with PCR are generally why diversity indices calculated from the results of PCR-based experiments may not be very robust.

17.5.1 DNA Fingerprinting

PCR fingerprinting is used to distinguish differences in the genetic makeup of microbial populations from different samples and can be accomplished by several different methods. The advantages of these techniques are that they are rapid and inexpensive and thus enable high sample throughput and can be used to target sequences that are phylogenetically or functionally significant (Fjellbirkeland et al. 2001). Depending on the primers chosen, PCR fingerprints can be used to distinguish between isolates at the strain level or to characterize target microbes at the community level. The more common PCR fingerprinting techniques in use today for characterizing soil microbial community composition are denaturing gradient gel electrophoresis (DGGE) or temperature gradient gel electrophoresis (TGGE) (Muyzer and Smalla, 1998) and terminal restriction fragment length polymorphism (T-RFLP) analysis (Liu et al. 1997; Marsh 2005; Thies 2007b). Both techniques can be used to separate PCR products that are initially the same length by employing additional methods to separate the amplicons into a greater number of bands or operational taxonomic units (OTUs) that are then used for community comparisons.

DGGE and TGGE are identical in principle. Both techniques impose a parallel gradient of denaturing conditions along a polyacrylamide gel. Double-stranded DNA PCR amplicons are loaded in wells at the top of the gel and, as the DNA migrates, the denaturing conditions of the gel gradually increase. In DGGE, the denaturants are typically urea and formamide; in TGGE it is temperature. Because native double-stranded DNA is a compact structure, it migrates faster than partially denatured DNA. The sequence of a fragment determines the point in the gradient gel at which denaturation will start to retard mobility. Sequence affects duplex stability by both percentage G+C content and neighboring nucleotide interactions (e.g., GGA is more stable than GAG). The resulting gel yields a ladder of bands in each lane characteristic of the DNA extracted and amplified from the original sample. There is not a direct correspondence between bands in the DGGE gel and organism diversity, however. Sequences amplified from the DNA of different organisms may have similar melting properties in the presence of the denaturant and thus occupy the same band in the denaturing gel. DNA fragments cloned from different bands may yield as many different sequences as clones analyzed. Since there is not a one-to-one correspondence between bands and taxa, the bands are referred to as OTUs. The OTUs form the basis of similarity and multivariate analyses of data derived from various soil communities.

While the power of DGGE and TGGE to detect PCR amplicon diversity within a single gel is high, the resolving power of these and other gel-based analyses, is limited by the number of bands capable of 'fitting' and being counted as individual bands on a single gel. In practice, no more than 80–100 distinct sequence types may be resolved despite the potential for single base-pair sensitivity. An important advantage that DGGE analysis has over T-RFLP (see below) is that PCR amplicons of interest that are resolved on a DGGE gel can be excised from the gel, re-amplified, cloned and sequenced, thereby obtaining taxonomic and/or phylogenetic information about amplifiable members of the soil community. For phylogenetic assignment of cloned sequences, variable regions within the SSU rRNA genes are amplified. An important disadvantage of the gradient gel approach is that the amplicon size must be restricted to under 600 base pairs in length to optimize separation within the gel matrix. Therefore, full length rRNA gene sequences cannot be recovered using these methods. DGGE and TGGE are now being applied frequently in soil microbial ecology to compare the structures of complex microbial communities and to study their dynamics. The basic method and applications were recently reviewed by Nakatsu (2007).

T-RFLP analysis, as in DGGE analysis, begins with amplifying soil community DNA using targeted primers, but with the key differences that one or both primers are labeled with a fluorochrome(s) and that resulting amplicons are hydrolyzed with restriction enzymes to create DNA fragments of varying size that are labeled with the fluorochrome at either the 5′ or 3′ end. These terminal fragments are then sized against a standard molecular size marker using automated DNA sequencing techniques. The resulting electropherogram (peaks representing the sizes of the terminal restriction fragments, TRFs) is used as a DNA fingerprint characteristic of the soil community sampled. Resulting TRF sizes are analogous to bands on a DGGE gel and are also referred to as OTUs, since any one terminal fragment size is not restricted to any taxonomic group per se (Marsh 2005). TRF profiles are compared subsequently between samples by use of similarity matrices and multivariate statistics.

With new capillary sequencers, up to 384 samples can be analyzed in a single run. T-RFLP also has a higher resolving power than DGGE, with often twice as many OTUs determined per sample (Jones and Thies 2007), making T-RFLP the preferred choice for a high throughput method to initially screen for differences between communities. Devare et al. (2004) applied the T-RFLP technique to compare rhizosphere bacterial communities colonizing transgenic and non-transgenic corn and found that communities clustered by sampling time and year, but not by corn hybrid. Other studies have used T-RFLP to evaluate the effects of soil management on fungal community composition (Edel-Hermann et al. 2004) and the effects of solarization and crop rotation on bacterial communities (Culman et al. 2006), among many other applications. Artursson et al. (2005) combined bromodeoxyuridine immunocapture with T-RFLP to examine the effects of mycorrhizal inoculation and plant species on the active soil bacterial metagenome. T-RFLP need not be restricted to studying the 16S rRNA gene. This technique can be used as a quick screen for any gene for which specific primers can be devised to examine differences between communities in environmental samples, such as *nif*H to compare populations of nitrogen-fixing bacteria or *amo*A to study ammonia oxidizing bacterial populations in soil. The main drawback of the use of this approach is the inability to further characterize TRFs or obtain sequence information as the sample is lost shortly after it is sized. However, once profiles are compared, the original PCR products from samples of interest can be used for cloning and sequencing experiments as described above. Alternatively, a gel-based approach called the 'physical capture method' of T-RFLP analysis can be employed when recovery of sequence information is desired (Blackwood and Buyer 2007). The technique is not as discriminating as the capillary approaches, but this level of resolution may not be needed for some applications or may not be possible in some research settings.

T-RFLP often yields a higher number of OTUs for use in comparative analyses than DGGE. However, all of these techniques yield numbers of OTUs that do not come close to the estimates of extant diversity in soil populations as estimated by DNA:DNA reassociation kinetics (discussed above). Hence, we are still viewing the tip of the iceberg as far as characterizing soil microbial diversity with these higher throughput DNA fingerprinting techniques.

A new DNA-based fingerprinting approach, two-dimensional polyacrylamide gel electrophoresis (2D-PAGE), can be used to separate PCR amplicons of the ITS regions first by size in a non-denaturing gel and then by melting characteristics in a second, denaturing gradient gel (Jones and Thies 2007). This approach yielded an order of magnitude higher number of OTUs than DGGE alone and three times the number of OTUs obtained by use of T-RFLP. Because the technique is gel-based, DNA spots of interest can be excised from the second dimension gel and then cloned and sequenced. Far fewer OTUs were found in spots on the second dimension gel than were recovered from corresponding bands on the first dimension sizing gel, thus the technique allows OTUs of interest to be recovered much more easily. The disadvantage of this technique is that it is more laborious, therefore it does not lend itself to high sample throughput. Yet, its improved ability to discriminate between soil communities and retrieve sequence information make it a powerful technique for elucidating key differences in community structure between studied samples. Jones and Thies (2007) used the technique to study changes in soil bacterial community composition in relation to a naturally occurring gradient of Zn and Cd content in a soil in upstate New York.

Several additional PCR fingerprinting techniques target the ribosomal gene sequences. Ribotyping makes use of differences in the chromosomal positions or structure of rRNA genes to identify or group isolates of a particular genus or species. Ribotyping has been shown to be reproducible and hence has gained popularity for isolate fingerprinting and has found use in bacterial source tracking and other studies where the similarity of isolates obtained from different samples needs to be compared. The most frequently used ribotyping method is to identify RFLPs of rRNA genes by probing a Southern transfer of genomic DNA that has been hydrolyzed with an endonuclease. In amplified ribosomal DNA restriction analysis (ARDRA), rRNA gene sequences are amplified. In automated ribosomal intergenic spacer analysis (ARISA), the ITS region is amplified. PCR amplicons resulting from use of both methods are hydrolyzed subsequently with restriction enzymes and the resulting variations in restriction fragment sizes are analyzed on a gel. Chelius and Lepo (1999) used RFLPs of PCR amplified *nif*H sequences to study the diversity of nitrogen-fixing bacteria in the rhizosphere of wetland plant communities. In these applications, bands in the gel are again termed OTUs and similarities and differences between the fingerprints from different samples are analyzed using multivariate techniques. Use of ARISA may yield more OTUs from a given sample, but as the number of bands on the gel increases, the more difficulty one has in resolving individual bands in the analysis.

17.5.2 *Quantitative and Real-Time PCR*

An advance in PCR analysis that allows specific gene targets to be quantified is quantitative PCR (qPCR), also called real-time PCR. qPCR is a method that employs fluorogenic probes or dyes to quantify the number of copies of a target DNA sequence in a sample. This approach has been used successfully to quantify target genes that reflect the capacity of soil bacteria to perform given functions. Examples include the use of ammonia monooxygenase (*amo*A), nitrite reductase (*nir*S or *nir*K), and particulate methane monooxygenase (*pmo*A) genes to quantify ammonia oxidizing (Hermansson and Lindgren 2001), denitrifying (Henry et al. 2004) and methanotrophic (Kolb et al. 2003) bacteria, respectively, in soil samples. qPCR coupled with primers to specific internal transcribed spacer (ITS) or rRNA gene sequences has also been used to quantify ectomycorrhizal (Landeweert et al. 2003) and endomycorrhizal fungi (Filion et al. 2003) as well as cyst nematodes (Madani et al. 2005) in soil.

17.5.3 Statistical Methods

The successful application of molecular techniques to population studies, particularly those based on the analysis of DNA or RNA in a gel matrix, relies heavily on the correct interpretation of the banding or spot patterns observed on electrophoretic gels. Gel images are typically digitized and band detection software is used to mark the band locations in the gel. The resulting band pattern is then exported to a statistical software package for analysis. Some analyses require that the fingerprint patterns obtained are first converted to presence/absence matrices; although average band density data are also used. The matrices generated are then compared using cluster analysis, multi-dimensional scaling, principal component analysis, redundancy analysis, canonical correspondence analysis, or additive main effects with multiplicative interaction model, among others. Each analysis will allow community comparisons, yet each has associated strengths and weaknesses. There are a number of software packages available that will enable one to compare and score PCR-fingerprints and produce similarity values for a given set of samples. Software packages, such as BioNumerics and GelCompar (Applied Maths, Kortrijk, Belgium), Canoco™ (Microcomputer Power, Ithaca, NY), PHYLIP (freeware via GenBank and the RDPII) and MatModel™ (Microcomputer Power) among others are used commonly. An advantage of using analysis programs, such as Bionumerics or GelCompar, is that fingerprints of communities generated from the use of several different markers can be combined. Generating a combined fingerprint in this way increases the robustness of similarity analyses based on PCR-fingerprints because it reduces the impact that one or two minor band differences has on the similarity matrices produced. The RDPII (Release 8.3) website provides analytical support for the analysis of T-RFLP data. The details of other analytical programs that support the analysis of data based on operational taxonomic units have also been published lately (Schloss and Handelsman 2005, 2006a,b).

The information that can be obtained from molecular characterization depends on the analysis technique. 16S rRNA gene sequencing can aid in assigning species into genera and can be used for determining relationships between genera, but the information is frequently unable to resolve differences between closely related species. To overcome this limitation, one could use additional genetic information contained within ITS regions either by sequencing or by RFLP to further discriminate between closely related species.

To add value to the study of soil community ecology, a technique must be robust, that is, yield specific information about communities at the level of resolution required; it must be rapid and allow high throughput in order for the large number of samples needed for landscape studies to be processed with moderate effort.

17.6   Conclusions

Molecular tools are offering unparalleled opportunities to characterize Bacteria, Archaea and Eucarya in culture and directly from field soils. These tools are allowing us to ask questions at much larger geographic scales than have been possible previously. We are now able to examine such issues as how microbial populations vary across soil types and climatic zones (Fierer and Jackson 2006), in association with plant roots and between various plant species (Cardon and Gage 2006; Costa et al. 2006), and in response to soil management (e.g., Culman et al. 2006) or soil pollution (Liu et al. 1997). Molecular approaches also provide improved tools for seeking new inoculant consortia that may provide benefit in cropping systems. Genotypes that enjoy high representation in the soil population are likely to be competent saprophytes and be well adapted to site conditions. Pre-adapted strains that are also highly effective and genetically stable would then be excellent target organisms for future inoculants.

Soil has been dubbed 'The Final Frontier'. Modern molecular techniques developed to study microbial populations finally allow us access to the very large proportion of organisms that are present in the soil that we are currently unable to culture under laboratory conditions (Handelsman and Smalla 2003). They are also allowing us to begin to link identify with function (Dumont and Murrell 2005), which will lead to a better understanding of how changes in soil management practices may be altering ecosystem dynamics. Continually evolving technical developments open new horizons of research and applications that are enabling a far more complete and less biased view of microbial biodiversity in soil and the rhizosphere.

References

Amann RI, Ludwig W (2000) Ribosomal RNA-targeted nucleic acid probes for studies in microbial ecology. FEMS Microbiol Rev 24:555–565

Amann RI, Ludwig W, Schleifer KH (1995) Phylogenetic identification and in situ detection of individual microbial cells without cultivation. Microbiol Rev 59:143–169

Aneja MK, Sharma S, Munch JC, Schloter M (2004) RNA fingerprinting – a new method to screen for differences in plant litter degrading microbial communities. J Microbiol Methods 59:223–231

Artursson V, Finlay RD, Jansson JK (2005) Combined bromodeoxyuridine immunocapture and terminal-restriction fragment length polymorphism analysis highlights differences in the active soil bacterial metagenome due to *Glomus mosseae* inoculation or plant species. Environ Microbiol 7:1952–1966

Binnerup SJ, Bloem J, Hansen BM, Wolters W, Veninga M, Hansen M (2001) Ribosomal RNA content in microcolony forming soil bacteria measured by quantitative 16 S rRNA hybridization and image analysis. FEMS Microbiol Ecol 37:231–237

Blackwood CB, Buyer JS (2007) Evaluating the physical capture method of terminal restriction fragment length polymorphism for comparison of soil microbial communities. Soil Biol Biochem 39:590–599

Bloemberg GV, Wijfjes AHM, Lamers GEM, Stuurman N, Lugtenberg BJJ (2000) Simultaneous imaging of *Pseudomonas fluorescens* WCS365 populations expressing three different autofluorescent proteins in the rhizosphere: new perspectives for studying microbial communities. Mol Plant-Microbe Interact 13:1170–1176

Bringhurst RM, Cardon ZG, Gage DJ (2001) Galactosides in the rhizosphere: utilization by *Sinorhizobium meliloti* and development of a biosensor. Proc Natl Acad Sci USA 98:4540–4545

Bruns MA, Buckley DH (2002) Isolation and purification of microbial community nucleic acids from environmental samples. In: Hurst CJ, Crawford RL, Knudsen GR, McInerney MJ, Stetzenbach LD (eds) Manual of environmental microbiology. American Society for Microbiology Press, Washington, DC

Cardon ZG, Gage DJ (2006) Resource exchange in the rhizosphere: molecular tools and the microbial perspective. Ann Rev Ecol Evol Syst 37:459–488

Chalfie M, Tu Y, Euskirchen G, Ward WW, Prasher DC (1994) Green fluorescent protein as a marker for gene-expression. Science 263:802–805

Chelius MK, Lepo JE (1999) Restriction fragment length polymorphism analysis of PCR-amplified *nif*H sequences from wetland plant rhizosphere communities. Environ Tech 20:883–889

Cole JR, Chai B, Farris RJ, Wang Q, Kulam SA, McGarrell DM, Garrity GM, Tiedje JM (2005) The Ribosomal Database Project (RDP-II): sequences and tools for high-throughput rRNA analysis. Nucleic Acids Res 33:D294–D296 doi:10.1093/nar/gki038

Costa R, Gotz M, Mrotzek N, Lottmann J, Berg G, Smalla K (2006) Effects of site and plant species on rhizosphere community structure as revealed by molecular analysis of microbial guilds. FEMS Microbiol Ecol 56:236–249

Courtois S, Frostegard A, Goransson P, Depret G, Jeannin P, Simonet P (2001) Quantification of bacterial subgroups in soil: comparison of DNA extracted directly from soil or from cells previously released by density gradient centrifugation. Environ Microbiol 3:431–439

Culman SW, Duxbury JM, Lauren JG, Thies JE (2006) Microbial community response to soil solarization in Nepal's rice-wheat cropping system. Soil Biol Biochem 38:3359–3371

Devare MH, Jones CM, Thies JE (2004) Effect of Cry3Bb transgenic corn and tefluthrin on the soil microbial community: biomass, activity, and diversity. J Environ Qual 33:837–843

Dumont MG, Murrell JC (2005) Stable isotope probing – linking microbial identity to function. Nat Rev Microbiol 3:499–504

Edel-Hermann W, Dreumont C, Perez-Piqueres A, Steinberg C (2004) Terminal restriction fragment length polymorphism analysis of ribosomal RNA genes to assess changes in fungal community structure in soils. FEMS Microbiol Ecol 47:397–404

Edwards RA, Rodriguez-Brito B, Wegley L, Haynes M, Breitbart M, Peterson DM, Saar MO, Alexander S, Alexander EC, Rohwer F (2006) Using pyrosequencing to shed light on deep mine microbial ecology. BMC Genomics 7:57

Ekins R, Chu FW (1999) Microarrays: their origins and applications. Trends Biotechnol 17:217–218

Errampalli D, Leung K, Cassidy MB, Kostrzynska M, Blears M, Lee H, Trevors JT (1999) Applications of the green fluorescent protein as a molecular marker in environmental microorganisms. J Microbiol Methods 35:187–199

Felske A, Engelen B, Nübel U, Backhaus H (1999) Direct ribosome isolation from soil to extract bacterial rRNA for community analysis. Appl Environ Microbiol 62:4162–4167

Fierer N, Jackson RB (2006) The diversity and biogeography of soil bacterial communities. Proc Natl Acad Sci 103:626–631

Filion Mm, St-Arnaud M, Jabaji-Hare SH (2003) Direct quantification of fungal DNA from soil substrate using real-time PCR. J Microbiol Methods 53:67–76

Fjellbirkeland A, Torsvik V, Ovreas L (2001) Methanotrophic diversity in an agricultural soil as evaluated by denaturing gradient gel electrophoresis profiles of *pmo*A, *mxa*F and 16S rDNA sequences. Antonie Van Leeuwenhoek 79:209–217

Fritze H, Pennanen T, Kitunen V (1998) Characterization of dissolved organic carbon from burned humus and its effects on microbial activity and community structure. Soil Biol Biochem 30:687–693

Frostegard A, Baath E, Tunlid A (1993) Shifts in the structure of soil microbial communities in limed forest as revealed by phospholipid fatty acid analysis. Soil Biol Biochem 25:723–730

Gans J, Wolinsky M, Dunbar J (2005) Computational improvements reveal great bacterial diversity and high metal toxicity in soil. Science 309:1387–1390

Griffiths RI, Whiteley AS, O'Donnell AG, Bailey MJ (2000) Rapid method for coextraction of DNA and RNA from natural environments for analysis of ribosomal DNA- and rRNA-based microbial community composition. Appl Environ Microbiol 66:5488–5491

Griffiths RI, Manefield M, Ostle N, McNamara N, O'Donnell AG, Bailey MJ, Whiteley AS (2004) (CO$_2$)-^{13}C pulse labelling of plants in tandem with stable isotope probing: methodological considerations for examining microbial function in the rhizosphere. J Microbiol Methods 58:119–129

Guschin DY, Mobarry BK, Proudnikow D, Stahl DA, Rittmann BE, Mirzabekov AD (1997) Oligonucleotide microchips and genosensors for determinative and environmental studies in microbiology. Appl Environ Microbiol 63:2397–2402

Handelsman J (2004) Metagenomics: application of genomics to uncultured microorganisms. Microbiol Mol Biol Rev 68:669–685

Handelsman J (2005) Metagenomics: application of genomics to uncultured microorganisms (vol 68, p 669, 2004). Microbiol Mol Biol Rev 69:195

Handelsman J, Smalla K (2003) Conversations with the silent majority. Curr Opin Microbiol 6:271–273

Handelsman J, Rondon MR, Brady SF, Clardy J, Goodman RM (1998) Molecular biological access to the chemistry of unknown soil microbes: a new frontier for natural products. Chem Biol 5:R245–R249

Henry S, Baudoin E, Lopez-Gutierrez JC, Martin-Laurent F, Baumann A, Philippot L (2004) Quantification of denitrifying bacteria in soils by *nir*K gene targeted real-time PCR. J Microbiol Methods 59:327–335

Hermansson A, Lindgren PE (2001) Quantification of ammonia-oxidizing bacteria in arable soil by real-time PCR. Appl Environ Microbiol 67:972–976

Hurt RA, Qiu XY, Wu LY, Roh Y, Palumbo AV, Tiedje JM, Zhou JH (2001) Simultaneous recovery of RNA and DNA from soils and sediments. Appl Environ Microbiol 67:4495–4503

Jiao JY, Wang HX, Zeng Y, Shen YM (2006) Enrichment for microbes living in association with plant tissues. J Appl Microbiol 100:830–837

Jones CM, Thies JE (2007) Soil microbial community analysis using two-dimensional polyacrylamide gel electrophoresis of the bacterial ribosomal internal transcribed spacer regions. J Microbiol Methods 69:256–267

Kolb S, Knief C, Stubner S, Conrad R (2003) Quantitative detection of methanotrophs in soil by novel *pmo*A-targeted real-time PCR assays. Appl Environ Microbiol 69:2423–2429

Landeweert R, Veenman C, Kuyper TW, Fritze H, Wernars K, Smit E (2003) Quantification of ectomycorrhizal mycelium in soil by real-time PCR compared to conventional quantification techniques. FEMS Microbiol Ecol 45:283–292

Lane DJ (1991) 16S/23S rRNA sequencing. In: Nucleic acid techniques in bacterial systematics. Wiley, Chichester, pp 115–175

Leake JR, Ostle NJ, Rangel-Castro JI, Johnson D (2006) Carbon fluxes from plants through soil organisms determined by field (CO$_2$)-^{13}C pulse-labelling in an upland grassland. Appl Soil Ecol 33:152–175

Lee N, Halkjaer N, Andreasen KH, Juretschko S, Nielsen JL, Schleifer KH, Wagner M (1999) Combination of fluorescent in situ hybridization and microautoradiography – a new tool for structure-function analyses in microbial ecology. Appl Environ Microbiol 65:1289–1297

Leininger S, Urich T, Schloter M, Schwark L, Qi J, Nicol GW, Prosser JI, Schuster SC, Schleper C (2006) Archaea predominate among ammonia-oxidizing prokaryotes in soils. Nature 442:806–809

Li Y, Dick WA, Tuovinen OH (2004) Fluorescence microscopy for visualization of soil microorganisms – a review. Biol Fert Soils 39:301–311

Liu WT, Stahl DA (2002) Molecular approaches for the measurement of density, diversity and phylogeny. In: Hurst CJ, Crawford RL, Knudsen GR, McInerney MJ, Stetzenbach LD (eds) Manual of environmental microbiology. American Society for Microbiology Press, Washington, DC

Liu WT, Marsh TL, Cheng H, Forney LJ (1997) Characterization of microbial diversity by determining terminal restriction fragment length polymorphisms of genes encoding 16S rRNA. Appl Environ Microbiol 63:4516–4522

Lu YH, Abraham WR, Conrad R (2007) Spatial variation of active microbiota in the rice rhizosphere revealed by in situ stable isotope probing of phospholipid fatty acids. Environ Microbiol 9:474–481

Madani M, Subbotin SA, Moens M (2005) Quantitative detection of the potato cyst nematode, *Globodera pallida*, and the beet cyst nematode, *Heterodera schachtii*, using real-time PCR with SYBR green I dye. Mol Cell Probes 19:81–86

Manefield M, Whiteley AS, Griffiths RI, Bailey MJ (2002a) RNA stable isotope probing, a novel means of linking microbial community function to phylogeny. Appl Environ Microbiol 68:5367–5373

Manefield M, Whiteley AS, Ostle N, Ineson P. Bailey MJ (2002b) Technical considerations for RNA-based stable isotope probing: an approach to associating microbial diversity with microbial community function. Rapid Comm Mass Spec 16:2179–2183

Margulies M, Egholm M, Altman WE, Attiya S, Bader JS, Bemben LA, Berka J, Braverman MS, Chen YJ, Chen ZT, Dewell SB, Du L, Fierro JM, Gomes XV, Godwin BC, He W, Helgesen S, Ho CH, Irzyk GP, Jando SC, Alenquer MLI, Jarvie TP, Jirage KB, Kim JB, Knight JR, Lanza JR, Leamon JH, Lefkowitz SM, Lei M, Li J, Lohman KL, Lu H, Makhijani VB, McDade KE, McKenna MP, Myers EW, Nickerson E, Nobile JR, Plant R, Puc BP, Ronan MT, Roth GT, Sarkis GJ, Simons JF, Simpson JW, Srinivasan M, Tartaro KR, Tomasz A, Vogt KA, Volkmer GA, Wang SH, Wang Y, Weiner MP, Yu PG, Begley RF, Rothberg JM (2005) Genome sequencing in microfabricated high-density picolitre reactors. Nature 437:376–380

Marsh TL (2005) Culture-independent microbial community analysis with terminal restriction fragment length polymorphism. Methods Enzymol 397:308–329

McDonald IR, Radajewski S, Murrell JC (2005) Stable isotope probing of nucleic acids in methanotrophs and methylotrophs: a review. Org Geochem 36:779–787

Mullis KB, Faloona FA (1987) Specific synthesis of DNA *in vitro* via a polymerase catalysed chain reaction. Methods Enzymol 155:335–350

Muyzer G, Smalla K (1998) Application of denaturing gradient gel electrophoresis (DGGE) and temperature gradient gel electrophoresis (TGGE) in microbial ecology. Antonie Van Leeuwenhoek 73:127–141

Nakatsu CH (2007) Soil microbial community analysis using denaturing gradient gel electrophoresis. Soil Sci Soc Am J 71:562–571

NCBI News (2006/2007) National Center for Biotechnology Information 15(2):2

Neufeld JD, Mohn WW (2005) Unexpectedly high bacterial diversity in arctic tundra relative to boreal forest soils, revealed by serial analysis of ribosomal sequence tags. Appl Environ Microbiol 71:5710–5718

Neufeld JD, Yu ZT, Lam W, Mohn WW (2004) Serial analysis of ribosomal sequence tags (SARST): a high-throughput method for profiling complex microbial communities. Environ Microbiol 6:131–144

Nicol GW, Schleper C (2006) Ammonia-oxidising Crenarchaeota: important players in the nitrogen cycle? Trends Microbiol 14:207–212

Ouverney CC, Fuhrman JA (1999) Combined microautoradiography–16S rRNA probe technique for determination of radioisotope uptake by specific microbial cell types in situ. Appl Environ Microbiol 65:1746–1752

Prasher DC, Eckenrode VK, Ward WW, Prendergast FG, Cormier MJ (1992) Primary structure of the *Aequorea-victoria* green-fluorescent protein. Gene 111:229–233

Radajewski S, McDonald IR, Murrell JC (2003) Stable-isotope probing of nucleic acids: a window to the function of uncultured microorganisms. Curr Opin Biotechnol 14:296–302

Rangel-Castro JI, Killham K, Ostle N, Nicol GW, Anderson IC, Scrimgeour CM, Ineson P, Meharg A, Prosser JI (2005) Stable isotope probing analysis of the influence of liming on root exudate utilization by soil microorganisms. Environ Microbiol 7:828–838

Rondon MR, August PR, Betterman AD, Brady SF, Grossman TH, Liles MR, Loiacono KA, Lynch BA, MacNiel IA, Minor C, Tiong CL, Gilman M, Osburne MS, Clardy J, Handelsman J, Goodman RM (2000) Cloning the soil metagenome: a strategy for accessing the genetic and functional diversity of uncultured microorganisms. Appl Environ Microbiol 66:2541–2547

Roslev P, Iversen N, Henriksen K (1998) Direct fingerprinting of metabolically active bacteria in environmental samples by substrate specific radiolabelling and lipid analysis. J Microbiol Methods 31:99–111

Rotthauwe J-H, Witzel K-P, Liesack W (1997) The ammonia monooxygenase structural gene *amoA* as a functional marker: molecular fine-scale analysis of natural ammonia-oxidizing populations. Appl Environ Microbiol 63:4704–4712

Saiki RK, Gelfand DH, Stoffel S, Scharf SJ, Higuchi R, Horn GT, Mullis KB, Erlich HA (1988) Primer-directed enzymatic amplification of DNA with a thermostable DNA-polymerase. Science 239:487–491

Schloss PD, Handelsman J (2005) Introducing DOTUR, a computer program for defining operational taxonomic units and estimating species richness. Appl Environ Microbiol 71:1501–1506

Schloss PD, Handelsman J (2006a) Introducing SONS, a tool for operational taxonomic unit-based comparisons of microbial community memberships and structures. Appl Environ Microbiol 72:6773–6779

Schloss PD, Handelsman J (2006b) Introducing TreeClimber, a test to compare microbial community structures. Appl Environ Microbiol 72:2379–2384

Sessitsch A, Gyamfi S, Stralis-Pavese N, Weilharter A, Pfeifer U (2002) RNA isolation from soil for bacterial community and functional analysis: evaluation of different extraction and soil conservation protocols. J Microbiol Methods 51:171–179

Smalla K, Sobecky PA (2002) The prevalence and diversity of mobile genetic elements in bacterial communities of different environmental habitats: insights gained from different methodological approaches. FEMS Microbiol Ecol 42:165–175

Thies JE (2007a) Molecular methods for studying the soil biota. In: Paul EA (ed) Soil microbiology, ecology and biochemistry, 3rd edn. Academic Press. Elsevier, Oxford, pp 85–118

Thies JE (2007b) Soil microbial community analysis using terminal restriction fragment length polymorphisms. Soil Sci Soc Am J 71:579–591

Thies JE, Grossman JM (2006) The soil habitat and soil ecology. In: Uphoff N, Ball AS, Fernandes E, Herren H, Husson O, Laing M, Palm C, Pretty J, Sanchez P, Sanginga N, Thies J (eds) Biological approaches to sustainable soil systems. CRC Press, Taylor and Francis Group, Boca Raton, pp 59–78

Torsvik VL (1980) Isolation of bactcrial-DNA from soil. Soil Biol Biochem 12:15 21

Torsvik V, Ovreas L (2002) Microbial diversity and function in soil: from genes to ecosystems. Curr Opin Microbiol 5:240–245

Torsvik V, Goksoyr J, Daae FL (1990) High diversity in DNA of soil bacteria. Appl Environ Microbiol 56:782–787

Torsvik V, Daae FL, Sandaa RA, Ovreas L (1998) Novel techniques for analysing microbial diversity in natural and perturbed environments. J Biotechnol 64:53–62

Treonis AM, Ostle NJ, Stott AW, Primrose R, Grayston SJ, Ineson P (2004) Identification of groups of metabolically-active rhizosphere microorganisms by stable isotope probing of PLFAs. Soil Biol Biochem 36:533–537

Treusch AH, Leininger S, Kletzin A, Schuster SC, Klenk HP, Schleper C (2005) Novel genes for nitrite reductase and Amo-related proteins indicate a role of uncultivated mesophilic Crenarchaeota in nitrogen cycling. Environ Microbiol 7:1985–1995

Tunlid A, White DC (1992) Biochemical analysis of biomass, community structure, nutritional status and metabolic activity of microbial communities in soil. Soil Biochem 7:229–262

Ward DM (1998) A natural species concept for prokaryotes. Curr Opin Microbiol 1:271–277

Wellington EMH, Berry A, Krsek M (2003) Resolving functional diversity in relation to microbial community structure in soil: exploiting genomics and stable isotope probing. Curr Opin Microbiol 6:295–301

Wintzingerode Fv, Gobel UV, Stackebrandt E (1997) Determination of microbial diversity in environmental samples: pitfalls of PCR-based rRNA analysis. FEMS Microbiol Rev 21:213–229

Woese CR (1987) Bacterial evolution. Microbiol Rev 51:221–271

Woese CR, Kandler O, Wheelis ML (1990) Towards a natural system of organisms: proposal for the domains Archaea, Bacteria and Eucarya. Proc Natl Acad Sci USA 87:4576–4579

Wu LY, Thompson DK, Li GS, Hurt RA, Tiedje JM, Zhou JZ (2001) Development and evaluation of functional gene arrays for detection of selected genes in the environment. Appl Environ Microbiol 67:5780–5790

Xi C, Lambrecht M, Vanderleyden J, Michiels J (1999) Bi-functional *gfp*-and *gus*A-containing mini-Tn5 transposon derivatives for combined gene expression and bacterial localization studies. J Microbiol Methods 35:85–92

Zhou JZ (2003) Microarrays for bacterial detection and microbial community analysis. Curr Opin Microbiol 6:88–294

Zhou JZ, Thompson DK (2002) Challenges in applying microarrays to environmental studies. Curr Opin Biotech 13:204–207

Zwirglmaier K (2005) Fluorescence in situ hybridisation (FISH) – the next generation. FEMS Microbiol Lett 246:151–158

Chapter 18
Morphotyping and Molecular Methods to Characterize Ectomycorrhizal Roots and Hyphae in Soil

Laura M. Suz(✉), Anabela M. Azul, Melissa H. Morris, Caroline S. Bledsoe, and María P. Martín

18.1   Introduction

At the interface between plants and soils, ectomycorrhizal (ECM) fungi explore soils, acquire resources, transfer resources to plants, and acquire carbon from plants. Mycorrhizas enhance plant survival, nutrition and growth and play key roles in ecosystems processes such as decomposition, nutrient cycling, soil carbon storage, productivity and sustainability. Mycorrhizas are critical for plant colonization of new soils (e.g. mine spoils, volcanic deposits, glacial moraines). ECM diversity ensures plant reestablishment after disturbance and can enhance survival and growth of trees in reforestation. ECM fungi can promote fine root development as well as produce antibiotics, hormones and vitamins. Mycorrhizal associations may help protect roots from pathogens and moderate effects of heavy metals and toxins. Many environmental problems may be alleviated by mycorrhizas – problems such as pollution, erosion, soil degradation, climate change, degradation of natural resources, and poor land use management.

Mycorrhizal abilities to carry out important functions are linked to diversity. ECM diversity is large and documented in many ecosystems, particularly coniferous ecosystems (Gehring et al. 1998; Goodman and Trofymow 1998; Kranabetter and Wylie 1998; van der Heijden et al. 1998; Stendell et al. 1999; Bidartondo et al. 2000). This ECM diversity has been based on surveys of fruiting bodies, but is now based on more recent methods – morphotyping (microscopic observations) and phylotyping (molecular characterization). An advantage of fruiting body surveys is ease of collection and identification based on morphology; a disadvantage is the assumption that fungi fruiting in an area also form ectomycorrhizas on nearby roots. Clearly identification of ectomycorrhizas on roots is preferable. However there are difficulties in ECM identification – complex sampling design and

L.M. Suz
Area Defensa del Bosc, Centre Tecnològic Forestal de Catalunya,
E-25280, Solsona, Lérida, Spain
e-mail: laura.martinez@ctfc.es

C.S. Nautiyal, P. Dion (eds.) *Molecular Mechanisms of Plant and Microbe Coexistence*. Soil Biology 15, DOI: 10.1007/978-3-540-75575-3
© Springer-Verlag Berlin Heidelberg 2008

extraction of roots from soils. We describe two complementary methods – morphotyping and phylotyping. This chapter focuses on these two methods, because they are effective and allow ECM fungi to be identified to genus and species. Use of both methods capitalizes on the benefits of each, while minimizing the disadvantages. Both methods are time consuming, although morphotyping requires time at a microscope while phylotyping requires time at a lab bench. Morphotyping is relatively inexpensive, but requires training to recognize key microscopic features. Morphotyping may not allow identification to genus or species in some cases. Phylotyping requires more expensive molecular reagents and materials but often allows identification of ECM root tips to genus and species.

18.2 Background

18.2.1 Taxa Forming ECM

ECM fungi form ectomycorrhizas with many woody plants such as plants in the Pinaceae, Fagaceae, Betulaceae, Myrtaceae and Ericaceae. There are about 5000–6000 ECM fungal species in the Ascomycotina (Ascomycetes) and Basidiomycotina (Hymenomycetes). Ascomycetous ECM belong almost exclusively to the orders Elaphomycetales, Leotiales, Pezizales and Pleosporales (Molina et al. 1992; de Roman et al. 2005) while basidiomycetous ECM include the class Hymenomycetes, subclass Hymenomycetidae (Agerer 1987–2006; Molina et al. 1992) which comprise the orders Boletales, Gomphales, Thelephorales, Agaricales (families Amanitaceae, Cortinariaceae and Tricholomataceae), Russulales (family Russulaceae) and Cantharellales (family Cantharellaceae) (Agerer 1987–2006). ECM fungal species can have narrow, intermediate and broad host ranges (Molina et al. 1992).

18.2.2 Description of ECM Structures in Roots and in Soils

Anatomical characteristics of ECM structures are conserved at the species and genus level (Agerer 1987–2006; Agerer and Rambold 2004–2007). Among the anatomical features, there are four key complexes that distinguish the ECM structure and allow recognition of the fungus involved in the symbiosis: (a) outer mantle layers in plan view, (b) rhizomorphs, (c) shape of cystidia, and (d) emanating hyphae (Agerer 2006). Other anatomical features that may be used as diagnostic features are the Hartig net, sclerotia, chlamydospores, laticifers and contents (Agerer 1987–2006).

Mantles are hyphal sheaths around roots and can be divided into two main groups according to hyphal distribution and organization: (1) plectenchymatous and (2) pseudoparenchymatous (Agerer 1987–2006, 1991, 1995; Agerer and Rambold 2004–2007). Plectenchymatous mantles have discernible hyphae, frequently loosely woven. Pseudoparenchymatous mantles do not have discernible individual hyphae; they have short-celled, inflated, densely packed hyphae, resembling a true parenchyma. Both mantle types include standard organizational patterns (Agerer 2006). In Fig. 18.1 the letters a–f correspond to different types of plectenchymatous mantles and the letters g–j correspond to pseudoparenchymatous mantles.

Among plectenchymatous mantles, mantle type B is considered the most primitive (Agerer 2006) and has hyphae randomly arranged with no discernible pattern (Fig. 18.1a). Mantle type A is ring-like and hyphae commonly grow together for a short distance and ramify at places where other hyphae join, forming loops (Fig. 18.1b). When loops of the joining hyphae become very massive and the connecting hyphae between the loops less distinct, mantle type A is called star-like. The ring-like mantle is well represented in the order Boletales while the star-like mantle is present in the family Bankeraceae (Agerer 1987–2006; Agerer and Rambold 2004–2007) and in some Thelephoraceae (Azul et al. 2006d) (Fig. 18.1c). Mantle type C is distinguished by the presence of a gelatinous matrix between hyphae (Fig. 18.1d), apparently as a water reservoir; hyphae are randomly distributed with no discernible pattern. Mantle type D is characterized by the occurrence of cystidia. This mantle type is used only when cystidia are present on the surface of plectenchymatous mantles, since they can be present on pseudoparenchymatous mantles surface (see Fig. 18.1i). Mantle type E is differentiated by the presence of multi-ramified hyphae with short, frequently y-shaped or almost rectangular branches, so-called squarrosely branched (Fig. 18.1e). Mantle type F is identified by the occurrence of inflated cells, globular terminal cells or other cells, above an undifferentiated mantle type. Mantle type G is typified by its star-like pattern, with hyphae compactly distributed and with no space between them, e.g. *Cenococcum geophilum* Fr. and *Quercirhiza flavocystidiata* (Fig. 18.1f) (Azul et al. 2006c). Mantle type H is rather similar to mantle type E and is characterized by the presence of hyphae somewhat inflated and loosely distributed, with interspaces between hyphae. Mantle type I is characterized by the presence of quite short, often slightly tortuous or irregularly bent perpendicular cells, forming a velvet-like structure on the mantle surface. These hyphal ends can be regarded as cystidia, since they are all stainable with sulfo-vanillin (Agerer 1986).

Among pseudoparenchymatous mantles there are mantles that exhibit angular to roundish cells: types K, L, O, and P (Fig. 18.1i,g) and mantles that show epidermoid, puzzle-like structures: type M and Q (Fig. 18.1h–j). Mantle type K is differentiated by the presence of angular to roundish cells sometimes arranged in rosettes. Mantle type O is differentiated by the presence of heaps of flattened cells. The majority of pseudoparenchymatous mantles bear a hyphal net, or small groups of globular or flattened cells, on the surface. Mantle type P corresponds to a pseudoparenchyma with angular cells bearing a hyphal net, and mantle type Q corresponds to a pseudoparenchyma with epidermoid cells bearing (Fig. 18.1j). Mantle type N, that has some cells containing oil droplets that stain in sulfo-vanillin (Agerer 1995), is now considered as an additional feature of mantle types O, P, and Q and not as a distinct mantle pattern (Agerer 2006). Although these standard mantle patterns can be used for recognition, there are transitions between some mantle types, making definitive attribution of a

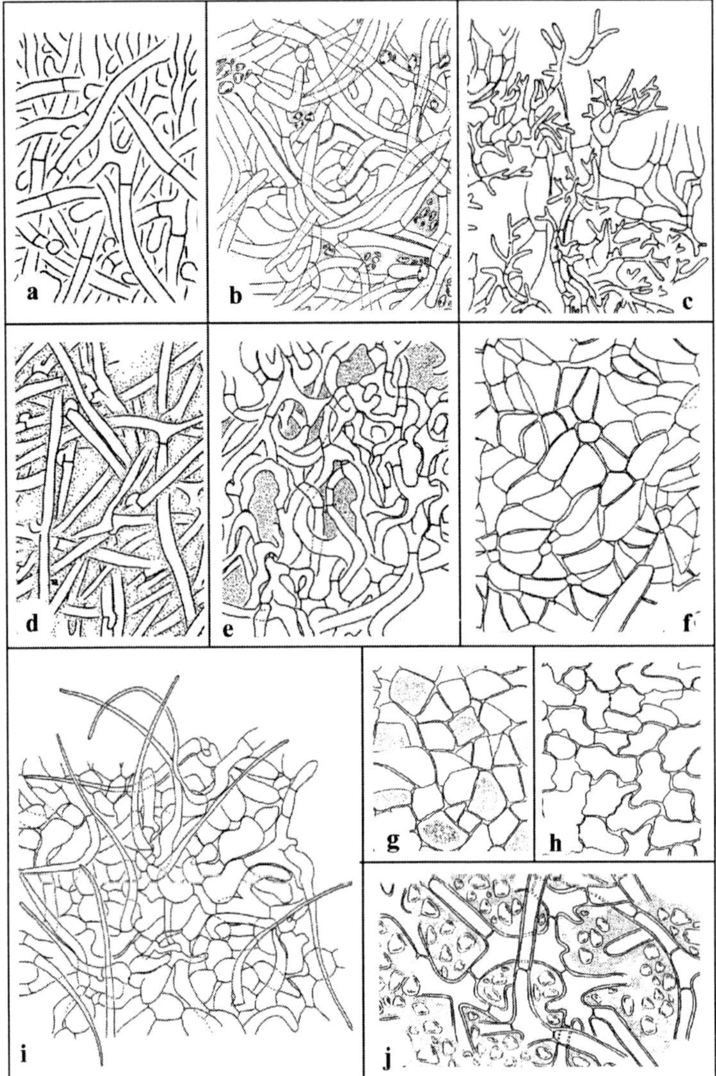

Fig. 18.1 Mantle types according Agerer 1987–2006. **a** Mantle type B: plectenchymatous, random hyphal arrangement, no discernible pattern (*Hysterangium stoloniferum*, Raidl and Agerer 1998). **b** Mantle type A: plectenchymatous, ring-like pattern (*Quercirhiza sclerotiigera*, Azul et al. 2001d). **c** Mantle type A: plectenchymatous, star-like arrangement, in this example bearing a loose, delicate hyphal net pattern (*Quercirhiza tomentellofuniculosa*, Azul et al. 2006d). **d** Mantle type C: plectenchymatous, no discernible pattern in a gelatinous matrix (*Gomphus clavatus*, Agerer et al. 1998). **e** Mantle type E: plectenchymatous, squarrosely branched hyphae (*Quercirhiza dendrohyphidiomorpha*, Azul et al. 2006a). **f** Mantle type G: plectenchymatous, star-like pattern (*Quercirhiza flavocystidiata*, Azul et al. 2006c). **g** Mantle type L: pseudoparenchymatous, angular to roundish cells (*Quercirhiza internangularins*, Azul et al. 2001b). **h** Mantle type M: pseudoparenchymatous, epidermoid cells (*Quercirhiza ectendotrophica*, Azul et al. 2001a). **i** Mantle type P: pseudoparenchymatous, angular to roundish cells with a hyphal net (*Quercirhiza auraterocystidiata*, Azul et al. 2006b). **j** Mantle type Q: pseudoparenchymatous, epidermoid cells with a hyphal net (*Quercirhiza ectendotrophica*, Azul et al. 2001a). All ECM mantle pictures have been utilized with author's permission

pattern difficult. From a phylogenetic point of view, hyphal organization in pseudoparenchymatous mantles is considered more advanced (Agerer 1995).

Rhizomorphs are extramatrical mycelia that may grow either as simple scattered hyphae from the mantle into the soil or may form 'multi-hyphal linear aggregates', the so-called rhizomorphs (Agerer 1999). Rhizomorphs may have evolved as efficient structures for water and nutrient transport. Rhizomorphs may be characterized according to their structure and ontogeny into seven types according to their internal organization (Agerer 1999). Type A corresponds to uniform-loose rhizomorphs (Fig. 18.2a) with normal vegetative hyphae loosely distributed. Type B has uniform-compact rhizomorphs, i.e., densely agglutinated hyphae of uniform shape. Type C, thelephoroid rhizomorphs, is slightly differentiated and has one peripheral hypha, rather different in diameter and in structure (Fig. 18.2b,c). Type D, ramarioid rhizomorphs, are internally differentiated and distinct due to ampullate inflations at the hyphal septum of the internal cells (Agerer 2006). Type E, russuloid rhizomorphs, are characterized by irregularly distributed thickened hyphae, frequently with incomplete septa, accompanied by thick-walled hyphae with several septa separated by short distance. Type F, phlegmacioid rhizomorphs, have thicker hyphae, sometimes with a large septal pore, with a random distribution and often embedded in a matrix. Type G, agaricoid rhizomorphs, are highly differentiated and possesses vessel-like central hyphae with partially or even completely dissolved septa. This type has not been observed on ECM fungi. Type H, boletoid rhizomorphs (Fig. 18.2d,e), corresponds to highly differentiated rhizomorphs with central vessel-like hyphae. The internal organization of highly differentiated rhizomorphs is key to recognition and delimitation of fungal relationships (Agerer 1999). The main difference between agaricoid and boletoid rhizomorphs is related to their ontogeny. Agaricoid rhizomorphs have simple, vessel-like hyphae, while boletoid rhizomorphs have vessel-like hyphae that fork close to their origin, one branch growing toward the rhizomorph base, the other, towards the tip. For both types, vessel-like hyphae originate in early ontogenetical stages. Node-like structures and split-type hyphal ramification can be observed in the Thelephoraceae, Amanitaceae and Tricholomataceae.

Some genera have distinctive rhizomorphs that allow easy identification *Cortinarius*, *Dermocybe*, *Sarcodon*, *Tomentella*, *Tricholoma* and *Xerocomus*. For example, *Cortinarius* and *Dermocybe* have abundant, distinctive rhizomorphs, while *Tricholoma* has diverse rhizomorphs. In the Boletales, highly differentiated "boletoid rhizomorphs" are observed in the Boletaceae, Gyroporaceae, Melanogastraceae, Paxillaceae, Rhizopogonaceae, Sclerodermataceae and Suillaceae (Agerer 2001). Unlike the Boletales, the Cortinariaceae exhibit undifferentiated or differentiated rhizomorphs, or no rhizomorphs at all. Rhizomorphs are particularly difficult to observe in some genera – *Russula*, *Lactarius*, *Hygrophorus*, or not seen in *Inocybe*, *Rozites* and *Tuber* (Agerer 1987–2006). Dimorphic rhizomorphs are less frequent in *Dermocybe crocea* (Schaeff.) M. M. Moser, *D. palustris* (M. M. Moser) M. M. Moser and *D. semisanguinea* (Fr.) M. M. Moser (Agerer 1995).

Organization and emanating hyphal structures of rhizomorphs allow categorization into different ECM exploration types: (a) contact, (b) short-distance, (c) medium-distance and (d) long-distance (Agerer 2001). The contact type, commonly

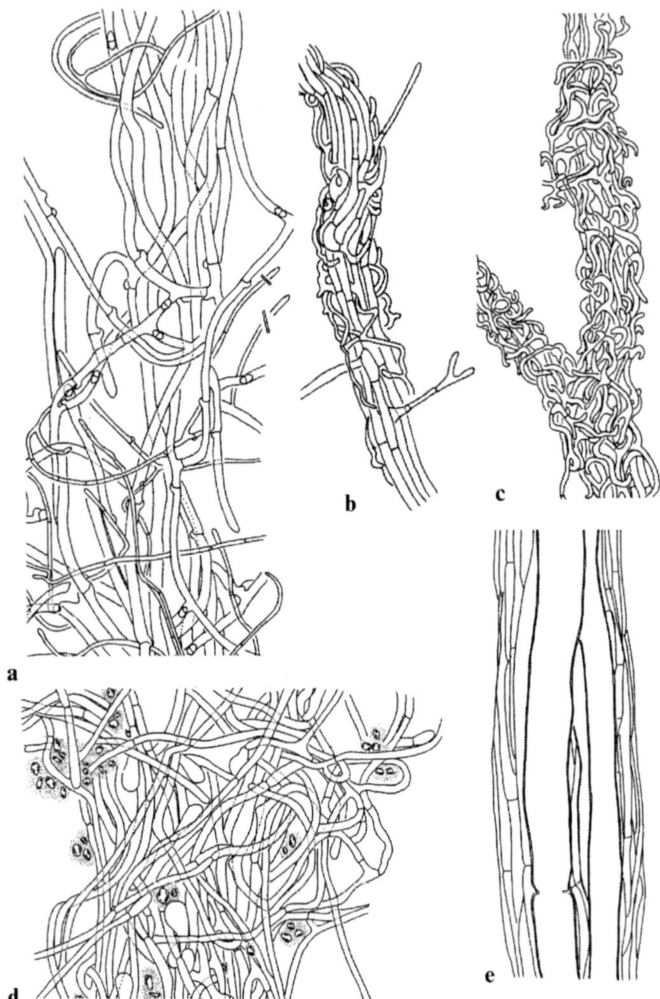

Fig. 18.2 Rhizomorph types according Agerer 1987–2006. **a** Mantle type A: uniform and loose hyphae (from *Quercirhiza auraterocystidiata*, Azul et al. 2006b). **b,c** Mantle type C: thelephoroid type, with peripheral hyphae thinner than central ones, (from *Quercirhiza tomentellofuniculosa*, Azul et al. 2006d). **d** Mantle type H: boletoid with vessel-like hyphae – surface (from *Quercirhiza pedicae*, Azul et al. 2001c). **e** Mantle type H: boletoid with vessel-like hyphae – inner layers with vessel-like hyphae with septa partially or completely dissolved (from *Quercirhiza sclerotiigera*, Azul et al. 2001d). All figures have been utilized with author's permission. All ECM rhizomorph pictures have been utilized with author's permission

hydrophilic, includes typical smooth mantles with infrequent emanating hyphae, which are frequently lost during removal of ECM from soil cores. The short-distance type has ectomycorrhizas with dense emanating hyphae that grows widely into surrounding soil. The medium-distance type matches with ECM fungi that develop rhizomorphs uniform-loose, uniform-compact, thelephoroid, or phlegmacioid (Agerer 2006) and grows often more than 30–50 mm. The long-distance type includes the highly differentiated rhizomorphs with vessel-like hyphae and may reach a length of several decimeters. Usually the long-distance exploration types include hydrophobic rhizomorphs.

Cystidia, very important diagnostic features, may be present on the ECM mantle and rhizomorph surfaces. They may also be observed on the cap skin, gills, and stipe of fruit bodies of Hymenomycetes. Cystidia may be diverse in structure and size (Agerer 1987–2006). Some are similar to normal hyphae, straight, bent, or hook-like; others present a distinct structure. They may be unramified or ramified at the proximal part with a monopodial ramification, or with dichotomous, tritomous or quadritomous ramifications.

Emanating hyphae may be present on mantles and rhizomorph surfaces. The main features of emanating hyphae are diameter, distance between septa, shape, color, cell wall thickness, surface, presence of crystals, presence of contents or drops of pigments, type of ramification, presence and type of anastomoses and presence and shape of clamps (Agerer 1987–2006). The type of anastomoses, the presence of intrahyphal hyphae and the shape of clamps are the most important features (Agerer 1995). Features of emanating hyphae are not as informative about fungal relationships as are mantles in plan view or rhizomorph organization.

The Hartig net, the zone of contact between plant and fungus, is produced by hyphae that penetrate intercellular spaces connecting outer cells of the root axis. This zone of contact is of central importance since it corresponds to the interface between symbionts. The Hartig net is usually formed from the inner mantle layers to the epidermal layers (in Angiosperms) or it can intrude into the cortical layers more deeply and toward the endodermis (in Gymnosperms). In most Angiosperms, the Hartig net may be limited to the anticlinal walls of the cortical cells – paraepidermal Hartig net – or may develop completely around the cortical cells – pariepidermal Hartig net (Godbout and Fortin 1983). Sometimes, very occasionally, haustoria-like intrusions may occur (Azul et al. 2001a).

Laticifers are typical of the genera *Lactarius* and *Russula* and correspond to latex-containing hyphae. They can be rather long, thick and/or scarcely branched, and are present in both plectenchymatous and pseudoparenchymatous mantles. Sclerotia are rather infrequent among the ectomycorrhizas. They have been observed in *Cenococcum geophilum* Fr. and in some members of the Boletaceae, Cortinariaceae, Paxillaceae and Pisolithaceae (Agerer and Rambold 2004–2007).

Chlamydospores are quite frequent in the Bankeraceae but also observed in the Thelephoraceae (Agerer 1987–2006). Mantle hyphae, rhizomorphs, cystidia and emanating hyphae may contain oil droplets, brownish contents, bluish contents, crystals or other appositions. For example, some members of the Thelephoraceae have blue granules on the surface of mantle hyphae and cystidia.

Chemical reactions are useful in identification of fruiting bodies and ectomycorrhizas (Table 18.1) (Agerer 1991). Reagents may be applied to whole ectomycorrhizas or to preparations of mantles and rhizomorphs. One of the most important chemicals is Melzer's reagent; the amyloid reaction is well represented in members of the Gomphidiaceae, Albatrellaceae and *Rozites*. Another distinctive chemical reaction is the

Table 18.1 Reagents and chemical reactions used to identify ectomycorrhizas

	Reagent	Preparation	Positive reaction
Very important	Melzer's reagent	0.5 g iodine, 1.5 g KI, 20 mL distilled water and 20 mL chloral hydrate	Amyloid reaction: stain blue. Dextrinoid reaction: stain slightly brownish
	Lactic acid	90%	Mantle colors become brighter. Reagent used to make permanent mantle preparations
	KOH	10% aqueous solution (w/v)	Three distinct reactions have been recognized: (a) cell walls with blue or brown color change to green, (b) blue granules change to green and (c) cell walls with brownish or brown color become darker or lighter
	Iron(II) sulfate	1 g Iron(II) sulfate in 10 mL distilled water and some drops of H_2SO_4	Mantle color changes to greenish, bluish or grayish. Care is needed to avoid misinterpretation of a similar reaction with root cells remnants on mantle preparation
	Guaiac	Solution of 1 g guaiac resin in 6 mL ethanol 70%	Patchy cells of bluish color, particularly in inner mantle layers. Very important to displace the reagent by water or lactic acid
Important	Brilliant cresyl blue		Two different reactions: (a) cell walls become bluish; (b) cell walls turn to reddish (called metachromatic reaction)
	Sulfo-vanillin	Crystals of vanillin, H_2SO_4	Cell contents turn to pink or black and hyphal walls turn to red. On *Lactarius* species, laticifers become all staining of the, and/or some cells of the mantle and all hyphal ends on a mantle
	Formol	40% aqueous solution (w/v)	Mantle color changes to grey or greenish
	Toluidine blue	1% (w/v) aqueous solution	Differently intense blue color of cell walls
	Cotton-blue-lactic-acid	0.05 g cotton blue solution in 30 mL lactic acid (90%)	Intense bluish color of cell walls after displacing the reagent by water or lactic acid
Optional	Ethanol	70% aqueous solution (w/v)	Elution of wall colors, contents or of pigment granules on the hyphal wall
	Phenol-aniline		Mantle cell changes to grayish color
	Sudan III	Dissolve 1 g of Sudan III in 500 mL ethanol 96% (in water bath) and add 500 mL glicerin	Lipids become reddish

sulfo-vanillin test. To a water-mantle preparation vanillin crystals and a drop of H_2SO_4 are added to one side of the coverslip and observed immediately (due to rapid reaction, laticifers may dissolve quickly). The sulfo-vanillin reaction is typical of *Lactarius* and *Russula*.

18.3   Study Design

We cannot overemphasize the importance of clear research hypotheses and related study design. While there are many research questions concerning ECM diversity and function, not all are testable. It is essential to form a question that is testable. Once a hypothesis or research question is refined and focused, the sampling design may be developed. A key element in sampling design is an understanding of the spatial and temporal distribution of ectomycorrhizas and roots in soil. Ectomycorrhizas are associated with plant roots that are not evenly distributed in soils but are aggregated based on patterns of soil fertility, soil texture, competing roots, etc. Since one often does not know the spatial and temporal distribution of roots and their associated mycorrhizas, preliminary sampling may be necessary. For example, one may collect soil cores to sample ECM diversity and determine "species effort curves". Detailed discussion of sampling design is beyond the scope of this chapter, so the reader is referred to Johnson et al. (1999). See also Sect. 18.5.2 in this chapter.

18.4   Morphotyping of ECM Roots

18.4.1   Introduction

The ability of ECM fungi to benefit their hosts is closely related to their structures. ECM fungi that colonize roots, and modify root color, shape and function, are often characterized by extensive external hyphal development. Few ECM species are well known and described in detail, particularly field-grown ECM fungi. Advances in molecular biology and morphotyping have demonstrated the poor correspondence between the species composition of fruiting bodies and of fungi colonizing roots. Identification of inconspicuous fruiting bodies (e.g. resupinates) highlights the overall diversity among ECM communities. These advances make it possible to study functional properties in nature by monitoring and by spatiotemporal analysis of below-ground ECM fungal communities. Studies on ectomycorrhizal structures are equally important to clarify functional properties in the relationship between different mycosymbionts (Agerer 2006).

   ECM morphology is useful but not considered as diagnostic because color, ramification and mycorrhizal system size depend on growth conditions and plant host (Agerer 1991). Since the first studies on ectomycorrhiza structure on *Castanea sativa* by Gibelli (1883) and on *Fagus sylvatica* and *Carpinus betulus* by Frank

(1885), over 343 species have been described in detail (de Román et al. 2005) and only about 6–7% of all presumably ECM fungi. Morphotyping has limitations for ECM identification, particularly in *Russula*, *Lactarius* and *Cortinarius*. Many fungal groups can be distinguished by morphological and anatomical characterization, a relatively inexpensive method allowing examination of large numbers of root tips. Studies of ECM structures allow better understanding of their spatial distribution (Tedersoo et al. 2003; Lilleskov et al. 2004; Baier et al. 2006).

18.4.2 *Extraction of Roots from Soils and Sample Storage*

To study ECM structure, morphological and anatomical characteristics must be intact as samples are removed from soil. Ectomycorrhizas can be sampled from roots or by tracing hyphal or rhizomorphs connections to fruiting bodies. In both situations samples should be taken with soil cores to ensure that connections between ECM roots, hyphae and rhizomorphs are not damaged or disrupted. If the research focuses on investigation of ECM fungi associated with fruiting bodies, the stipe of the fruiting body should be carefully cut and marked, especially in species with dark stipes. It is necessary to balance the number of samples and the size of the soil core to enhance retention of ECM natural features. Unfortunately, many ecological studies involve large numbers of samples with low numbers of ECM tips and no voucher specimens. Soil core samples must have unique collection numbers and can be stored at 4°C in plastic until processing within two weeks if at all possible.

18.4.2.1 Protocols: Extraction and Assessment of ECM

Procedure

List of Materials Dissecting microscope, large Petri dishes, needles, pipettes, fine forceps.

Cleaning Procedure and ECM Extraction from Soil (1) Immerse the soil core in water and soak carefully in water until saturated; (2) Wash roots gently with pipettes to limit damage to the ectomycorrhizas; (3) Place cleaned roots (<1 mm diameter) in a Petri dish with filter paper soaked with water.

Notes (a) Petri dishes with filter paper prevent color changes of ectomycorrhizas and hyphal growth; (b) ECM tips can be stored in the refrigerator up to seven days after sampling.

ECM Abundance and Richness To evaluate changes in ECM abundance and richness in soil horizons or under different soil conditions, root tips must be counted.

ECM Abundance Total ECM abundance corresponds to the total number of living ectomycorrhizas per 100 cm^3 soil volume. Relative abundance can be expressed as either (a) number of living ectomycorrhizas per meter of fine roots, or

(b) number of tips of a given genus or species divided by total number of living ECM root tips in the same soil core.

ECM Richness ECM diversity within soil horizons and/or study areas may be estimated using distinct diversity descriptors: (a) species richness, i.e., number of ECM species observed (S); (b) Shannon-Wiener (H) and Simpson (λ) diversity indexes; (c) Pielou evenness (H'); (d) Margalef (D), Log α (S) and Jack-knife richness indexes; and (e) Whittaker β-diversity index (Magurran 1988).

Notes (a) The effects of soil horizons, soil types, or land use systems on ECM abundance may be determined by univariate analyses – ANOVA (Zar 1996); (b) The relationship between ECM diversity descriptors and study areas conditions can be further determined upon multivariate analyses, e.g. by using CANOCO 4.5 software (Ter Braak and Smilauer 2002).

18.4.2.2 Protocols: Description of ECM

Procedure

List of Materials Dissecting microscope, microscope (NIC), pipettes, needles, fine forceps, Petri dishes, chemicals (see Table 18.1).

ECM Morphology (1) Observe ECM root tips in water under a dissecting microscope (6×, 12×, 25×) using a black background and lamps of daylight quality; (2) Isolate morphotypes by morphology, i.e., color, ramification type, systems, size and texture, presence of emanating hyphae, cystidia, rhizomorphs, and/or sclerotia (see Agerer and Rambold 2004–2007); (3) Take photos of mycorrhizal systems, maintaining the black background and lamps of daylight quality (Agerer 1991); (4) Use some living ectomycorrhizas to check the chemical reactions (see Table 18.1).

Notes (a) Ectomycorrhizas may exhibit high diversity of colors (Agerer 1987–2006). The color of the ectomycorrhizas is very useful for the first isolation after extraction from soil and can be an important feature for identifying some species, since ectomycorrhizas color often mirrors the color of the fruiting body cap; (b) Dimensions of ECM systems and unramified ends should be taken into account despite not being a distinctive character. The dimensions are affected by the host species and by the physical properties of the substrate in which the ectomycorrhizas have grown. The values always should be presented considering the different range of measurements.

ECM Anatomy (1) Prepare slides by using mantle squashes and mantle peels from fresh material; (2) Use remaining root tissue of the ectomycorrhizas to prepare slides by squashing for confirming presence of and structure of the Hartig net; (3) Observe slides prepared with water at 400× and 1000× magnification, with a Normarski interference contrast microscope, to register the presence of cytoplasmic contents, such as oil droplets; (4) Add lactic acid to the slide after observing mantle peels in water to study anatomical features: mantle plan view, rhizomorphs, cystidia, emanating hyphae, Hartig net, chlamydospores, sclerotia (Agerer 1991, 1995; 1987–2006; Agerer and Rambold 2004–2007); (5) Take photos and drawings of anatomical features (Agerer 1991; Agerer and Rambold 2004–2007); (6) Store

ECM samples and reference vouchers in ethanol 50%, FAA, or 2.5% CTAB. Preserve slides with lactic acid as reference vouchers as well.

Other Considerations

See Deemy (an information system for characterization and DEterminatin of EctoMYcorrhizae): http://www.deemy.de/

18.5 Molecular Identification of ECM Roots and Hyphae in Soil

18.5.1 Introduction

In recent years, use of molecular methods to identify fungal species has provided new insights into the below-ground fungal community and a more precise approach to fungal diversity studies. In ECM fungi, molecular identification is especially important for root tips and hyphae whose morphological identification can be difficult or impossible. In this section, we provide DNA-based protocols to identify ECM fungi in different fungal and environmental samples. Different molecular techniques can be chosen depending on the aims of the study, research hypotheses and the fungal material. Following DNA extraction and PCR, various techniques such as RFLP, T-RFLP, cloning and sequencing resolve identification of individual fungi or develop fungal community profiles. When the aim is to quantify the amount of DNA from a certain fungus in a pure or complex DNA sample, one can use a quantitative PCR method (Real-time PCR).

18.5.2 Guidelines for Sampling Design and Collection of Samples

ECM fungi can be present in the environment in four different stages: spores, ecto-mycorrhizas or root tips, mycelium and fruiting bodies. This section focuses on procedures to detect and identify ectomycorrhizas (both single and pooled root tip samples) and mycelia (hyphae) in soil. However, these procedures can also be used for identification of fruiting bodies and spores. Sampling area, sample size, distances among samples, spatial and vertical distribution of fungi and complexity of samples (single vs pooled) are factors that should be considered in sampling design. Presence or absence of species in soil can be influenced by sample size, since diversity can change within 1 cm in soil. It is recommended to keep distances between samples greater than size of samples (or soil cores) collected (Taylor 2002). Some ECM species seem to develop preferentially in the organic layer while other species grow in mineral layers of the soil profile (Stendell et al. 1999; Taylor and Bruns 1999). Moreover, the apparent non-random distribution of species complicates sampling of ECM communities to assess ECM richness (Landeweert et al. 2005). When studying root tip diversity, it is also important to be cautious in the interpretation of diversity data, especially when comparing two ECM communities (Taylor 2002). Single root tip DNA extractions are not realistic when analysing thousands of root tips and therefore extracted root tips are sometimes pooled prior to DNA extraction (Zhou and Hogetsu 2002). Before DNA isolation, soil and root tip samples can be stored at 4 °C for a few days. However, samples should be processed as soon as possible to reduce contamination from other fungi and DNA degradation. For long-term storage, samples should be frozen (at −20 °C or −80 °C for longer periods) or freeze-dried and then frozen. Samples can also be ground into a fine powder after freeze-drying. Root tip samples can be stored in CTAB buffer.

18.5.3   DNA Extraction

18.5.3.1   Introduction

Efficient isolation of DNA is essential in order to perform techniques to identify ECM fungi. DNA extraction protocols are subject to modifications depending on the origin of the samples and the quality and length of DNA needed for posterior analyses. These modifications usually consist of improving homogenization (e.g. in soil samples or mummified root tips), including additional cleaning steps or using extra reagents (e.g. polyvinylpyrrolidone (PVP) or Proteinase K) to remove substances that could inhibit the PCR reaction such as polyphenols and humic acids in soils and tannins in root tips. Apart from DNA, RNA can also be isolated and in higher amounts than DNA, since it is more abundant in cells. RNA requires special procedures to be extracted and cannot be stored for long periods of time because it is much more susceptible to degradation and contamination than DNA. This chapter includes only descriptions of DNA-based techniques.

18.5.3.2   Protocol: Root Tip DNA Extraction

This CTAB-based protocol of DNA extraction from single to several root tips is based on protocols previously developed by Rogers and Bendich (1985), Doyle and Doyle (1990) and Henrion et al. (1992). It can also be use to isolate DNA from fruiting bodies and mycelium from pure culture of ECM fungi.

Procedure

Equipment and Plasticware Water bath, 1.5-mL microcentrifuge tubes, pellet pestles, micropipettes, microcentrifuge.
Reagents 2% CTAB lysis buffer, chloroform, isopropanol, ice-cold ethanol (70%), sterile deionized water (or TE buffer).
  Mechanical and Chemical Lysis (1) Place one or several root tips in a 1.5-mL microcentrifuge tube; (2) Add 600 µl of 2% CTAB lysis buffer to each sample and grind with a pellet pestle.
  Incubation (3) Place microcentrifuge tubes in a water bath at 65 °C for 40 min to 1 h.

DNA Purification (4) Centrifuge the tube at 13,000 rpm for 5 min; (5) Transfer the upper phase into a new microcentrifuge-tube; (6) Add 600 µl of chloroform. Mix well by hand until the suspension is colloid; (7) Centrifuge at 13,000 rpm for 15 min; (8) Transfer the upper phase to a new 1.5-mL microcentrifuge-tube.

DNA Precipitation (9) Precipitate the DNA with 1.25 volumes of isopropanol (≈750 µl). Mix well by agitating the microcentrifuge-tube and keep it at −20 °C for 30 min. Samples can remain in the freezer overnight. (10) Centrifuge samples at 13,000 rpm for 30 min; (11) Discard the upper phase (by pipette or pouring).

DNA Washing (12) Add 200 µl of 70% ice-cold ethanol (−20 °C) to wash the DNA; (13) Centrifuge samples at 7,000 rpm for 5 min; (14) Discard the upper phase and let the DNA-pellet dry (about 5 min on the heating block).

DNA-Pellet Solubilization (15) Solubilize the DNA-pellet in 50 µl of sterile double deionized water or TE buffer.

Notes (a) Before incubation, 5–7 µl of Proteinase K (20 mg/mL) can be added to the lysis buffer; (b) To inhibit polyphenol oxidisation processes, 0.2% of β-mercaptoethanol can be added after incubation (step 3) followed by incubation for another 30 min after the purification steps.

Other Considerations

Troubleshooting Low DNA yields may result from different reasons: (a) Incomplete tissue homogenization: use liquid nitrogen prior to grinding to get fine powder from each sample before incubation in lysis buffer; (b) Insufficient lysis buffer: increase its volume in homogenization step; (c) Old material: add proteinase K during incubation as described above; (d) Incomplete solubilization of the final DNA pellet: warm the tubes at 65 °C for a few minutes, vortex briefly if needed; (e) DNA degradation: root tips were not correctly stored or frozen before DNA isolation; (f) Presence of proteins, salts, etc. DNA was not sufficiently washed; pellet the DNA again and repeat the protocol from washing step.

DNA Quantification The quality and purity of the extracted DNA is essential for PCR amplification of the target DNA. Once crude DNA is obtained and solubilized, there are two procedures to quantify the DNA: (1) with electrophoresis: run an aliquot of DNA in an agarose gel (usually 0.8%) and measure fluorescence emitted by an added nucleic acid gel stain (ethidium bromide or SYBR® Green I dye) against serial dilutions of a known amount of DNA such as calf thymus DNA (Ranjard et al. 1998, 2003); (2) with a spectrophotometer, measure absorbance of the DNA sample at 260 nm and apply the formula: 1 O.D.=50 µg DNA/mL to estimate the quantity of DNA (include the dilution factor). The first procedure has the advantage that it provides together with the quantification of DNA an estimation of the presence of contamination by RNA and an estimation of the quality of DNA. However, DNA in very low quantities may not be visible in a gel but can be amplified by PCR.

Other Protocols Henrion et al. (1994) developed one of the most frequently used protocols for DNA isolation in fungi, based on modifications of the protocols of Henrion et al. (1992) and of Gardes and Bruns (1993). Another SDS protocol developed by Edwards et al. (1991) for DNA isolation from plants has been extensively used for ECM fungi (Paolocci et al. 1999; Baciarelli-Falini et al. 2006). Lee and Taylor (1990) also developed an SDS protocol for fungal mycelia and single spores, using β-mercaptoethanol.

Commercial Kits To isolate DNA from fungi, the following fungal DNA commercial kits are often used: DNA E.Z.N.A.® Fungal DNA miniprep kit (Omega Bio-Tek; Martín and García-Figueres 1999; Aguín-Casal et al. 2004) and Ultra Clean Microbe DNA Isolation Kit (Mo Bio Laboratories; Koide et al. 2005b). The DNeasy® Plant Mini Kit (Qiagen) for plants has been used by Genney et al. (2006), Hortal et al. (2006) and Gagné et al. (2006).

18.5.3.3 Protocol: Soil (Hyphal) DNA Extraction

When working with soil samples, one may use either direct or indirect methods to extract DNA. With the direct method, cells are lysed while they remain in the soil matrix (Ogram et al. 1987). With the indirect method, cells are recovered from the soil matrix before lysis (Balkwill et al. 1975). The direct method is more commonly used because it yields more DNA and a less biased sample of the microbial community diversity (Holben et al. 1988; Miller et al. 1999). The quantity and quality of DNA extracted per gram of soil will depend on the method used and on the properties of the substrate sampled (Martin-Laurent et al. 2001). When the objective is to detect hyphae in soil, it is essential to examine soil samples with a dissecting microscope for the presence of spores or ECM mantle tissues before DNA extraction. Soil samples can be sieved to remove small roots, stones and other debris. This CTAB-based direct DNA extraction protocol is based on the root tip DNA extraction protocol described above and adapted to soil conditions by Suz et al. (2006). Only modifications are described.

Procedure

Equipment and Plasticware Micropipettes, 50-mL polypropylene tubes, 1.5-mL microcentrifuge tubes, mortar/mortar pestle, pellet pestle, water-bath or thermoblock, microcentrifuge, centrifuge for 50-mL tubes.

Reagents 2% CTAB–1%PVP, lysis buffer, chloroform-isoamyl alcohol, isopropanol, ice-cold ethanol (70%).

Mechanical and Chemical Lysis (1) Weigh 7 g of soil and place it in a porcelain mortar; (2) Add 40 mL of 2% CTAB containing 1% PVP and grind the sample with a mortar pestle.

Incubation (3) Transfer sample to a 50-mL propylene tube and incubate for 1 h at 65 °C in a water bath.

DNA Purification (4) Centrifuge the tubes at 9800 g for 5 min at 10 °C; (5) Transfer 600 µl of the supernatant to a 1.5-mL.microcentrifuge tube; (6) Add 600 µl of chloroform-isoamyl alcohol (24:1). Mix well by hand until the suspension is colloid; (7) to (11) Follow protocol for root tip DNA extraction.

DNA Washing (12) to (14) Follow Protocol for root tips. Repeat once.

DNA-Pellet Solubilization (15) Solubilize the DNA-pellet as in protocol for root tips. Warm the 1.5-mL microcentrifuge tubes in a heating block at 65 °C for few minutes if pellet is hard to solubilize.

Other Considerations

DNA Purity To estimate DNA purity, ratios of absorbance at 260/280 should be measured. Ratios up to 1.8 indicate that the DNA extract is relatively free of proteins. When this ratio is lower, commercial kits are available to purify DNA extracts, e.g. Wizard DNA clean-up system (Promega; Smalla et al. 1993; Landeweert et al. 2003a,b) and Gene Clean® II kit (Bio 101; Bertini et al. 1998; Amicucci et al. 1998).

Other DNA Isolation Protocols DNA extraction protocols for soil samples were initially designed to extract DNA from bacterial communities; more recently they have been used for fungal communities. Some protocols involve grinding samples with liquid nitrogen (Volossiouk et al. 1995; Zhou et al. 1996). Other protocols include homogenization with beads (Yeates et al. 1998; Landeweert et al. 2003a,b [based on Smalla et al. 1993]) or freeze-thaw lysis (Miller et al. 1999). Some protocols use SDS as a detergent (Martin-Laurent et al. 2001; Ranjard et al. 2003) while others use CTAB (Anderson et al. 2003, from Griffiths et al. 2000).

Commercial Kits Several commercial kits are available to isolate DNA from soil: Ultraclean™ Soil and PowerSoil™ DNA Isolation kits (MoBio; Dickie et al. 2002; Chen and Cairney 2002; Koide et al. 2005a; Hortal et al. 2006; Parladé et al. 2007) and Fast DNA Spin kit for soil (Bio101; Griffiths et al. 2000; Guidot et al. 2003). These kits are often used for complex samples of root tips (Bergemann and Garbelotto 2006) or even for individual root tips (Tedersoo et al. 2006).

18.5.4 *Polymerase Chain Reaction (PCR)*

18.5.4.1 Introduction

Nucleic acid extraction coupled with PCR (Polymerase Chain Reaction) (Mullis and Faloona 1987) has significantly improved DNA-based detection and identification of ECM fungi. This technique allows in vitro amplification of a specific DNA fragment (target DNA) through successive temperature cycles (25–40). Thus the PCR method involves repeated cycles of three phases: (1) DNA denaturation, (2) annealing or primer hybridisation, and (3) DNA extension (Fig. 18.3). This results in exponential amplification of a target DNA sequence. A standard PCR reaction contains the DNA target sequence (10–100 ng), two primers (0.1–0.5 µM) – forward and reverse – that are complementary to the sequence of target DNA, nucleotides (dNTPs, 20–200 µM each), a thermal-stable DNA polymerase (0.5–2.5 units) with its corresponding buffer (1/10 of the final volume) and cofactor ($MgCl_2$, 0.5–5 mM, typically 1.5 mM) and sterile distilled water to a final volume of 25–100 μL (Edel 1998). In theory 'n' PCR cycles will correspond to 2^{n-1} copies of the target DNA.

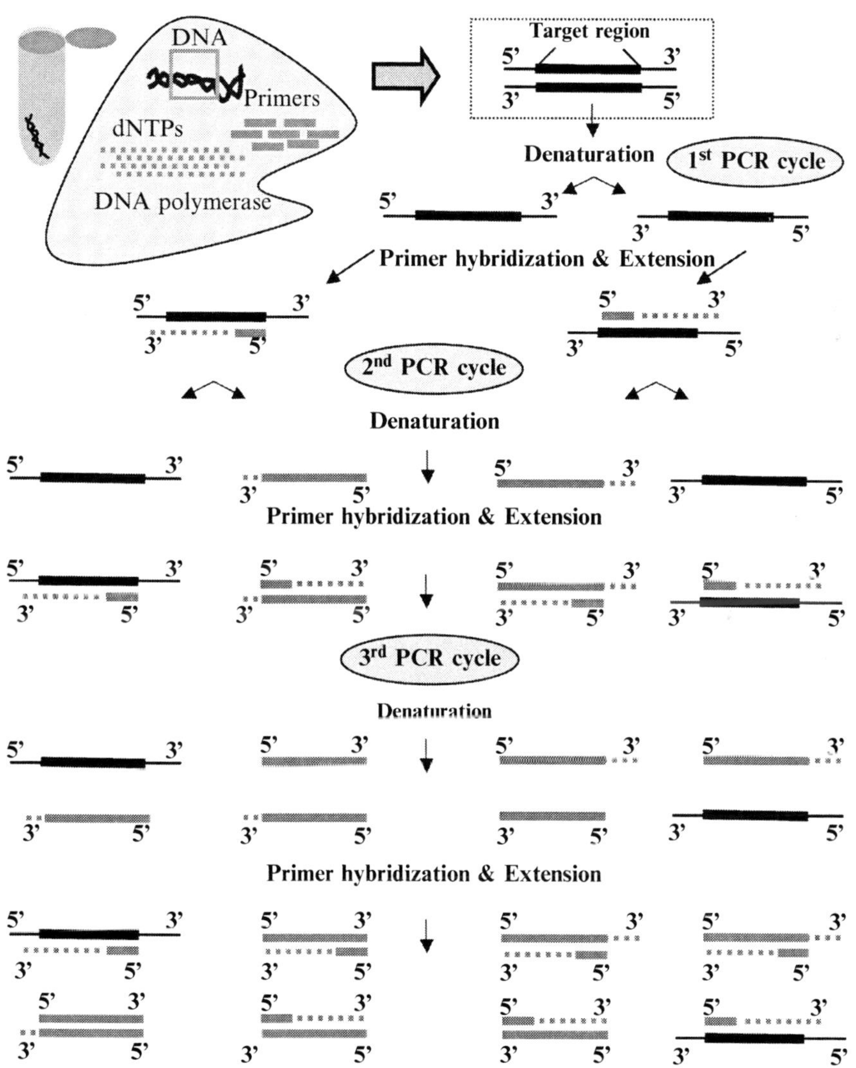

Fig. 18.3 Diagram showing three consecutive cycles of DNA amplification by PCR in three steps: DNA denaturation, Primer hybridization and DNA extension (based on Martín, 2000)

18.5.4.2 Selection of the Target Region

Since ECM fungi are not monophyletic, there are no primers designed to amplify this group. The ribosomal RNA operon (rRNA), present both in nuclei and mitochondria, is the most widely targeted region for PCR in ECM fungi. The number of copies repeated in tandem varies among fungal species (Hibbett 1992). Each rRNA gene unit repeat contains highly conserved rRNA genes 18 S, 5.8 S, 28 S and 5 S (in some taxa) and variable spacer or non-coding regions (ITS: Internal transcribed spacer region, and IGS: Intergenic spacer regions) allowing the comparison and discrimination of ECM fungi at different taxonomic levels. The 18 S (small subunit, SSU) and 28 S (large subunit, LSU) rRNA gene sequences allow identification to genus and family level (Bruns et al. 1992). Both ITS and IGS regions are more variable than other regions of the rRNA gene, presenting few or no homologies between divergent genomes. Polymorphisms in the ITS region allow identification between ECM species (Kårén et al. 1997; Pritsch et al. 1997). The largest database sequence information for molecular identification of fungi corresponds to ITS regions. In some species, the IGS region is more variable than the ITS region and is used to identify intraspecific polymorphisms. However, it may not be useful for isolates of the same species (Erland et al. 1994).

18.5.4.3 Selection of the Primer Set

Primers are short DNA sequences 15–30 bp in length. Efficiency and specificity should be considered when selecting primers. Primers should have ~50% CG content. Complementarity between forward and reverse primers and within each primer should be avoided to minimize formation of primer-dimers because they decrease the product yield. There are specific software programs to design and confirm the availability of primers. The earliest set of primers designed for ITS region amplification was ITS1/ITS4 (White et el. 1990) but they also amplify DNA from plant and algae. The fungal-specific primer ITS1F (Gardes and Bruns 1993) and the universal reverse primer ITS4 (White et al. 1990) have been extensively used in ECM fungi studies (Dickie et al. 2002, 2004; Chen and Cairney 2002; Genney et al. 2006). Since the majority of ECM fungi are basidiomycetes, the basidiomycete-specific reverse primer ITS4B has been coupled with ITS1F to amplify this group (Landeweert et al. 2003a,b; Gagné et al. 2006). Larena et al. (1999) developed a reverse primer specific to ascomycetes called ITS4A. Mitchell and Zuccaro (2006) reviewed published primers that amplify fungal sequences from nuclear SSU, LSU and ITS regions including 5.8 S rRNA. Primers designed to hybridize in complementary sequences of the DNA regions described above are detailed in several web sources: http://www.biology.duke.edu/fungi/mycolab/primers.htm; http://www. lutzonilab.net/pages/primer.shtml.

Primers to amplify the IGS region have also been designed (Henrion et al. 1992; Gardes and Bruns 1993). Martin and Rygiewicz (2005) designed a set of primers to discriminate between plant and fungal sequences, very useful in root tip identification. Primers specific to a certain ECM species have been designed for genera such as *Tuber* (Amicucci et al. 1998), *Armillaria* (Sicoli et al. 2003), *Lactarius* (Hortal et al. 2006) and *Rhizopogon* (Kennedy et al. 2007).

18.5.4.4 Cycling Conditions

In DNA denaturation, temperature to separate both strands is usually 90–95 °C. Annealing temperature is typically 40–60 °C. In DNA extension temperature is typically 72 °C but duration will depend on the length of the PCR products. PCR products <500 bp require ~30 s in this phase. Each 500-bp increase in length corresponds to an additional 30 s in the extension phase (1 min per each 1 Kb).

18.5.4.5 Tips

To avoid nucleic acid contamination, use sterile pipette tips with filters, autoclave all buffers, clean work surfaces with bleach followed by sterile water and prepare PCR reactions in a separate place in the laboratory. Keeping the components of the PCR in small aliquots helps avoid contamination from continuously pipetting from the same tube and degradation by continuous cycles of freeze-thawing.

18.5.4.6 Protocol: PCR

Procedure

Equipment and Plasticware 0.2-mL PCR tubes, 1.5-mL microcentrifuge tubes, thermocycler, micropipettes, transilluminator, vortex.
Reagents Sterile deionized water, *Taq*-polymerase/appropriate buffer, $MgCl_2$, dNTPs, direct and reverse primers, template DNA, agarose, ethidium bromide or SYBR® Green I dye, molecular weight marker (DNA ladder), TAE1X or TBE1X, load buffer.

(1) Calculate the total amount of each reagent needed for the total number of samples in the PCR-mix (see note (b) below).Concentration and volumes of the different components of the PCR reaction will depend on the thermo-stable polymerase chosen (check manufacturer's instructions). See example in Table 18.2 below; (2) Turn on thermocycler (lid needs to be 10 °C over the highest temperature of the PCR program before starting the PCR reaction); (3) Place on ice all components for PCR reaction except *Taq*-polymerase; (4) When reagents have thawed, prepare in a 1.5-mL microcentrifuge tube the PCR-mix for a final volume per sample of 20 µl in this order: sterile deionized water, enzyme buffer,

MgCl$_2$ (in the example, it is included in the Phusion buffer), dNTPs and primers; (5) Remove *Taq*-polymerase from the freezer and add the units per reaction recommended by the manufacturer to the PCR-mix (e.g. 0.4 units/μL of Phusion™ High-Fidelity DNA Polymerase in the example); (6) Vortex the mixture for 5 s to homogenize the reagents; (7) Distribute equal volumes of PCR-mix in each 0.2-mL microcentrifuge tube (in the example, 17 μl); (8) Vortex each DNA extraction briefly and add the DNA template to each PCR-tube; (9) Cover the PCR-tubes and place them in the thermocycler. Program cycling conditions (e.g. for ITS1F/LR3 amplicon: initial denaturation at 98 °C for 30 s, followed by 35 cycles of denaturation at 98 °C for 10 s, annealing at 56 °C for 30 s and extension at 72 °C for 1 min, with a final extension at 72 °C for 7 min). Variations on time and length of each step will depend on the *Taq*-polymerase chosen. (10) Resolve PCR products by gel electrophoresis in agarose 1.5% stained with ethidium bromide or SYBR® Green I dye (use gloves) and run samples in the gel with an appropriate molecular weight marker; (11) Visualize PCR products under UV light.

Notes (a) If thermocycler lacks a thermal lid, it is necessary to add oil to the tubes to avoid evaporation; (b) Make enough PCR-mix for three extra reactions (one negative control, one positive control and one for any pipetting errors); (c) Changes in volume of the different PCR-components will determine the volume of water added.

18.5.4.7 Protocol: Nested PCR

One challenge of working with root tips and hyphae in soil is that low amounts of DNA or DNA of low quality are obtained after DNA extraction, leading to weak or non-amplifications. Furthermore, plant DNA is co-extracted with fungal DNA in root tip samples. In soil samples, DNA from different organisms may be co-extracted with fungal DNA, resulting in low quality and low amounts of target DNA. To solve these problems it is possible to use a technique called "Nested-

Table 18.2 Concentrations of components needed for PCR amplification of the ITS regions and part of the 28 S of rRNA gene with primers ITS1F and LR3 (Gardes and Bruns 1993; Hopple and Vilgalys 1994). DNA extraction was performed with the Ultraclean™ DNA Isolation kit (MoBio) using pooled samples of ~100 lyophilized roots tips

Reagent	Initial concentration	μL/sample (1X)	Master mix (μL) (10X)
Milli-Q water		8.4	84.0
Phusion Buffer	5X	4.0	40.0
dNTPs	200 μM	0.4	4.0
ITS1F	5 μM	2.0	20.0
LR3	5 μM	2.0	20.0
Phusion *Taq*[a]	2 units/μl	0.2	2.0
Template DNA	Variable	3.0	
Total volume		20.0	

[a]Phusion™ High-Fidelity DNA Polymerase (New England, BioLabs)

PCR", that consists of two consecutive PCR reactions. The first PCR reaction uses an external primer pair, while the second PCR reaction uses an internal (nested) primer pair, or one of the external primers and an internal one (semi-nested PCR). Thus, the product of the first PCR is used as a template for the second PCR, increasing the sensitivity of the technique and allowing the detection of the target DNA, that would be undetectable or in very low amounts with only one PCR reaction. In soil DNA extractions, nested-PCR provides a lower threshold of sensitivity and permits much higher levels of dilutions of the DNA template (Volossiouk et al. 1995; Anderson and Parkin 2007).

Procedure

Choose two pairs of primers. Primers for the first PCR reaction should hybridize in external sequence parts of the sequence of interest. The product obtained from this PCR will be the template in the second PCR. After following procedure described in the previous PCR protocol: (12) Depending on the amount of product obtained by the first PCR using the first primer set, dilute the PCR product 10- to 1000-fold with sterile deionized water; (13) Follow the PCR process described in the previous protocol, using the second pair of primers (see note below) and the dilution series of the DNA template. The negative control (no DNA template) of the first amplification should be used as template in the second one. Thus, two negative controls are set up in the second amplification; (14) Place tubes in the thermocycler and run the appropriate program; (15) Resolve PCR products by gel electrophoresis in agarose 2% stained with ethidium bromide or SYBR® Green I dye (use gloves) using an appropriate molecular weight marker; (16) Visualize PCR products under UV light.

Note When first PCR is carried out with ITS1F and LR3 primers, primers that hybridize on 5.8 S ribosomal subunit, such as the forward primer ITS3 and the reverse one ITS2 (White et al. 1990) can be used for seminested PCR reactions coupled with the former primers. Also ITS1F coupled with ITS4 or ITS4B can be useful for ECM fungal identification.

Other Considerations

Use of 1–10% (w/v) dimethylsulfoxide (DMSO) decreases the melting temperature needed for separation of both strands of DNA. Addition of bovine serum albumin (BSA) (10–100 µg/mL) improves yield of the PCR product, binding fatty acids and phenolic compounds that can inhibit the PCR reaction.

PCR-Beads Use of PCR kits such as puRe Taq Ready-To-Go PCR beads (GE Healthcare) significantly reduces the number of pipetting steps, decreases handling errors and increases reproducibility. This kit has been used in several studies about ECM fungi (Martín and Calonge 2000; Tedersoo et al. 2006; Suz et al. 2006).

Troubleshooting Weak or no amplification could be due to: (1) Reagents are not in the required concentration or one of them is missing: check the protocol; (2) Low amount of DNA template: increase the volume of DNA template and correspondingly decrease the volume of deionized water, if improvement is not shown: try to perform a nested-PCR with inner primers to the DNA target; (3) Presence of PCR inhibitors: dilute DNA template or purify the DNA extraction. Add BSA or DMSO in the PCR-mix, to increase DNA yield. (4) Too short extension times: Increase 30 s for each 500 bp of PCR product. You can also increase $MgCl_2$ concentration: yield of amplified products will increase but specificity will decrease. When excessive background amplification is present, possible reasons could be: (1) Very high amount of DNA template: dilute it for PCR reaction; (2) Too many PCR cycles can increase the amount of non-specific background products; (3) High concentration of primers; (4) Poor quality of DNA template; (5) Too low annealing temperature. When non-specific amplification is shown, reasons could be: (1) High concentration of $MgCl_2$: decrease $MgCl_2$ concentration; (2) Contamination of reagents with amplifiable DNA: discard old reagents and prepare a new stock; (3) Primers hybridize to a secondary site in DNA template: think about designing new primers; (4) Annealing temperature is low; (5) Excessive number of cycles.

18.5.5 PCR-Based Techniques

18.5.5.1 Introduction

PCR-based techniques permit the study of environmental samples without the need to culture organisms, leading to a better understanding of the role of ECM fungi in complex environments. They also provide methods to identify ECM fungi at different life stages when morphological characters are ambiguous or missing. These PCR-based techniques allow investigation of relationships between closely related species and populations of single species. In direct PCR-based techniques such as RFLP, T-RFLP, sequencing and cloning, the primers hybridize to known sequences of the fungal genome (ITS regions, LSU and SSU of the rRNA gene, etc.). Selection of the appropriate PCR-based technique depends on the research objectives, the fungi of interest and the type of data analyses planned (Anderson and Cairney 2004).

18.5.5.2 PCR – Restriction Fragment Length Polymorphism (RFLP)

After checking by gel electrophoresis that single intense bands were obtained from each PCR reaction, the product can be analyzed by Restriction Fragment Length Polymorphisms (RFLPs). This analysis involves digestion of PCR products with a number of restriction enzymes that cleave DNA molecules at specific nucleotide sequences. Restriction fragments are resolved by gel electrophoresis; smaller DNA fragments travel greater distances towards the positive electrode than larger fragments. After a successful restriction digestion, a series of DNA fragments will appear in the gel. Pattern similarity can be used to differentiate species, strains and other taxonomic categories. To identify ECM and to carry out population studies, it is important to compare patterns obtained with those from local PCR-RFLP databases of fruiting bodies or pure cultures. In most projects, 100–500 root tips are sampled and 3–6 enzymes are used in RFLP analysis. With such large number of samples, automatic processing of data is required (see below: Taxotron® software package). One of the best tools for identification of those RFLP patterns that are not associated with a known pattern, is the direct sequencing of ITS regions followed by BLAST search (Altschul et al. 1997).

18.5.5.2.1   Protocol: RFLP

Procedure

Equipment and Plasticware Micropipettes, 1.5-mL microcentrifuge tubes, water rakes, vortex, microcentrifuge, water bath, camera.

Reagents Enzymes/buffer for each enzyme, sterile deionized water, molecular weight marker (DNA ladder).

Distribute 5 µl PCR-product to 1.5-mL microcentrifuge tubes, according to the number of restriction enzymes. Keep on ice.

Master Mix (15 µl/Sample/Enzyme) (2) Prepare a digestion mix for each enzyme according to manufacturer's instructions. Include an extra volume to each 10 samples. For each sample, mix the reactants in the following order: (a) Two microliters of buffer (1/10 of final volume). Be careful, each enzyme has its own buffer, which should be completely defrosted before use; (b) Sterile deionized water to final volume of 15 µl. Mix gently (vortex); (c) Two units of restriction enzyme (volume will vary depending on enzyme concentration). Mix gently (vortex) and spin down.

Incubation (3) Add 15 µl of master mix to each sample. Mix gently (vortex) and spin down; (4) Place tubes in a water bath at the temperature require for each enzyme, for at least 1 h/enzyme unit. After incubation, tubes should be stored at −20 °C.

Analysis of RFLP Patterns (5) Visualize DNA fragments by 2–2.5% agarose gel electrophoresis. According to the range of length of the analyzed fragments, different DNA size standards can be used. Record results with polaroid or digital pictures. Tip: run a DNA ladder with each two or three digested samples to obtain a more accurate measure of fragment lengths; (6) Determine fragment size using the software chosen (e.g. Taxotron® software system, see below).

Other Considerations

All recognition sequences are palindromes: both strands read the same in both directions. If the PCR buffer is not compatible with the restriction enzyme, purify the PCR product by the standard procedure (phenol-chloroform and ethanol) or by a commercial kit. The Taxotron® software system, especially developed for RFLP data processing (Institute Pasteur, Paris) includes: RestrictoScan® that allows detection of lanes, band and migration values in a digitalized picture and RestrictorType® that calculates molecular size, using the function of Schaffer and Sederoff (1981) to estimate fragment lengths. Authors have published RFLP data from different species of *Rhizopogon* (Martín et al. 1998, 2000b), *Macowanites* (Martín et al.

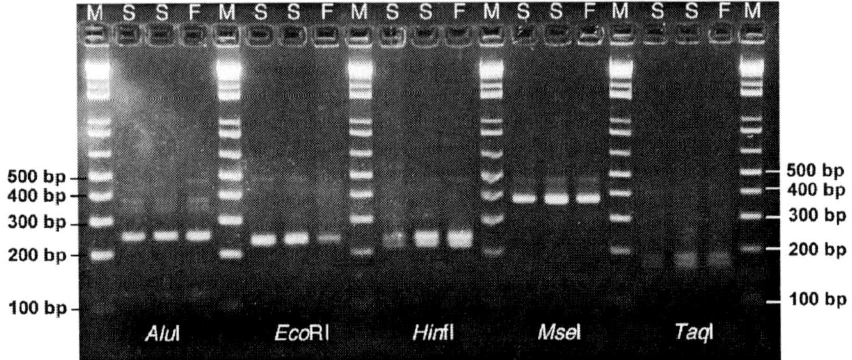

Fig. 18.4 RFLP patterns obtained alter digestion with five restriction enzymes (*AluI*, *EcoRI*, *HinfI*, *MseI* and *TaqI*) of a 465-bp PCR product corresponding to *Tuber melanosporum* from soil mycelia (S), and fruiting bodies (F). M: 1 Kb plus DNA Ladder. Fragments were resolved in 2% agarose gel electrophoresis for 40 min at 100 V

1999), *Russula* (Martín et al. 1999), *Tuber* (Suz et al. 2006, see Fig. (18.4), *Terfezia* and *Tirmania* (Martín et al. 2000a).

18.5.5.3 Terminal-RFLP (T-RFLP)

Terminal-Restriction Fragment Length Polymorphism (T-RFLP) is a molecular fingerprinting method commonly used to characterize ECM fungal communities (Burke et al. 2005; Koide et al. 2005a,b). This method facilitates increased throughput compared with gel-based fingerprinting techniques because it uses fluorescence electrophoresis and automated DNA sequencing technologies (Anderson and Cairney 2004). T-RFLP utilizes fluorescently labeled forward and/or reverse primers resulting in terminal restriction fragments (TRFs) containing labeled primers. Separation and size detection of fluorescently labeled TRFs is performed with a DNA sequencer where labeled fragments are recognized by the fluorescence detector.

18.5.5.3.1 Protocol: T-RFLP

Procedure

Equipment and Plasticware Thermocycler, incubator, capillary DNA sequencer.
Reagents Fluorescently labeled primers, dNTPs, *Taq*-polymerase, PCR buffer (sterile), Milli-Q filtered water, restriction enzymes and buffers, sizing standard, formamide, PCR purification method (e.g. QIAquick PCR purification kit). (1) Perform PCR (50-µL reaction volumes) using fluorescently labeled forward and reverse primers (0.2 µM of each primer); (2) Run PCR products on agarose gel; (3) Purify PCR product using a commercial kit such as QIAquick PCR purification kit (Qiagen) or ChargeSwitch®

PCR clean-up kit (Invitrogen); (4) Digest PCR product with restriction enzymes and recommended buffers according to manufacturer's protocol (see RFLP protocol above); (5) Incubate at recommended temperature depending on enzymes used; (6) Add 1.0 µL digestion product (diluted to optimum concentration) to 9 µL formamide and 0.5 µL of GS-500 ROX size standard (Applied Biosystems); (7) Denature at 95 °C for 5 min; immediately chill on ice; (8) Perform capillary gel electrophoresis on an ABI 3100 genetic analyzer (Applied Biosystems) or equivalent system.

Other Considerations

Amplification efficiency of fluorescently labeled primers tends to be low compared to unlabeled primers, which frequently leads to lower yields of PCR product. Since the output of T-RFLP is digital, specialized software is needed for determining TRF sizes and intensities. Peak Scanner™ and GeneMapper® are software programs available from Applied Biosystems that can be used with Applied Biosystems Genetic Analyzers. One major limitation of T-RFLP analysis is that sequence data cannot be produced from T-RFLP peaks, thus making it difficult to identify unknown taxa. In addition, in order to match unknown T-RFLP peaks to previously identified T-RFLP peaks, a robust local database must be created. As with RFLPs, enzyme selection is critical since closely related species can generate similar TRFs. In mixed-template environmental samples, rare taxa may be overlooked by T-RFLP because they are represented by relatively low amounts of DNA (Burke et al. 2005). Despite some limitations, T-RFLP is a relatively efficient and effective method for characterization of ECM communities.

18.5.5.4 DNA Sequencing

DNA sequencing allows determination of the nucleotide sequence of a given DNA segment. Sequencing is the most accurate technique to identify fungal species. Sequencing is applied to PCR products either directly or after cloning. The most popular sequencing method, 'chain termination' method, was developed by Sanger et al. (1977) and is based on use of labeled-dideoxinucleotides (ddNTPs) as DNA chain terminators, since they lack an OH group on the 3 carbon atom, necessary to form the linkage with other nucleotide. These fluorescently labeled ddNTPs are added together with non-labeled dNTPs to the sequencing reactions. Each time the chain incorporates a labeled ddNTP, this nucleotide constitutes the end of the chain. Thus, the reaction results in fragments of DNA with different lengths that only differ by one base from each other (Fig. 18.5). These fragments are heat denatured and separated by polyacrylamide gels. A laser within an automated DNA sequencing machine is used to analyze DNA fragments produced. Therefore, the composition and order of the whole sequence can be obtained, reading all single-stranded sequences from the fixed point to the last specific base. It is now possible to perform the reaction using a commercial kit that provides the required reagent compo-

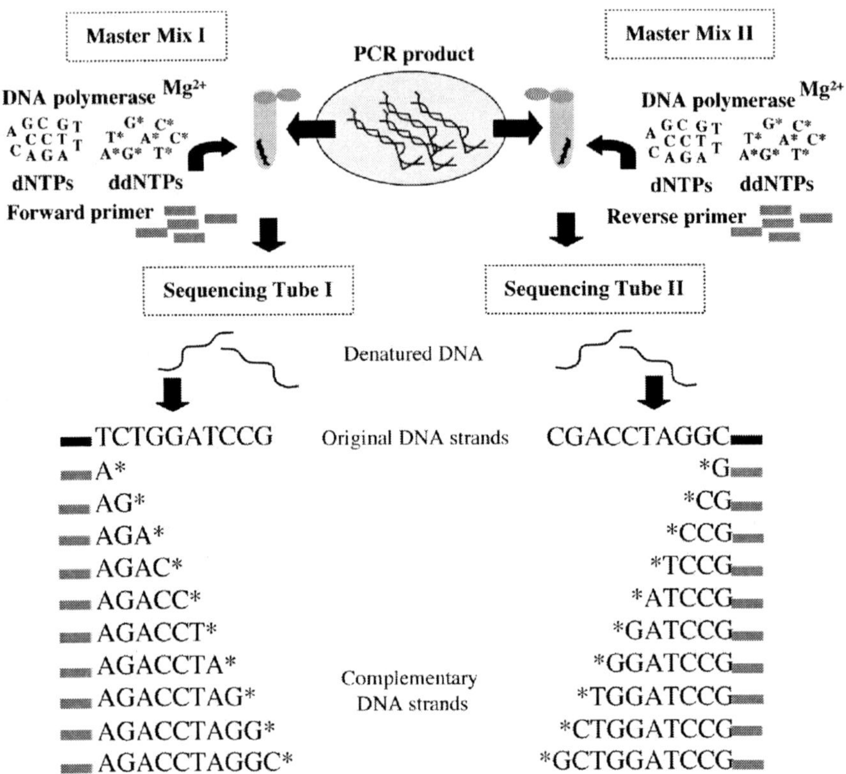

Fig. 18.5 Schematic diagram of the 'Chain Termination' sequencing method (Sanger et al. 1977). Two reactions per sample corresponding to both strands of the DNA PCR product are detailed

nents for the sequencing reaction (cycle sequencing) in a pre-mixed format. The user only provides the DNA template and the template-specific primer/s.

18.5.5.4.1   Protocol: DNA Sequencing from a PCR Product Using the BigDye Terminator v3.1 Cycle Sequencing Kit (Applied Biosystems)

Procedure

Equipment and Plastic ware Instrument Platform (ABI PRISM® 3700 DNA Analyzer; ABI PRISM® 3100 Genetic Analyzer; ABI PRISM® 3100-Avant Genetic Analyzer ABI PRISM® 310 Genetic Analyzer or all models of ABI PRISM® 377 DNA Sequencer), thermocycler (GeneAmp® PCR Systems 9700, 9600, 2700 or 2400), 1.5-mL microcentrifuge tubes/96-well plates (depending on the thermocycler used), micropipettes, vortex.

Reagents DNA template, primers (forward-reverse) complementary to the sequence of the DNA template, BigDye Terminator v3.1, Cycle Sequencing Kit.

Purification of the PCR Product (1) Purify PCR product to sequence using a commercial kit such as QIAquick PCR purification kit (Qiagen) or EZNA® Cycle Pure kit (Omega Bio-Tek). If nonspecific PCR products are detected when the PCR product is run out on agarose gel, the PCR product purification can be performed by excising the band from the gel and using a gel purification kit such as QIAquick Gel PCR purification kit (Qiagen). Single PCR products can be purified using Exo-Sap enzymes (Sigma).

DNA-Quantification of the PCR Product (2) Quantify purified DNA by gel electrophoresis or by spectrophotometry at 260 nm (Sect. 5.3; DNA quantification). The amount of DNA necessary depends on the length of the PCR product to sequence. It should range around 2 ng/µL per 100 bp of PCR product.

Sequencing Reaction (3) Distribute the DNA template volume into 1.5-mL microcentrifuge tubes. Keep on ice; (4) Take out from the freezer the components of the kit. Prepare the master mix as indicated in manufacturer's instructions (4 µL of Ready reaction Premix 2.5X, 2 µL of BigDye Sequencing Buffer 5X, 3.2 pmol of primer, 2 ng/µL each 100 bp of PCR product and water to a final volume of 20 µL); (5) Mix well and spin briefly; (6) Add the required volume of master mix to each DNA simple. Mix gently; (7) Place tubes in the thermocycler and follow the corresponded cycling conditions. E.g.: cycling conditions for ITS primers: 25 cycles of denaturation at 96 °C for 10 s, annealing at 50 °C for 5 s and extension at 60 °C for 4 min.

Purification of Extension Products Incorporated dye terminators should be removed before electrophoresis. There are some recommended purification protocols such as ethanol/EDTA, ethanol/EDTA sodium acetate precipitation and plate and spin column purification; (8) Choose the purification protocol supplied by the manufacturers depending on the desired particular application; (9) Send the resultant reactions to an automated sequencer. Each sample will result in a chromatogram of peaks of four colors, each corresponding to one labeled-ddNTP.

Other Considerations

Sequencing options after cloning are described in Sect 5.5 (Cloning technique). The reagents provided by the kit described above are suitable for performing fluorescence-based cycle sequencing reactions on single- or double-stranded DNA templates, on PCR fragments and on large templates (e.g. BAC clones). Users can adjust the quantities of reagents to a final volume < 20 µL.

Troubleshooting When the chromatogram shows peaks on top of peaks or peaks with a weak signal, it is probably due to the low quality of the DNA template or to the presence of contaminants: include a control DNA template in each run of sequencing to determine whether failed results are due to DNA quality or to a failure in the sequencing reaction. High concentrations of EDTA from the TE buffer used to resuspend the DNA in its isolation inhibit the BigDye reaction. Do not use too much EDTA in the DNA samples and follow recommended procedures detailed in the user manual for this kit. If the thermocycler used is not from Applied Biosystems you may need to optimize cycling conditions. DNA sequencing protocols will vary depending on the sequence systems and equipment used. Sequencing results can be compared to a number of publicly available fungal databases, including NCBI GenBank (http://www.ncbi.nlm.nih.gov/), UNITE (http://unite.ut.ee/) and mor (http://mor.clarku.edu/).

18.5.5.5 Cloning

While direct sequencing can be used for single-species PCR products, cloning or creation of clone libraries is a technique that can be used for environmental samples that produce mixed PCR products. In samples with multiple species of ECM fungi, the mixed product generated by PCR requires separation by cloning into a suitable vector. Several methods exist for cloning PCR-derived DNA fragments. The T/A cloning method is used because of convenience and efficiency. T/A cloning depends on the terminal transferase activity of *Taq*-polymerase, which adds a single deoxyadenosine (A) to the 3′ ends of PCR products. Individual PCR products are then inserted into linearized vectors with single overhanging 3′ deoxythymidine (T) residues. Next vectors are integrated into bacterial cells resulting in bacterial colonies that contain different fragments of fungal DNA. Since an efficient cloning reaction produces hundreds of colonies, profiling techniques (e.g. RFLP or DGGE/TGGE) can be used to screen clones and reduce the total amount of sequencing needed. Two cloning systems commonly used for construction of clone libraries in ECM studies are the TOPO TA Cloning® Kit (Invitrogen) and the pGEM®-T vector systems (Promega).

18.5.5.5.1 Protocol: Cloning (This Modified Protocol Uses Invitrogen's TOPO TA Cloning® Kit for Sequencing)

Procedure

Equipment and Plasticware Sterile toothpicks, 96-well microtiter plates, 0.5-mL microcentrifuge tubes, micropipettes, microcentrifuge, thermoblock 42 °C, 37 °C shaking and non-shaking incubator, thermocycler.

Reagents PCR Product, plasmid vector, *E. coli* competent cells, S.O.C. media (2% Tryptone, 0.5% Yeast Extract, 10 mM NaCl, 2.5 mM KCl, 10 mM $MgCl_2$, 10 mM $MgSO_4$, 20 mM glucose), LB (Luria-Bertani) agar plates with 50 μg/mL of ampicillin, LB broth with 50 μg/mL of ampicillin (add ampicillin just before use), sterile Milli-Q filtered water.

PCR (1) Add a 10-min extension time at 72 °C at the end of PCR cycling to facilitate addition of 3′A-overhangs on the PCR products.

Ligation (Day 1) (2) Incubate LB plates at 37 °C; (3) Set up 6-μL cloning reaction: (a) add 4.0 μL of fresh (see note below) PCR product to a 0.5-mL-microcentrifuge tube, (b) add 1.0 μL of sterile Mili-Q filtered water, (c) Add 1.0 μL of plasmid vector (pCR®4-TOPO®); (4) Mix gently and centrifuge briefly; (5) Incubate reaction for 20 min at room temperature and then put on ice.

Transformation (Day 1) (6) Thaw One Shot® TOP10 chemically competent cells on ice; (7) Add 2 µL of cloning reaction to each tube with TOP10 cells; (8) Mix gently and incubate on ice for 30 min; (9) Heat shock at 42 °C for 30 s; then transfer immediately to ice; (10) Add 260 µL of room temperature S.O.C. medium to each tube; 11) Place in incubator at 37 °C on shaker (200 rpm) for 1–1.5 h; (12) Spread 100 µL and 150 µL of each transformation onto prewarmed LB plates; (13) Invert plates (to prevent condensation) and place in a 37 °C incubator overnight (~12 h).

Picking Colonies (Day 2) (14) Dispense 100 µL of LB broth with ampicillin into each well of a 96-well microtiter plate; (15) Gently touch tip of sterile toothpick to a single colony (without dipping into agar); (16) Tap or swirl toothpick into well with LB broth and then carefully remove toothpick; (17) Incubate plate at 37 °C overnight.

Analyzing Clones (Day 3) (18) Perform PCR using <0.5 µL of LB culture mixture as template; (19) Run PCR products on 1.5% agarose gel; (20) Use profiling techniques to screen PCR products and then sequence representative clones.

Note Fresh PCR products should be used for cloning because terminal deoxyadenosines are susceptible to cleavage from repeated freezing and thawing.

Other Considerations

PCR amplification of environmental samples with complex DNA mixtures can result in generation of chimeric sequences (O'Brien et al. 2005; Jumpponen 2003). Chimeric sequences occur when a DNA fragment of one gene anneals with a homologous template to prime the next cycle of DNA synthesis. Formation of chimeric sequences can result in the identification of non-existent fungal species and overestimates of species diversity. Reducing the number of PCR cycles (fewer than 20) and increasing PCR extension times can minimize formation of chimeras (Suzuki and Giovannoni 1996; Qiu et al. 2001; Acinas et al. 2005). Computer programs such as Recombination Detection Program (Martin and Rybicki 2000) and Chimera Check from the Ribosomal Database Project II (Maidak et al. 2001) can be used to detect possible chimeric sequences. PCR bias and cloning bias can also influence the relative frequencies of DNA fragments recovered from cloning of mixed-template reactions. Cloning bias results from preferential cloning of certain DNA fragments while PCR bias is a result of unequal amplification of certain templates. The following modifications are recommended in order to minimize PCR bias in construction of environmental clone libraries: (1) combine several independent replicate PCR amplifications; (2) use low numbers of PCR cycles and (3) use low annealing temperatures (Suzuki and Giovannoni 1996; Qiu et al. 2001; Acinas et al. 2005). The number of clones to analyze per sample depends on the objectives of the study, the diversity of the sample, the importance of detecting rare species and the availability of time and financial resources. Landeweert et al. (2003a) determined that analyzing 30 clones per soil sample was sufficient for detecting the most common ECM species. Sequencing 50 clones per sample resulted in reasonable detection of fungal diversity in soil samples from a natural grassland (Anderson et al. 2003). Screening of clones with a profiling technique such as RFLP or denaturing gradient gel electrophoresis (DGGE) (Middleton et al. 2004) can greatly reduce the total amount of sequencing needed. For example, Smith et al. (2007) and Morris (2006) used RFLPs to screen 48 clones from samples of pooled EM root tips. Once it has been determined which clones will be sequenced there are two options for sequencing. Plasmid DNA can be extracted from the *E. coli* cells and sequenced directly or PCR can be performed with the clone culture mixture and then the PCR products purified and sequenced. Sequencing can be performed using primers that flank the vector cloning sites or primers used in the original PCR.

18.5.5.6 Real-Time PCR – Quantitative PCR

Real-time PCR measures the quantity of nucleic acid target in a DNA sample. Using this technique it is possible to estimate copy number or the amount of DNA target in the exponential phase of the PCR reaction rather than at the end, when increment of fluorescence truly correlates to the amount of DNA target. DNA synthesis is monitored using DNA-binding dyes such as fluorophore SYBR® green I (Invitrogen, Life Technologies). This dye binds to double-stranded DNA, but not to single stranded. As more PCR product is produced, more fluorescence is registered from SYBR® green. Another way to screen formation of copies of the target DNA through the reaction is by registering the fluorescence generated by fluorogenic target-specific probes, such as Real-time TaqMan™ PCR system (also known as fluorogenic 5′ nuclease chemistry). Copies of the DNA target emit fluorescence that is recorded in each cycle, and this fluorescent signal increases with number of copies. To determine the starting amount of the DNA target in an unknown sample, it is necessary to measure its Ct (cycle number at which fluorescence crosses the threshold) and interpolate this value in a determined standard curve. Confirmation of the identity of the PCR product is performed by examining the thermal denaturation plots or melting curves (Fig. 18.6), since the melting temperature of the PCR product mainly depends on the nucleotide composition. Formation of the PCR product can be monitored throughout the reaction, thus, it is possible to adjust the exact number of PCR cycles needed. Moreover, Real-time PCR does not need post-PCR processing, saving time. The availability of a technique such as Real-time PCR has opened a door for quantification of specific ECM fungi in environmental samples, allowing evaluation of their ecological and functional importance in different ecosystems. This technique is able to detect twofold changes in concentration of the target DNA, being very useful to estimate changes in fungal biomass.

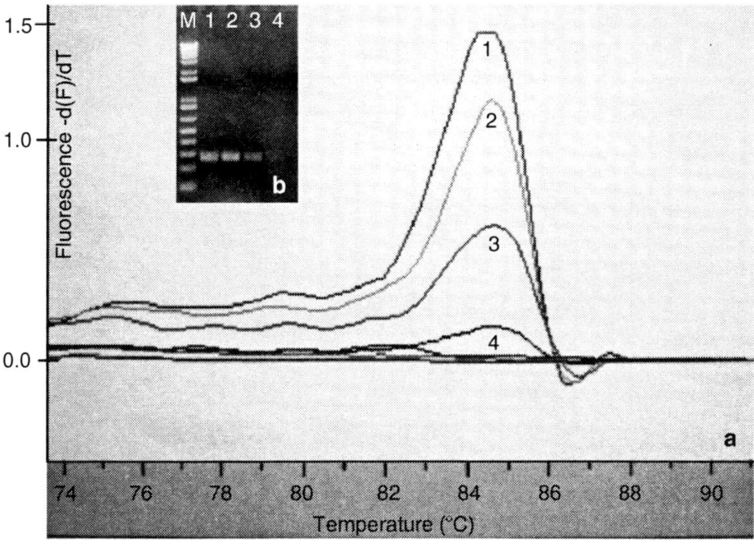

Fig. 18.6 Diagrams of **a** melting curves of tenfold dilution series of DNA from *T. melanosporum* obtained after Real-Time PCR with the LightCycler (Roche) system using SYBR® Green dye I, and **b** gel analysis after the PCR reaction: (1): undiluted simple; (2) to (4) dilutions 10^1, 10^2 and 10^3; (M) 1 kb plus DNA ladder. The lower the DNA concentration, the lower melting peaks (fluorescence emitted) and band intensities detected

18.5.5.6.1 Protocol: Real-Time PCR

Procedure (Using ABI Prism 7700 Equipment; Applied Biosystems)

Equipment and Plasticware ABI Prism® 7700 equipment, 96-wells plates, Sequence Detection System (Applied Biosystems), software version SDS 1.9.1 and Dissociation Curves 1.0, 1.5-mL microcentrifuge tubes, micropipettes, vortex.

Reagents Sterile Milli-Q water, primers, template DNA, known and unknown samples, agarose, molecular weight marker (DNA ladder), TAE1X or TBE1X, load buffer.

Establishment of a Calibration Curve (1) Establish a calibration curve with serial dilutions of a known amount of DNA template, a known number of copies of target DNA (standard) or plasmid DNA. The system determines the calibration curve and plots the Ct of these serial dilutions obtained after a certain number of PCR cycles, vs $\log_{10}$ of the quantity of target DNA ($\log_{10}$ (quantity)). Each calibration standard should be tested in duplicate or triplicate in each different run. The amount of target DNA in each unknown sample is calculated by interpolating the Ct value in the standard curve. The r^2 of the standard curve should be >0.99. Specificity of the designed primers, linearity of the reactions and sensitivity of the Real-time PCR should be checked.

Tips Serial dilutions of the standard should contain the correct amount of DNA target. Check the absolute quantity of DNA in each standard. Prepare aliquots of each dilution and freeze them at −80 °C, thaw only once before use.

Real-Time PCR Reaction (2) Prepare the PCR cocktail, preferably one cocktail per group of sample replicates. Calculate the quantity of each PCR component for all samples, containing: 0.2 µM of each primer, 12.5 µL of SYBR® Green PCR Master Mix and 10.5 µL of sterile Milli-Q water. Include DNA-free, negative and positive controls for each cocktail in each plate, vortex to homogenize the mixture; (3) Distribute 24 µL of the PCR cocktail in each well; (4) Add 1 µL of DNA template previously vortexed; (5) Cover the plate, place in thermocycler and program, e.g. as follows: (a) incubation step at 95 °C for 10 min; (b) DNA amplification for 20–40 cycles of 95 °C for 30 s, 58 °C for 30 s and 72 °C for 1 min. Annealing temperature will depend on the primers used and extension time will depend on the length of the PCR product. Number of cycles can be adjusted after monitoring the formation of the PCR product all over the reaction; (6) Program the melting curve temperature profile by, e.g. 95 °C for 1 min, 60 °C for 1 min and heating to 95 °C in 20 min; (7) Analyze data with the software version SDS 1.9.1 and Dissociation Curves 1.0.

Additional Confirmation of PCR Product (8) Confirm product identity by electrophoresis in a 2% agarose gel; (9) Visualize the gel on a transilluminator.

Notes Recommended primer concentration in SYBR® Green reactions is 50 mM, products should be ~150 bp on length (the shorter, the better) and the primers should not include more than two G or C bases in their last five bases. The PCR product obtained can be used for sequencing.

Other Considerations

Application of Real-Time PCR to ECM Studies Real-time PCR has recently been applied to ECM fungal community studies not only for fungal quantification (Landeweert et al. 2003b; Anderson and Parkin 2007) but also for ecological and functional studies such as the examination of gene expression (Miozzi et al. 2005), investigation of spatial distribution and temporal persistence of mycelia (Kennedy et al. 2007) and determination of ECM fungal competition (Kennedy et al. 2007; Parladé et al. 2007).

Optional Real-Time PCR Systems Other Real Time systems can be used such as the LigthCycler™ (Roche Molecular Biochemicals) (Landeweert et al. 2003b), RotorGene 3000 centrifugal amplification system (Corbett Research) (Martin and Rygiewicz 2005) or Bio-Rad iCycler iQ Multi-Color Real Time PCR Detection System (Guescini et al. 2003).

Troubleshooting if amplification does not appear, dilute or purify the DNA extracts. Losses of DNA occur after purification. If primer-dimmers are formed, raise the annealing temperature or the temperature when primers are not in a double strand. If unsuccessful, add lower concentrations of primers or design a new pair. The fluorophore SYBR® Green I dye does not distinguish target DNA from non-target DNA, thus, it is essential to use a good specific primer pair and to examine the melting curves to confirm the PCR product. When working with ITS regions, it is necessary to perform a calibration for each species. Quantification of a biomarker gene present in a known number of copies is another possibility (Raidl et al. 2005).

Acknowledgements This work was supported in part by the Departament de Medi Ambient i Habitatge from the Generalitat de Catalunya and by the scholarship 2002FI-00711 (DURSI-GENCAT) (Spain) to L.M. Suz, by the European Community Program "Structuring the European Research Area" under Synthesis at the RJB-CSIC and the grant SFRH7BPD/5560/200 from FCT-MCTES (Portugal) to Dr. A.M. Azul, by the National Science Foundation Grant DEB-9981711 to Dr. C.S. Bledsoe, and by the grant CGL2006-12732-C02-01 from the Ministerio de Educación y Ciencia (Spain) to Dr. M.P. Martin.

References

Acinas SG, Sarma-Rupavtarm R, Klepac-Ceraj V, Polz MF (2005) PCR-induced artifacts and bias: insights from comparison of two 16S rRNA clone libraries constructed from the same sample. Appl Environ Microbiol 71:8966–8969

Agerer R (1986) Studies on ectomycorrhizae. II. Introducing remarks on characterization and identification. Mycotaxon 26:473–492

Agerer R (1987–2006) Colour atlas of ectomycorrhizae: 1st–13th delivery. Einhorn-Verlag, Schwäbisch Gmünd

Agerer R (1991) Characterization of ectomycorrhiza. In: Norris JR, Read DJ, Varma AK (eds) Techniques for the study of mycorrhiza. Methods Microbiol 23:25–73

Agerer R (1995) Anatomical characteristics of identified ectomycorrhizas: an attempt towards a natural classification. In: Varma K, Hock B (ed) Mycorrhiza: structure, function, molecular biology and biotechnology. Springer, Berlin Heidelberg New York, pp 685–734

Agerer R (1999) Never change a functionally successful principle: the evolution of Boletales s.l. (Hymenomycetes, Basidiomycota) as seen from below-ground features. Sendtnera 6:5–91

Agerer R (2001) Exploration types of ectomycorrhizae. A proposal to classify ectomycorrhizal mycelial systems according to their pattern of differentiation and putative ecological importance. Mycorrhiza 11:107–114

Agerer R (2006) Fungal relationships and structural identity of their ectomycorrhizae. Mycol Progr 5:67–107

Agerer R, Rambold G (2004–2007) [first posted on 2004-06-01; most recent update: 2007-05-02]. DEEMY – An Information System for Characterization and Determination of Ectomycorrhizae. www.deemy.de – München, Germany

Agerer R, Beenken L, Christan J (1998) *Gomphus clavatus* (Pers.: Fr.) S. F. Gray + *Picea abies* (L.) Karst. Descr. Ectomyc. 3:25–29

Aguín-Casal O, Sáinz-Osés MJ, Mansilla-Vázquez JP (2004) *Armillaria* species infesting vineyards in northeastern spain. Eur J Plant Pathol 110:683–687

Altschul SF, Madden TL, Schäffer AA, Zhang J, Zhang Z, Miller W, Lipman DJ (1997) Gapped BLAST and PSI-BLAST: a new generation of protein database search programs. Nucleic Acids Res 25:3389–3402

Amicucci A, Zambonelli A, Giomaro G, Potenza L, Stocchi V (1998) Identification of ectomycorrhizal fungi of the genus *Tuber* by species-specific ITS primers. Mol Ecol 7:273–277

Anderson IC, Cairney JWG (2004) Diversity and ecology of soil fungal communities: increased understanding through the application of molecular techniques. Environ Microbiol 6:769–779

Anderson IC, Parkin PI (2007) Detection of active soil fungi by RT-PCR amplification of precursor rRNA molecules. J Microbiol Methods 68:248–253

Anderson IC, Campbell CD, Prosser JI (2003) Potential bias of fungal 18 S rDNA and internal transcribed spacer polymerase chain reaction primers for estimating fungal biodiversity in soil. Environ Microbiol 5:36–47

Azul AM, Agerer R, Freitas H (2001a) "*Quercirhiza ectendotrophica*" + *Quercus suber* L. Descr Ectomyc 5:67–72

Azul AM, Agerer R, Freitas H (2001b) "*Quercirhiza internangularis*" + *Quercus suber* L. Descr Ectomyc 5:79–83

Azul AM, Agerer R, Freitas H (2001c) "*Quercirhiza pedicae*" + *Quercus suber* L. Descr Ectomyc 5:85–91

Azul AM, Agerer R, Freitas H (2001d) "*Quercirhiza sclerotiigera*" + *Quercus suber* L. Descr Ectomyc 5:99–105

Azul AM, Agerer R, Freitas H (2006a) "*Quercirhiza dendrohyphidiomorpha*" + *Quercus suber* L. Descr Ectomyc 9/10:87–91

Azul AM, Martín MP, Agerer R, Freitas H (2006b) "*Quercirhiza auratercystidiata*" + *Quercus suber* L. Descr Ectomyc 9/10:81–86

Azul AM, Martín MP, Agerer R, Freitas H (2006c) "*Quercirhiza flavocystidiata*" + *Quercus suber* L. Descr Ectomyc 9/10:93–97

Azul AM, Martín MP, Agerer R, Freitas H (2006d) "*Quercirhiza tomentellofuniculosa*" + *Quercus suber* L. Descr Ectomyc 9/10:127–134

Baciarelli-Falini L, Rubini A, Riccioni C, Paolocci F (2006) Morphological and molecular analyses of ectomycorrhizal diversity in a man-made *Tuber melanosporum* plantation: description of novel truffle-like morphotypes. Mycorrhiza 16(7):475–484

Baier R, Ingenhaag J, Blaschke H, Gottlein A, Agerer R (2006) Vertical distribution of an ectomycorrhizal community in upper soil horizons of a young Norway spruce (*Picea abies* [L.] Karst.) stand of the Bavarian Limestone Alps. Mycorrhiza 16:197–206

Balkwill DL, Labeda DP, Casida LE (1975) Simplified procedures for releasing and concentrating microorganisms from soil for transmission electron-microscopy viewing as thin-sectioned and frozen-etched preparations. Can J Microbiol 21(3):252–262

Bergemann SE, Garbelotto M (2006) High diversity of fungi recovered from the roots of mature tanoak (*Lithocarpus densiflorus*) in northern California. Can J Bot 84:1380–1394

Bertini L, Potenza L, Zambonelli A, Amicucci A, Stocchi V (1998) Restriction fragment length polymorphism species-specific patterns in the identification of white truffles. FEMS Microbiol Lett 164:397–401

Bidartondo M, Kretzer AM, Bruns TD (2000) High root concentration and uneven ectomycorrhizal diversity near *Sarcodes sanguinea* (*Ericaceae*): a cheater that stimulates its victims? Am J Bot 87:1783–1788

Bruns TD, Vigalys R, Barns SM, González D, Hibbett DS, Lane DJ, Simon L, Stickel S, Szaro TM, Weisburg WG, Sogin ML (1992) Evolutionary relationships within the fungi: analyses of nuclear small subunit rRNA sequences. Mol Phylogenet Evol 1:231–241

Burke DJ, Martin KJ, Rygiewicz PT, Topa MA (2005) Ectomycorrhizal fungi indentification in single and pooled samples: terminal restriction fragment length polymorphism (TRFLP) and morphotyping compared. Soil Biol Biochem 37:1683–1694

Chen DM, Cairney JWG (2002) Investigation of the influence of prescribed burning on ITS profiles of ectomycorrhizal and other soil fungi at three Austrian sclerophyll forest sites. Mycol Res 106:532–540

de Román M, Clavería V, de Miguel AM (2005) A revision of the descriptions of ectomycorrhizas published since 1961. Mycol Res 109:1063–1104

Dickie IA, Xu B, Koide RT (2002) Vertical niche differentiation of ectomycorrhizal hyphae in soil as shown by T-RFLP analysis. New Phytol 156:527–535

Dickie IA, Guza RC, Krazewski SE, Reich PB (2004) Shared ectomycorrhizal fungi between a herbaceous perennial (*Helianthemum bicknellii*) and oak (*Quercus*) seedlings. New Phytol 164:375–382

Doyle JJ, Doyle JL (1990) Isolation of plant DNA from fresh tissue. Focus 12:13–15

Edel V (1998) Polymerase chain reaction in mycology: an overview. In: Bridge PD, Arora DK, Reddy CA, Elander RP (eds) Applications of PCR in mycology. CAB International, New York, pp 1–20

Edwards K, Johnstone C, Thompson C (1991) A simple and rapid method for the preparation of plant genomic DNA for PCR analysis. Nucleic Acid Res 19:1349

Erland S, Henrion B, Martin F, Glover LA, Alexander IJ (1994) Identification of the ectomycorrhizal basidiomycete *Tylospora Fibrillosa* Donk by RFLP analysis of the PCR-amplified ITS and IGS regions of ribosomal DNA. New Phytol 126:525–532

Frank AB (1885) Über die auf Wurzelsymbiose beruhende Ernährung gewisser Bäume durch unterirdische Pilze. Ber Dtsch Bot Ges 3:128–145

Gagné A, Jany J-L, Bousquet J, Khasa DP (2006) Ectomycorrhizal fungal communities of nursery-inoculated seedlings outplanted on clear-cut sites in northern Alberta. Can J For Res 36:1684–1694

Gardes M, Bruns TD (1993) ITS primers with enhanced specificity for basidiomycetes-application to the identification of mycorrhizae and rusts. Mol Ecol 2:1–6

Gehring CA, Theimer TC, Whitham TG, Keim P (1998) Ectomycorrhizal fungal community structure of pinyon pines growing in two environmental extremes. Ecology 79:1562–1572

Genney DR, Anderson IC, Alexander IJ (2006) Fine-scale distribution of pine ectomycorrhizas and their extramatrical mycelium New Phytol 170:381–390

Gibelli G (1883) Nuovi studii sulla malattia del Castagno detta dell' inchiostro. Mem R Acad Sci Ist Bologna 4:287–314

Godbout C, Fortin JA (1983) Morphological features of synthesized ectomycorrhizae of *Alnus crispa* and *A. rugosa*. New Phytol 102:429–442

Goodman DM, Trofymow JA (1998) Distribution of ectomycorrhizas in micro-habitats in mature and old-growth stands of Douglas-fir on southeastern Vancouver Island. Soil Biol Biochem 30:2127–2138

Griffiths RI, Whiteley AS, O'Donnell AG, Bailey MJ (2000) Rapid method for coextraction of DNA and RNA from natural environments for analysis of ribosomal DNA- and rRNA-based microbial community composition. Appl Environ Microbiol 66:5488–5491

Guescini M, Pierleoni R, Palma F, Zeppa S, Vallorani L, Potenza L, Sacconi C, Giomaro G, Stocchi V (2003) Characterization of the *Tuber borchii* nitrate reductase gene and its role in ectomycorrhizae. Mol Gen Genomics 269:807–816

Guidot A, Debaud JC, Effosse A, Marmeisse R (2003) Below-ground distribution and persistence of an ectomycorrhizal fungus. New Phytol 161:539–547

Henrion B, Le Tacon F, Martin F (1992) Rapid identification of genetic variation of ectomycorrhizal fungi by amplification of ribosomal RNA genes. New Phytol 122:289–298

Henrion B, Chevalier G, Martin F (1994) Typing truffle species by PCR amplification of the ribosomal DNA spacers. Mycol Res 98:37–43

Hibbett DS (1992) Ribosomal RNA and fungal systematics. Trans Mycol Soc Jpn 33:533–556

Holben WE, Jansson JK, Chelm BK, Tiedje JM (1988) DNA probe method for the detection of specific microorganisms in the soil bacterial community. Appl Environ Microbiol 54:703–711

Hopple JS, Vilgalys R (1994) Phylogenetic-relationships among coprinoid taxa and allies based on data from restriction site mapping of nuclear rDNA. Mycologia 86:6–107

Hortal S, Pera J, Galipienso L, Parladé J (2006) Molecular identification of the edible ectomycorrhizal fungus *Lactarius deliciosus* in the symbiotic and extraradical mycelium stages. J Biotechnol 126:123–134

Johnson NC, O'Dell TE, Bledsoe CS (1999) Methods for ecological studies of mycorrhizae. In: Robertson GP, Coleman DC, Bledsoe CS, Sollins P (eds) Standard soil methods for long-term ecological research. Oxford University Press, New York. Chap 18, pp 378–436

Jumpponen A (2003) Soil fungal community assembly in a primary successional glacier forefront ecosystem as inferred from rDNA sequence analysis. New Phytol 158:569–578

Kårén O, Högberg N, Dahlberg A, Jonsson L, Nylund J-E (1997) Inter- and intraspecific variation in the ITS region of rDNA of ectomycorrhizal fungi in Fennoscandia as detected by endonuclease analysis. New Phytol 136:313–325

Kennedy PG, Bergemann SE, Hortal S, Bruns TD (2007) Determining the outcome of field-based competition between two *Rhizopogon* species using real-time PCR. Mol Ecol 16:881–890

Koide RT, Xu B, Sharda J (2005a) Contrasting below-ground views of an ectomycorrhizal fungal community. New Phytol 166:251–262

Koide RT, Xu B, Sharda J, Lekberg Y, Ostiguy N (2005b) Evidence of species interaction with an ectomycorrhizal fungal community. New Phytol 166:305–316

Kranabetter JM, Wylie T (1998) Ectomycorrhizal community structure across forest openings on naturally regenerated western hemlock seedlings. Can J Bot 76:189–196

Landeweert R, Leeflang P, Kuyper TW, Hoffland E, Rosling A, Wernars K, Smit E (2003a) Molecular identification of ectomycorrhizal mycelium in soil horizons. Appl Environ Microbiol 69:327–333

Landeweert R, Veenman C, Kuyper TW, Fritze H, Wernars K, Smit E (2003b) Quantification of ectomycorrhizal mycelium in soil by real-time PCR compared to conventional quantification techniques. FEMS Microbiol Ecol 45:283–292

Landeweert R, Leeflang P, Smit E, Kuyper T (2005) Diversity of an ectomycorrhizal fungal community studied by a root tip and total soil DNA approach. Mycorrhiza 15:1–6

Larena I, Salazar O, González V, Julián MC, Rubio V (1999) Design of a primer for ribosomal DNA internal transcribed spacer with enhanced specificity for ascomycetes. J Biotechnol 75:187–194

Lee SB, Taylor JW (1990) Isolation of DNA from fungal mycelia and single spores. In: Innis MA, Gelfand DH, Sninsky JJ, White TJ (ed) PCR protocols. A guide to methods and applications. Academic Press, San Diego, pp 282–287

Lilleskov EA, Bruns TD, Horton TR, Taylor DL, Grogan P (2004) Detection of forest stand-level spatial structure in ectomycorrhizal fungal communities. FEMS Microbiol Ecol 49:319–332

Magurran AE (1988) Ecological diversity and its measurement. Princeton University Press, New Jersey

Maidak BL, Cole JR, Lilburn TG, Parker CT, Saxman PR, Farris RJ, Garrity GM, Olsen GJ, Schmidt TM, Tiedje JM (2001) The RDP-II (Ribosomal Database Project). Nucleic Acids Res 29:173–174

Martin D, Rybicki E (2000) RDP: detection of recombination amongst aligned sequences. Bioinformatics 16:562–563

Martin KJ, Rygiewicz P (2005) Fungal-specific PCR primers developed for analysis of the ITS region of environmental DNA extracts. BCM Microbiol 5:28–39

Martín MP (2000) Protocols: DNA isolation, PCR and RFLP analyses. In: Martín MP (ed) Methods in root-soil interactions research. Protocols. Slovenian Forestry Institute, Ljubljana, Slovenia, pp 35–44

Martín MP, Calonge FD (2000) *Rhizopogon aromaticus* (Boletales, Basidiomycotina) a new species found in Spain. Mycotaxon 75:425–429

Martín MP, García-Figueres F (1999) *Colletotrichum acutatum* and *C. gloeosporioides* cause anthracnose on olives. Eur J Plant Pathol 105:733–741

Martín MP, Högberg N, Nylund J-E (1998) Molecular analysis confirms morphological reclassification of the genus *Rhizopogon*. Mycol Res 102:855–858

Martín MP, Högberg N, Llistosella J (1999) *Macowanites messapicoides*, a hypogeous relative to *Russula messapica*. Mycol Res 103(2):203–208

Martín MP, Díez J, Manjón J-L (2000a) Methods used for studies in molecular ecology of ectomycorrhizal fungi. In: Martín MP (ed) Methods in root-soil interactions research. Protocols. Slovenian Forestry Institute, Ljubljana, Slovenia, pp 25–28

Martín MP, Kårén O, Nylund J-E (2000b) Molecular ecology of hypogeous mycorrhizal fungi: *Rhizopogon roseolus* (Basidiomycotina). Phyton 40(4):135–141

Martin-Laurent F, Philippot L, Hallet S, Chaussod HR, Germon JC, Soulas G, Catroux G (2001) DNA extraction from soils: old bias for new microbial diversity analysis methods. Appl Environ Microbiol 67:2354–2359

Middleton SA, Anzenberger G, Knapp LA (2004) Denaturing gradient gel electrophoresis (DGGE) screening of clones prior to sequencing. Mol Ecol Notes 4:776–778

Miller DN, Bryant JE, Madsen EL, Ghiorse WC (1999) Evaluation and optimization of DNA extraction and purification procedures for soil and sediment samples. Appl Environ Microbiol 65:4715–4724

Miozzi L, Balestrini R, Bolchi A, Novero M, Ottonello S, Bonfante P (2005) Phospholipase A2 up-regulation during mycorrhiza formation in *Tuber borchii*. New Phytol 167:229–238

Mitchell JI, Zuccaro A (2006) Sequences, the environment and fungi. Mycologist 20:62–74

Molina R, Massicotte H, Trappe JM (1992) Specificity phenomena in mycorrhizal symbioses: community-ecological consequences and practical implications. In: Allen MF (ed) Mycorrhizal functioning, an integrative plant–fungal process. Springer, Berlin Heidelberg New York, pp 357–423

Morris MH (2006) Diversity, composition and structure of ectomycorrhizal fungal communities on roots of *Quercus* spp. in California and Mexico. PhD thesis, University of California, Davis

Mullis KB, Faloona FA (1987) Specific synthesis of DNA in vitro via a polymerase-catalyzed chain reaction. Methods Enzymol 155:335–350

O'Brien HE, Parrent JL, Jackson JA, Moncalvo JM, Vilgalys R (2005) Fungal community analysis by large-scale sequencing of environmental samples. Appl Environ Microbiol 71:5544–5550

Ogram A, Sayler GS, Barkay T (1987) The extraction and purification of microbial DNA from sediments. J Microbiol Methods 7:57–66

Paolocci F, Rubini A, Granetti B, Arcioni S (1999) Rapid molecular approach for a reliable identification of *Tuber* spp. ectomycorrhizae. FEMS Microbiol Ecol 28:23–30

Parladé J, Hortal S, Pera L, Galipienso L (2007) Quantitative detection of *Lactarius deliciosus* extraradical soil mycelium by real-time PCR and its application in the study of fungal persistence and interspecific competition. J Biotechnol 128:14–23

Pritsch K, Boyle H, Munch JC, Buscot F (1997) Characterization and identification of black alder ectomycorrhizas by PCR/RFLP analyses of the rDNA internal transcribed spacer (ITS). New Phytol 137:357–369

Qiu X, Wu L, Huang H, McDonel PE, Palumbo AV, Tiedje JM, Zhou J (2001) Evaluation of PCR-generated chimeras, mutations, and heteroduplexes with 16s rRNA gene-based cloning. Appl Environ Microbiol 67:880–887

Raidl S, Agerer R (1998) *Hysterangium stoloniferum* Tul. & Tul. + *Picea abies* (L.) Karst. Descr Ectomyc 3:31–35

Raidl S, Bonfigli R, Agerer R (2005) Calibration of quantitative real-time Taqman PCR by correlation with hyphal biomasa and ITS copies in mycelia of *Piloderma croceum*. Plant Biol 7:713–717

Ranjard L, Poly F, Combrisson J, Richaume A, Nazaret S (1998) A single procedure to recover DNA from the surface or inside aggregates and in various size fractions of soil suitable for PCR-based assays of bacterial communities. Eur J Soil Biol 34:89–97

Ranjard L, Lejon DPH, Mougel C, Scheher L, Merdinoglu D, Chaussod R (2003) Sampling strategy in molecular microbial ecology: influence of soil sample size on DNA fingerprinting of fungal and bacterial communities. Environ Microbiol 5:1111–1120

Rogers SO, Bendich AJ (1985) Extraction of DNA from milligrams amounts of fresh, herbarium and mummified plant tissues. Plant Mol Biol 5:69–76

Sanger F, Nicklen S, Coulson AR (1977) DNA sequencing with chain-terminating inhibitors. Proc Natl Acad Sci USA 74:5463–5467

Schaffer HE, Sederoff RR (1981) Improved estimation of DNA fragment lengths from agarose gel. Anal Biochem 115:113–122

Sicoli G, Fatehi J, Stenlid J (2003) Development of species-specific PCR primers on rDNA for the identification of European *Armillaria* species. For Pathol 33:287–297

Smalla K, Cresswell N, Mendonca-Hagler LC, Wolters A, van Elsas JD (1993) Rapid DNA extraction protocol from soil for polymerase chain reaction-mediated amplification. J Appl Bacteriol 74:78–85

Smith ME, Douhan GW, Rizzo DM (2007) Ectomycorrhizal community structure in a xeric *Quercus* woodland based on rDNA sequence analysis of sporocarps and pooled roots. New Phytol 174:847–863

Stendell ER, Horton TR, Bruns TD (1999) Early effects of prescribed fire on the structure of the ectomy-corrhizal fungal community in a Sierra Nevada ponderosa pine forest. Mycol Res 103:1353–1359

Suz LM, Martín MP, Colinas C (2006) Detection of *Tuber melanosporum* DNA in soil. FEMS Microbiol Lett 254:251–257

Suzuki MT, Giovannoni SJ (1996) Bias caused by template annealing in the amplification of mixtures of 16S rRNA genes by PCR. Appl Environ Microbiol 62:625–630

Taylor AFS (2002) Fungal diversity in ectomycorrhizal communities: sampling effort and species detection. Plant Soil 244:19–28

Taylor DL, Bruns TD (1999) Community structure of ectomycorrhizal fungi in a *Pinus muricata* forest: minimal overlap between the mature forest and resistant propagule communities. Mol Ecol 8:1837–1850

Tedersoo L, Koljalg U, Hallenberg N, Larsson K-H (2003) Fine scale distribution of ectomycor-rhizal fungi and roots across substrate layers including coarse woody debris in a mixed forest. New Phytol 159:153–165

Tedersoo L, Hansen K, Perry BA, Kjøller R (2006) Molecular and morphological diversity of Pezizalean ectomycorrhiza. New Phytol 170:581–596

Ter Braak CJF, Smilauer P (2002) CANOCO manual and CanoDraw for Windows user's guide: software for canonical community ordination (version 4.5). Microcomputer Power, Ithaca, NY

van der Heijden M (1998) Mycorrhizal fungal diversity determines plant biodiversity, ecosystem variability and productivity. Nature 396:69072

Volossiouk T, Robb EJ, Nazar RN (1995) Direct DNA extraction for PCR-mediated assays of soil organisms. Appl Environ Microbiol 61:3972–3976

White TJ, Bruns T, Lee S, Taylor J (1990) Amplification and direct sequencing of fungal ribosomal RNA genes for phylogenetics. In: Innis MA, Gelfand DH, Sninsky JJ, White TJ (eds) PCR protocols. A guide to methods and applications. Academic Press, San Diego, pp 315–322

Yeates C, Gillings MR, Davison AD, Altavilla N, Veal DA (1998) Methods for microbial DNA extraction from soil for PCR amplification. Biol Proc Online 1:40–47

Zar JH (1996) Biostatistical analysis, 3rd edn. Prentice Hall International, London

Zhou J, Bruns MA, Tiedje JM (1996) DNA recovery from soils of diverse composition. Appl Environ Microbiol 62:316–322

Zhou ZH, Hogetsu T (2002) Subterranean community structure of ectomycorrhizal fungi under *Suillus grevillei* sporocarps in a *Larix kaempferi* forest. New Phytol 154:529–539